Sensors and their Applications VII

Sensors Series

Series Editor: **B E Jones**

Other books in the series

Solid State Gas Sensors
Edited by P T Moseley and B C Tofield

Techniques and Mechanisms in Gas Sensing
Edited by P T Moseley, J O W Norris and D E Williams

Hall Effect Devices
R S Popović

Thin Film Resistive Sensors
Edited by P Ciureanu and S Middelhoek

Biosensors: Microelectrochemical Devices
M Lambrechts and W Sansen

Sensors VI: Technology, Systems and Applications
Edited by K T V Grattan and A T Augousti

Intelligent Sensor Systems
J E Brignell and N M White

Automotive Sensors
M H Westbrook and J D Turner

Thermal Sensors
Edited by G C M Meijer and S A W Van Herwaarden

Advances in Actuators
Edited by A P Dorey and J H Moore

Sensors and their Applications VII

Proceedings of the seventh conference on
Sensors and their Applications,
held in Dublin, Ireland, 10–13 September 1995

Edited by

A T Augousti

Kingston University

Institute of Physics Publishing
Bristol and Philadelphia

1995

British Library Cataloguing-in-Publication Data

A catalogue record for this book is available from the British Library.

ISBN 0 7503 0331 X

Library of Congress Cataloging-in-Publication Data are available

Series Editor: **Professor B E Jones**, Brunel University

Published by Institute of Physics Publishing, wholly owned by The Institute of Physics, London
Institute of Physics Publishing, Techno House, Redcliffe Way, Bristol BS1 6NX, UK
US Editorial Office: Institute of Physics Publishing, The Public Ledger Building, Suite 1035, Independence Square, Philadelphia, PA 19106, USA

Printed in the UK by J W Arrowsmith Ltd, Bristol

Preface

As the 20th Century draws to a close the importance of sensor systems as essential constituents of the technological infrastructure which supports modern civilization is widely recognized. These sensor systems find application in every area of human activity, from transport and construction, through marine and built environment applications, to biomedical and manufacturing applications, to name but a few.

The UK is fortunate in being at the forefront of the extensive activity which is taking place worldwide in the development of these new sensor systems. This is reflected by the full and varied meeting programmes of the national professional groups concerned with measurement and instrumentation, and the formation in recent years of a substantial number of special interest sensor groups. The pace of developments in the sensing area shows no signs of diminishing, and the continued interaction of a range of technologies, and in particular the growing processing power available at the sensor head, is continuously producing more powerful sensors which can be deployed in an increasingly wide range of conditions. The pressure for this growth is sustained both by the natural evolution of the relevant technologies, and the emergence of new ones, but more recently by the effect of national and European directives also, as well as by the enforcement of global agreements.

Sensors and their Applications VII aims to capture some of the excitement of these developments. It records the Proceedings of the Seventh Conference on Sensors and their Applications, held in Dublin, Ireland on 10–13th September 1995, organized by the Instrument Science and Technology Group of the UK Institute of Physics. The contributions to this volume amply illustrate the range and high quality of the work which is taking place within the UK and the EC, as well as containing several notable contributions from farther afield. This volume is the third of the conference proceedings published in the Sensors Series by Institute of Physics Publishing.

It is my pleasure to acknowledge the important contributions made by the Organizing Committee and the Technical Programme Committee, both in their support with regard to the arrangements for the Conference, and in particular for their diligence in refereeing the abstracts in order to maintain the high quality of presentations at the conference and hence the material which is included in this volume. I would particularly like to thank Ms Jacki Butler of the Meetings Department at the Institute of Physics, for her invaluable assistance and efficient service throughout the long period of planning for the conference itself.

Every effort has been made to ensure that the contributions to this volume have been properly prepared for publication, in order to maximize its utility to a wide audience. I am confident that the interested reader will find this a useful overview of the national sensor scene in the UK in 1995.

A T Augousti
Kingston University

Contents

Section B: Sensor Arrays and Silicon Sensors

Section C: Acoustic and Ultrasonic Sensors

Section D: Optical Sensors

Section E: Flow Measurement and Process Tomography

Section F: Biomedical Sensors

Section G: Sensor Modelling and Interfacing

Section H: Sensor Applications

Late Paper

Section A

Gas and Environmental Sensors

Environmental sensors: Present practices and future prospects

R W Bogue

Robert Bogue & Partners, Kingston House, Tuckermarsh, Bere Alston, Devon PL20 7HB, UK

Abstract. This paper reviews some of the methods and sensors employed in environmental monitoring practices, with a particular emphasis on the differing approaches adopted when monitoring pollutants in water and air. It subsequently discusses certain technological trends and recent developments and illustrates the scale of future prospects by presenting a selection of market forecasts. It also aims to dispel a number of misconceptions.

1. Introduction

1.1 Background

If future historians need a leitmotiv for the 1980s, the "decade of the environment" seems most appropriate. The decade witnessed "green" issues emerging from relative obscurity to a centre-stage position within the political arena and public debate. This groundswell of concern, which involves both macro-environmental issues such as global warming and ozone depletion, and more localised considerations such as drinking water and urban air quality, has led to the emergence of a multi-million pound market for environmental sensors and instruments.
An "environmental" sensor is defined as one that is used to control or monitor the state of the environment and many exploit novel techniques and technologies and respond to hitherto un-sensed quantities. The majority serve one of the following purposes:

- to control pollution abatement processes;
- monitoring pollutants as they enter the environment (at, or close to, the point of discharge);
- assessing the state of the external environment.

This latter use is, perhaps, the most varied as the "external environment" includes in-land water courses (rivers, lakes, reservoirs), estuaries and beaches, oceans, aquifers, ambient air, the upper atmosphere and derelict industrial and contaminated land.

1.2 Legislation

Most environmental monitoring is undertaken as a consequence of legislation but contrary to the often voiced view, such legislation is not new. Laws governing the release of pollutants have existed in the UK for many years. However, much present-day monitoring is performed as a direct consequence of recent EU laws; the origins of which lie with the 1972 European Summit

Conference in Paris, where a decision was made to introduce a Community-wide raft of environmental policies and laws. Since then, numerous directives have been issued by the CEC which form part of the statuatory laws in most EU countries. These cover, for instance, the emission of toxic and environmentally threatening gases to atmosphere, the chemical and microbiological quality of drinking water, the composition of industrial discharges, bathing water quality, waste disposal, incineration, and many other issues.

It is often argued that, once a substance is implicated by legislation, sensors will be required to monitor it but at best, this is only partly true. Species in the liquid phase can, more often that not, be determined accurately and cost-effectively by existing laboratory analytical methods, and sensors are only required if it is deemed necessary to monitor the species at the point of discharge (if sufficiently localised) or in the external environment.

Moreover, legislators only stipulate continuous monitoring if realistic technologies exist. Dioxins illustrate this point, as, although emitted by incinerators and perceived to pose a threat to public health, there are no methods that can determine these compounds in real time in the ngm^{-3} concentrations that are typical.

2. Environmental monitoring practices

2.1 Quantities monitored

Perhaps the most important point to note is that a very large number of chemical and other quantities are implicated by legislation but no generic measurement techniques, sensing principles or technologies are available for their determination. Some of the quantities implicated by environmental legislation are listed in Table 1.

Table 1 - Some quantities implicated by environmental legislation

Air/atmosphere	In-land water	Other
Carbon monoxide	Ammonia	Noise
Carbon dioxide	Nitrates	Ionising radiations
Sulphur dioxide	Phosphates	PCBs, PCPs (ground)
Ozone	Metals	Trace metals (sea)
Methane	Halogenated organics	Radon (buildings)
Nitrogen oxides	Suspended solids	Methane (landfill)
1,3 butadiene	Cyanide	Oil (sea)
VOCs	Pesticides	Radioisotopes
Dioxins	Fuels and oils	
Particulates	Biochemical oxygen demand	
Metals	Chemical oxygen demand	
Chlorine	Total organic carbon	
Hydrogen chloride	Algae and algal blooms	
Hydrogen sulphide	Bacteria and viruses	
Ammonia	Phenols	
CFCs	Chlorine	
	Dissolved oxygen	
	pH	

2.2 Aquatic monitoring

Much aquatic monitoring is undertaken by non-sensor means. The NRA examines water samples for 200-300 different analytes and of the 4.3 million determinations made annually, approximately 95% are achieved by sampling and laboratory analysis. Analytical techniques such as AAS, GC, HPLC, MS and GCMS are widely used and even taking into account new sensor and instrument developments, the NRA estimates that field measurements with sensors will not exceed 10% of the total. This is because:

- contrary to popular misconception, laboratory analysis can be very cost-effective, particularly when large numbers of samples are involved;
- trace level determinations are often required (sub-ppb) which exceed the detection limits of most available chemical sensors;
- for compliance monitoring, approved high accuracy analytical methods are stipulated by legislation.

Also, sensors are not available for most of the analytes involved and sensor-based field measurements in the aquatic environment are restricted presently to a very limited range of analytes such as pH, DO, ammonia, conductivity, some metals, turbidity, etc.
Conversely, fixed instruments, which are used widely to monitor pollution abatement processes and analyse aqueous discharges, can, by virtue of their size and higher costs, exploit relatively complex techniques such as wet reagent-aided photometry and voltammetry. Thus, they are able to determine a far wider range of ions and cations, as well as more "difficult" analytes such as TOC and COD etc. However, these instruments are costly to purchase and operate and achieve often less than acceptable field reliability.
A point to note is that several of the quantities measured are not themselves pollutants but indicators of pollution or the health of the aquatic environment. An example of such a surrogate measurement is dissolved oxygen, where low DO levels may indicate the presence of a substance with a high BOD or COD. In the case of the recent pollution of the Carnon River in Cornwall by metal-laden flood water from Wheal Jane, the NRA monitored metal concentrations by measuring the water's conductivity.

2.2 Air and atmospheric monitoring

The situation regarding the determination of air quality and the composition of gaseous emissions is very different, as most measurements are made with sensors. This is because:

- air monitoring is a relatively recent practice and many sensors and analysers have been developed around modern technology;
- many major sources of atmospheric pollution are localised allowing (and often demanding) measurements at source;
- far fewer species are involved than in the aqueous environment and some near-generic sensing techniques are emerging.

This last point warrents comment. Several extremely accurate, sensitive and selective, quasi-generic optical gas sensing techniques are emerging. For instance, methods such as non-dispersive infra-red absorption, gas filter

correlation spectroscopy, Fourier transform IR spectroscopy and Raman scattering LIDAR are, between them, able to monitor any species that absorbs in the UV/vis/NIR part of the electromagnetic spectrum, often both at source and in the external environment. Other optical methods such as chemiluminescence, UV fluorescence and microwave spectroscopy add to this powerful armoury of techniques. Many are of recent origin, having been made possible by developments in optics and optoelectronics.
As yet, no such technological revolution has occurred in the aqueous sensing context.
The dramatically differing situations regarding water and air monitoring are summarised in Table 2.

Table 2 - Summary of air and water monitoring practices

	Air	Water
Number of analytes	Small (tens)	Very large (hundreds)
Monitoring practices	Recent	Long established
Use of sensors	Extensive	Limited
Sensing technologies	Recently developed	Many are long established
Nature of technologies	Mostly optical	Electrochemical etc.

2.3 Existing markets

Prior to considering future prospects, it is useful to gain an insight into the scale of the environmental sensing business. A selection of data are illustrated in Table 3.

Table 3 - Markets for environmental sensing products, 1994, in US$

Region	Products	Market
W. Europe	Stack gas emission monitors (NO_x, SO_2, CO etc.)	74m
W. Europe	Vehicle emissions analysers (HCs, NO_x, CO)	170m
W. Europe	Airborne particulate monitors	21m
W. Europe	Ambient water quality monitors	67m
W. Europe	Aqueous discharge monitors	234m
USA	Air quality monitors (all types)	675m
W. Europe	Laboratory environmental analysis services	1.1bn (1993)

(All sources: Frost & Sullivan)

The last figures illustrate very well the heavy reliance placed on laboratory analysis.

3. Future prospects

3.1 Factors stimulating future prospects

Prospects for novel environmental sensors result principally from the following factors:

1) Where they offer significant technical benefits over those sensors now employed;
2) Where they offer significant economic benefits over those sensors now employed;
3) Where sampling and laboratory analysis is too slow and/or too costly;
4) Where legislation stipulates, directly or indirectly, new monitoring regimes (species, applications, etc.).

All manner of technologies are being investigated in the environmental sensing context. Examples include fibre optic sensors, fibre-coupled remote analysers, integrated optical sensors, silicon ISFETs, biosensors, solid-state electrodes and microelectrodes, and multi-sensor arrays etc.

3.1 Technical benefits

Progress in sensor technology continues to yield improvements to sensitivity, specificity and response times and in some instances, allows hitherto impossible measurements to be made.
The ever-growing body of work on fibre optic sensors for waterborne analytes illustrates well the application of new technology to overcome problems associated with traditional sensors. Examples include the work in the UK by M Squared Ltd which aims to develop a family of sensitive yet highly selective sensors for ammonia, which is part-funded by the NRA, and the fluorescence quenching effect fibre optic sensor for dissolved oxygen, developed at Joanneum Research in Austria.
Soil analysis is an example of where innovative technology has yielded novel sensing capabilities. This application is of particular importance in the US, where investigating the nature of soil pollution at former military and other sites is of growing importance.
A group from the Naval Research Laboratory, Washington, is developing a fibre-coupled optical sensor that exploits infra-red reflectance spectroscopy and responds to oils in soil. It utilises the fact that certain oils exhibit a clear peak at 3.58 microns. Trials have shown that diesel marine fuel levels as low as 0.0023% by weight in sand can be detected. The sensor is linear over the range 0.02-0.1%.
Metals are also a major contaminator of derelict sites and a team at the Los Alamos National Laboratory are experimenting with fibre-coupled LIBS (laser-induced breakdown spectroscopy), in which an IR laser vapourises the soil and the light produced is transmitted by the fibre to a remote spectrometer. Detection limits range from 2 ppm to 60 ppm, depending on the metal involved.
Whilst LIBS cannot determine metals in-situ, a technique recently commercialised by the US company Scientific Ecology Group, offers prospects to achieve this by using a novel method termed prompt gamma neutron activation analysis (PGNAA). This uses a 14 MeV deuterium-tritium accelerator to create the neutrons which interact with metal atoms which in turn release gamma rays. These are detected with a liquid nitrogen cooled germanium detector, as used in conventional gamma spectroscopy. The instrument has determined metals such as Hg at concentrations of around 100 ppm, several inches below ground. Future developments will increase the technique's sensitivity and depth of penetration.

3.2 Economic benefits

Several opportunities for novel sensing products derive from the need to reduce the ownership costs of fixed water quality monitors. A recent survey of the 5-year life costs of a selection of on-line monitors by Severn-Trent Water (Fowles, 1994) revealed that these usually equal or often exceed the purchase price. The total 5-year life costs (consumables, reagents and servicing) for four on-line aluminium monitors, expressed as percentages of the purchase price, were 121%, 73%, 161% and a massive 282%. Interestingly, the instrument with the highest ownership cost had the lowest purchase price, whilst that with the highest price had the lowest ownership cost. Such ownership costs, which in these cases vary between £10K and £20K, are unacceptable to many users and suggest strong prospects for novel on-line chemical sensors and analysers.

Researchers often focus their attention on developing sensors with low unit costs but as the above illustrates, in the case of on-line devices, this may well be a spurious aim. The users need low life costs and some of the analytes for which such improved on-line sensors are needed, include TOC, COD, BOD, metals (heavy and process), halogenated organics, nitrates, phosphates and ammonia.

Another potential cost benefit will emerge from multi-parameter sensors. An example is "CENSAR", a thick-film chemical sensor array being developed jointly by Siemens Plessey and the University of Southampton, that simultaneously determines the DO, pH, redox, temperature and conductivity of a water sample. Future work aims to add sensors for metals and chlorine to the array.

A further and very active, cost-related field of research, is the development of disposable sensors. Several academic groups and companies are working on disposable thick-film sensors for the determination of metals and other waterborne analytes. Palintest Ltd has recently launched two such sensors, for lead and copper, which emerged from a collaboration with the University of Cranfield. These offer an alternative to wet chemical test kits or lab analysis and are able to determine compliance with the levels set by EU drinking water legislation. Opportunities exist also for disposable sensors for pesticides, phenols and many other waterborne analytes.

3.3 Improvements over sampling and laboratory analysis

As a specific example, complex organics are routinely determined in the laboratory by techniques such as GCMS but in contrast to lab analysis of "simple" analytes, these measurements are often extremely time consuming and costly. For example, the cost of analysing a sample for a dioxin such as 2,3,7,8-tetrachlorodibenzo-p-dioxin (2,3,7,8-T_4CDD), is around £400-600 on a batch basis and a single analysis can cost up to £1K (Bogue, 1994). Such analyses often take a week or more to complete, due to the very complex extraction, purification and separation techniques involved. Developments in test kit-based immunoassays are addressing these issues and Ensys Inc has a kit for 2,3,7,8-T_4CDD that can resolve 190 ppt, and with a preconcentration procedure it can reach 1 ppq (10^{-12}). Immobilising the immunoassay onto a biosensor is the obvious next step and has been achieved already for at least one pesticide. Ohmicron (US) has marketed a family of IA-based pesticide kits for some years and earlier this year, launched a disposable, field biosensor for atrazine featuring anti-atrazine antibodies and a platinum detection electrode.

From January 1st 1997, emissions of dioxins and furans from Europe's hazardous waste incinerators will be limited to 0.1 ngm^{-3}. Sampling and lab analysis would be prohibitively costly and excessively time-consuming and a challenge to sensor technologists will be developing an instrument that can determine these levels, ideally in-situ and in real-time.

These developments will also increase the speed with which data are obtained but nowhere is this more important than in the case of BOD (biochemical oxygen demand). The quantity has been used since before the first World War and is a vital measure of an effluent's capacity to de-oxygenate a water course. Presently, the five-day incubation procedure is the standard method and much research aims to yield real-time BOD data. Laboratory BOD monitors based on live cell biosensors have been launched by Lange (Germany) and at least one Japanese company, and an on-line device which employs an immobilised colony of Trichosporon cutaneum has been launched by Medingen (Germany). As highlighted above, the ownership costs of this will need to be low if it is to gain widespread acceptance.

3.4 Legislatively driven opportunities

Many technological innovations are originally stimulated by legislation. The laws governing the leakage of petroleum fuels from underground storage tanks in the US provide a good example. In response to this the US company FCI Environmental embarked on an R&D programme which culminated in the "Petrosense", perhaps the world's first fibre optic chemical sensors for industrial applications. This exploits a hydrocarbon-responsive fibre coating whose refractive index alters reversibly in the presence of the analyte, leading to an attenuation of the light in the fibre. The sensor can determine BTEX over the range 0-20,000 ppm, with an accuracy of around +/- 3 ppm.

The 1990 Clean Air Act Amendments (CAAA) implicate 189 chemical species and groups as health threatening but a recent survey (Kelly, 1995) showed that no ambient data existed for 74 of these. More importantly, an examination of the methods available for sampling and measuring these compounds revealed that directly applicable methods only exist for 126 of the total. Metals such as As, Sb, Be, Cd, Cr, Co, Pb, Hg and Ni etc. are amongst the species implicated and a need exists to quantify these in stack emissions; a problem being addressed by a research group at the US Sandia National Laboratories. The technique under development is optical fibre-coupled laser spark spectroscopy, whereby a laser creates a plasma of excited metal ions. On cooling, light is emitted and taken by the fibres to a remote grating spectrometer. Detection limits are typically in the 2-40 mgm^{-3} region, although beryllium levels of 0.1 mgm^{-3} can be detected with the technique.

The market for vehicle exhaust emission analysers is a good example of one created as a direct consequence of recent legislation. EU legislation stipulated a more strict vehicle emission test and as previously noted, the Western European market for this equipment was worth US$ 170m in 1994.

Legislation which ultimately stimulates a need for novel sensing technologies is often preceeded by progress in medical science. For example, airborne particulates have been implicated as posing a health threat for many years and monitoring equipment now constitutes a significantly sized market (see Table 3, above). Recent epidemiological research has revealed that particle size as well as overall concentration is critical (the "PM10" issue, see Bogue 1994), and more recently still, research has shown that particle composition is of equal or perhaps greater importance, as carcinogenic VOCs can be adsorbed onto certain particulate materials such as diesel soot. Thus emerge the capabilities likely to be required from future generations of particulate monitors: the ability to quantify, categorise by size and chemically speciate the particulate materials in a sample.

3.5 Summary of prospects

Table 4 lists a small selection of the many prospects that exist for new or improved environmental sensors.

Table 4 - Some prospects for new and improved environmental sensors

Product	Analytes
- On-line sensors with reduced costs of ownership	BOD, TOC, COD, metals, ammonia, nitrate, phosphate, etc.
- Low cost disposable field sensors	pesticides, metals, halogenated organics, ammonia, phenols, etc.
- Stack emission analysers	dioxins, metals, speciated VOCs.
- Ambient air quality monitors	NO_2, speciated VOCs, particle size and composition.
- Sensors for soil analysis	metals, oils, organics, etc.

3.6 Market forecasts

A selection of market forecasts are illustrated in Table 5.

Table 5 - Forecasts for environmental sensing products in Western Europe in US$ million

	1995	1996	1997	1999	2000
Stack gas analysers	83	-	101	103	-
Particulate monitors	24	-	30	30	-
Vehicle exhaust gas analysers	183	-	214	171	-
All air quality monitors	376	-	446	422	-
Ambient water quality instruments	76	-	100	-	-
Aqueous discharge monitors	275	-	378	-	-
Environmental biosensors (World)	-	15	-	-	100

(Sources: Frost & Sullivan, Cranfield Biotechnology)

This table serves to dispel another misconception: that environmental sensing markets are all growing. Whilst many market sectors are indeed developing rapidly, some will saturate and diminish thereafter as users complete their purchasing programmes. The vehicle exhaust gas analyser and particulate monitor markets illustrates this well.

The most pertinent point is that, whilst legislation certainly drives the markets, it also moderates them.

4. Concluding comments

Some of the more important points raised in this paper are summarised below.

1) Whilst legislation is the driving force behind much environmental monitoring, it is not new, having existed in the UK for over a century. Furthermore, a substance or quantity being implicated by legislation is no guarantee that sensors will be required to monitor it.

2) Most monitoring of aquatic pollutants is undertaken by sampling and laboratory analysis and this practice will continue for the forseeable future. Laboratory analysis is often an extremely cost-effective method and markets for laboratory analytical services far outstrip those for sensors.

3) Prospects for novel sensors for aquatic monitoring do exist but only in specific instances where they offer clear-cut economic or operational benefits. Multi-sensor arrays and disposable sensors are examples.

4) The key to the successful development of on-line environmental sensors is low overall ownership cost rather than low unit costs.

5) Strong prospects exist for sensors offering novel capabilities. Examples include the determination of metals in stack gases, pollutants in soil, ambient levels of speciated VOCs, and real-time particulate analysis (size and composition).

6) Whilst many environmental sensing markets are developing rapidly, some will saturate as users implicated by legislation complete their purchasing programmes. For many sensing products, there is a narrow window of opportunity lasting only a few years. Legislation creates markets but also moderates them.

References

Bogue R W, Dioxins in the Environment. Environmental Sensors, October 1994.

Bogue R W, Environmental dust and particulate monitoring. Environmental Sensors, May 1994.

Fowles G, Lifetime costs of process measurements. Proc. SWIG conference, Water quality Measurement: Impact on the bottom line, Cambridge, June 1994.

Kelly T J and Gordon S M, Identification of ambient sampling and analysis methods for the 189 Clean Air Act hazardous air pollutants. Proc. Pittcon'95, New Orleans, March 1995.

An Overview of Electronic Noses and their Applications

R.J. Elliott-Martin[1], P.N. Bartlett[1], J.W. Gardner[2], and T.T. Mottram[3]

[1] Department of Chemistry, University of Southampton, Southampton, SO17 1BJ, UK
[2] Department of Engineering, University of Warwick, Coventry, CV4 7AL, UK
[3] Silsoe Research Institute, Wrest Park, Silsoe, Bedford, MK45 4HS, UK

Abstract. A review on the developments in the field of artificial olfaction is presented. The engineering aspects of such a device designed to mimic the human sense of smell show how the technology has reached a stage where instruments are now available on the market. Several applications are outlined including the recent work on the feasibility of disease diagnosis in dairy cattle using an electronic nose.

1. Introduction

1.1 What are Odours?

Odorant molecules are typically hydrophobic and polar in nature with molecular masses of up to about 300 Daltons. Compounds larger than this tend to be insufficiently volatile to pass through the nasal cavity and across the olfactory epithelium. A simple odour is a single molecule some examples of which are given in Fig. 1, together with the odour type and their olfactory thresholds [1]. Most natural smells are complex in that they are a result of a mixture of perhaps many tens, and possibly many hundreds of different individual volatile chemicals, e.g. >600 compounds have been identified as playing some role in the flavour of coffee. The size, shape and polar properties of the molecule determine its odour characteristics [2], however the rules which govern this are poorly understood.

1.2 The Human Nose

The human olfactory system uses a large number of non-specific receptors which show broad patterns of response. These receptors are located at the surface of the olfactory hairs extending from the olfactory cells and into the epithelium (Fig. 2(a)). Secondary messengers produced at the olfactory cells generate signals which travel down axons to the glomeruli nodes located in the olfactory bulb. Further processing takes place at the mitral cells before finally traversing the granular cells where the signals then pass into the brain.

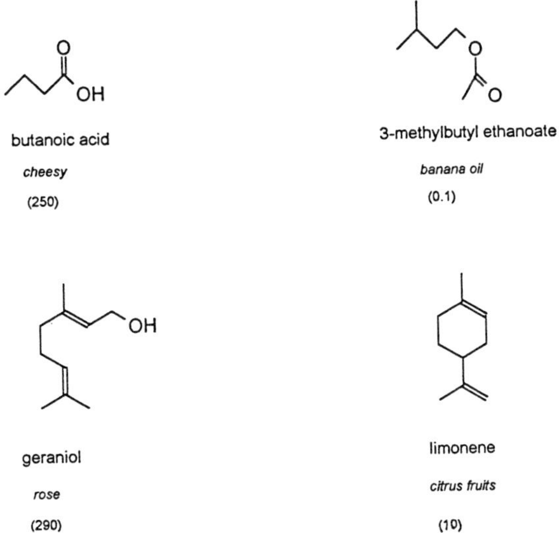

Fig. 1 Simple Odorants. Olfactory thresholds in water in ppb are given in parenthesis

It appears that more than 100 different receptor proteins of low specificity and overlapping sensitivities exist in the nose. The subsequent neural processing in the brain however allows discrimination between several thousand simple and complex odours.

1.3 The Need for an Electronic Instrument

In the evaluation of the smell and flavour of various industrially important products the human nose is the primary instrument for quality control. Trained individuals are required, the drawback being that these individuals can only work for short periods of time and that there are problems associated with variation in performance depending upon age, health and diet. Other techniques which may have limited use in odour classification, e.g. gas chromatography combined with mass spectrometry are time consuming and expensive. The aim of the electronic nose concept is to provide low cost and rapid sensory information by mimicking the discriminatory powers of the mammalian nose.

2. Engineering Aspects of Electronic Nose Technology

2.1 Sensors

An electronic nose (Fig. 2(b)) is an instrument based upon an array of electronic, non-specific chemical sensors, e.g. thin film conducting polymer gas sensors [3], thick film tin oxide semiconductors [4], and metal-oxide-semiconductor field-effect transistors (MOSFETs) [5].

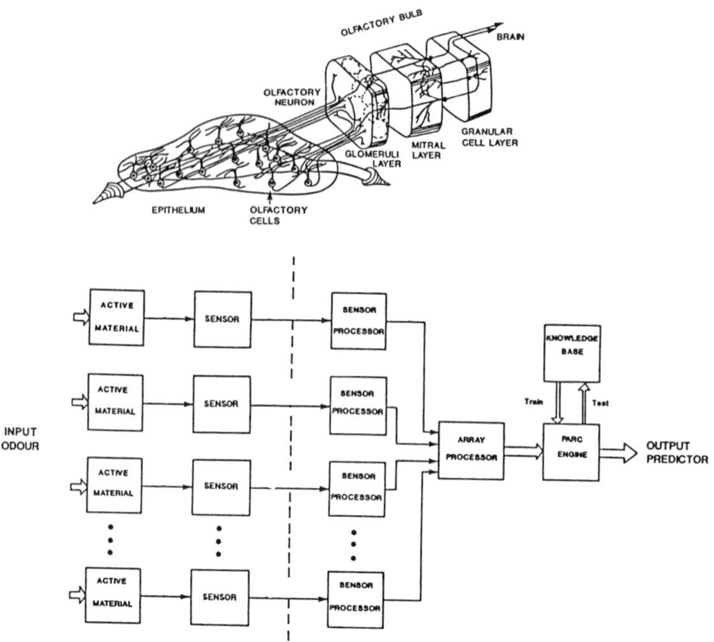

Fig. 2 (a) Schematic diagram of human olfactory system
(b) Architecture of electronic nose

By analogy to the human nose it is necessary to be able to identify many odours that may contain hundreds of individual chemical components. Many of the gas sensors listed above are ideal in this context in that they have partial sensitivity, i.e. they respond to a range of gases rather than a specific one.

Tin oxide semiconductor gas sensors are available commercially and are quite sensitive to combustible and reducing materials such as CH_4 and CO. By virtue of their operating temperature (300-500 °C) these sensors respond to odorants on the basis of a combustion mechanism. Conducting polymers have the advantage of being chemically functionalizable at the precursor stage offering a wider range of materials that can be tailored to meet the needs of the particular olfactory application. Sensors based on conductive polymers are also easier to process than the more conventional inorganic materials as they can be fabricated electrochemically or deposited by spin coating. Figure 3 shows a sensing head comprising an array of 12 discrete sensors using modified poly(pyrrole)s and poly(aniline)s [6]. These sensors are manufactured by electrochemical growth of the active material across a 10 μm gap between two gold electrodes evaporated onto a silicon substrate. Each device has an approximate coverage of 1 mm^2 polymer film.

Other sensing materials used include lipid layers, siloxanes and phthalocyanines. The former have been incorporated into arrays of surface acoustic wave devices [7] and thus an odour is detected by the event of a mass change brought about by chemisorption.

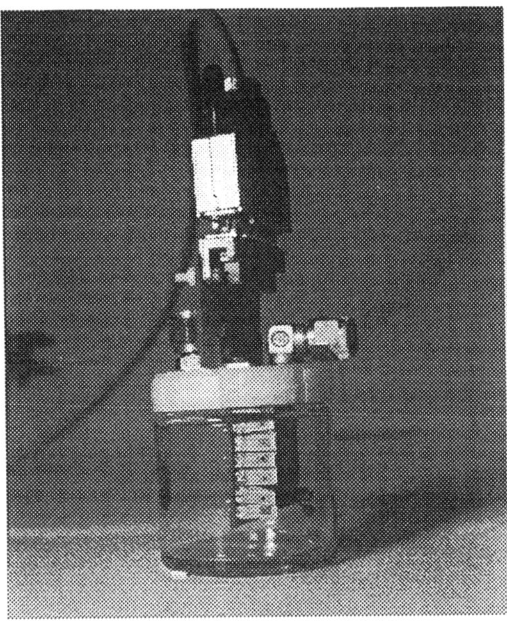

Fig. 3 Conducting polymer sensor array used in the Southampton-Warwick electronic nose

After an odour is presented to the active material , which converts a chemical input into an electrical signal the information is processed by interface electronics (Fig. 2(b)), e.g. conducting polymer resistances can be converted into an analogue voltage suitable for input into a microprocessor. The data can then be processed using simple algorithms before being transferred to a pattern recognition machine (section 2.3).

2.2 Sampling of Odours

The choice of sampling method and sample presentation depends upon the particular application, the form of the odour sample and the type of the sensors used. This is important in the acquisition of high quality and reproducible results. Two approaches are the use of a static headspace system or a flow injection system. In the former system a measured volume of odorant can be injected into a closed vessel containing the sensor array. In flow injection systems a carrier gas is used to carry the odour from the sample to the sensors. In this respect the arrangement is more like the mammalian system in which the flow of air carries the odour to the olfactory epithelium. However, the flow rate is an important consideration in such a system as metal oxide, conducting polymer and surface acoustic wave devices are sensitive to both temperature and humidity of the carrier gas. The static instrument has an advantage in that control of the temperature at which the measurement is carried out is relatively easy. This is important since temperature can affect both the release of volatile from the sample and the response of the individual sensors.

2.3 Pattern Recognition Techniques

Upon presentation of the sample to the electronic nose each sensor usually responds to a different degree to the next resulting in an individual pattern of sensor array responses for a given odour. It is this pattern of responses which forms the basis of discrimination between different odours. Pattern recognition (PARC) machines can take the form of either a multivariate statistical technique or artificial neural network [8,9].

A variety of methods have been adopted from the field of chemometrics, two such examples are discriminant function analysis (DFA) and principal component analysis (PCA). The main aim of DFA is to predict group membership from a set of dependent variables. PCA reduces the high dimensionality of a multivariate problem (12 dimensions in the case of the electronic nose depicted in Fig. 3), where the variables are partly correlated and allows the data to be displayed in a 2 or 3 dimensional scattergram. Other methods include template matching and cluster analysis [8,9].

Artificial neural networking techniques such as back-propagation [10] are quite attractive because they can handle non-linear data (as would be the case at high odour concentrations), and they are tolerant to sensor drift or noise. This then mimics, to a certain extent the human olfactory system.

3. Applications of Electronic Noses

3.1 Commercial Instruments

Odour monitors in the form of hand held devices have been available on the market for several years. However, although they measure odour intensity, they only contain one sensor and thus have no capability of distinguishing between different smells. Instruments that do meet the criteria for definition of an electronic nose are manufactured in the UK (AromaScanner - *Aromascan*), (NOSE - *Neotronics*) and France (Intelligent Nose - *Alpha MOS*). The latter system (Fig. 4) consists of an array of metal oxide sensors and operates as a flow injection sampler.

3.2 Some Typical Results

The most commonly reported application of electronic noses is their use to classify the aroma of various foods and beverages, e.g. coffees [11], beers [3], whiskies, cheese and fish [12]. Figure 5 shows the DFA results for the headspace odours of three commercial coffees obtained

Fig. 4 Commercial electronic nose

from a 12 element tin oxide array [11]. When comparing the group centroids, C the first discriminant function, Z_1 best separates out the coffee types labelled (a) and (b) while the second discriminant function, Z_2 best separates out coffee types (a) and (c).

Recently, applications in the agricultural environment have taken place. Odours from livestock wastes have been assessed by an electronic nose based on poly(pyrrole) sensors [13]. In this work a photoionization detector and an electronic nose were used to investigate odours arising from pig and chicken slurries. The photoionization detector proved to be the more sensitive technique whilst the electronic nose was more readily able to discriminate between different odours.

Another use has manifested itself in the potential monitoring of health in dairy cattle [14]. Ketosis, an important disorder in cows is traditionally diagnosed at the clinical stage by the veterinarian smelling the cow's breath. It may be possible to detect the condition before it reaches this stage by presenting breath samples to an electronic nose. Figure 6 shows the PCA results for breath samples of livestock with known condition of health [15,16], where normal breath samples are labelled as (H) and ketotic samples are labelled as (K). These preliminary findings suggest that it is possible to discriminate between healthy and sick cows with an 89% success rate of classifying ketosis samples. However, more detailed studies are required for obtaining repeatability of sample, as the sensing in this case is complicated by many biological and environmental factors.

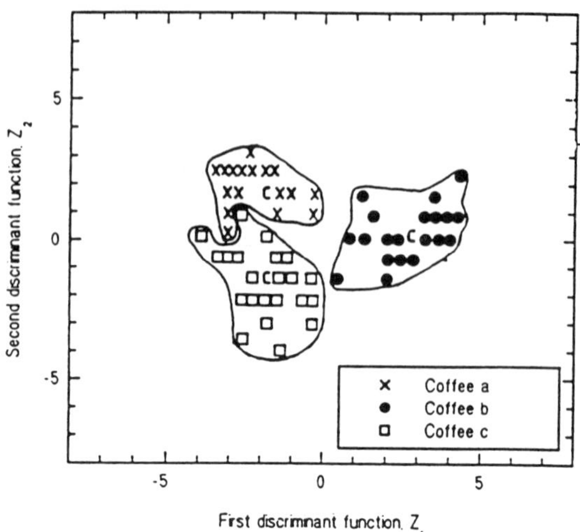

Fig. 5 Discriminant function analysis plot of three commercial coffee odours

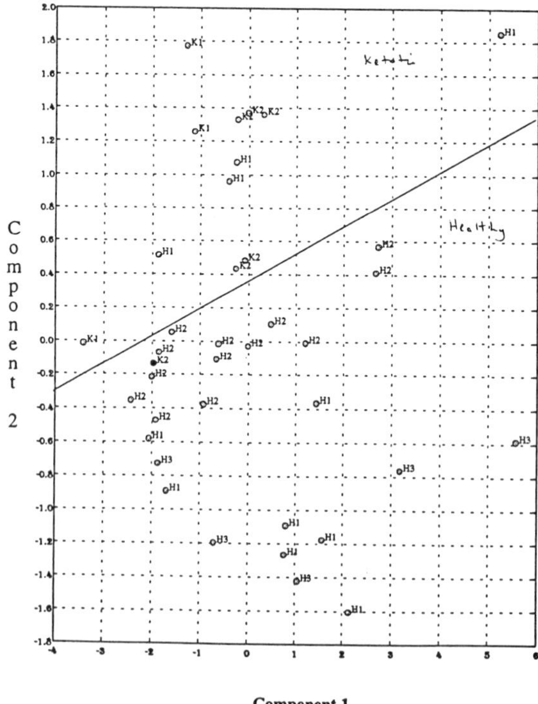

Fig. 6 Discrimination of healthy and diseased cow breath using principal components analysis

4. Conclusions

It is clear that, under laboratory conditions, arrays of chemical sensors linked to appropriate pattern recognition techniques are able to mimic some of the discriminatory capabilities of the human nose. Besides the assessment of various foodstuffs and beverages, there is considerable scope in the field of environmental monitoring. At present there is not a universal nose that can solve all odour sensing problems. Instead the technology is application specific, which means developing the appropriate sensor structures, materials and PARC methods.

Acknowledgements

The authors wish to acknowledge the BBSRC for funding a linked research group and MAFF for T.T. Mottram's part in this work.

5. References

[1] G.H. Dodd, P.N. Bartlett and J.W. Gardner, in J.W. Gardner and P.N. Bartlett (eds.), *Sensors and Sensory Systems for an Electronic Nose, NATO ASI Series E: Applied Sciences,* Vol. 212, Kluwer, Dordrecht, (1992), 1-13.

[2] M.G.J. Beets, *Structure-Activity Relationships in Human Chemoreception*, Applied Science Publishers, London, (1978)

[3] T.C. Pearce, J.W. Gardner, S. Friel, P.N. Bartlett, and N. Blair, *Analyst*, 118, (1993), 371-377

[4] H.V. Shurmer, J.W. Gardner, and P. Corcoran, *Sensors and Actuators B1*, (1990), 256-260

[5] F. Winquist, E. G. Hörnsten, H. Sundgren, and I. Lundström, *Meas. Sci. Technol.*, 4, (1993), 1493-1500

[6] J.W. Gardner, P.N. Bartlett, and J.M. Elliott, unpublished results

[7] M. Ohnishi, T. Ishibashi, Y. Kijima, C. Ishimoto, and J. Seto, *Sensors Mater.*, 4(1), (1992), 53-60

[8] J.W. Gardner and P.N. Bartlett, in P. Moseley, J. Norris, and D. Williams (eds), *Techniques and Mechanisms in Gas Sensing*, Adam Hilger, Bristol, (1991), 347-380

[9] J.W. Gardner and P.N. Bartlett, in J.W. Gardner and P.N. Bartlett (eds), *Sensors and Sensory Systems for an Electronic Nose, NATO ASI Series E: Applied Sciences,* Vol. 212, Kluwer, Dordrecht, (1992), 161-179

[10] J.W. Gardner, E.L. Hines, and M. Wilkinson, *Meas. Sci. Technol.*, 1, (1990), 446-451

[11] J.W. Gardner, H.V. Shurmer, and T.T. Tan, *Sensors and Actuators B,* 6, (1992), 71-75

[12] R. Ólafsson, E. Martinsdottir, G. Ólafsdottir, Š. I. Sigfusson, and J.W. Gardner, in J.W. Gardner and P.N. Bartlett (eds), *Sensors and Sensory Systems for an Electronic Nose, NATO ASI Series E: Applied Sciences*, Vol. 212, Kluwer, Dordrecht, (1992), 257-272

[13] P.J. Hobbs, T.H. Misselbrook, B.F. Pain, *J. Agric. Engng. Res.*, 60, (1995), 137-144

[14] T.T. Mottram, (1992), *UK Patent*, GB 9224404.5

[15] R.J. Elliott-Martin, T.T. Mottram, J.W. Gardner, and P.N. Bartlett, *Proceedings of the Symposium on Olfaction and Electronic Nose,* Toulouse, (1995)

[16] R.J. Elliott-Martin, T.T. Mottram, P.J. Hobbs, J.W. Gardner, and P.N. Bartlett, *J. Agric. Engng. Res.*, to be submitted

Hazardous Gas Sensing Using Near Infrared Laser Diodes

V. Weldon, J. O'Gorman, J. Hegarty and P. Phelan[1]

Optronics Ireland, Physics Department, Trinity College, Dublin 2, Ireland.

Tel. + 353-1-6081987. **Fax.** + 353-1-6798412

and

T. Tanbun-Ek

AT&T Bell Laboratories, New Jersey 07974, U.S.A.

Abstract

Optical sensors using narrow linewidth near infrared laser diodes provide high detectivity and selectivity in the area of trace gas sensing. We report on the sensing of the environmentally important gases CH_4, CO_2 and H_2S using distributed feedback laser diodes emitting at a wavelength (λ) around $\lambda = 1.5$ μm.

Introduction

Increased awareness of environmental issues has focused attention on the need for constant air quality monitoring of an increasing number of hazardous gases. Laser based optical absorption techniques have applications in the area of trace gas sensing. They are capable of high selectivity since they can probe specific vibrational features in the optical absorption spectrum of the target species. The detectivity achievable using optical techniques primarily depends on the magnitude of the absorption coefficient of the particular vibrational transition.

Intensive research and development of near-infrared laser diodes (LDs) for telecommunications applications has resulted in the ready availability of low threshold and narrow linewidth devices which are operable around room temperature. Concurrently there has been significant development of related technologies, e.g. efficient photodetectors and low cost laser drivers and controllers. However, near-infrared LDs (emitting in the

---

[1]*Present address:*
Opto Electronica, INESC, Rue Jose Falcao 110, 4000 Porto, Portugal

wavelength (λ) range 0.8 $\mu m \leq \lambda \leq$ 1.7 μm) also have applications in the area of near-infrared spectroscopy and optical absorption trace gas sensing since many gases of industrial and environmental interest have vibrational overtone and or combination absorption features in this wavelength range[1]. Absorption band strengths at these wavelengths, although significantly less than in the fundamental absorption wavelength region 2 $\mu m \leq \lambda \leq$ 30 μm, are of sufficient magnitude to achieve parts per million (ppm) sensitivity when laser based optical absorption gas sensing is used in conjunction with frequency or wavelength modulation spectroscopy (WMS) and synchronous detection signal recovery techniques.

In this paper we review our recent work[2 - 5] using telecommunications type distributed feedback (DFB) laser diodes for detection and monitoring of hazardous gases at low concentrations. The fundamental transitions of most gases of environmental interest, especially hydrocarbons and hydrosulphides, occur at long wavelengths, typically in the wavelength range 2 $\mu m \leq \lambda \leq$ 30 μm. These transitions arise from molecular vibrational modes, such as, in the case of H_2S, symmetric stretch mode (ν_1), bending mode (ν_2) and asymmetric stretch mode (ν_3) where ν_1, ν_2 and ν_3 refer to the transition frequencies.

Fortunately, many environmentally significant gases have vibrational overtone and combination absorption features in the range 0.8 $\mu m \leq \lambda \leq$ 1.7 μm. These features arise because anharmonicity of vibrational modes relaxes the selection rules governing transitions between energy levels. In general, overtone features associated with a fundamental vibration frequency υ_1 occur at approximately $2\upsilon_1$, $3\upsilon_1$, etc. Similarly, combination features will occur at frequencies approximately equal to the sum of two or more fundamental frequencies. Absorption band strengths at these wavelengths are significantly less than in the fundamental absorption region. For hydrogenic gases however, absorption is of sufficient magnitude to achieve parts per million (ppm) sensitivity when used in conjunction with frequency or wavelength modulation spectroscopy (WMS) and synchronous detection techniques. More importantly, these overtone and combination absorption features are accessible with near infrared laser diodes developed for the telecommunications industry and which, especially in comparison with lead salt laser diodes operating at wavelengths in the fundamental absorption region, are relatively inexpensive and operate around room temperature. In particular, single wavelength or single mode InGaAsP distributed feedback laser diodes with emission wavelengths around $\lambda = 1.3$ μm and λ=1.55μm have been used for spectroscopic monitoring of hazardous gases which are of environmental and industrial interest [6,7]. An obvious advantage of optical sensors, particularly at this wavelength, is their compatibility with optical fibres. This feature extends their potential use to distributed sensing where many locations, remote from the laser source, can be monitored simultaneously. A further advantage of such distributed optical sensing is that locations which are electrically noisy and chemically hostile may be effectively monitored in this way.

Experimental

In this paper we present a unified review of our recent work which targets the detection of CH_4, CO_2 and most recently H_2S, using an optical absorption system based on single mode, narrow linewidth DFB laser diodes which are routinely used in telecommunications applications. The most essential requirements for laser diodes for gas sensing applications are single frequency operation with a narrow linewidth and a wide and continuous wavelength tuning range. The laser should also exhibit (above lasing threshold) a linear light output power verses applied current in order to satisfy the requirements for the most sensitive signal recovery schemes such as WMS and synchronous detection.

In Figure 1 we show the light output power verses current for a DFB laser diode emitting at around a wavelength (λ) = 1.5 μm, while in Figure 2(a). we show the laser emission spectrum exhibiting a single frequency performance. Figure 2(b) shows the variation with bias current of the laser spectral linewidth measured using a delayed self heterodyne technique[8]. Clearly, for conditions under which we operate the laser the linewidth ($\delta\upsilon$) is < 5 MHz. which is significantly less than the typical low pressure gas absorption feature linewidth of about 400 MHz. In Figure 3 we show a schematic of the experimental set-up used to demonstrate the application of DFB laser diodes to trace gas sensing. Temperature control and stabilisation of the laser diode to 0.1°C was achieved using a thermoelectric module and a standard commercial temperature control unit. A multipass gas cell was used to allow open path absorption measurements in a controlled way. With this arrangement, pressure, concentration and absorption length can be easily varied and controlled and also baseline information obtained by evacuation. Coarse control of the laser diode emission wavelength was achieved by adjustment of the laser temperature. Fine tuning of the laser diode emission wavelength was achieved by varying the injection current. The laser light transmitted through the gas was detected with commercial InGaAs photodetector which provided good noise and dark current performance. Absorption measurements were made by detecting the change in intensity of the laser beam transmitted through the gas as the laser emission wavelength is swept across the absorption feature of interest.

Strong signals obtained with a high concentration of gas in the cell were detected by modulation of the laser intensity with a chopper and synchronous lock-in detection of the signal transmitted through the gas cell (Direct Detection). For weak signals, obtained with low gas concentrations, the chopper was removed and a technique known as Wavelength Modulation Spectroscopy (or harmonic detection) was employed. This technique is described in detail elsewhere [9]. It suffices to say here that as the laser emission wavelength is scaned through the absorption feature, the change in optical transmission (due to the modulation of the DFB emission wavelength by an applied A.C. current) leads to a non-linearity in the detected response. Consequently, a weak absorption signal may be synchronously detected at the second harmonic of the modulation frequency, 2F. This technique allows high sensitivity detection with excellent signal to noise ratio (SNR).

Results

Figure 4 shows the absorption spectrum obtained for a H_2S sample of 99.96% purity at a pressure of 300mB over a 1 metre path length. The absorption line at around $\lambda = 1576.1$ nm was targeted to find the low detection limit for H_2S in this wavelength region. The detection limit was not directly measured, but extrapolated from the magnitude of the absorption signal obtained using wavelength modulation spectroscopy, shown in Figure 5, when the H_2S concentration was 1.5×10^5 ppm and the corresponding signal to noise ratio. From the corresponding measured noise signal at 1576.1nm (not shown) the peak to peak SRN is calculated to be approximately 3600. Hence the minimum detectable H_2S concentration path length product is [(1.5 x 10^5 * L)/(3.6 x 10^3)] ppm.m. Since the path length (L) is 1 metre the detection limit is approximately 42 ppm. Therefore, a path length greater that approximately 5 metres provides a low detection limit of less than 10 ppm. Similar measurements on other hazardous gases have shown detection limits around the ppm level. For example we have monitored the CH_4 rotational absorption line (R8) in the $2\upsilon_3$ vibrational overtone at a wavelength around 1.64μm and obtained a detectivity of 1 ppm over a 1 metre path length [2]. Targeting a CO_2 absorption line at λ=1.5769μm, within the tuning range of the same DFB laser diode used for H_2S detection, yields a detectivity of 100ppm over a path length of 40 metres.

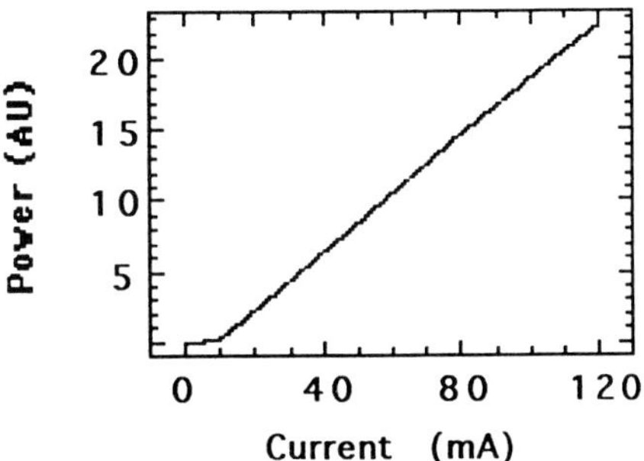

Figure 1. Output optical power verses current for a DFB laser diode.

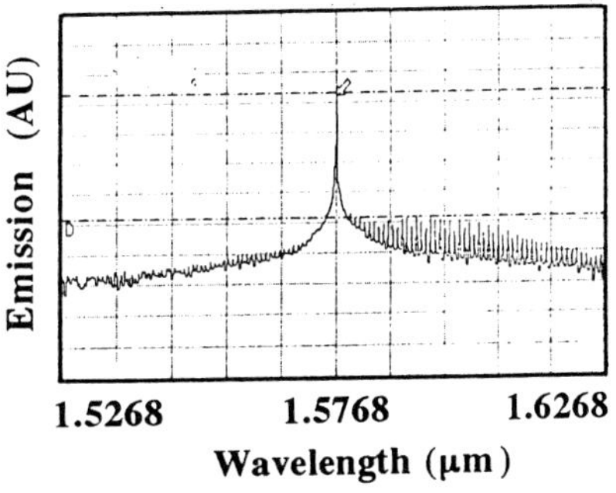

Figure 2(a). Spectral emission for a DFB laser diode showing single frequency performance.

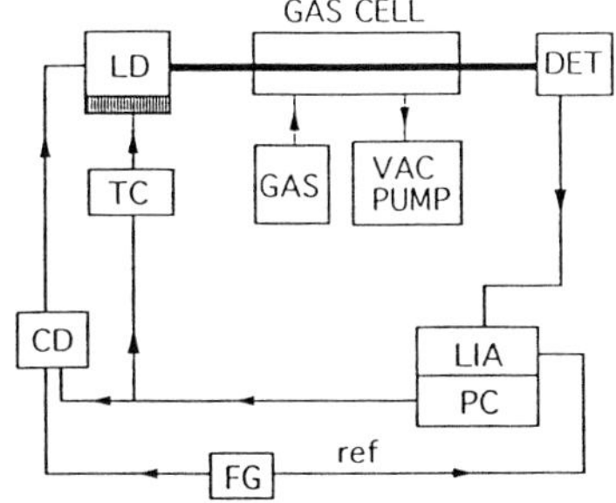

Figure 3. Schematic of experimental set-up. **CD** Current Driver, **TC** Temperature Controller, **LD** Laser Diode, **LIA** Lock-in Amplifier, **FG** Function Generator, **DET** Detector.

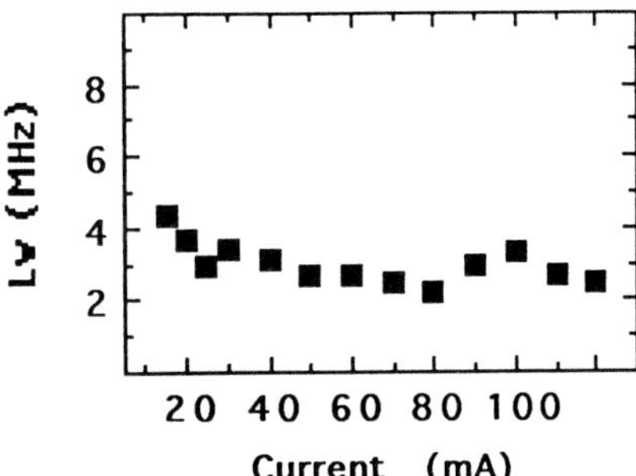

Figure 2(b). Spectral linewidth for a DFB laser diode as a function of bias current.

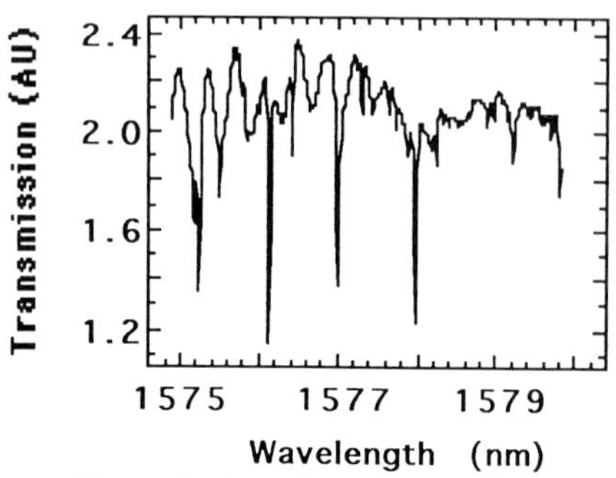

Figure 4. Absorption spectrum of H_2S at a pressure of 300mB in the wavelength range1575 nm $\leq \lambda \leq$ 1580 nm.

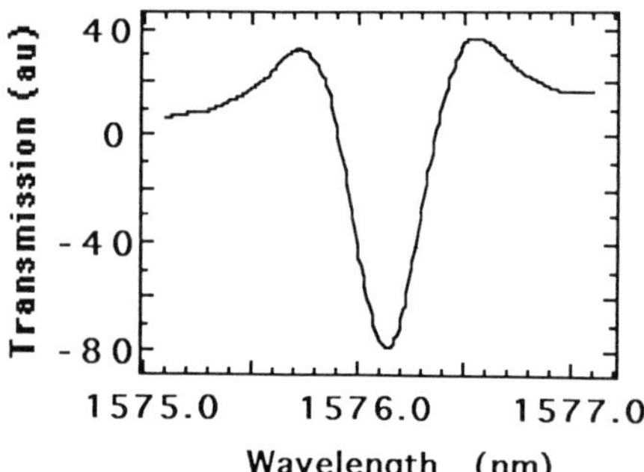

Figure 5. Absorption spectrum of a H_2S sample of 1.5×10^5 ppm. concentration buffered by Nitrogen gas at atmospheric pressure obtained by WMS.

Conclusion

In conclusion we have investigated the sensing of hazardous gases at low concentrations using optical absorption methods based on standard components developed for the telecommunications industry. We have shown that such systems are capable of detection around the ppm level. Our detection limits are below the commonly accepted hazardous threshold levels for the gases targeted. While our limits are dependent on path length, the primary attraction of such optical systems is for applications such as perimeter line monitoring of industrial sites where monitoring using such long path lengths is not only feasible but *it is required.*

References

1. Bomse,D.S., Hovde,D.C., Oh,D.B., Silver,J.A., Stanton,A.C.,"Diode Laser Spectroscopy for on-line Chemical Analysis", SPIE, Vol. 1681, 138-148 (1992).
2. Weldon,V., Phelan,P., Hegarty,J., "Use of rotational sideband absorption for high sensitivity methane detection with a 1.64 μm DFB laser" Electronics Letters, Vol.28, No.22, October 1992.
3. Weldon,V., Phelan,P., Hegarty,J., "Methane and Carbon Dioxide sensing using a DFB laser diode operating at 1.64 μm," Electronics Letters, Vol.29, No.6 March 1993.
4. Weldon,V., O'Gorman,J., Phelan,P., Hegarty,J., Tanbun-EkT,T., "H_2S and CO_2 Gas Sensing using DFB Laser Diodes Emitting at 1.57 μm" to be published in Sensors and

Actuators B, proc. of 2nd European Conference on Optical Chemical sensors and Biosensors, Firenze, Italy, April,1994.

5. Weldon,V., O'Gorman,J., Phelan,P., Tanbun-Ek,T.,"Gas Sensing with $\lambda = 1.57\mu m$ DFB Laser Diodes using Overtone and Combination Band Absorption", SPIE Optical Engineering Vol. 3 No. 12 pp 3867-3870. Dec. 1994.
6. Uehara,K., Tai,H.,"Remote detection of Methane using a 1.66mm diode laser", Appl. Opt., **31** (1992) 809-814.
7. Tai,H., Yamamoto,K., Osawa,S., Uehara,K.,"Remote detection of Methane using a 1.66μm diode laser in combination with optical fibres", Inst. Radio and Elect. Eng., Proc. of the 7th OFS Conference, Sydney, Australia, Dec.1990, 51-54.
8. Okoshi,T., Kikuchi,K., Nakayama,A., "Novel method for high resolution measurement oflaser output spectrum" Electronics Letters, Vol.16, No.16, July 1980.
9. Bromse,D.S., Stanton,A.C., Silver,J.A., "Frequency modulation and Wavelength modulation Spectroscopies:comparison of experimental methods using Lead-Salt diode lasers", Appl. Opt.,. **31**, (1992) 718-731.

OPTICAL GAS RESPONSE OF AN ORDERED AZOBENZENE-POLYSILOXANE FILM

A. Newton, L.S. Miller, I.R. Peterson♦

Centre for Molecular and Biomolecular Electronics, School of N.E.S.
Coventry University, Priory Street, Coventry CV1 5FB, UK

ABSTRACT: The time-dependence of reflectance of 90 nm thick Langmuir-Blodgett films of a polysiloxane with azobenzene sidechains was monitored as a function of temperature. The activation energy of the response was 0.63 eV, which is presumed to be the binding-site depth. The recovery had an activation energy of 0.21 eV suggesting that it was limited by other mechanisms. Absorption measurements on a very thick film revealed additional absorption due to the gas, peaking at 480 nm and with an oscillator strength of about 4% of the total, consistent with the estimate for binding site depth obtained from temperature variation.

1. Introduction

Over the last few years it has become apparent that the optical parameters of thin films provide a very interesting principle for gas sensing[1,2]. A number of different configurations have been proposed for accurately converting changes of these parameters to light intensities. The optical signal can then be converted accurately to electrical form over a very large dynamic range. Typical film thicknesses are comparable to the wavelength of light, so that the time for diffusion of gas molecules to active sites in the film can be usefully small.

Recently, a system of this sort has been demonstrated for the detection of NO_2, in which the sensing film is fabricated by the Langmuir-Blodgett deposition of a siloxane polymer with substituted azobenzene sidegroups•. This system typically responds to the gas in <10 s, with a detection threshold down to 1 ppm, and essentially complete recovery over times essentially equal to the response times. The selectivity of response is excellent, but a slow degradation of the sensitivity on prolonged exposure is observed.

In the present work we provide a more detailed characterisation of the films to clarify the sensing mechanism as a prelude to optimisation of the system. Reported measurements include the optical absorption spectrum and the response as a function of temperature, each as a function of gas concentration. This enables some conclusions to be drawn about the nature of transport and binding mechanisms.

♦ Also at NIMA Technology Ltd, The Science Park, Coventry CV4 7EZ.

• L. S. Miller, M. Simpson, D. A. Parry, D. J. Walton, A. M. McRoberts, Azobenzene polysiloxane optical gas sensor: Sixth Sense **23**, Autumn 1994 and at The Manufcaturing Week, N.E.C., U.K., 1994.

2. Experimental

The substituted siloxane polymer was synthesised[3] by the route of Carr et al.[4] and displayed a melting point of 193-194 °C. The polymer was made up into a spreading solution in chloroform (BDH, AnalaR grade) at a chromophore concentration of 0.002 M and stored in the dark under refrigeration (5 °C) until use.

The Langmuir-Blodgett deposition facilities at Coventry have been described previously[5]. Monolayers were spread on a pure-water subphase and compressed to a surface pressure of 30 mN/m. At a temperature of 20 °C, multilayers were deposited onto silanated glass and silicon substrates by successive immersion and withdrawal. The optical absorption spectrum and interference-enhanced[2] reflectance responses of the films were measured using the system described elsewhere[6]. During the recording of the spectrum, the sample temperature and the composition of the gas mixture in contact with it were computer controlled.

3. Results and discussion

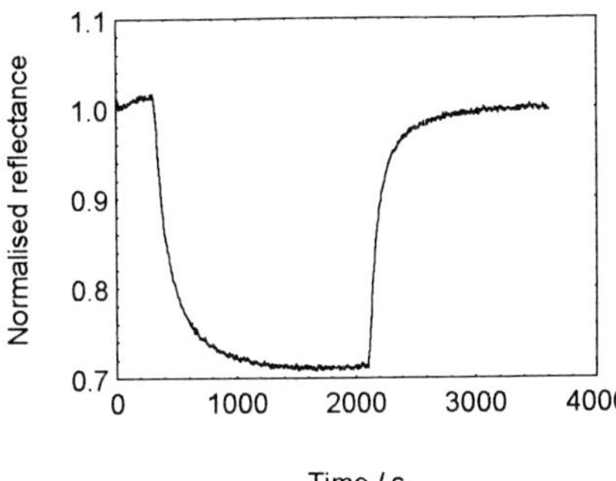

Figure 1: reflectance vs time for a 42 layer sample on silicon.

Figure 1 shows a typical fractional reflectance change. This has been normalised to the reflectance in air. The sample was initially in dry air, which was switched to 100 ppm of NO_2 in dry air and then back to dry air. This procedure was repeated over a range of temperatures between -11°C and +34°C. Figure 2 shows the maximum fractional change in reflectance as a function of the reciprocal temperature. Interpretation as a simple exponential (fractional change of reflectance = $A \exp[-E_b/kT]$) leads to an activation energy E_b of 0.63 eV.

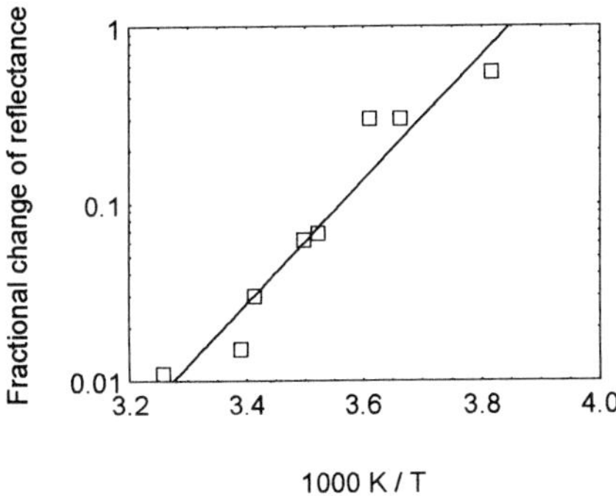

Figure 2: fractional change of reflectance vs reciprocal temperature.

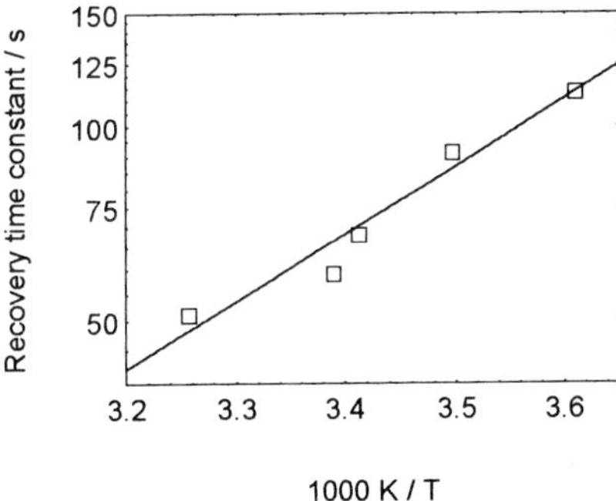

Figure 3: recovery time constant vs reciprocal temperature.

The recovery times in figure 1 (which were actually longer than for most samples) were least-squares fitted to single exponentials with adjustable amplitudes and time constants. The deduced time-constants are plotted in figure 3 as a function of reciprocal temperature; in this case the assumption of Arrhenius behaviour leads to an activation energy for desorption of 0.21 eV. Not all the curves could be well-fitted to a single exponential, which results in fewer points than for figure 2, and alternative procedures for fitting the recovery can lead to a variation of the finally-fitted activation energy to a maximum value of around 0.3 eV.

The discrepancy between the two activation energies of 0.63 eV and 0.21 eV derived from the two types of measurement is significant and suggests that the binding process responsible for sensing is not the limiting factor in recovery.

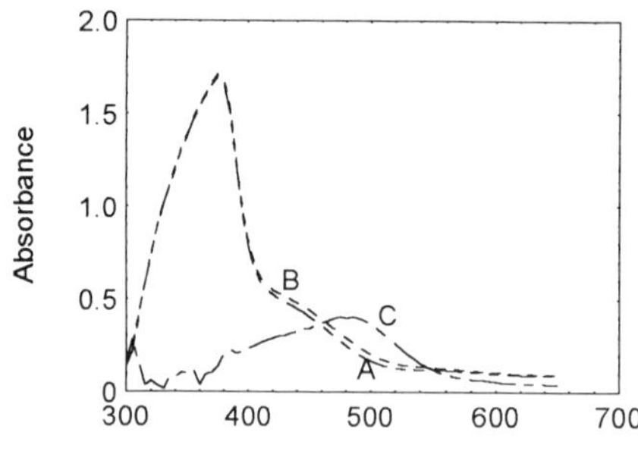

Figure 4: absorbance of a 414-layer film before and during exposure to NO_2 and the difference × 10.

Figure 4 shows the absorbance spectrum at room temperature of a 414 layer film (A) in contact with dry air, (B) after 20 min. exposure to a concentration of 100 ppm NO_2 in dry air and (C) their difference (on a 10x expanded scale).
It can be seen that the additional absorbance has a maximum at about 480 nm. The area under this curve is about 4% of the original total absorbance integral.

The absorption peaks of NO_2 and N_2O_4 gases occur at 400 nm and below, ruling them out as the chemical species causing the change in optical properties. In any case unbound gas would not be expected to be present in the films at volume concentrations much higher than that in the gas phase, which would give negligible absorption. This indicates the intimate involvement of the azobenzene moieties in the sensing process.

Since the gas sensing action is observed to be rapidly reversible, and the apparent activation energy of the process is smaller than that of typical chemical bonds, this sensing must involve the formation of an adduct between an NO_2 molecule and some site on an azobenzene. The simultaneous occurrence of a chemical reaction or *trans/cis* isomerisation cannot be excluded at this stage. To give rise to a change in optical parameters, the optical behaviour of the adduct must be different from that of the parent azobenzene. The total oscillator strength after exposure to the gas increases implying that the oscillator strength of the adduct is greater than that of the original azobenzene. Hence the fraction of adduct is probably less than the 4% change in total absorbance.

On the assumption that the binding sites are independent and have a well-defined binding energy, it is possible to relate the fraction of occupied sites to the concentration in the gas phase. For low concentrations:

$$\theta = C \times (h^2/2\pi MkT)^{3/2} \times \exp(-E_a/kT),$$

where θ is the fraction of occupied sites, C the concentration of the active species in the gas, h is the Planck constant, M is the mass of an NO_2 molecule, k is the Boltzmann constant and T absolute temperature. Using the above estimate of <4% for the adduct fraction, this equation leads to an upper limit on the binding energy of about 0.58 eV, in satisfactory agreement with the value obtained from the temperature dependence of the sensing signal.

4. Conclusions

1. A polysiloxane with azobenzene sidechains has been deposited on silicon as a thin (about 90 nm) film by the Langmuir-Blodgett technique and studied as an optical NO_2 sensor based on interference-enhanced reflectometry. The temperature was varied from -11°C to +34°C. The absorbance spectrum of a film around 900 nm in thickness has also been taken in the presence or absence of the gas.
2. The optical sensing action is found to have an activation energy of about 0.63 eV, which indicates the binding sites to be of this depth. The increased absorption of the thick film in the presence of the gas indicates an upper limit to the binding energy of 0.58 eV, which is in reasonable agreement.
3. The recovery of this material, when fitted to a single exponential, displays an activation energy of 0.21 eV, suggesting that the recovery is limited by other processes.

5. References

1. A. Mandelis and C. Christofides, Solid State Gas Sensor Devices, *Chemical Analysis* **125**, Wiley 1993.
2. L. S. Miller, A. L. Newton, C. G. D. Sykesud and D. J. Walton, Optical Gas Sensing Using Langmuir-Blodgett Films, *Sensors, technology, systems and applications*, Adam Hilger 1991.
3. L. S. Miller, D. J. Walton, P. J. W. Stone, A. M. McRoberts and R. S. Sethi, Langmuir-Blodgett films for nonlinear optical applications, *J. Mat. Sci: Mats. in Electronics*, **5** (1994) 75-82.
4. N. Carr, M. J. Goodwin, A. M. McRoberts, G. W. Gray, R. Marsden and R. M. Scrowston, *Makromol. Chem., Rapid Comm.* **8** (1987) 487.
5. L. S. Miller, D. E. Hookes, P. J. Travers and A. P. Murphy, *J. Phys E: Sci. Inst.* **21** (1988) 163.
6. L. S. Miller, A. M. McRoberts, D. J. Walton, D. A. Parry and A. L. Newton, Optical Gas Sensing Using Langmuir-Blodgett Films, ICMBE, Goa 1994 and submitted Mat. Sci. Eng. C: Biomimetic Mats., Sensors and Systems (conference edition).

The use of tin dioxide based thick films as room temperature humidity sensors

Geraint Williams and Gary S.V. Coles

Department of Electrical and Electronic Engineering, University of Wales Swansea, Singleton Park, Swansea, SA2 8PP, U.K.

Abstract: Sintered films of tin oxide mixed with bismuth oxide held at room temperature experience considerable changes of up to 4 orders of magnitude in both capacitance and d.c. resistance under conditions of varying relative humidity. Studies have shown that varying the initial sensor composition can markedly change the room temperature humidity response characteristics. For example, the capacitance of a sensor prepared from pure SnO_2 shows no variation over the whole relative humidity test range while a film of $Sn_2Bi_2O_7$ responds only to relative humidities of 50% or higher. The greatest humidity sensitivities are exhibited by SnO_2 based films which have an initial Bi_2O_3 composition ranging between 15 and 25% w/w.

1. Introduction

Semiconductor gas sensors based on tin dioxide are usually operated at elevated temperatures and respond via resistance changes to the presence of low levels of combustible gases in the surrounding atmosphere. Water vapour can adsorb on the tin dioxide surface to form surface hydroxyl species with the subsequent release of electrons to the semiconductor conduction band. Consequently an increase in ambient humidity leads to decreased sensor resistance and thus interferes with measurements carried out in reducing gas containing environments. Researchers have attempted to utilise this usually undesirable property as the basis of a stannic oxide humidity sensor operated at 350°C [1]. However, the significant reducing gas cross sensitivity and temperature dependence of the device require that filters and compensation circuits must be employed for satisfactory operation. In this paper, results showing that tin dioxide thick films which incorporate Bi_2O_3 can be used as room temperature humidity sensors will be presented. These films behave in a manner similar to porous ceramic humidity sensors where the ionic conductivity of the sensing element, usually a metal oxide, increases as water vapour is physisorbed on the surface and within the micropores of the material [2,3]. Proton transfer between neighbouring molecules of the condensed water is thought to be the principal mechanism responsible for conduction within the ceramic, especially at moderate to high humidities.

2. Experimental Details

Initial results were obtained using sensors prepared from a mixture of SnO_2 powder (Keeling and Walker Superlite) and 17% w/w Bi_2O_3 (B.D.H. Analar grade) which was screen printed over the contact array of a planar alumina substrate and subjected to heat treatment at 800°C in air. X-ray diffraction analysis of the sensor material has shown that all the Bi_2O_3 undergoes a solid state reaction with the SnO_2 to form $Bi_2Sn_2O_7$ [4] so that the sensing film is actually a mixture of tin dioxide and the stannate (28% w/w in this case). A sensor composed wholly of $Bi_2Sn_2O_7$ can be prepared via the heat treatment of a mixture of

SnO_2 and Bi_2O_3 (61% w/w) powders. Sensors of varying composition in the range 0 - 61% w/w Bi_2O_3 were prepared and tested by monitoring the capacitance and resistance changes observed when placed in the headspace above a series of aqueous saturated salt solutions held within a constant temperature chamber.

3. Results and Discussion

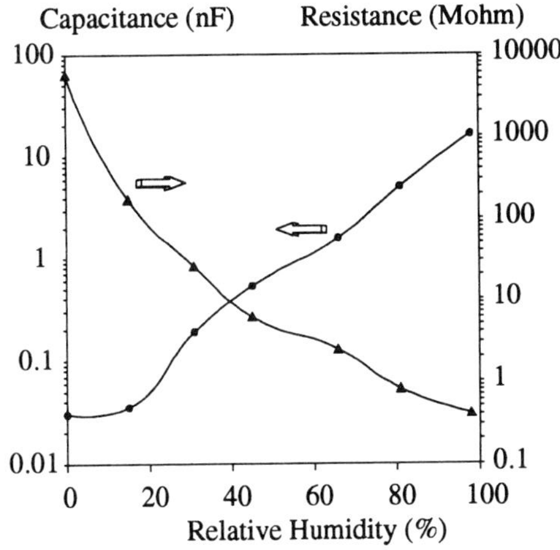

Figure 1: *Effect of changes in relative humidity on the capacitance and resistance of a thick film sensor composed of SnO_2+Bi_2O_3 (17% w/w) held at 20° C.*

Figure 1 shows the changes observed in both capacitance and d.c. resistance of a SnO_2/Bi_2O_3 (17% w/w) thick film held at 20°C as the relative humidity of the surrounding environment is varied between 0 and 98%. Capacitance increases of up to 3 orders of magnitude are matched by even greater decreases in d.c. resistance. The nature of the response curves plotted in figure 1 indicate that for actual applications, thick film capacitance should be monitored for relative humidities in excess of 40%, while changes in d.c resistance should be used for R.H. < 40%. Experiments carried out to determine the response time of the films revealed that constant capacitance or resistance values were reached within minutes of increasing or decreasing the relative humidity (see Figure 2).

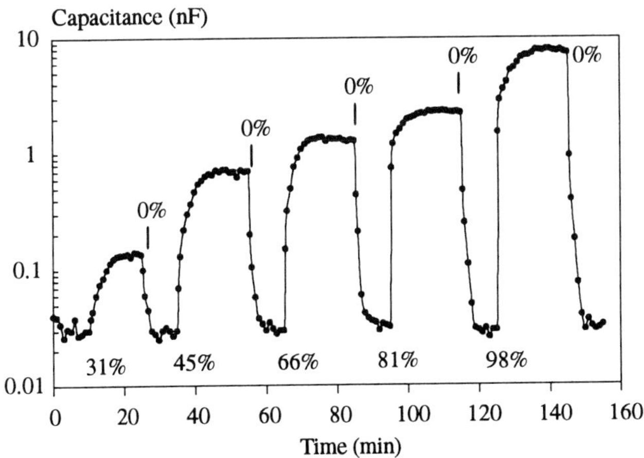

Figure 2: *The dynamic capacitance response of a thick film SnO_2-Bi_2O_3 (17% w/w) sensor film held at 20° C to changes in relative humidity.*

As would be expected for semiconductor based materials, the sensors exhibit a temperature dependence of the humidity response in the10 - 30°C range. Figure 3 shows that an increase in ambient temperature leads to a drop in the size of the overall resistance changes observed in the 0 - 98% humidity range and an enhancement in the capacitance modulations measured under the same conditions.

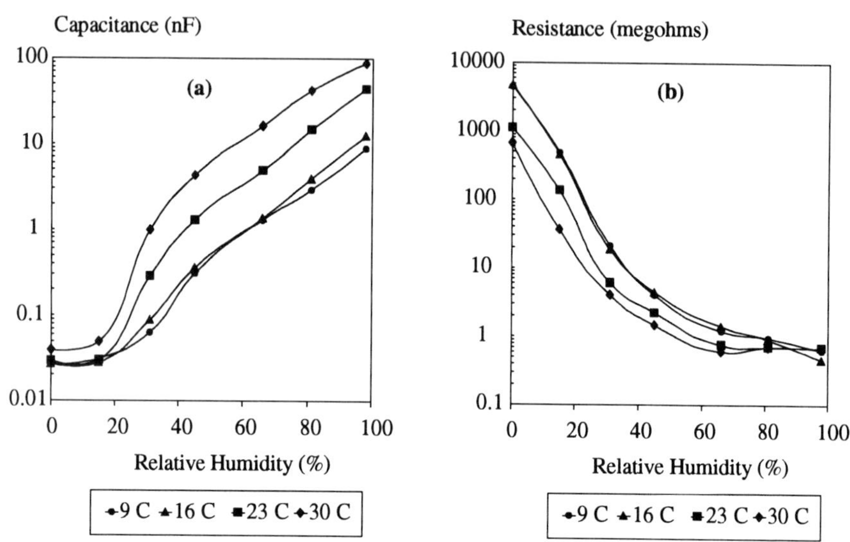

Figure 3: *The effect of ambient temperature variation on (a) capacitance and (b) resistance changes observed in different humidities for a thick film sensor composed of* SnO_2-Bi_2O_3 *(17% w/w).*

Studies have also shown that varying the initial SnO_2-Bi_2O_3 composition can markedly change the room temperature humidity response characteristics. For example, the capacitance of a sensor prepared from SnO_2 only shows no variation over the whole relative humidity test range. However, the incorporation of bismuth oxide causes increased resistivity and decreased capacitance and confers a significant humidity response whose characteristics can be modified depending on the initial quantity of Bi_2O_3 added. A summary of the results obtained for a series of sensors of varying composition is shown in Figure 4. Sensors composed purely of $Bi_2Sn_2O_7$ (*i.e.* having an initial 61% w/w Bi_2O_3 content) respond only to changes in humidity above levels of 60% or greater, while sensors possessing a stannate content of between 13 and 28% w/w display humidity sensitivity at levels of 20% R.H. or greater.

After a period of several weeks storage under ambient conditions the films become insensitive to changes in humidity and retain a high resistance and low capacitance over the whole R.H. test range. However, the original humidity response of the films can be regenerated by simply heating the films to 300°C for 1h in air. This type of "heat refreshment" to regain humidity response characteristics lost when irreversible adsorption occurs is common in porous metal oxide ceramic sensors [2,3,5] which depend on capillary condensation of water vapour causing ionic conductivity within the sensing material. Sensor passivation can also be avoided by returning the films to a dry environment when they are not in use.

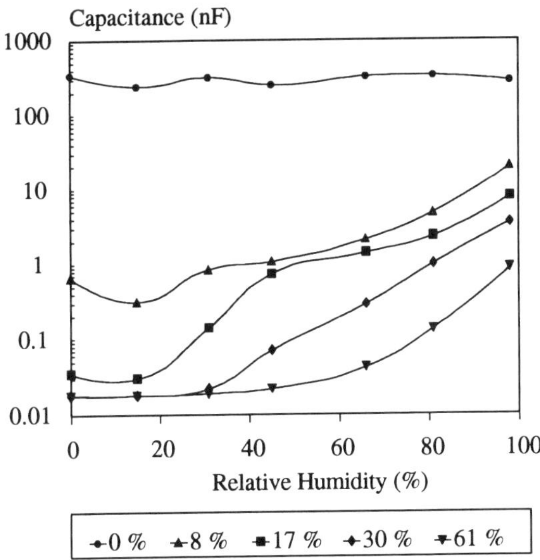

Figure 4: *Capacitance changes observed at 20° C as a function of increasing R.H. for a series of SnO_2/Bi_2O_3 based thick films. Compositions are given in weight percentage.*

When maintained at elevated temperatures in the range 200 - 500°C, the resistance of SnO_2 thick films, both in the presence and absence of the Bi_2O_3 additive, decreases with increasing relative humidity. However, the magnitude of the resistance changes observed are small in comparison with those displayed by the SnO_2 - Bi_2O_3 films held at 20°C. For example, the resistance changes observed for an undoped SnO_2 and a SnO_2/Bi_2O_3 (17% w/w) sensor operated at 300°C are 5 fold and 3 fold respectively upon testing over the 0 - 98% relative humidity range. Compare these results with the findings obtained at room temperature where the resistance/capacitance of the undoped SnO_2 film remains unchanged over the same test range while the Bi - added sensor exhibits a resistance capacitance change of several orders of magnitude. Therefore it is evident that thick films composed of a sintered SnO_2 - Bi_2O_3 mixture can act both as ionic and electronic type ceramic humidity sensors. At elevated temperatures the resistance changes observed are caused by the donation of electrons from water molecules to the conduction band of the SnO_2 as chemisorption of H_2O to form surface hydroxyl species occurs. The room temperature humidity sensitivity of the material arises as a result of capillary condensation of water vapour within the micropores of the tin dioxide - bismuth stannate system.

4. Conclusion

Changes in relative humidity give rise to considerable variations in both the capacitance and resistance of thick sintered films of tin dioxide mixed with bismuth oxide held at room temperature. Experiments have shown that stable sensor readings are obtained within approximately 5 min of varying the R.H., while all the changes in film capacitance and resistance are reversible. As with most semiconductor based ceramics, the humidity sensitivity is temperature dependent, while passivation of response occurs over a period of weeks when the films are stored in ambient conditions, though this can easily be overcome by heat cleaning at regular intervals.

References

1. S. Mukode and H. Futata, A semiconductive humidity sensor, Sensors and Actuators, **16** (1991) 1-11.
2. T. Seiyama, N. Yamazoe and H. Arai, Ceramic humidity sensors, Sensors and Actuators, **4** (1983) 85-96.
3. N. Yamazoe and Y. Shimizu, Humidity sensors: Principles and Applications, Sensors and Actuators, **10** (1986) 379-398).
4. G.S.V. Coles, S.E. Bond and G. Williams, Metal Stannates and their role as potential gas sensing elements, J. Mater. Chem., **4(1)** (1994) 23-27.
5. Y. Yokomizo, S. Uno, M. Harata, H. Hideaki and K. Yuki, Microstructure and humidity sensitive properties of $ZnCr_2O_4$-$LiZnVO_4$ ceramic sensors, Sensors and Actuators, **4** (1983) 559-606.

New substrates for thick and thin-film gas sensors

Matthew P. Elwin , Gary S. V. Coles and J. Watson

Department of Electrical and Electronic Engineering, University of Wales Swansea, Singleton Park, Swansea SA2 8PP, U.K.

Abstract: The majority of solid state gas sensors comprise two main parts, the active layer and a substrate which usually consists of a heater and a contact array through which the active material can be interrogated. Predominantly, research in this field has focused on the semiconductor layer. However there is much scope for improvement in the substrate which can increase the performance and efficiency of the sensor. A review of current substrate design and performance is presented and then recent and proposed in house work is described.

1. Introduction

The increasing demand for gas sensors in domestic and industrial environments has lead to the development at Swansea of four selective sensors. All are based on a layer of tin dioxide deposited onto a substrate [1,2]. The importance of the substrate is evident since it not only provides the mechanical support and electrical contact to the active material, but also produces the high temperatures required to optimise the oxidation process at the semiconductor surface which is the basis of the gas sensing mechanism. Besides selectivity there are other properties which commercially viable devices should possess. These are briefly described in table 1. The majority of these properties are in some way influenced by the substrate on which the sensor is produced. Selectivity is often achieved by the addition of dopants, but the temperature of the substrate is also a crucial parameter. Sensitivity is to some extent dependent on the surface area of active material and thus on the size and shape of the substrate. It will

Table1:- Desirable properties of gas sensors

Selectivity	Sensors should respond to just one gas in the presence of others.
Sensitivity	Sensors should be able to detect low gas concentrations.
Fast response speed	Giving read out or alarm as soon as the danger arises.
Low ambient temperature and humidity variations	Enabling sensor operation in variable environments.
Negligible drift	Preventing the need for regular calibration or characterisation.
Longevity	Sensors should maintain their characteristics years after installation.
Power consumption	This should be kept a minimum especially for battery operated devices.
Cost	Sensors should be cheap especially for domestic use.

also depend on the contact array; for example if a highly resistive compound is used then the sensor should be prepared using a substrate with interdigitated contacts in order to reduce the base resistance to an easily measurable value (see figure 1c.I). Alternatively, if the compound is of low resistivity then a simple contact array would suffice (figure 1c.II). The response speed of the sensor will depend on the substrate temperature since the oxidation process occurs much more rapidly at high temperatures. However there is an upper temperature limit resulting from thermal excitation of electrons from the valence band to the conduction band which swamps the conductance change due to the presence of contaminant gas. Sudden changes in ambient temperature due to drafts will tend to affect the overall temperature of the substrate. These effects can be reduced by suitable substrate design and sensor housing. Control circuits incorporating a wheatstone bridge and differential amplifier are also used to keep the resistance and hence temperature of the heater reasonably constant [3] .

The majority of these characteristics can be improved by better substrate design, though it is the power consumption and cost which are of primary interest since they are heavily dependent on the substrate configuration. The power consumption of a semiconductor gas sensor is due almost entirely to the current being passed through the heater track, the remainder being drawn by control and monitoring circuitry. The cost of these sensors is heavily influenced by the amount of highly technical equipment required to produce the substrates (e.g. clean room facilities, sputtering rigs, screen printers, laser cutters and the housing). However when manufactured in large quantities these costs can be significantly reduced.

2. Present Substrates

There are a number of substrates currently in use by the semiconductor gas sensor industry and research labourites. A selection are reviewed here, all of which are based on alumina. Companies such as Figaro produce complete gas sensors whilst other companies such as Rosemount manufacture the substrate alone. However, due to high cost, several research institutions are developing their own substrates. Two examples are shown in figure 1d & e. All of these substrates are illustrated in figure 1 and their properties are described below. In an experiment to determine the power consumption of these substrates at given temperatures a thermocouple was mounted on the centre of the alumina tile using a room temperature curing ceramic paste. A variable supply was then used to heat the substrate by passing a current through

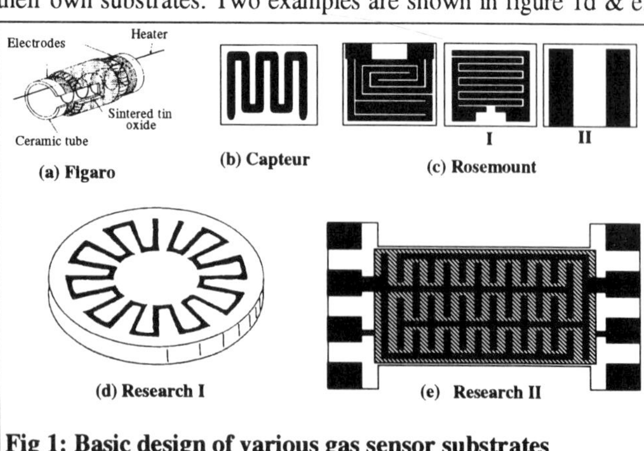

Fig 1: Basic design of various gas sensor substrates

the heater track. The results are presented in figure 2. The temperature variation across the surface was obtained by repeating the above experiment with the thermocouple on the edge of the substrate. Figure 3 shows the results obtained for a research II substrate.

Figaro substrates are probably best developed on the market. In this device the heater is a simple coil of wire which passes through the centre of a ceramic tube. Its tubular shape ensures that the majority of the heat must enter the active material. Also, small size mean it has a comparatively low power consumption. However it still has a fairly large temperature gradient between the centre and ends of the tube. The Capteur and Rosemount substrates are very similar although Capteur has a higher power consumption which may be due to the fact that its heater is not as well developed and also because it is mounted on its side resulting in more heat escaping directly from the heater by convection rather than passing through the substrate. Both are made from small alumina squares with the heater and contact array screen printed on opposite sides. Research substrate I also use this configuration, though the alumina tile is larger, with heater and contact array designs being circular in an attempt to reduce temperature gradients across the surface. However, as with all designs of this type there is a temperature distribution across the substrate with the hottest place at the centre. Research II substrates are also constructed from a thin ceramic tile but in this case the heater has been printed onto the same side as the contact array and the two have been separated by a thin electrically insulating layer. The advantage with this method is that less heat escapes, so that the substrate is more efficient. However due to the size of this substrate the power consumption is very high and there are again temperature gradients across the surface.

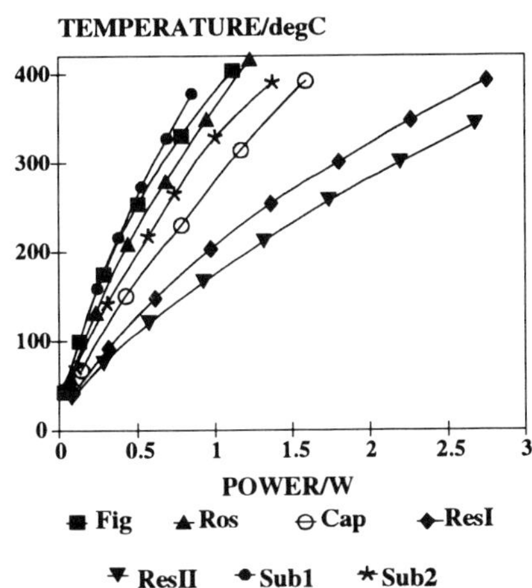

Fig 2: Substrate power consumption

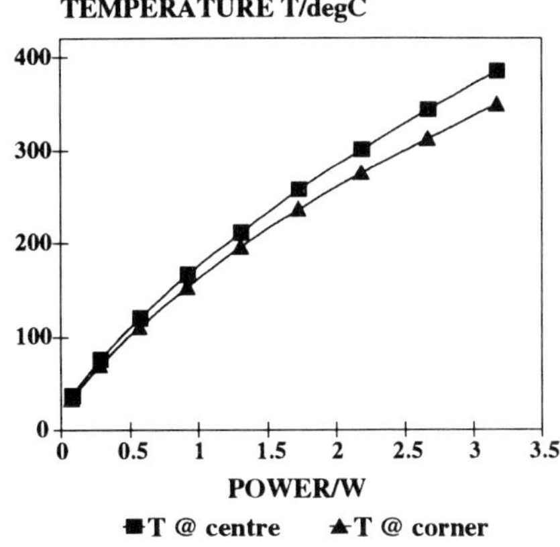

Fig 3: Temperature gradients of research II substrate

3. Substrate Design and Construction

Preliminary work at Swansea focused on the development of an alumina substrate, with the aim of producing a cheaper, lower powered alternative to current commercial devices. In designing alternatives some important properties to be considered were: the substrate size, thermal expansion coefficients, and thermal conductivity of all materials used to achieve maximum efficiency of heat transfer from substrate to the gas sensing material. Also it should be noted that to enable operation with a conventional 5 volt supply, the heater resistance needs to be approximately 10Ω. The construction of substrates was carried out with the aid of clean room technology. Computer aided design (CAD) was used to develop heater and contact array patterns which were then laser printed and photographed to produce the required masks. Substrates were then ultrasonically cleaned and the designs transferred onto them using photolithography and r.f. sputtering techniques. A number of designs have been tested, two of which are shown in figure 4. Figure 4a has heater and contact arrays on opposite sides of a 0.5mm thick alumina tile and its relatively small size means it has a power consumption comparable with the Figaro device (see figure 2 Sub1). Figure 4b shows a larger substrate which thus has a higher power consumption (figure 2 Sub2). However it has several advantages. The heater and contact array are on the same side and thus can be produced in one deposition process, hence reducing the time and cost of manufacture. Also the heater surrounds the contact array thus reducing the temperature gradients across the active region.

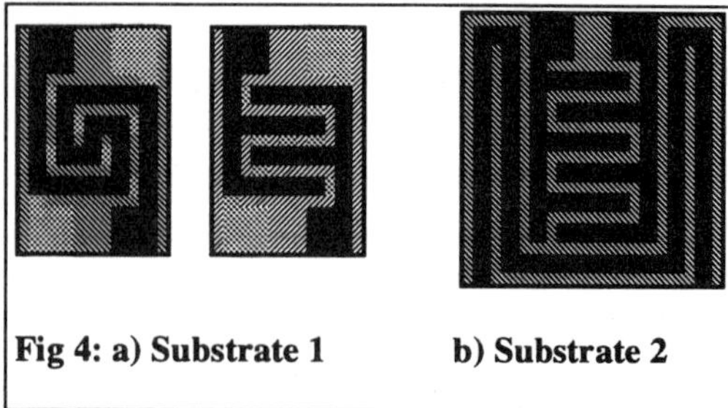

Fig 4: a) Substrate 1 b) Substrate 2

3.1 Further development

Further development of these substrates has resulted in producing possible dual sensors. For substrate 1 an electrically insulating layer of Al_XO_Y is reactively sputtered over the heater track (by sputtering Aluminium in a 5% oxygen, 95% argon atmosphere) followed by a second contact array. The design of substrate 2 means that a second contact array with or without a second heater can be sputtered directly onto the reverse side of the alumina tile. The idea is that two materials can be deposited, for example one selective to methane and one selective to carbon monoxide [1,2]. The methane selective compound usually performs better at higher temperatures and so this would be deposited over the insulated heater, the carbon monoxide selective material would be deposited on the cooler opposite side. Such a sensor, mains operated, would be an ideal device for domestic use since the above mentioned gases can be found in the home from sources including faulty boilers and gas fires.

Table 2 summarises problems which have arisen with the sputtered materials, along with possible solutions. With the exception of increasing the magnetron power (sputtering rig is limited to 50 watts, because of difficulty cooling the magnetron) the solutions for the heater tracks have been implemented. In particular the use of high purity alumina (99.6%), with a smoother surface, has resulted in better adhesion and lower resistance. This was also found to improve resistance stability as was the use of adhesion layers. Platinum (Pt) was used as the heater and contact array and chromium, titanium and tantalum (Ta) were used to aid adhesion. The best results, to date, were found for the Ta/Pt tracks, which showed adhesion improvement to the rougher 96% purity alumina.

Table 2:- Some problems with thin film heater tracks and possible solutions.		
Problem	Reasons	Solutions
Long sputtering time to produce 10Ω heater tracks.	Poor film formation. Low sputter yield rate.	Use smoother substrates. Increase magnetron power and make shorter, fatter tracks.
Abnormally high increase in heater resistance at elevated temperatures.	Rough substrate surface. Poor adhesion of metal and substrate.	Use smoother substrates. Lay down an adhesion layer between track and substrate.
Abnormally low increase in heater resistance at elevated temperatures. (Mainly when an adhesion layer is used)	Track eutectics. Intermetallic layer formation.	Use high purity substrates. Make adhesion layer thinner to minimise reaction or interdiffusion [4].
Breakdown of electrical insulation layer upon sputtering of contact array.	Film to thin. Energetic sputtered atoms tunnelling through film.	Lengthen sputtering time or increase magnetron power. Use an adhesion layer which has a low sputtering kinetic energy [5].

3.2 Micropower substrates

Although the substrates tested above are all based on alumina and are very convenient for the characterisation of thick and thin-film gas sensing compounds they are really only suited to mains operation. For battery powered devices the sensor will have to be constructed on a micro power substrate. This has been achieved to some extent using microelectronic technology, where sensors having power consumption's of the order of 100mW have been developed [6-8]. However this is still high for battery operation where power consumption's in the region of 10mW are desirable. One proposed method of reducing the power further is to produce a silicon substrate which incorporates a buried Zener diode as the heater. The advantage with using a Zener diode is that the opposite temperature coefficients of two mechanisms involved in Zener operation mean that the temperature of the sensor should remain relatively constant without the need of complicated control circuitry. A similar silicon substrate has been developed where the heater is simply a diffused resistor and control is maintain by an adjacent buried diode [9]. However such devices are limited to temperatures of approximately 250°C. This temperature could be increased if larger band gap semiconductors were used.

4. Conclusions

These micropower sensors are more suited to thin film technology and are still in the early stages of development. As a result there are no devices presently available which can compete with thick film devices, such as the figaro sensor. Thus future work at Swansea will involve development of not only micropwer substrates, but also on devices based on ceramics, which can incorporate either thin or thick film materials and can compete with present markets. Indeed such devices can be produced relatively inexpensively and for many applications where mains is available (such as many industrial applications and domestic use), low power consumption is not a priority.

5. References

1. G.S.V. Coles, G. Williams, B. Smith, Selectivity studies on tin oxide-based semiconductor gas sensors, Sensors and Actuators B, **3** (1991) 7-14.

2. G.S.V.Coles, Making the sensitive selective:- A review of semiconductor gas sensor technology, In Electronic materials for the 21st century. Ed. by Marshall and Kirov ISBN 981-02-1431-6, 161, 1993.

3. M. Benammar and W. C. Maskell, Temperature control of thick-film printed heaters, J.Phys. E: Sci. Instrum. **22** (1989) 933-936.

4. J. M. Poate, K. N. Tu and J. W. Mayer (eds.), Thin Films - Interdiffusion and Reactions, Wiley, New York, 1978, pp.1

5. J. A. Thornton, J. L. Lamb, Substrate heating rates for planar and cylindrical post magnetron sputtering sources, Thin Solid Films, **119** (1984) 87-95.

6. V. Demarne and A. Grisel, An integrated low-powered thin-film CO gas sensor on silicon, 2nd International Meeting on Chemical Sensors, Bordeaux, France, July 7-10, 1986.

7. U. Dibbern, A substrate for thin-film gas sensors in microelectronic technology, Sensors and Actuators B, **2** (1990) 63-70.

8. P. Corcoran, H. V. Shurmer and J. W. Gardner, Integrated tin oxide sensors of low power consumption for use in gas and odour sensing, Sensors and Actuators B, **15-16** (1993) 32-37.

9. Qinghai Wu, Kwang-Man Lee and Chung-Chiun Liu, Development of chemical sensors using microfabrication and micromachining techniques, Sensors and Actuators B, **13-14** (1993) 1-6.

Long-term drift in silicon nitride chemfets

S. Ko, M.K. Andrews & M.B. Moore

New Zealand Institute for Industrial Research, P.O. Box 31-310, Lower Hutt, New Zealand. E-mail: m.andrews@irl.cri.nz

Abstract: Much has been written about the response times and hysteresis effects of chemfets, but there is less information on long term drift performance, which is an aspect which will significantly affect their general acceptance. To investigate this, quasi-static CV measurements were done on model gate systems comprising buffer/nitride-oxide/silicon over times of hundreds of hours. We show that the flat-band voltage of this system drifts systematically positive at 1-2 mv/day for both p and n-silicon, principally as a result of water contact. HF contact produces similar but more extreme changes. In both cases, etch experiments show that the drift is caused by the build-up of a negative charge layer very close to the liquid-nitride interface. In some cases, the charge can be removed by low temperature annealing and therefore shows some similarities with water-related oxide traps in silicon dioxide.

Introduction

It is more than 20 years since Bergveld's original paper relating to chemical sensing transistors (Bergveld, 1970) and still they have not found general acceptance. In a recent review, Janata (1994) observed that the food industry should be a potential area of application of chemfet pH sensors because of their non-glass construction. It is however an area where issues of reliability and calibration frequency become more important than miniaturisation and response speed. The literature contains much about the latter, and Bousse and co-workers in particular (e.g. Bousse et al 1990) have extensively investigated step responses on a time scale of hours.

Only recently have reports appeared suggesting an ongoing drift might be present. Yu Dun et al (1991) examined the output from n-channel chemfets and identified a drift of 2.6mV/day over an 18 hour period. Hein and Egger (1993) compared n- and p-channel chemfets under operation and in storage and reported positive drifts ranging from 1 to 9mv/day. The drift in storage (buffer) was suggested to be a surface chemical reaction. Cabruja et al (1991) measured flatband voltage shifts of chemfet gate structures using quasi static CV methods and again reported a positive drift rate over a 24 hour period. Here we use a similar quasi static technique to investigate drifts over periods of several hundred hours. The technique allows different structures to be investigated without the need to fabricate entire transistors, and avoids complications in interpretation which might arise in extended testing when voltage shifts might be caused by progressive encapsulation failure.

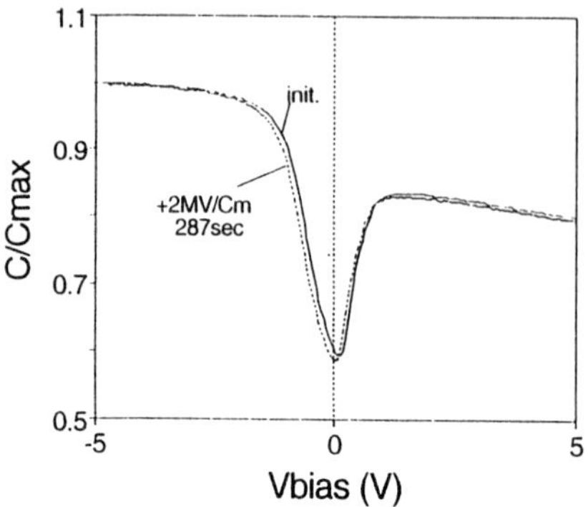

Fig.1 HF CV curve for a metal gate FET before and after a voltage stress was applied. This test, and others with negative stress, indicate that charge is being injected across the metal-nitride boundary. The rise in capacitance at positive bias is an artifact caused by the transistor turning on.

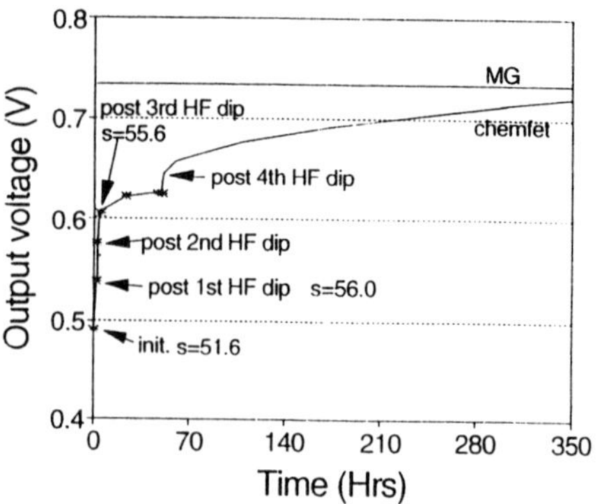

Fig.2 Time history of a chemfet output signal in pH6 buffer when given repeated 30sec dips in dilute HF before being logged for several days. The initial treatment increases the sensitivity S (in mV/pH) and raised the output voltage; subsequent dips only shift the output level. The final drift rate was 4.1 mV/day. During this time, the metal gate transistor did not drift.

We establish that a long term drift occurs and show that there are similarities between it and the threshold shift caused by the HF dip commonly used to initially condition the nitride gate. Etching experiments show that the shift is produced by a layer of negative charge very close to the nitride surface; in some cases the charge has been substantially reduced by low temperature anneals.

Experiments

We have conducted experiments over several years with n-channel chemfets whose gate consists of 40nm oxide overlaid with 80nm of LPCVD nitride. Standard chips contain two chemfets and one identical transistor with an aluminium gate. Experiments extended in time clearly show a steady increase in threshold voltage; further, the fact that the increase is similar at both chemfet sites points to a fundamental mechanism, rather than a local anomaly.

Fig 1 shows a high frequency CV curve generated on the metal gate transistor of a chemfet, before and after a voltage bias stress. The parallel shift and its sign show that the Si-insulator interface is sound, but that charge can be injected across the nitride-metal surface. In Fig 2, we track the behaviour of a chemfet through its initial immersion in pH6 buffer, subsequent brief etches in dilute HF, to logging of the drift over a an extended period. As expected, the initial HF etch raises the sensitivity from 51 to about 55mV/pH; however this is accompanied by a threshold voltage shift. Subsequent HF dips further raise the threshold voltage but do not change the pH sensitivity. The drift rate after

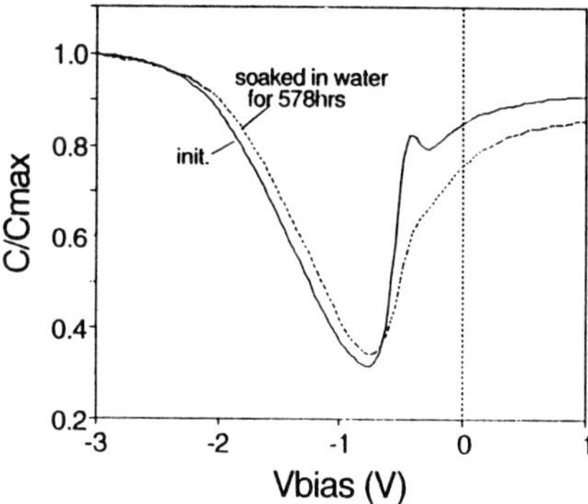

Fig.3 Quasi-static CV measurements taken on p-type silicon showing parallel movement of the curves between the initial measurement in water and when measured after 578 hours immersion. The inversion (right hand) end of the curves is unreliable and not used.

several hundred hours is about 4mv/day. The threshold voltage of the metal gate transistor did not change over this time indicating that the phenomenon is related to the liquid immersion, and is not a problem of the bulk gate dielectric.

It is not possible to perform conventional HF CV tests on devices requiring connection via a reference electrode/electrolyte. Flat band voltage drifts of the gate structure of chemfets were therefore studied by quasistatic CV tests on silicon wafers coated with appropriate gate materials. The silicon samples were sealed with O-rings to plastic cylinders to make containers into which buffer and a reference electrode were inserted. This system has the advantage that insulation failure of the chemfet is not an issue, different materials can be studied without the need to fabricate a transistor, and annealing experiments can be done on the samples if necessary.

Fig 3 shows an example of the basic data obtained, in this case the shift of the CV curve to the right (+ve) after the sample (here p-type) was soaked in distilled water for 576 hours. The important part of the curves for establishing chemfet drift is the depletion region, here between -1 and -2 volts. From curves such as these, the movement of the flatband voltage, taken as 0.7 times the capacitance in accumulation, has been examined for p and n samples in acid and alkali buffers, and in pure water. The results are shown in Fig 4 and Fig 5, and are summarised as follows;

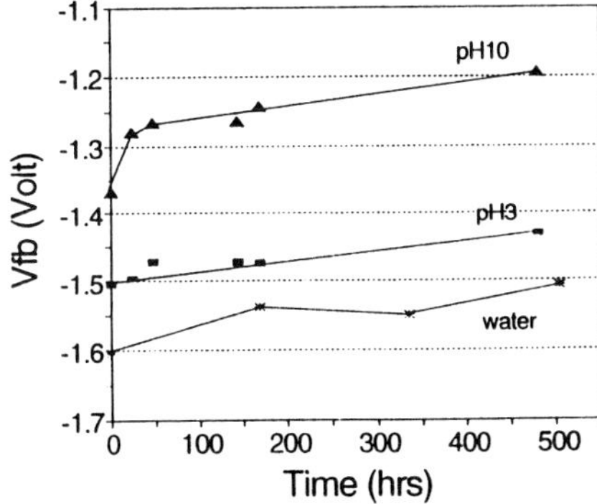

Fig.4 Flatband voltage movement of p-wafers coated with nitride-oxide dielectric, stored under different liquid conditions.

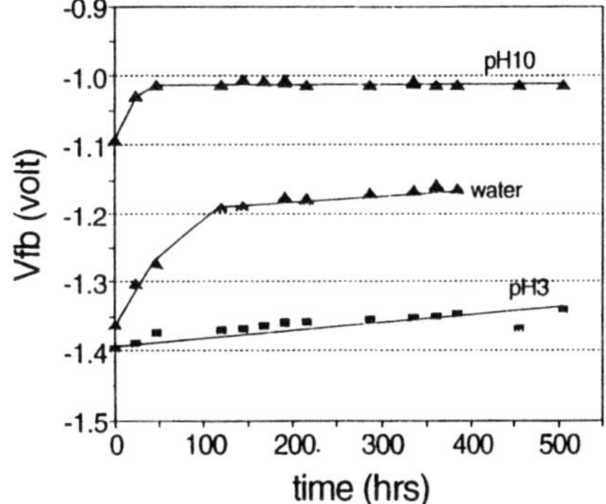

Fig.5 Flatband voltage movement of n-wafers coated with nitride-oxide dielectric, stored under different conditions.

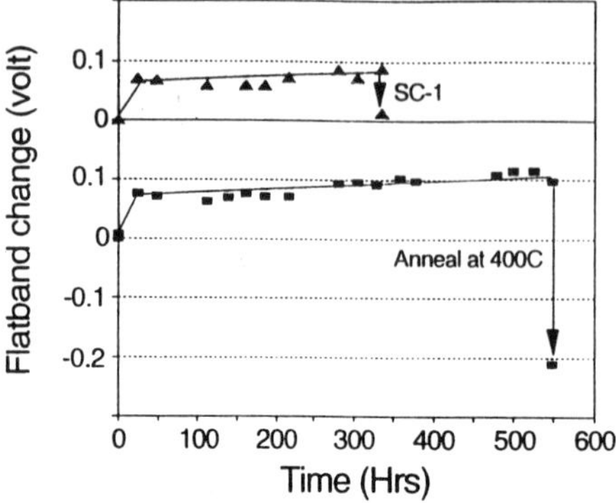

Fig.6 The effect on the flatband voltage of two n-type samples of soaking, followed by an SC-1 clean in one case and a forming gas anneal in the other.

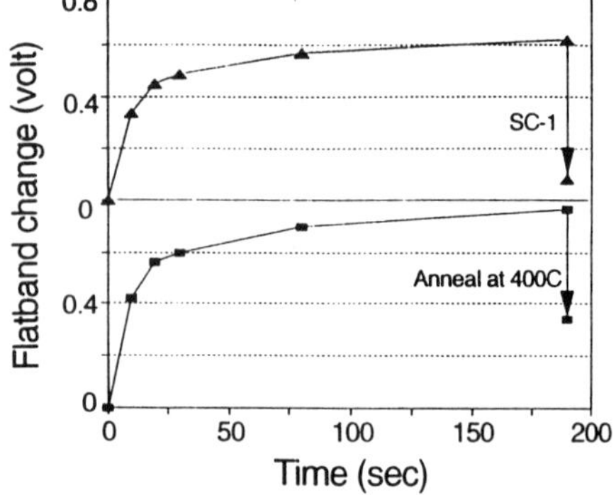

Fig.7 The effect of immersion time in buffered HF on the flatband voltage of two samples. One sample was then annealed in forming gas, and the other given an SC-1 clean.

1. All flatband values show a positive drift, including those soaked only in water.
2. The magnitude of the drifts are very comparable with those measured directly on chemfets, giving confidence that we are measuring the same phenomenon.
3. Samples immersed in alkali always showed a fast initial drift greater than that found in acid, but the basic drift is water related.
4. There is some evidence that the drift of n wafers is less than that of p-wafers, but this needs to be checked. In the case shown, the n-wafer drift was near zero at pH10, and only 1.3mv/day at pH3.

We have made some preliminary efforts to explore the charge responsible for these drifts by etching and annealing experiments, and have identified a similarity between the on-going positive drift and the extreme flatband voltage jumps caused by HF-dipping, illustrated in Fig 2. Fig 6 shows the rising trend of the flatband voltage for two n-silicon samples soaked at pH10; one sample was then subjected to an RCA-1 clean (Kern, 1983), designated SC-1, which is known to be a mild etch (Watanabe et al, 1983), while the other was given an anneal in forming gas at 400C. In both cases, when the samples were remeasured, the flatband voltage had recovered towards its original value. SC-1 cleans always produced downward shifts, but the downward movement produced by the anneal is not always seen. Similar but more repeatable effects are associated with the HF treatment. Fig 7 shows that the flatband voltage rises dramatically with the dip time but reaches something of a plateau after 2-3 minutes. Both the SC-1 clean and the forming gas anneal substantially restore the flatband voltage to its original value.

Discussion and conclusions

These results confirm that the drift in nitride-oxide gate chemfets, not surprisingly, is related to surface effects at the liquid/nitride interface rather than the bulk of the dielectric. We confirm the results of Hain and Egger (1993) that p and n silicon both drift in a positive direction; the general similarity of drift in water or buffers indicates that the effect is primarily water related. Contrary to Hain and Egger, we have found no significant difference in drift rate of chemfets whether soaking powered-up or not. The form of the CV curves indicates a progressive build-up of negative charge with time. The fact that removing a few angstrom from the surface by the SC1 clean lowers the flatband voltage is very significant and shows that the charge is located close to the insulator surface. In this regard, the effect of HF dipping is interesting. This also etches silicon nitride slowly, but clearly the principal effect is a reaction, again in the few angstrom near the liquid interface (since SC1 cleans remove the effect), which loads this zone with negative charge. We have demonstrated that annealing HF-treated samples at 400C in an N2/H2 ambient goes some way to restoring the original flatband voltage. Annealing previously soaked samples at 400C in N2/H2, and anneals in moist ambients at 200C frequently (but not always) lowered the flatband voltage. There are therefore some similarities between the drift behaviour and its recovery, and the phenomenon of water-related traps in silicon dioxide recorded by Nicollian and Brews (1983). In this regard, we note that LPCVD silicon nitride does contain silicon-oxygen bonds, but the number is process-dependent. The rather low drift rates exhibited by the n-silicon of between zero and 1.3 mV/day in Fig 5 are encouraging, but the samples were not prepared in the same reactor as the samples on p-silicon. While the results indicate that chemfets based on silicon nitride can have low drifts, a conclusion that superior drift characteristics is obtainable from chemfets based on n-silicon is not yet warranted.

Acknowledgements

We are grateful to Hyundai Electronics Corporation for the provision of nitride coated n-silicon samples.

References

Bergveld P. IEEE Trans. Biomed. Eng. BME-19, (1970), 70-71.

Bousse L., D. Hafeman and N. Tran, Sensors and Actuators, B1, 1990, 361-367.

Cabruja E., A. Merlos, C. Cane, M. Lozano, J. Bausells and J. Esteve, Surface Sci., 251/252 (1991), 364-368.

Hein P., and P. Egger, Sensors and Actuators, B13-14 (1993), 655-656.

Janata J. The Analyst, 119, (1994), 2275-2278.

Kern W., RCA Engineer, 28(4), 1983, 99-105.

Nicollian E.H. and J.R. Brews, p532 in "MOS Physics and Technology", John Wiley, New York, 1982.

Watanabe N., M. Harazono, Y. Hiratsuka and T. Edamura, Electrochem. Soc. Ext. Abstr., 83(1), (1983), 221-222.

Yu Dun, Wei Ya-Dong and Wang Gui-hua, Sensors and Actuators, B3, (1991), 279-285.

Fish Freshness Determination with a Metallo-Porphyrins Coated QMB Sensor Array

C. Di Natale, J.A.J. Brunink, F. Bungaro, F.A.M. Davide, A. D'Amico
Dept. of Electronic Engineering, University of Rome "Tor Vergata"
via della Ricerca Scientifica, 00133 Roma; Italy

R. Paolesse, T. Boschi
Dept. of Chemical Science and Technology, University of Rome "Tor Vergata"
via della Ricerca Scientifica, 00133 Roma; Italy

M. Faccio, G. Ferri
Dept. of Electrical Engineering, University of L'Aquila
Monteluco di Roio, 67040 L'Aquila; Italy

Abstract

The monitoring of fish freshness by sensor arrays has been attempted in the past by several research groups employing a variety of sensors.
In this paper the results of a study concerned with the determination of fish freshness with a sensor array of quartz microbalances (QMB) coated with different metallo-porphyrins are presented and discussed. The array should be sensitive and more or less selective to the different compounds, e.g. amines and alcohols exhalated by fish upon decomposition, for this purpose the different metallo-porphyrins were applied.
The preliminary results show that the sensors exhibit different sensitivities with respect to a number of representatives, e.g. methanol and diethylamine, for the different compounds exhalated by stored fish. From this it may be concluded that the sensor array can be applied for multicomponent analysis.
This has been proven by the application of the sensor array for the classification of cod-fish stored for a varying number of days. In this way it should be usable for a simple routine qualification of the freshness of fish.

Introduction

The quality control of food represents one of the most important applications of sensors. In particular the freshness of fishes has been taken into consideration in many papers (see for example Ref.1 and 2) adopting a number of different sensors. This particular application can be achieved by monitoring some specific compounds or studying the chemical pattern emitted by the fish through a multicomponent approach, known to be correlated with the ageing of the fish. In the first case a preliminary study with the kind of substances arising from the ageing fish is necessary in order to individuate those particular markers whose concentration can be correlated with the degree of freshness. Once the compounds have been selected it is necessary to search for sensors which can be used for their detection with high selectivity. Of course these sensors have to exhibit a strong selectivity towards these compounds. In this approach fish freshness is usually related to the amount of present amines, such as trimethylamine, and great effort has been devoted to the research of sensors showing an high selectivity towards trimethylamine [3, 4]. Nevertheless, the formation of amines due to

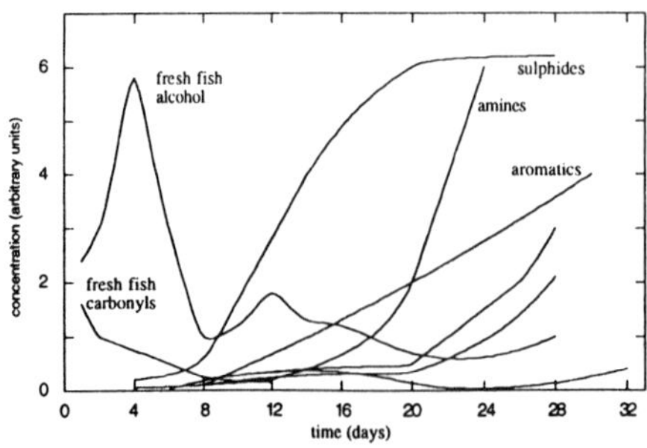

Fig. 1: Variation in time of the concentration of certain compounds realted with fish freshness [2].

the decomposition starts only some days after the fishing.
Chemical investigations pointed out that a lot of substances, in particular alcohols, are produced during the first days of the fish storage [4, 5], see Fig. 1. So, in order to measure the freshness of fish during the first days a multicomponent approach is required.
The multicomponent approach is based on the utilization of a number of sensors (configured in a sensor array) each endowed with different selectivity properties [6]. By means of proper data processing the sensor outputs can be correlated, in this particular case, with the degree of freshness of fish.
In this work an array of quartz microbalance (QMB) sensors has been adopted to monitor the freshness of fish. In the following details concerning the sensors fabrication and testing will be given. Particular attention has been devoted to the choice of the sensors according to their selectivity in order to cover as much as possible, the range of substances shown in Fig.1.
In the last section the classification properties of the array in recognizing the number of conservation days of cod-fish are reported.

SENSORS FABRICATION

Porphyrins represent one of the most widespread classes of molecules present in nature: they consist of four pyrrole rings linked by methenyl groups (see Fig. 2), and act as dianionic ligands; the four nitrogen atoms form a frame that can complex most of the metals in the Periodic Table.
The coordination sphere of the bonded metals are completed by ligand donors linked at the residual axial positions. The natural metallo-porphyrins have their biological functions associated with axial binding phenomena: for example hemoglobin binds dioxygen molecule by an iron (II) porphyrin moiety.

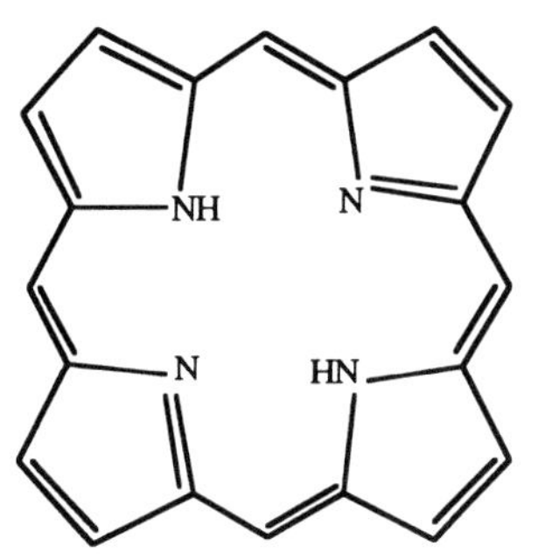

Fig. 2: The structure of generic porphyrin.

The properties of the axial bonding have been applied in order to construct sensor devices. The selectivity of the sensing mechanism depends primarily on the central metal complexed in the porphyrin pocket. So, the same porphyrin with different central metals gives sensors with clearly different selectivities. This features makes this class of compounds very suitable to be employed for the fabrication of sensor arrays.
The metallo-porphyrins used in this preliminary study were Ru-TPP(CO), Rh-TPPCl, Co-TPP and Mn-TPPCl, where TPP is the dianion of the *meso*-tetraphenylporphyrin; these complexes were prepared by literature methods [7].
The sensors were prepared by dissolving the metallo-porphyrins in $CHCl_3$ (100 mg in 25 ml of solvent) and two drops of the resulting solution were slowly evaporated on the surfaces of AT-cut quartz crystals. The fundamental frequency of the quartzes was around 5 MHz.

TEST MEASUREMENTS

The suitability of each sensor to be employed as component of an array for the determination of the freshness of fish has been proved by measuring the response of the sensors to some compounds that can be considered as representative for the classes of compounds shown in Fig. 1.
The sensors have been tested with the following compounds: triethylamine, diethylamine, methanol and tiophene. The responses of the sensors to the four compounds are shown in Fig. 3 - 6.

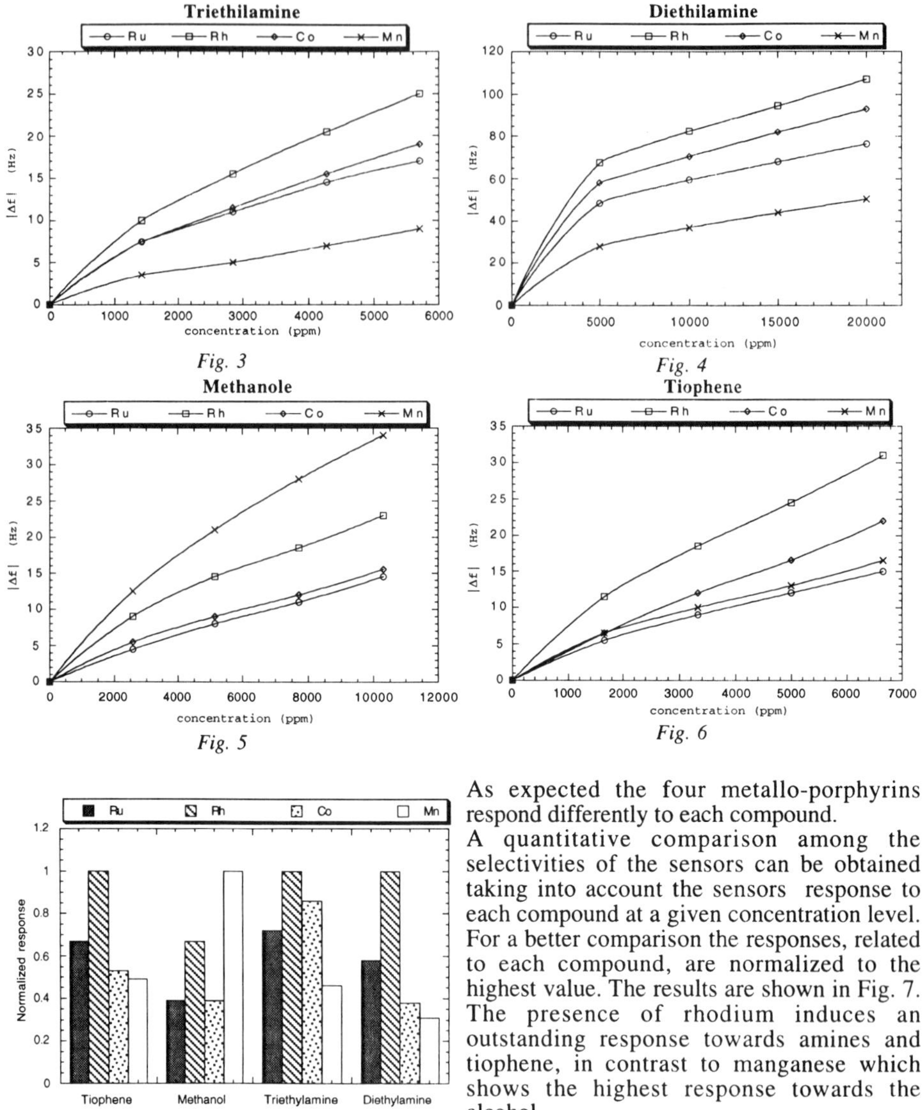

Fig. 3

Fig. 4

Fig. 5

Fig. 6

Fig. 7: Comparison of the normalized responses of the sensors to the test compounds.

As expected the four metallo-porphyrins respond differently to each compound.
A quantitative comparison among the selectivities of the sensors can be obtained taking into account the sensors response to each compound at a given concentration level. For a better comparison the responses, related to each compound, are normalized to the highest value. The results are shown in Fig. 7. The presence of rhodium induces an outstanding response towards amines and tiophene, in contrast to manganese which shows the highest response towards the alcohol.

COD-FISH CLASSIFICATION

The behaviour of the sensor array to the discrimination of the fish quality have been tested preliminarily, measuring the exhalation from cod-fish stored at room temperature. The cod-fish was stored at room temperature in order to enhance the exhalation rate of compounds, due to decomposition, significantly. In this way the sensor characteristics, i.e. sensitivity and selectivity, of the different components of the array could be studied within a short period.
Cod-fish was stored at room temperature into sealed bottles. For the measurement a quantity of gas inside the bottles was extracted by means of a syringe, and injected in a measurement chamber. From the same bottle different concentrations were measured.

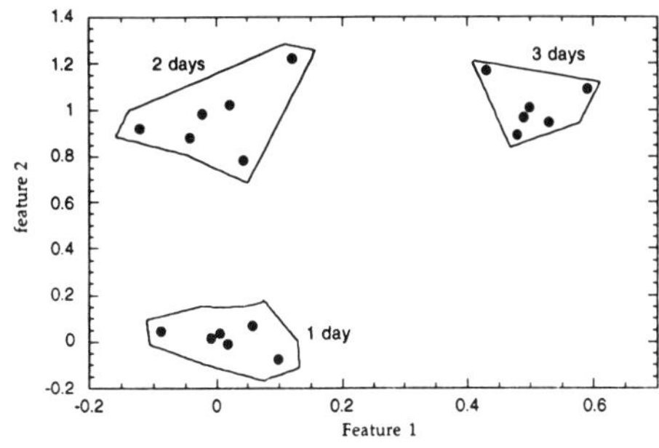

Fig.8: Classification results shown in the feature space.

The data was processed by a feed forward neural network trained by back-propagation algorithm. The results are displayed in a suitable feature space shown in Fig. 8.
As can be seen a net separation among the classes has been achieved.

CONCLUSIONS

In this preliminary study it was proven that it is possible to apply a sensor array of quartz microbalances modified with Ru^{3+}, Rh^{3+}, Co^{2+} and Mn^{2+} *meso*-tetraphenylporphyrine to qualify the freshness of fish.
First it was shown that these sensors respond with distinct sensitivities to a number of representatives for the various classes of compounds exhalated by fish upon storage. Based on this result the sensor array was exposed to the gasses emitted by cod-fish stored at room temperature for a varying number of days. The responses of the sensor array ware analyzed by an artificial neural network and resulted in an undoubtedly classification related to the number of days stored.
From this it can be concluded that such a sensor array can be applied, in a simple routine way, to study the degree of freshness of fish.

REFERENCES

[1] R. Olafsson, E. Martinsdottir, G. Olafsdottir, S. Sigfusson and J. Gardner; Monitoring of fish freshness using tin oxide sensors; in J. Gardner and P. Bartlett (eds) *Sensors and sensory systems for an electronic nose*, Kluwer, Dordrecht, 1992.
[2] M. Schweizer-Berberich, S. Vahinger and W. Göpel; Characterisation of food freshness with sensor array, *Sensors and Actuators B*, 18-19 (1994) 282-290.
[2] M. Egashira, Y. Shimizu and Y. Takao; Trimethylamine sensor based on semiconductive metal oxides for detection of fish freshness, *Sensors and Actuators B*, 1 (1990) 108-112.
[3] H. Nanto, H. Sokooshi and T. Kawai; Aluminum doped ZnO thin film gas sensor capable of detecting freshness of sea food, *Techn. Digest of the 4th International Meeting on Chemical Sensors, Tokyo (Japan) Sept. 13-17, 1992* 810-811.
[4] D.B. Josephson, R.C. Lindsay and G. Olafsdottir; Measurement of volatile aroma constituents as a means for following sensory deterioration of fresh fish and fishery products, in D.F. Kramer and L. Liston (eds) *Seafood quality determination Symp., Nov 10-14 , 1986*, Elsevier, Amsterdam, 1986.
[5] N.J.C. Strachan and F.J. Nicholson; Gill air analysis as an indicator of cod freshness and spoilage, *Int. J. Food Sci. Techn.*, 27 (1990) 43-47.
[6] C. Di Natale and F. Davide; Multicomponent analysis in chemical sensing, in A. D'Amico and G. Sberveglieri (eds) *Proceedings of the 1st European school on sensors*, World Scientific, Singapore, 1995.
[7] J. Buchler; Synthesis and properties of metallo-porphyrins, in D. Dolphin (ed) *The porphyrins Vol 1A*, Academic Press, New York, 1978.

Development and Applications of an Electronic Nose based on Arrays of Piezoelectric Sensors

M.P. Byfield[1], I.P. May[1], L.F. Wunsche[2], and C.R. Vuilleumier[2]

[1] GEC-Marconi Ltd., Hirst Research Centre, Elstree Way, Borehamwood, Herts WD6 1RX, United Kingdom. Tel: +44 181 732 0237 , Fax: +44 181 732 0100

[2] Firmenich SA, Corporate Research, CH-1211 Geneva 8, Switzerland. Tel: +41 22 780 3258 , Fax: +41 22 780 3334

Sensor array technology based on piezoelectric sensors has been developed. The sensors respond rapidly and reversibly to a wide range of analytes and are well-suited to real-time continuous monitoring either in the form of an array or as individual sensors. Applications in the fragrance and petrochemical industries and for environmental monitoring have been demonstrated.

1. Introduction

Electronic noses are sophisticated chemical sensing systems based on the combination of an array of broad-band sensors with information processing techniques. The analysis of sensor array data with signal processing and pattern recognition algorithms can be used to derive information on complex multi-component samples, atmospheres or environments and may be regarded as the conceptual analogue of the mammalian olfactory mechanism.

The development of a robust and reliable electronic nose is critically dependent on the qualities and performance of the sensor technology. A number of technologies for gas sensing are available but many of these suffer from problems due to baseline-drift, long term instability, irreversiblity of response, high sensitivity to moisture or high operating temperatures. Piezoelectric chemical sensors, such as those based on quartz crystal resonators, represent a stable, reliable and extremely versatile gas sensor technology with a number of advantages including their reliable and robust nature, operation at room temperature, low power requirements, miniature size, relatively simple associated electronics and low cost.

This paper describes the main features of the technology and presents some applications in the fragrance, flavouring and petrochemical industries

2. Experimental

In this study arrays of 10MHz AT-cut quartz crystals were coated with a range of materials based on polysiloxanes and other similar polymers. The films had a thickness of 1-2 microns. The supporting electronics including all the drive and mixer circuitry was designed and developed in-house. Using this circuitry the noise level of each 10MHz sensor was in the range 1-2Hz. Experiments were carried out at 25°C using a purpose-built automated vapour generation system controlled by a PC equipped with a counter-timer card to allow real-time recording of the sensor array responses. A portable prototype sensor array instrument was also used in this work.

3. Results

For the majority of applications the sensors respond linearly to increasing concentrations of target gases or vapours (Figure 1a) so reducing the complexity of the signal processing requirements. The sensor dynamics are rapid and fully reversible (Figure 1b) and may be used to yield valuable additional information on the analyte(s) which is independent of that derived from the steady-state response values. Multi-parameter models have been developed for extracting this information.

Sensor array technology is well-suited for industrial applications where the detection and measurement of small differences between samples or products is required. Using an array of piezoelectric sensor with broad chemical specificity relatively subtle discrimination can be accomplished. Figure 2 illustrates the ability of a sensor array to distinguish between different grades of fuel; flow dilutions of the fuel vapours were identical. This sensor could form the basis of systems designed to prevent errors in the delivery of fuel to underground storage tanks at petrol stations or at the forecourt pump for delivery into vehicle fuel tanks.

A series of proprietary fragrance formulations could readily be discriminated based using an array of piezoelectric sensors (Figure 3). To avoid delays in reaching equilibration that can arise due to the presence of heavier, less volatile constituents in perfumes, analysis of the sensor response dynamics was carried out. This is illustrated in Figure 4 where dynamic analysis was used for the rapid detection of the presence of a small amount of an adulterant in a perfume prior to equilibration.

Further experiments were carried out to assess the performance of the sensor arrays in quality control analysis in the fragrance and flavour industry. There is potentially a large market for this technology for checking both raw materials and finished formulations. Representative examples are the detection of a rogue batch of mango flavour lacking 1% of a specific component (Figure 5a), and identification of a small but very significant, in olfactory terms, formulation error in a proprietary fragrance formulation (Figure 5b).

In addition to demonstrating an ability to detect missing components in complex samples, the sensor array proved to be well-suited for the detection of adulteration or contamination of products. The contamination of rivers and other water courses with toxic organic pollutants is causing much concern at present. Experiments carried out showed that traces of petrol could be detected in water with high sensitivity (Figure 6)

4. Conclusions

Piezoelectric sensor technology has been developed for the rapid analysis of complex products or samples across a wide range of industries. It provides a capability for portable real-time monitoring which can complement existing off-line methods.

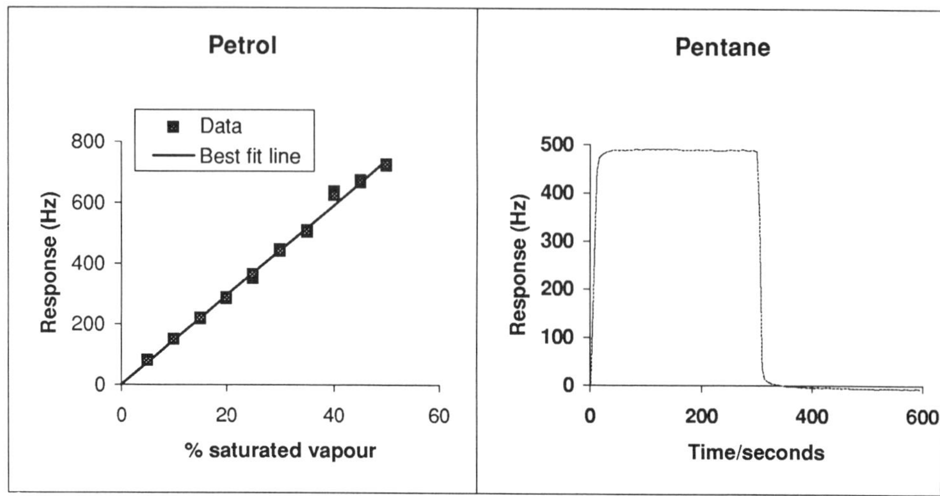

Figure 1a. **Typical linear response of an individual sensor to increasing amounts of analyte, in this case petrol vapour.**

Figure 1b. **Response of an individual sensor to pentane vapour Exposed to pentane at t = 0, removed at t = 300 seconds**

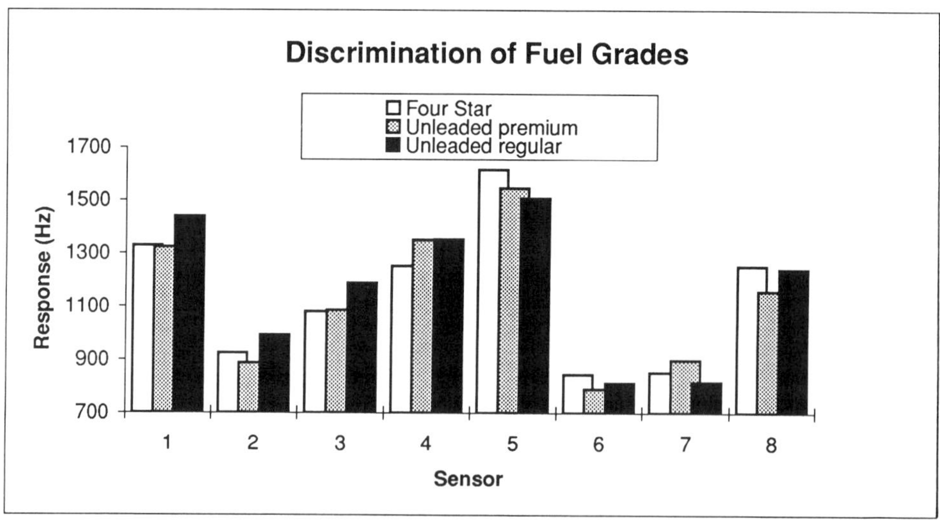

Figure 2. **Discrimination between different grades of automotive fuel using an array of piezoelectric sensors. The sensor noise was 1-2Hz.**

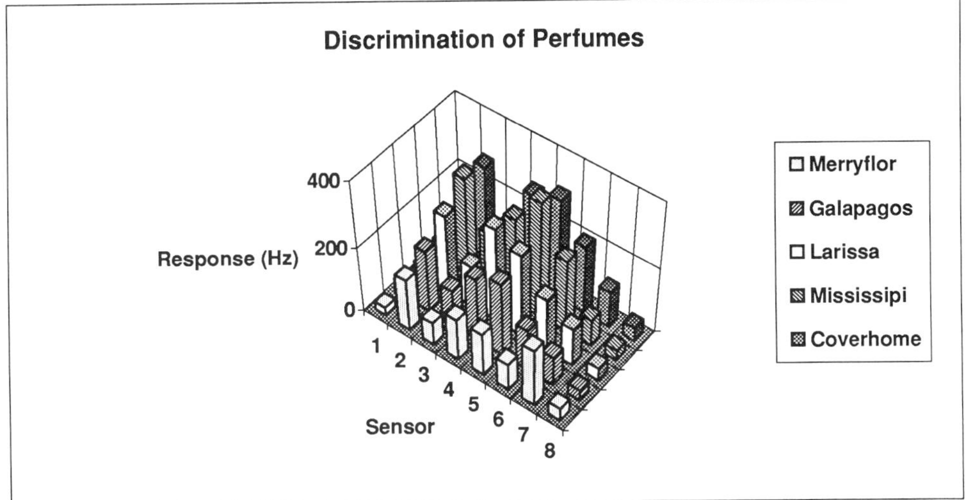

Figure 3. **Discrimination between 5 proprietary fragrance formulations. The names refer to internal codes.**

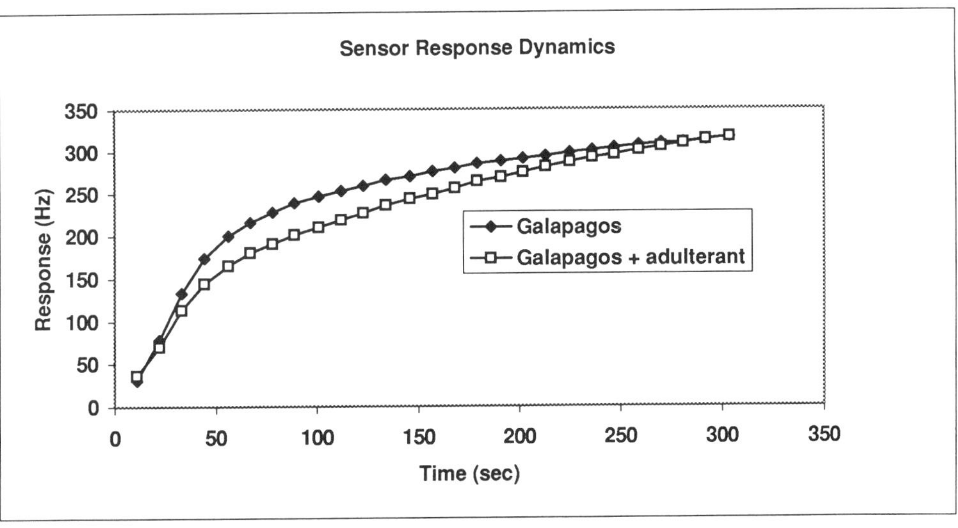

Figure 4. **Analysis of the dynamics of the response of an individual sensor to a genuine fragrance formulation and to an adulterated sample of the 'same' formulation The curves represent the fit of a multiparameter model to the data. Using this model it is possible to identify adulterated samples on the basis of their array response dynamics. 'Galapagos' is an internal code name.**

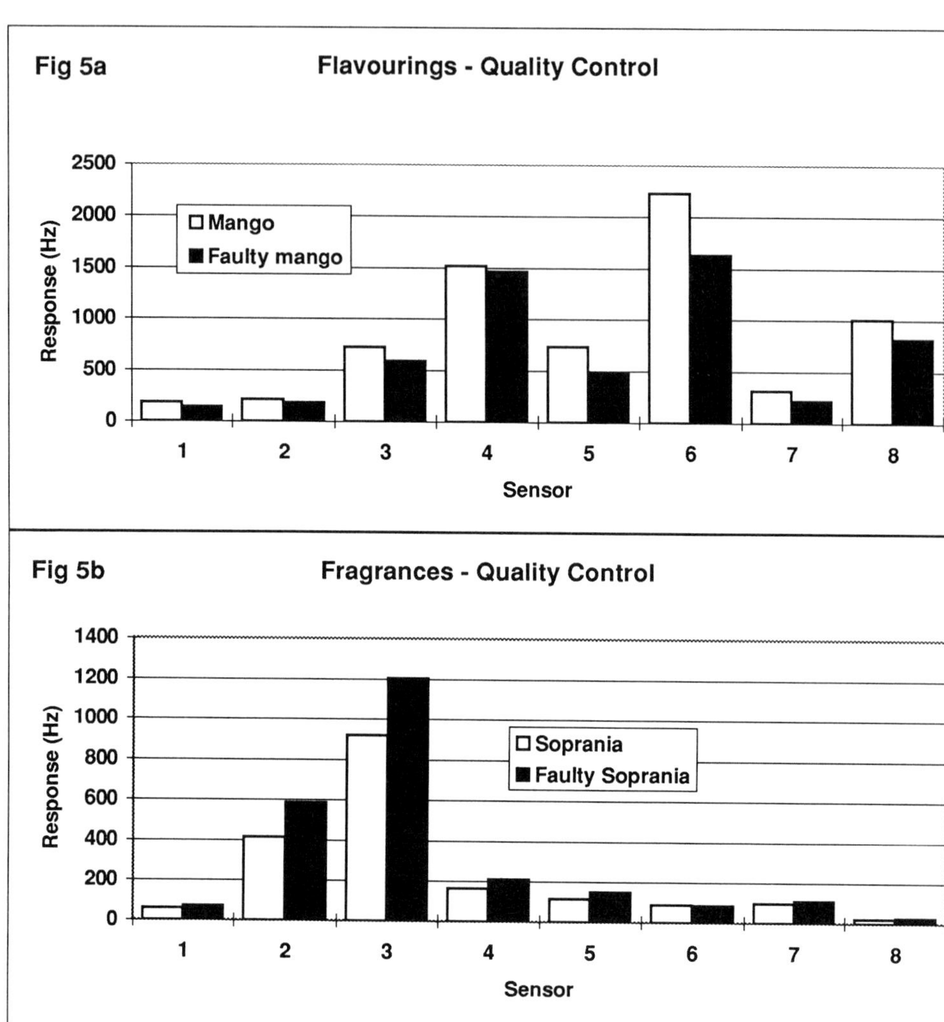

Figure 5 Use of an array of piezoelectric sensors to carry out quality control analysis for (a) a flavour, mango, and (b) a fragrance, Soprania. The faulty batch of mango was lacking 1% of a specific constituent whilst the faulty Soprania contained 2.3% of one particular constituent rather than 1.3% in the correct formulation.

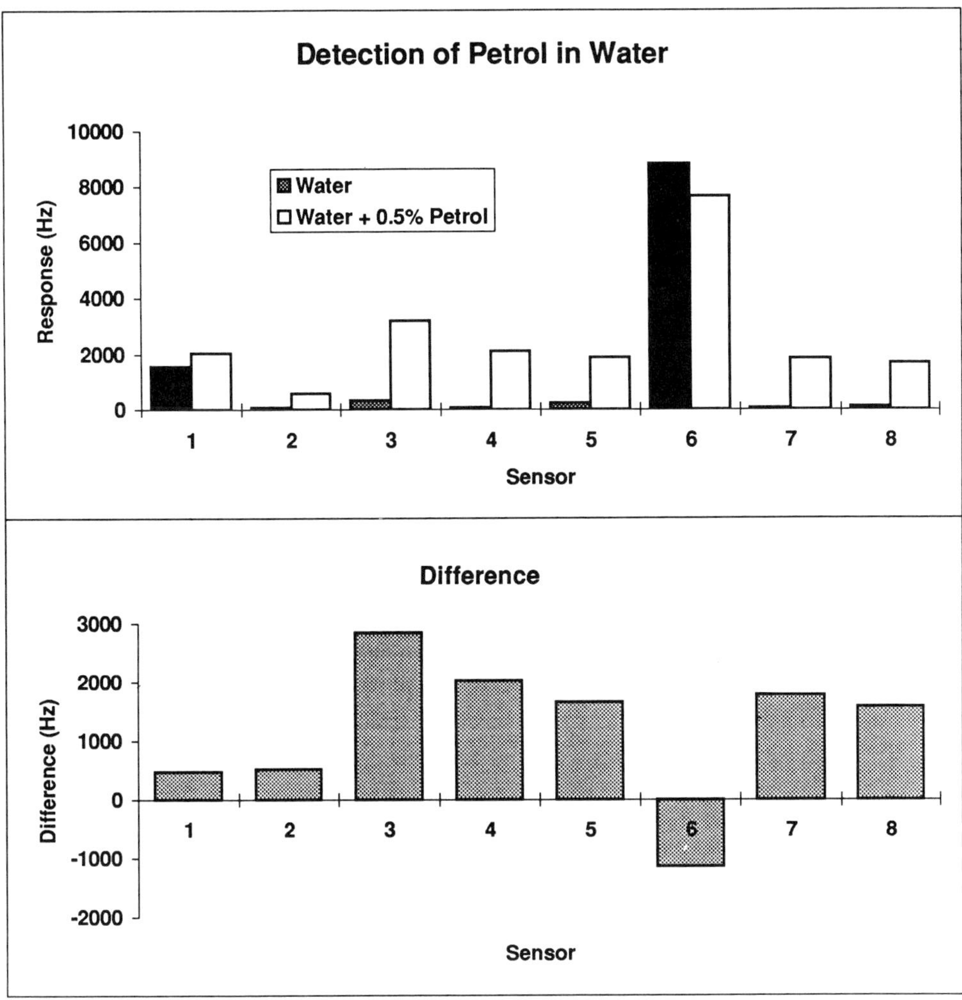

Figure 6. Detection of 0.5% petrol contamination in water using an array of piezoelectric sensors

Nanoengineered dual sensor device for intelligent monitoring of gases

Andy Pike and Julian W. Gardner

Department of Engineering, University of Warwick, Coventry CV4 7AL, UK.

Abstract. Despite the considerable effort that has been directed towards the development of chemical sensors over the past decade or so [1], the performance of solid-state gas sensors has failed to achieve the expectations of the marketplace. The problems associated with these sensors are well known, namely, substantial drift in the baseline and sensitivity due to ageing or poisoning of the active material. These problems also severely reduce the calibration period or working life of the sensor. There are two approaches to solving these problems: either develop superior active materials by changing the chemistry appropriately or compensate through the use of an intelligent sensor system [2]. We have chosen the latter approach here and so this paper describes the design of a novel dual sensor structure. The theory of the dual chemoresistive sensor has been described previously [3,4].

1. Theoretical considerations

The expected steady-state response of a single sensor (i.e. fractional change in conductance, X) has been derived for co-planar conductometric gas sensors, in terms of the sensor geometry and the nature of the reaction process [4]. The response can be simplified into two limiting cases, where the electrode separation is either much smaller or much greater than the thickness of the active film. The dual sensor device employs a pair of such sensing elements lying under the same active film. A characteristic response diagram can be drawn by plotting the ratio of individual sensor responses in the dual device, see Figure 1. The concentration-dependence of the dual device depends upon the distribution of the gas molecules within the active film, and hence on the physical and chemical properties of the active material. Conducting polymers and phthalocyanines [5] exhibit a Type I response (as do semiconducting oxides with either low activity (or density) of reaction sites), while a Type II response is expected for conventional semiconducting oxides run at high temperatures such as Pd-doped SnO_2. The characteristic response of a dual sensor device for a Type I material can be used as a simple real-time diagnostic for film poisoning, whereas the response from a Type II material can be used both as a superior sensing parameter and as a poisoning diagnostic, see Table 1.

2. Structure of dual sensor device

Figure 2 shows the design of the dual sensor device. Our novel device comprises two chemoresistors one with a 10 μm gap fabricated using conventional micromachining techniques and the other with a 100 nm gap nanoengineered using focused ion beam milling. The aspect ratio used for the devices with 10 μm and 0.1 μm interelectrode gaps are 25 and 2500, respectively. Active materials often require an elevated operating temperature for optimal performance. Therefore, a platinum heater has been positioned below the active area in order to obtain the appropriate operating temperature (up to 600°C). The thin silicon nitride membrane reduces the heat loss to less than 60 mW at 300°C. A similar structure exhibited a rapid thermal response (*ca.*

Table 1. Application of Type I and Type II materials.

Sensor Type	Typical Active Material	Application of Dual Sensor Response
I	Conducting polymers Phthalocyanines	Ratio is independent of gas concentration and a simple poison diagnostic parameter. Use narrow gap device as sensor when poisoning observed.
II	Semiconducting oxides	Ratio is related to concentration and provides a sensor parameter with common-mode effects eliminated (e.g. temperature/ageing). Also curve is a poison diagnostic.

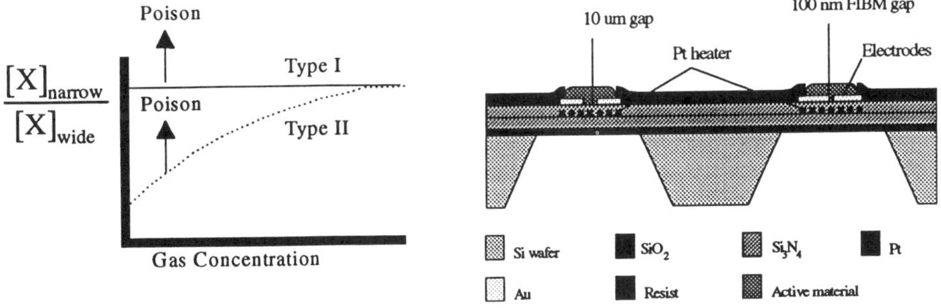

Figure 1. Characteristic behaviour of sensor type I and type II dual sensor devices.

Figure 2. A schematic of the dual sensor structure.

10 ms), and a stable power consumption (drift less than 0.3 % in 3 months) [6]. This structure is thus suitable for the use of thin active layers such as gas-sensitive conducting polymers Type I) or doped tin oxide (Type II).

3. Active materials and Experiments

Poly(pyrrole) and poly(aniline) conducting polymers are deposited electrochemically on the devices at the University of Southampton, and thin metal oxide films are deposited from a liquid suspension. Results are taken using two automated mass flow rigs, one set up to analyse the response of the dual sensor device to various ppm levels of hazardous gases (such as CO and NO_2) and the other one set up to analyse the response to volatile organic vapours (e.g. n-octane, benzene and toluene). We also report on the effect of poisoning on the dual sensor response - simulated through the introduction of high levels of acidic gases which are known to degrade these active materials.

4. Acknowledgments

We would like to thank Prof. dr. de Rooij and Mrs. M. Koudelka-Hep at the Institute of Microtechnology (University of Neuchâtel) for the fabrication of the sensors and Prof. Bartlett (University of Southampton) for the deposition of the conducting polymers.

5. References

1. W Göpel, TA Jones, M Kleitz, I Lundström,and T Seyama (eds.) 1991 *Sensors: A comprehensive survey, Chemical and Biochemical Sensors*, Vol. 2/3, VCH Publishers, Weinheim, Germany, 716 pp.
2. J Brignell and N White 1994 *Intelligent Sensors*, IOP Publishing Ltd, UK, 272 pp.
3. P McGeehin. PT Moseley and DE Williams, Self-diagnostic solid-state gas sensors employing both novel and model materials, *Sensors 93, Nuremberg, Germany, October 1993.*
4. JW Gardner, Intelligent gas sensing using an integrated sensor pair, *Eurosensors IX, Toulouse, July 1994* (to be published in Sensors & Actuators).
5. D Kohl, University of Geißen, Germany, *Private Communication*, July 1994.
6. JW Gardner, A Pike, NF de Rooij, M Koudelka-Hep, PA Clerc, A Hierlemann and W Göpel, Integrated chemcial sensor array for detecting organic solvents, *Eurosensors IX, Toulouse, July 1994* (to be published in Sensors & Actuators).

An improved REMPI detector for monitoring atmospheric NO_x samples

M Campbell[1], R Zheng[1] and K W D Ledingham[2]

[1]Department of Physical Sciences, Glasgow Caledonian University, Glasgow G4 0BA, UK
[2] Department of Physics & Astronomy, University of Glasgow, Glasgow G12 8QQ, UK

Abstract. It has been known for some time that the presence of NO plays an important part in the degradation of organic pollutants and in the formation of ozone and nitric acid in the troposphere. In recent years, there has been a significant increase in the concentration of atmospheric NO_x and a consequent decrease in air quality. The authors believe that there is now a need for a computer-controlled, near-real time monitoring system for atmospheric gas samples. This paper describes an improved laser-based ionisation chamber technique which could form the basis of such a system. It involves a [2+1] REMPI procedure at 383nm based on the $C^2\Pi(v=0) \leftarrow X^2\Pi(v=0)$ transition in NO. The resulting threshold sensitivity is better than 1ppb for NO and 10ppb for NO_2.

1 Introduction

Trace levels of NO normally exist in the upper atmosphere due to the photo-dissociation of nitogen dioxide into nitric oxide and oxygen. The concentration of NO is critical to the complex chemistry of the upper atmosphere and is known to have a significant effect on the formation of tropospheric ozone and the degradation of organic pollutants [1)]. Enhanced NO levels in our environment, due to increased emissions of NO_x gases from the combustion of fossil fuels, cannot be understated as they can have a profound impact on the aforementioned processes. At present, the mean concentration of NO_2, and therefore [2)] of NO, within cities in the UK and USA is approximately 35ppb which is within the allowed maximum of 65ppb (135μg m^{-3}) set by current European legislation. Nevertheless, it is known that, on occasions, concentrations can rise to over 150ppb. It would seem, therefore, that there is a need for a state-of-the-art analyser for atmospheric pollutants such as NO_x gases. Two important classes of sensing system, chemical-based and laser-based, will now be discussed.

1.1 Chemical based sensing systems

Several chemical methods are currently available for the detection of NO_x gases but while each method has advantages they also have distinct disadvantages. For example, metal phalocyanine films are relatively inexpensive and easy to operate but are not particularly sensitive (<ppm) [3,4)]. Ion chromatography techniques have better sensitivity (~ppb) but require long collection times (~24hrs) [5)]. Chemiluminescence, on the other hand, is very sensitive (~ppt) but suffers from interference effects [6)].

1.2 Laser-based sensing systems

The processes by which multiphoton absorption takes place in NO_x gases have been extensively studied in the past [7-9)]. Laser induced fluorescence (LIF) and resonance enhanced multi-photon ionisation (REMPI) techniques, with and without jet cooling, have been successfully applied to laboratory gas mixtures in both vacuum time-of-flight spectrometers and atmospheric pressure ionisation chambers [10-12)]. The underlying principle of REMPI detection involves ionising the molecular species of interest in the presence of an electric field. By scanning the laser light over a wide range of wavelengths it is possible to obtain a characteristic resonance ionisation "fingerprint" for any given molecule in the presence of other molecules. The technique has given rise to a series of prototype detectors for gases and vapours based on an ionisation chamber, charge-sensitive pre-amplifier and tuneable, pulsed uv laser [13-14]. One of these detectors was specifically developed for the detection of NO_x gases [15] using the [1+1] resonance excitation-ionisation channel for NO, $A^2\Sigma(v=0)$<--$X^2\Pi(v=0)$, at 226nm. The threshold sensitivity was, however, limited to 150ppb for NO molecules.The purpose of this paper is to report improvements in the threshold sensitivity of the system using a near uv [2+1] resonance excitation-ionisation channel for NO at 383nm.

2 Theory

Multiphoton ionisation (MPI) occurs when a molecule absorbs sufficient numbers of photons of arbitrary wavelength such that the first ionisation potential is exceeded. Various REMPI mechanisms for the excitation-ionisation of NO are shown in figure 1.

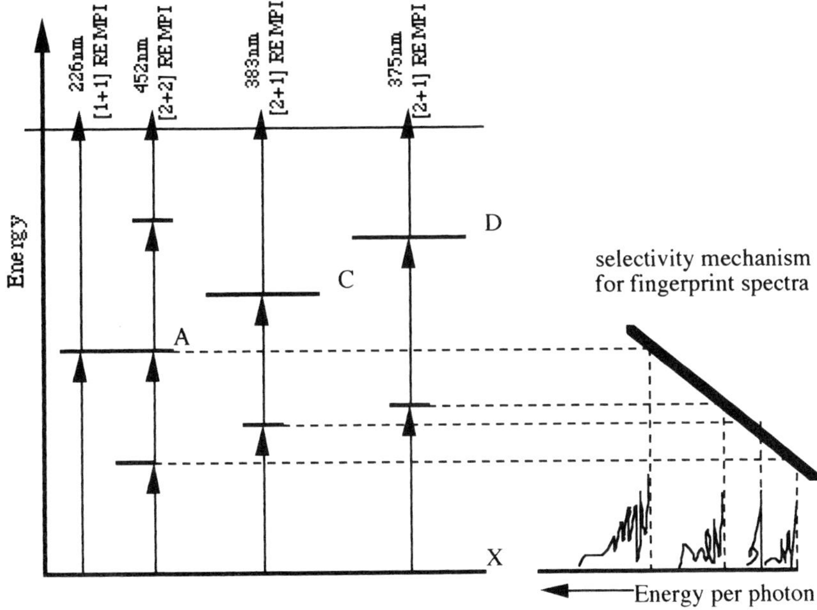

Figure 1 Electronic states of NO and the wavelength selectivity mechanism

If the photon wavelength is tuned to correspond to an allowed transition between ro-vibronic states of the molecule then ionisation is enhanced at that particular wavelength by several orders of magnitude. The requirement here is that the energy of the excitation step to a real intermediate level is greater than or equal to half the ionisation energy. In the original REMPI [1+1] system at 226nm, the $A^2\Sigma(v=0)$<--$X^2\Pi(v=0)$ transition tended to produce NO spectra in the presence of strong molecular oxygen peaks and this limited the threshold sensitivity. An alternative [2+1] transition at 383nm, $C^2\Pi(v=0)$<--$X^2\Pi(v=0)$, was selected in preference to the [1+1] transition since it is free of these interference effects. Although this transition relies on the simultaneous absorption of three photons at 383nm, any disadvantage due to low cross-section is offset by the fact that dye laser photon frequency does not have to be doubled, a process which is usually less than 20% efficient.

3 Experimental details

The gas mixtures used in these experiments consisted of air + 1ppm NO_2 or nitrogen + 1ppm NO. Pulsed lasers were used to provide the necessary large photon fluxes which permit multiphoton absorption by molecules present in the laser beam volume. The number of photons required to ionise each molecule depends on the energy of the individual photons, the ionisation potential of the molecule concerned and on whether or not dissociation of the molecule is required prior to ionisation, as in the case of NO_2. Ionisation potentials for NO and NO_2 are 9.25eV and 9.79eV respectively. The laser source consisted of a Lumonics EPD 330 dye (BBQ) laser, pumped by a 308nm (XeCl) Lumonics TE 860M-3 excimer laser operating at frequencies of between 10Hz and 15Hz. The beam consisted of intense (3mJ) light pulses of 6ns duration in the wavelength region 380nm -385nm which resulted in a peak power density of 160MW cm^{-2} and a pulse bandwidth of 0.015nm. The detector is an ionisation chamber which consists of two parallel stainless steel plates 25mm long with a separation of 15mm which are housed within a cast metal box. A variable high voltage power supply, normally operated in the range (100-400)V, was connected across the plates via an Ortec 142H preamplifier. The position of the laser beam within the chamber was chosen to be 1mm from the anode since the collection efficiency for ions decreases rapidly when the beam moves away from this electrode. During this type of experiment, it is desirable to maintain the beam intensity within the chamber. This condition was achieved by measuring the intensity continuously during a wavelength scan using a Molectron J3-05 joulemeter and maintaining pulse amplitude at a constant level by means of an attenuator. The data acquisition consisted of two SRS gated boxcar integrators and a Macintosh IIfx computer. The system is shown in block diagram form in figure 2.

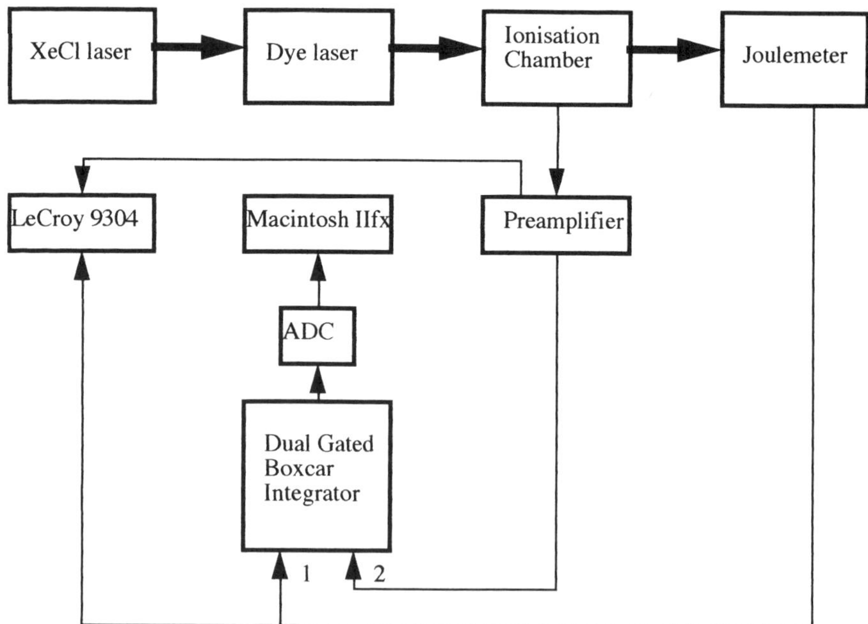

Figure 2 Block diagram of the experimental arrangement

The time-of-flight mass spectrometer and its associated instrumentation are not included in figure 2 since the characterisation of the system, in terms of the signals obtained using the ionisation chamber at specific wavelengths and the corresponding mass data, has established that the signals are solely due to NO ions (mass 30).

4 Results

The results obtained for 1ppm NO in N_2 gas and subsequent dilutions in pure nitrogen as summarised in figure 3a. Each spectrum was obtained by averaging 10 laser shots. By plotting the peak intensity at 383.75nm as a function of gas concentration, it can be shown that the response of the ionisation chamber is fairly linear as indicated in figure 3b.

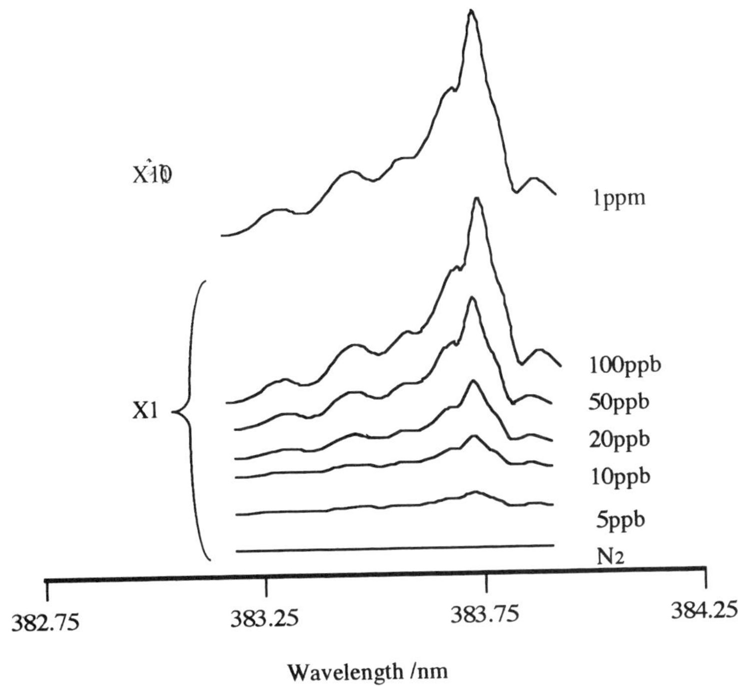

Figure 3a Spectra for NO at various concentrations in N_2

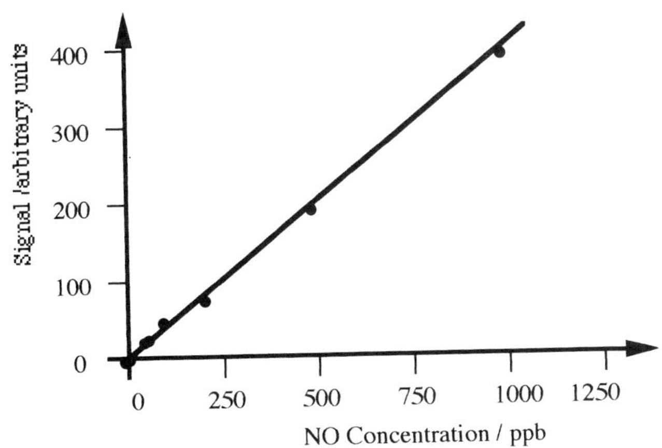

Figure 3b Signal size as a function of NO concentration

5 Discussion and conclusions

These results indicate that NO can be detected in ordinary samples of atmospheric air to a threshold value of 1ppb using this 3 photon technique. NO_2, on the other hand, is limited to 10ppb at present due to the fact that a further photon is required for the initial dissociation of nitrogen dioxide into nitric oxide and oxygen. It is felt that these sensitivity values can be reduced even further by averaging over thousands of laser shots.

The authors believe that this system could easily form the basis of a commercial city-centred pollution analysis and control centre of the type shown in figure 4. The present system sensitivity and the advent of new tuneable solid state laser systems could mean that user-friendly, computer-controlled REMPI systems could simultaneously measure atmospheric levels of NO_x, SO_x, and CO_x gases on a city-wide, 2-hourly basis. Such systems, operated by non-specialists and utilising disposable ionisation chambers, could transform the air quality of cities by monitoring pollutants and controlling the movements of their sources.

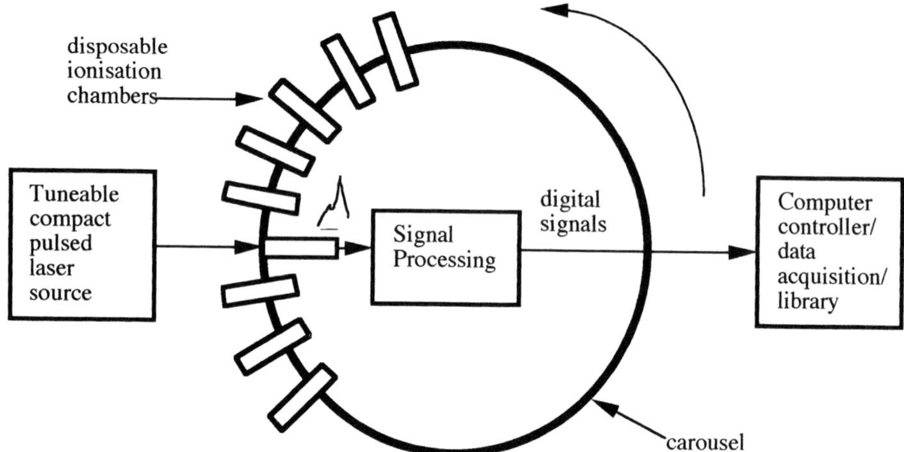

Figure 4 Towards a commercial atmospheric pollution detection system

6 References

1) Johnston H, *Science* **173** (1971) 517-522
2) Logan J A, *J. Geophys. Res.* **88** (1983) 10785-10807
3) Rigby G P, Wilson A and Wright J D, *Sensors: Technology, Systems and Applications, Ed. K T V Grattan, Adam Hilger, Bristol* (1991) 121-126
4) Sadaoka Y, Jones T A and Gopel W, *Sensors and Actuators* **B1** (1990) 148-153
5) Sickles J E, Grohse P M, Hodson L L, Salmons C A, Cox K W, Turner A R and Estes E D, *Anal. Chem.* **62** (1990) 338-346
6) Zafiriou O C and True M B, *Environ. Sci. Technol.* **20** (1986) 594-596
7) Uselman W M and Lee E K C, *J. Chem Phys.* **65(5)** (1976) 1948-1955
8) Morrison R J, Rockney B H and Grant B R, *J. Chem. Phys.* **75(6)** (1982) 2643-2651
9) Slanger T G, Bischell W K and Dyer M J, *J. Chem. Phys.* **79(5)** (1983) 2231-2240
10) Hippler M, Yates A J and Pfab J, *Inst. Phys. Conf. Ser. 113* Section **9** (1990) 303-306
11) Guizard S, Chapoulard D, Horani and Gauyacq D, *Appl. Phys. B.* **48** (1989) 471-477
12) Miller J C, *Anal. Chem.* **58** (1986) 1702-1705
13) Marshall A, Clark A ,Ledingham K W D, Singhal R P and M. Campbell, *Sensors: Technology, Systems and Applications, Ed: KTV Grattan, Adam Hilger*, (1991) 151-155
14) Marshal A, Clark A ,Jennings R, Ledingham K W D, Sander J and Singhal R P, *Int. J. Mass Spectrom. Ion Proc.* **116** (1992) 143-156
15) Clark A , Deas R M, Kosmidis C, Ledingham K W D, Marshall A, Sander J, Singhal R P, Campbell M and Zheng M, *Sensors VI, Technology, Systems and Applications", Ed:K.T.V. Grattan and A.T. Augousti*, IOP Publishing, 57-61, 1993.

The development of a novel area modulated gas sealed microelectrode electrochemical sensor.

A McNaughtan, R O Ansell, J R Pugh

Department of Physical Sciences, Glasgow Caledonian University, Cowcaddens Road, Glasgow. G4 0BA.

Abstract.
A novel area modulated gas sealed microelectrode electrochemical sensor has been developed. The use of a gas seal has enabled the problems of leakage currents and poor seals, often a limiting factor with conventional encapsulated microelectrodes, to be overcome. The novel area modulation technique enables the size of the electrode to be controlled and also offers the possibility of a readily renewable electroactive surface. This is of particular importance in real industrial or environmental applications where electrode fouling can be a major problem. The electrode has been demonstrated in a linear ramp cyclic voltammetric measurement system using the well-characterised oxidation of potassium ferrocyanide.

INTRODUCTION.

Electrochemical sensors are widely used in analytical investigations. Under controlled mass transport conditions the current associated with the oxidation or reduction of a substance at an electrode is proportional to the concentration of the substance in solution. At an inlaid disc microelectrode of a few micrometers diameter, the mass transport of the substance to the electrode is controlled by diffusion to the electrode and a steady state diffusion limited current is rapidly established. The microelectrode electrochemical sensor therefore offers the possibility of rapidly determining the concentration of a substance in solution.

The range of application of microelectrode sensors is vast and the many advantages of the microelectrode have been well documented[1,2,3,4]. The microelectrode can be used in simple carefully designed measurement systems and is small, inexpensive and disposable. A near to ideal steady state response can be achieved using a simple two terminal electrochemical cell configuration. It is particularly suitable for in situ, portable and remote sensing applications. However, the main disadvantages of the microelectrode compared to conventional sized electrodes are the very small currents that have to be measured and the difficulty in cleaning small electrodes which becomes more severe as the electrode size is reduced.

The main drawback which has limited the use of microelectrode sensors in real applications is the ready fouling of the electrode[4]. Cleaning the electrode by polishing tends to degrade the seal between the electrode surface and the insulating surround and results in a non-ideal response which limits the detection at low concentration levels. The current density at the inlaid disc microelectrode is highest at the interface between the electrode surface and the insulating surround and a high quality seal is crucial for low concentration measurements.

Many designs for microelectrode sensors have been published using a wide range of methods and materials. At Glasgow Caledonian University platinum inlaid disc microelectrodes are routinely fabricated by sealing a 25μm diameter high purity platinum wire (Goodfellow Metals, Cambridge) in a soft soda glass capillary tube using a microflame butane burner and polishing the end to expose the platinum wire cross section. Investigations using cyclic voltammetry and ac impedance spectroscopy with these electrodes have indicated that the limit of detection is due to leakage current across the glass insulator surface[5]. However, the soft glass produces a good seal and the electrode can be easily polished. In contrast, commercially available electrodes of a similar design have been found to exhibit a lower leakage current but require a more elaborate polishing procedure and are more prone to degradation of the seal. As a result the limit of detection of the microelectrode sensor has been found to be generally in the range of 10^{-6} mol l^{-1} to 10^{-8} mol l^{-1} for most analytes and the theoretically predicted lower detection limit has yet to be achieved.

Difficulties in fabricating microelectrodes with highly resistive surrounds and difficulties in maintaining electrodes with high quality seals is at present limiting their use in real applications and in ultra trace investigations[4]. Although many designs for microelectrodes have been published, as yet, no single design is able to simultaneously provide a good seal, low leakage currents and an electrode which can be readily cleaned.

The novel microelectrode which has been developed at Glasgow Caledonian University provides a high quality highly resistive seal by using an inert gas rather than a solid insulator and while the electrode cannot be polished, it does offer the possibility of a readily renewable electroactive surface. In addition, alternative electrode materials which are difficult to encapsulate in resistive surrounds can be readily used with this design.

ELECTRODE DESIGN AND MEASUREMENT SYSTEM.

The microelectrode sensor comprises a novel gas sealed wire electrode which has a method of varying the level of immersion of the wire in solution in an oscillating manner. This enables a small surface area of the electrode to be exposed intermittently to the solution. The varying level of immersion of the wire electrode in solution may be achieved by providing a means of varying the position of the electrode and maintaining the position of the solution fixed, or by maintaining the position of the electrode fixed and providing a means of varying the position of the solution. The latter method has been adopted and is shown in figure 1.

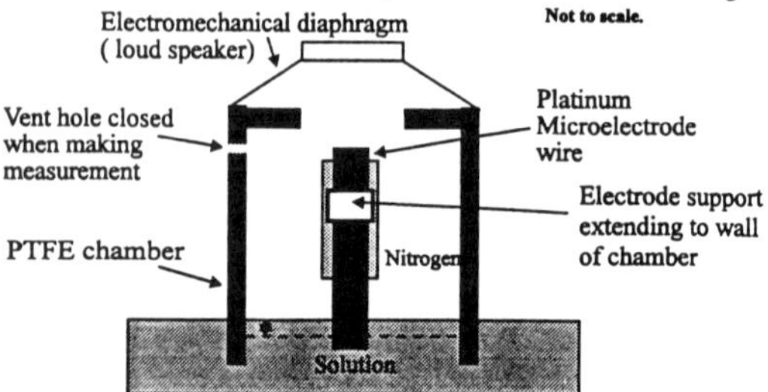

Figure 1. Area modulated gas sealed microelectrode design.

An electromechanical diaphragm (loudspeaker) was used to modulate the pressure in the chamber. The enclosed space around the wire was filled with nitrogen to reduce the possibility of oxygen reacting with the electrode. The change in pressure causes a small change in the area of the wire immersed in the solution. The component of the faradaic current associated with the small change in area was then measured using a low noise current to voltage convertor and a lock-in amplifier. The use of an ac detection system is essential in order to discriminate against the faradaic component associated with the portion of the wire permanently immersed in solution.

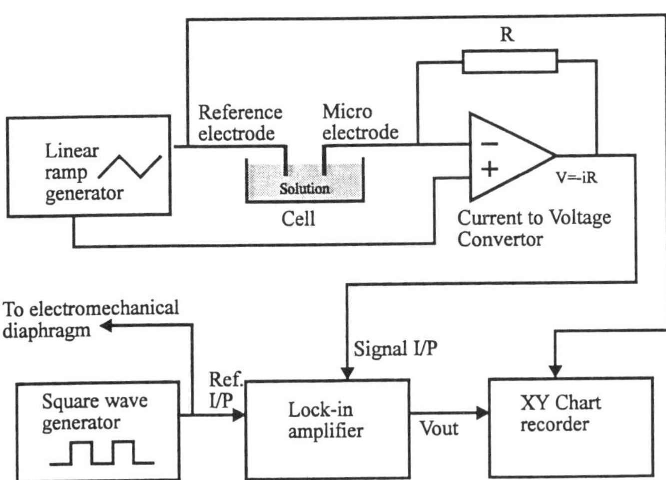

Figure 2. Block diagram of the measurement system.

A block diagram of the measurement system is shown in figure 2. An EG & G Brookdeal 9503-SC lock-in amplifier was used to discriminate between the ac current component associated with the change in area of the electrode and the current component associated with the portion of the wire permanently immersed in the solution. A square wave signal with a frequency of 14Hz was used to modulate the electromechanical diaphragm and also provide the reference signal for the lock-in amplifier. The frequency was chosen to allow a steady state diffusion current to be established. Details of the custom-built current to voltage convertor are given elsewhere in the literature[6]. The reference electrode consisted of a silver/silver chloride wire immersed in the solution. A custom-built linear ramp function generator provided the bias potential used to drive the faradaic reaction.

RESULTS.

The microelectrode sensor has been used in a linear scan cyclic voltammetric measurement system. The well characterised potassium ferrocyanide reaction was used to verify the operation of a conventionally encapsulated platinum microelectrode sensor. Figure 3 shows the voltammogram obtained using the area modulated microelectrode. The potential of the electrode was swept between 0V and 0.5V at a scan rate of $2mVs^{-1}$. The voltammogram of the modulated electrode is close to the ideal response and shows a very flat diffusion limited plateau which is characteristic of an electrode with a highly resistive high quality seal.

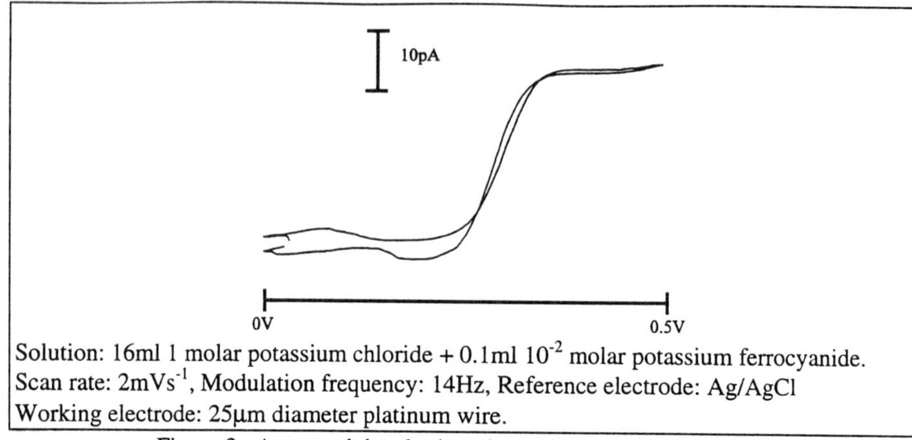

Figure 3. Area modulated microelectrode cyclic voltammogram.

DISCUSSION.

A novel gas sealed area modulated microelectrode sensor has been developed which exhibits very low leakage currents and a high quality seal which results in a near ideal shaped voltammogram. By using a gas seal the electrode offers the possibility of using alternative electrode materials such as gold which is difficult to seal in an insulating surround. Also, by changing the level of immersion of the wire in the solution a fresh new section of wire could be immersed in the solution to give a new electroactive surface.

Mass transport effects such as hydrodynamic modulation of the wire in solution, depletion of the analyte around the portion of the wire permanently in solution and surface tension effects are still under investigation. However, initial investigations have shown that the electrode offers the possibility of overcoming the main limitations of the encapsulated microelectrode sensor although further work is required in order to fully characterise the electrode.

REFERENCES.

[1] S Pons, M Fleischmann, (1987), The Behaviour of Microelectrodes, Analytical Chemistry, 59, pp1391A-1399A.

[2] R M Wightman, (1981), Microvoltammetric Electrodes, Analytical Chemistry, 53, pp1125A-1134A.

[3] R M Wightman, D O Wipf, (1989), Voltammetry at Ultramicroelectrodes, (ed A J Bard), Marcel Dekker, New York, pp268-344.

[4] A M Bond, (1994), Past, Present and Future Contributions of Microelectrodes to Analytical Studies Employing Voltammetric Detection, Analyst, 119, R1-R21.

[5] A McNaughtan, R O Ansell, J R Pugh, (1991), Determination of the Detection Limit of Micro-electrode Chemical Sensors, (ed KTV Grattan), The Adam Hilger Series on Sensors, pp49-54

[6] A McNaughtan, R O Ansell, J R Pugh, (1994), The Measurement of Microelectrode Sensor Characteristics Using Impedance Spectroscopy, Measurement Science and Technology, 5, pp789-792.

Thin tin dioxide films for combustible gas sensors

Geraint Williams and Gary S.V. Coles

Department of Electrical and Electronic Engineering, University of Wales Swansea, Singleton Park, Swansea SA2 8PP, U.K.

Abstract: Factors which influence the magnitude of the resistance response of thin film SnO_2 sensors to gases such as H_2, CO and CH_4 have been investigated. Considerable variations in sensitivity are observed upon changing the film thickness, processing temperature and substrate temperature during deposition for sensors prepared via several different r.f. magnetron sputtering pathways. The use of ultra thin activator layers of Pd, Ag or Pt (0-10 nm) deposited on the SnO_2 film surface leads to marked increases in sensitivity especially in the presence of hydrogen.

1. Introduction

Although methods of preparing thin films of tin dioxide were developed over forty years ago [1], it is only very recently that such techniques have been applied to the production of gas sensors. Initial reports of the use of SnO_2 thin films prepared by radio frequency magnetron sputtering as gas sensitive resistors exploited their sensitivity towards oxidising gases such as NO or NO_2 [2] with relatively little interference from low levels of reducing gases such as CO, H_2, C_3H_8 and CH_4. Subsequent improvements in performance, especially in terms of selectivity have been achieved by modifying the deposition method [3]. Researchers are now attempting to develop SnO_2 films for the purpose of detecting reducing gases such as those listed above. Recent developments in the field are given in a review by Sberveglieri [4]. Interest has been fuelled by the emergence of silicon microfabrication technology and the ease with which thin film deposition methods can be allied with such techniques [5]. The work described here aims to show that the fabrication route and processing of sputtered SnO_2 thin films greatly influences the reducing gas sensitivity of the material. The role of ultra thin activator films of Pd or Ag deposited onto the surface of the SnO_2 has also been investigated with the ultimate aim of producing a sensor of defined selectivity.

2. Experimental Details

Sensors were prepared by depositing thin tin dioxide films over the contact array of planar alumina substrates (supplied by Apex Ltd.) via the following methods.

(a) R.f. magnetron sputtering of a metallic tin layer in an argon environment and subsequent thermal oxidation of Sn to SnO_2 in an oxygen flow [3]. In several cases the metal was deposited on to a heated substrate held at a temperature exceeding the melting point of tin. Molten tin condenses on the substrate surface and surface tension effects cause the formation of Sn microspheres. The room temperature resistivity of these films are high since there is little contact between the Sn spheres. However, thermal treatment in oxygen causes an increase in the volume of the metal particles of up to 35% as oxidation proceeds forming a continuous semiconducting SnO_2 film of greatly increased surface area [4].

(b) Reactive r.f magnetron sputtering employing a pure Sn target in an environment containing 5% v/v oxygen in argon. The majority of the SnO_2 films were produced at room temperature, although in one experiment sensors were fabricated at several different substrate temperatures in the range 150 - 500°C. The substrate target distance employed was 10 cm, while the a gas pressure of 2×10^{-3} mbar was maintained within the sputtering chamber.

Surface doping of the sensors was achieved by sputtering thin metallic films of either Pd, Pt, Ag or Ni over the SnO_2 films. Deposition rates for all the materials used were estimated by measuring the weight of films sputtered over a known area (usually 10 cm^2). By using known density values and assuming uniform deposition over the entire area, absolute film thicknesses could be calculated for a number of different sputter times leading to the generation of calibration plots for each material studied (see Table 1)

Table 1: *Calculated deposition rates for sputtered films*

Material deposited	Power level (W/cm^2)	Deposition rate (nm/min)
Sn	1.32	5.8
SnO_2	1.32	7.4
Pd	9.87	20
Ag	9.87	39
Pt	9.87	18
Ni	4.39	11

3. Results and discussion

3.1 The influence of film thickness

A range of reactively sputtered thin film sensors in the thickness range 40 nm - 0.85 µ m were prepared and subjected to testing at different operating temperatures (200 - 600°C) in dry air containing 1% v/v of H_2, CO or CH_4. Each film was initially heated to 600°C in air for 5h along with 2h heating and cooling ramps prior to evaluating its gas sensing properties. All the sensors displayed maximum H_2 and CO sensitivity at approximately 480°C while the greatest CH_4 response was observed at 530°C. Plots of maximum sensitivity versus film thickness are shown in figure 1, where it is evident that optimum performance is obtained by employing a film thickness of 100 nm.

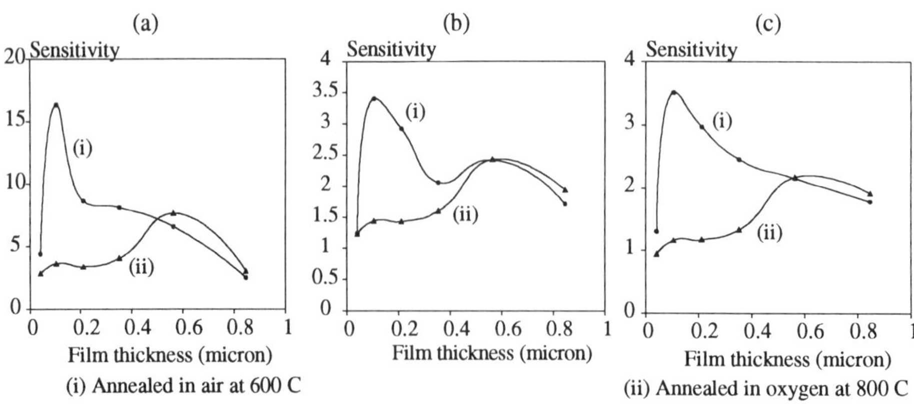

Figure 1: *The effect of varying SnO_2 film thickness on the sensitivity of reactively sputtered sensors exposed to 1% v/v concentrations of (a) H_2 ,(b) CO and (c) CH_4 in dry air at operating temperatures of approximately 480° C. Sensitivity is represented by the ratio R_{air}/R_{gas} , where R_{air} is the sensor resistance in dry air and R_{gas} is the sensor resistance in a reducing gas containing environment.*

In a bid to increase sensitivity by changing the stoichiometry of the tin dioxide films, the sensors were heated to 800°C for 5h under oxygen and then re-tested at their optimum temperatures. The results obtained are also displayed in figure 1. With the exception of films having a thickness of 0.6 μm or greater, after high temperature treatment in oxygen all sensors are considerably less sensitive to all three gases. It is obvious therefore that either the oxygen content of the annealing environment or the increased temperature or a combination of both has a profound effect on sensor properties. More detailed studies were performed on sensors of the same thickness heated to the same temperature in either oxygen or air prior to testing. For 100 nm SnO_2 films, it appears that annealing in air gives rise to the highest H_2 and CO sensitivity, while in the case of 420 nm films, the situation is reversed where sensors annealed in oxygen display the greatest resistance response. The methane sensitivity of the films remains unaffected by a change in the annealing atmosphere.

3.2 The effect of processing temperature

Studies were carried out on films of constant thickness (420 nm) heated under oxygen for 5h prior to testing. A summary of the effects of annealing temperature on the optimum sensitivity towards H_2, CO or CH_4 are displayed in figure 2. Maximum hydrogen and CO sensitivity is observed after annealing at 700°C, while CH_4 sensitivity is highest at lower temperatures (500 - 600°C). The resistivity of the films remain relatively constant up to processing temperatures of 800°C, however any further increase causes marked rises in resistivity signifying that the stoichiometry of the SnO_2 film is being irreversibly changed.

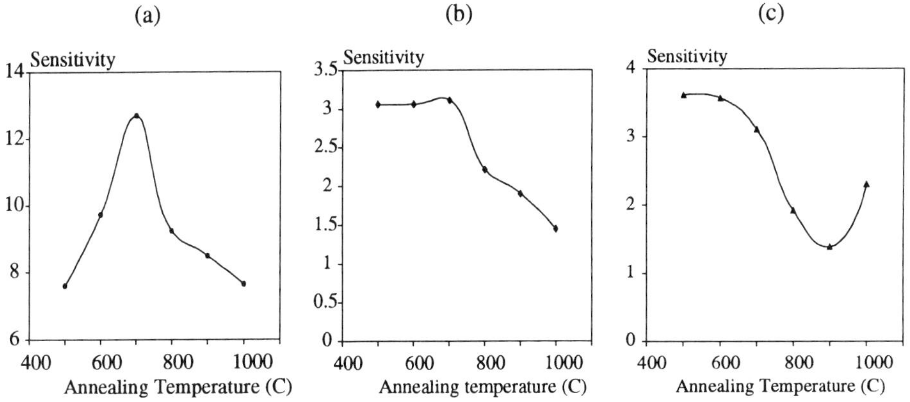

Figure 2: *The variation of sensitivity with annealing temperature for thin film SnO_2 sensors (0.42 micron thickness) operated at 480° C in the presence of 1% v/v concentrations of (a) hydrogen, (b) CO and (c) methane. Sensors were heated in an oxygen flow for 5h prior to testing.*

3.3 Different film preparation methods

A comparison of the gas sensing properties of SnO_2 thin films of 420 nm thickness produced by different r.f magnetron sputtering pathways is given in table 2. Films prepared by methods (a) (see section 2) where the metallic Sn deposition takes place at room temperature appear to be the least sensitive of all.. This is hardly surprising considering these form the basis of selective NO_2 sensors of low resistivity which display minimal cross sensitivity to reducing gases [3]. A significant improvement in performance is achieved by sputtering the preliminary metallic Sn layer onto a substrate held at a temperature above the tin melting point (232°C), though both CO and CH_4 sensitivities remain disappointingly low. As discussed previously the sensitivity of films of this thickness can be improved by pre-treatment under oxygen at 700°C. However, considerable sensitivity increases could be achieved by reactive sputtering onto a substrate heated to 250°C during the deposition procedure.

Table 2: *Characteristics of thin tin dioxide films prepared by different sputtering methods.*

Method of SnO_2 film preparation	**Annealing conditions**	**Optimum operating temp (°C)**	R_{air}/R_{H2}	R_{air}/R_{CO}	R_{air}/R_{CH4}
Reactive sputtering at room temperature	No heat treatment prior to testing	480	8.68	2.70	2.79
Reactive sputtering at room temperature	700°C in oxygen for 5h.	460	13.2	2.52	5.35
Reactive sputtering on to a substrate heated to 250°C	No heat treatment prior to testing	430	40.6	6.90	10.4
Reactive sputtering on to a substrate heated to 250°C	700°C in oxygen for 5h.	430	15.5	2.24	4.81
Deposition of metallic Sn film at room temperature	Thermal oxidation of Sn at 600°C in O_2 for 5h.	520	5.10	1.39	1.46
Deposition of metallic Sn film onto a substrate held at 250°C	Thermal oxidation of Sn at 600°C in O_2 for 5h.	480	22.9	3.11	4.06

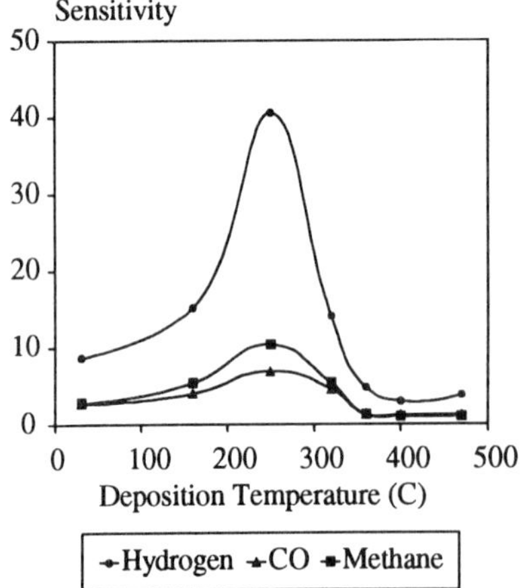

Figure 3: *The effect of deposition temperature on the sensitivity of 0.42 μ m reactively sputtered SnO_2 films towards 1% v/v concentrations of H_2, CO and CH_4 in dry air. The films were annealed in dry air at 550° C for 1h prior to testing*

In light of these findings, an experiment was carried out to optimise the substrate deposition temperature with respect to H_2, CO or CH_4 response. The results obtained for reactively sputtered films are summarised in figure 3 opposite. Any increase of the deposition temperature beyond the optimum of 250°C causes a marked drop in sensitivity to each of the combustible gases under test. A second series of sensors prepared under the same conditions were subjected to thermal treatment under oxygen at 700°C prior to testing. In this case, films deposited at room temperature were equally as sensitive as those prepared using a substrate temperature of 250°C. For both sets of sensors, deposition temperatures of 400°C or greater yielded highly resistive sensors which remained largely unresponsive to all the reducing gases tested.

3.3 Surface doping using ultra thin metallic films

Preliminary studies were carried out by depositing 2 nm films of metals such as Pd, Pt, Ag and Ni over the surface of a batch of 420 nm reactively sputtered SnO_2 sensors. The results are summarised in table 3. In general, each of the activator films increases sensitivity towards all three test gases, though there are some exceptions. For example, the use of silver increases both H_2 and CO sensitivity while methane response remains unaffected, while conversely, surface doping with Ni does not influence the magnitude of hydrogen sensitivity.

Table 3: *Reducing gas sensitivity of a series of surface doped SnO_2 films upon exposure to 1% v/v contaminant concentrations in dry air.*

Surface activator film	Optimum temperature (°C)	R_{air}/R_{H2}	R_{air}/R_{CO}	R_{air}/R_{CH4}
None	455	10.1	1.81	3.07
Pd	380	17.3	3.43	6.63
Pt	450	23.0	2.71	8.31
Ag	460	20.3	2.61	2.88
Ni	420	10.5	2.90	5.33

More detailed investigations of the most promising systems have concentrated on establishing the optimum activator film thickness while keeping the SnO_2 sensing layer constant. A batch of reactively sputtered films of 420 nm thickness were prepared and annealed under oxygen at 700°C for 5h. These sensors films were then surface doped with sputtered layers of either Pd or Ag in the range 0.3 - 20 nm and subsequently exposed to 1 % v/v mixtures of H_2, CO and CH_4 in air over a wide range of operating temperatures. The results obtained for both systems is summarised in figure 4. In general, Pd surface deposits of 2 nm or less cause a marked increase in sensitivity towards each of the test gases.

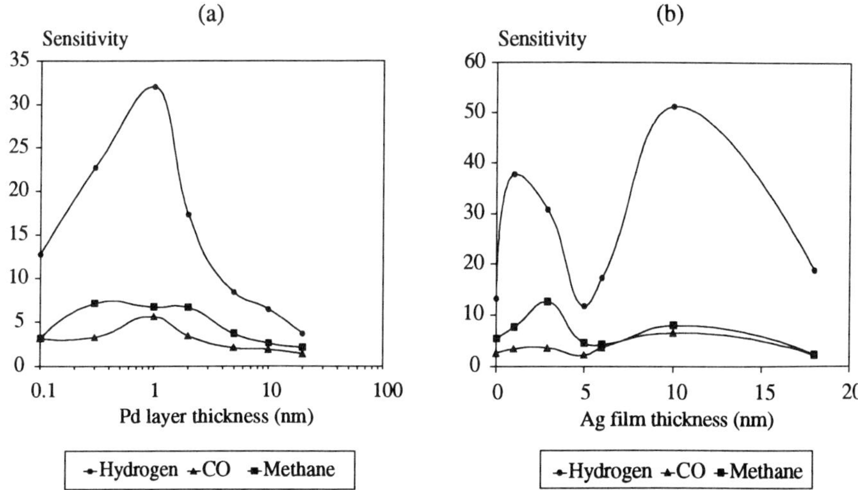

Figure 4: *The influence of activating layer thickness on the maximum sensitivity of 0.42 μ m reactively sputtered SnO_2 films surface doped using (a) Pd and (b) Ag in the presence of 1% v/v concentrations of H_2, CO and CH_4.*

Pd surface doping also causes a significant decrease in the optimum operating temperature from approximately 500°C for undoped SnO_2 to under 400°C for sensors employing Pd activator layers of 1 nm or greater. In the case of Ag, there appear to be two regions in the sensitivity versus activator film thickness plots where enhanced sensor response is exhibited. It may be possible that the two distinct regions arise as a consequence of a change in the mechanism whereby the surface dopant gives rise to enhanced sensitivity. For example, in the case of Pd, increased sensitivity is thought to be caused by spillover of hydrogen catalysed by islands of the metal on the SnO_2 surface. It is unsurprising therefore that the maximum sensitivity is achieved by using 1 nm Pd deposits which do not display electrical continuity. By analogy, the same may be true of Ag under similar conditions, while at increased thicknesses the influence of electronic interactions between SnO_2 and the Ag^+/Ag^0 redox couple [6] become more important.

4. Conclusions

The influence of deposition parameters such as film thickness and annealing conditions on the performance of reactively sputtered tin dioxide thin films has been extensively studied. From the results presented it appears that sensors fired in air yield the best results when a processing temperature of 600°C and a film thickness of 100 nm is employed. However, for sensors annealed under O_2, a temperature of 700°C and film thicknesses in the range 400 - 600 nm display the highest sensitivity. A marked enhancement in sensitivity can also be achieved by heating the substrate to 250°C during the film deposition process for sensors produced by both metallic Sn sputter/thermal oxidation and reactive sputtering pathways. A series of optimisation studies have established that surface doping the SnO_2 films by sputtering ultra thin activator layers of Pd and Ag over the sensing material leads to significant improvements in sensitivity. An optimised Pd-doped sensor exhibits high hydrogen and methane sensitivity while the use of an Ag activating layer of 10-20 nm leads to marked improvements in H_2 response but with little interference from CO and CH_4. Additional studies are now underway to determine how the changes in film processing and deposition parameters affect the microstructure of the SnO_2 and to ascertain whether any trends observed can be linked with the gas sensing behaviour of the material.

References

1. Z.M. Jarebski and J.P. Marton, Physical properties of SnO_2 materials. I. Preparation and defect structure, *J. Electrochem. Soc.*, **123(7)** (1976) 199C-205C.
2. S.C. Chang, Thin-film semiconductor NO_x sensor, IEEE Trans., Electron. devices, ED-26 (1979) 1875-1880.
3. G. Williams and G.S.V. Coles, NO_x response of tin dioxide gas sensors, *Sensors and Actuators B*, **15-16** (1993) 349- 353.
4. G. Sberveglieri, Classical and novel techniques for the preparation of SnO_2 thin film gas sensors, *Sensors and Actuators B*, **6** (1992) 239-247.
5. Q. Wu, K.M. Lee and C.C. Lui, Development of chemical sensors using microfabrication and micromachining techniques, *Sensors and Actuators B*, **13-14** (1993) 1-6.
6. N. Yamazoe, Y. Kurokawa and T. Seyiama, Effects of additives on semiconductor gas sensors, *Sensors and Actuators*, **4** (1983) 283-289.

Mid-infrared Fibre Sensor for the In-situ Monitoring of Chlorinated Hydrocarbons.

J.E. Walsh and B.D. MacCraith.
School of Physical Sciences, Dublin City University, Glasnevin, Dublin 9, Ireland.
M. Meaney, J.G. Vos and F. Regan.
School of Chemical Sciences, Dublin City University, Glasnevin, Dublin 9, Ireland.

Abstract.

An infrared fibre optic sensor which operates in the 4 to 16 μm region of the spectrum, has been developed for the *in-situ* monitoring of chlorinated hydrocarbons (CHCs) in water. The sensing element consists of a silver halide ($AgCl_xBr_{1-x}$) fibre optic, coated with a polymer which both enriches the analyte in the evanescent wave region of the fibre and minimises water interference. Water absorbs strongly in the mid-infrared region which makes more conventional transmission spectrometry of water borne chemical species difficult. Using trichloroethylene (TCE) as a representative pollutant, evanescent wave spectrometry in the mid-infrared (MIR) region is shown to provide good performance down to single ppm levels. Furthermore, we show that the technique can be applied to multi-analyte samples.

Introduction.

The increasing demands of national and European directives concerning the quality of water and the environmental protection of their resources are having a significant impact on the development of advanced sensors. There is an increasing need for more detailed quality information concerning the presence and extent of chemical species such as CHCs in water to ensure acceptable standards. Current requirements demand accurate measurement in the ppm-ppb range [1,2]. There is also an increasing need to develop sensors that can provide this information on-line from the point of intake and discharge. Off-line analysis of analyte samples is accurate but can be time consuming and expensive. Fourier transform infrared (FTIR) spectrometry is a powerful and well established analytical method for monitoring and identifying chemical species, especially organic vapours and liquids which have a spectral "finger print" in the mid-infrared [3]. However, this technique is somewhat limited when dealing with aqueous solutions of an analyte due to strong background absorption of water in the mid-infrared [4]. Fibre optic evanescent wave spectroscopy (FEWS) is a technique which allows *in-situ* mid-infrared absorption spectroscopy in aqueous environments [5,6,7].

In difficult or hazardous environments FEWS analysis systems can be separated from the analyte or sample point to facilitate on-line monitoring.

Theory and design of FEWS.

The evanescent wave of guided radiation in an optical fibre decays exponentially in amplitude with distance from the core/cladding interface (figure 1) [8]. An analyte in the cladding region will absorb the evanescent wave at analyte specific wavelengths and the degree of absorption is directly related to the analyte concentration. Furthermore, if an analyte enriching, hydrophobic polymer is used as the fibre cladding greater sensitivity can be achieved. Other factors which determine the sensitivity of FEWS include the length, diameter and shape of the sensor fibre as well as the input illumination conditions, such as angle and numerical aperture [7,9].

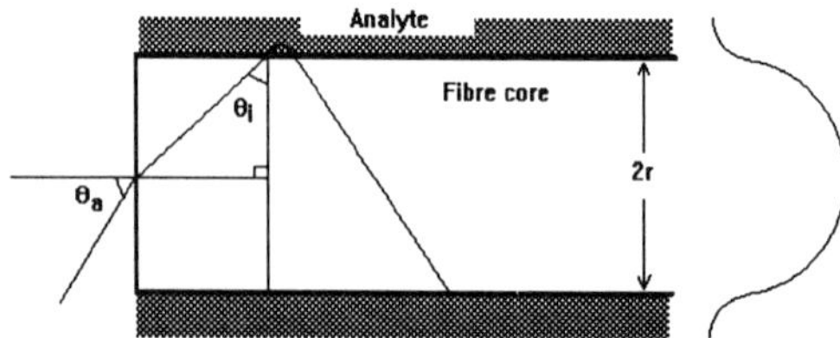

Figure 1. Internal and evanescent rays in an optical fibre with corresponding intensity profile.

The development of infrared fibre optics, operating in the 4 to 16 μm region of the spectrum, has facilitated the development of evanescent wave sensors for the in-situ detection of chlorinated hydrocarbons in water [5,6,7]. The sensing element consists of a silver halide optic fibre, coated with a polymer, such as polyisobutylene (PIB), which enriches CHCs in the evanescent wave region of the fibre. In order to increase the evanescent signal, and therefore the sensitivity of the sensor, a number of novel launch designs and fibre configurations are being examined.

Set-up.

Figure 2 shows the current instrument set-up. Using a fibre optic interface accessory, infrared light from the FTIR's black-body source is launched into a 1.5 m length of 860/1000 μm core/clad mid-infrared (MIR) fibre optic cable. Light from this cable is coupled into a 0.15 m length of unclad 1000 μm MIR fibre optic cable which is mounted, using septa, in a flow-through sample cell. The other end of this measurement fibre is coupled into another 1.5 m length of 860/1000 μm core/clad which returns the light directly to the FTIR's detector. The type of detector used is a liquid nitrogen cooled mercury-cadmium-telluride (MCT) photoconductor. The measurement fibre is coated with the appropriate polymer within the sample cell.

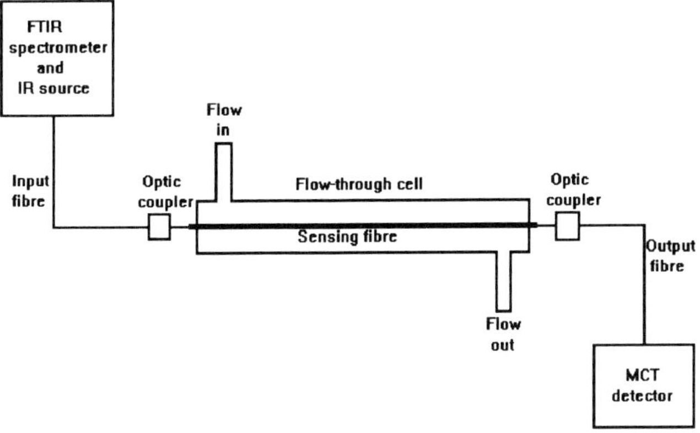

Figure 2. Layout of mid-infrared fibre optic sensor.

Results and discussion.

The absorption spectrum of the PIB coating can be calculated using the signal from the uncoated fibre as a reference. The absorption spectrum for a 10 μm coating of PIB which has not been exposed to an analyte solution is shown in figure 3. One of the advantages of PIB is the low number of absorption features in the spectral region of interest for CHC detection which is above 10 μm [?]. A typical absorption spectra for TCE recorded with the PIB coated evanescent wave sensor are shown in figure 4. Each analyte has a unique absorption signature in the mid-infrared whose intensity varies with analyte concentration. In figure 5 absorption spectra for increasing concentrations of trichloroethylene (TCE) are shown and the resulting calibration curve is produced in figure 6. The current limit of detection for TCE is 3 ppm. Finally, an absorption curve for multiple analytes, shown in figure 7, demonstrates the potential power of the sensor to quantify multi-analyte data. Each analyte has a unique absorption feature allowing it to be detected in the presence of the other analytes.

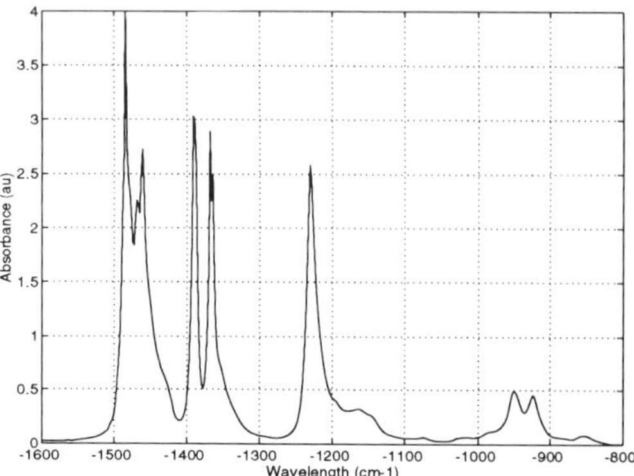

Figure 3. Absorption spectrum of PIB.

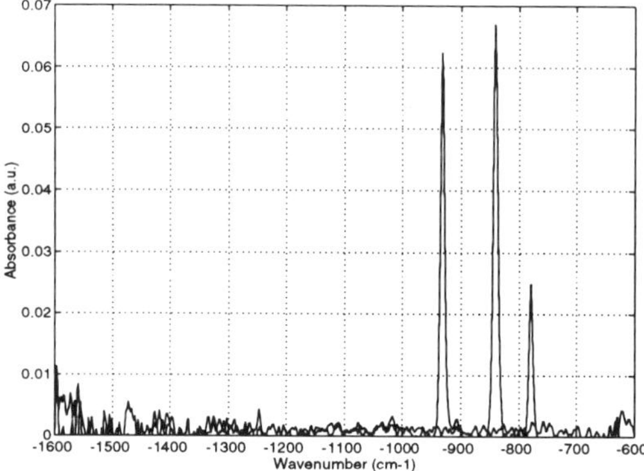

Figure 4. Absorption spectrum of 1000 ppm TCE.

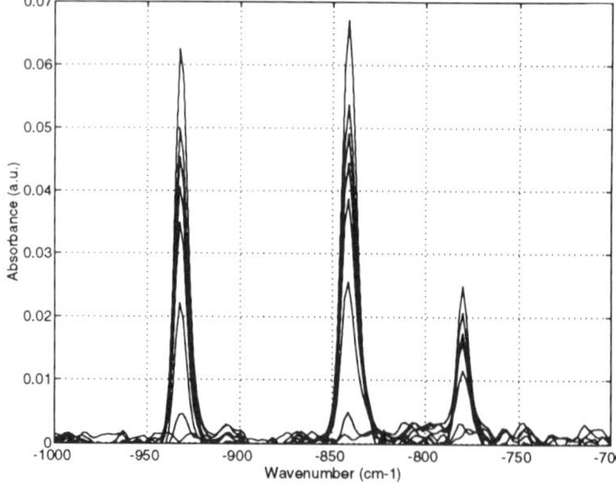

Figure 5. Absorption spectra for various concentrations of TCE.

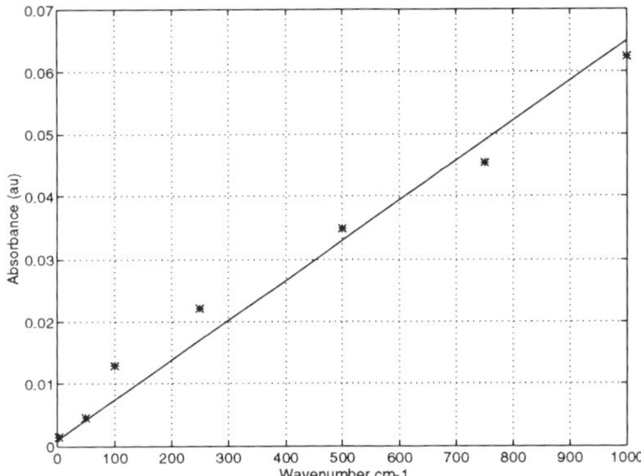

Figure 6. Calibration curve for TCE absorption peak at 840 cm^{-1}.

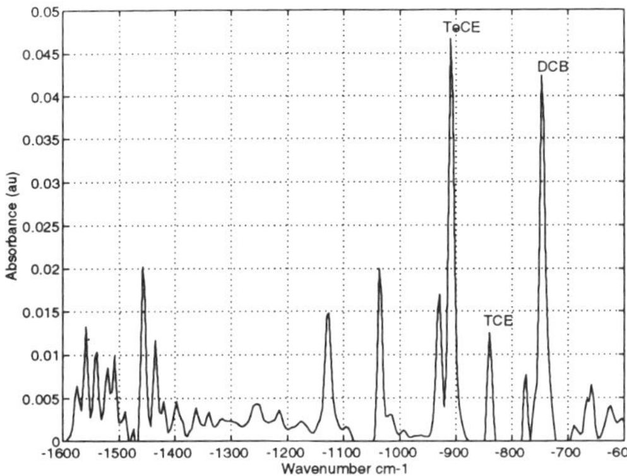

Figure 7. Absorption spectrum of multiple CHC sample including trichloroethylene (TCE), tetrachloroethylene (TeCE) and dichlorobenzene (DCB).

Conclusions.

A fibre optic evanescent wave sensor for the *in-situ* detection of CHCs, using PIB coated silver halide fibres, has been demonstrated. Limits of detection in the ppm range for TCE have been achieved and a calibration curve produced. Finally, the potential of the system for monitoring multi-analyte samples has been demonstrated.

In order to increase the evanescent signal, and therefore the sensitivity of the sensor, a number of novel launch designs and fibre configurations are being examined. These include illuminating the sensor fibres at an angle, increasing the length of the fibre in the analyte region, tapering and/or bending the fibres in the analyte region, improving the fibre coatings to further enrich the analyte in the evanescent region and doping the coating to increase the evanescent signal.

In addition, data analysis techniques such as principle components analysis are being applied. These techniques allow the multiple spectral component variations, due to combinations of chemical species, to be quantified in the presence of background variations such as the water content of the polymer coating.

References.

1. EEC Council Directive (No. 76(464)) of 4 May 1976 on pollution caused by certain dangerous substances discharged into the aquatic environment of the community.
2. *Guidelines for drinking-water quality,* World Health Organisation, Vol. 1, Recommendations (WHO, Genf, 1984).
3. *Improving the infrared detection of analytes in aqueous solutions using polymer coated ATR elements,* C.W. Meuse and S.A. Tomellini, Anal. Lett., 22 (9), 2065-2073, 1989.
4. *Principles and applications of Fourier transform infrared (FTIR) process analysis,* W.M. Doyle, Process Control and Quality, 2, 11-41, 1992.
5. *New IR fibre-optic chemical sensor for in-situ measurements of chlorinated hydrocarbons in water,* R. Krska, K. Taga and R. Kellner, Appl. Spec., 47 (9), 1484-1487, 1993.
6. *Evanescent wave infrared spectroscopy of liquids using silver halide optical fibres,* S. Simhony, I. Schnitzer, A. Katzir and E.M. Kosower, J. Appl. Phys., 64 (7), 3732-3734, 1988.
7. *Fiber optic sensor for chlorinated hydrocarbons in water based on infrared fibers and tuneable diode lasers,* R. Krska, R. Kellner, U. Schiessl, M. Tacke and A. Katzir, Appl. Phys. Lett., 63 (14), 1868-1870, 1993.
8. *Internal Reflection Spectroscopy,* N.J. Harrick, 1987, (Harrick Scientific Corporation, New York).
9. *Demonstration of an optimised evanescent field optical fibre sensor,* Z.M. Hale and F.P. Payne, Anal. Chim. Acta., 293, 49-54, 1993.

Section B

Sensor Arrays and Silicon Sensors

Piezomagnetic materials for use in microelectromechanical systems

M R J Gibbs

Sheffield Centre for Advanced Magnetic Materials and Devices, Department of Physics, The University of Sheffield, Sheffield, S3 7RH, UK

Abstract. The aim of this paper is to introduce the principles of piezomagnetism within the context of the latest developments in thin film materials preparation. It is demonstrated that the properties of modern piezomagnetic films are competitive for consideration as the active elements in a range of microelectromechanical systems. Amorphous ferromagnetic thin films and magnetic multilayer systems are discussed in detail. A discussion is given of the potential benefits to be derived from the use of piezomagnetic materials. A demonstrator pressure sensor based on a resonant microbridge is described.

1. Principles of piezomagnetism

1.1 Domain structure and the magnetisation loop

It has been established [1] that the optimum piezomagnetic response in a ferromagnet comes from the coherent rotation of the magnetisation through 90°. A material should therefore be chosen with a well defined uniaxial anisotropy. Such a material would have a domain structure consisting of 180° domain walls separating uniformly magnetised domains (see Fig.1).

Fig.1
A schematic diagram of the ideal domain structure for a piezomagnetic material

As the applied magnetic field, H, is increased from zero, the direction of magnetisation within the domains rotates through an angle θ towards the field direction.
The zero field direction of the domain magnetisation is defined by the directional of the uniaxial anisotropy, K. Rotation of the magnetisation from the easy direction results in a torque on the magnetic moments, Γ_K, given by

$$\Gamma_K = 2K\cos\theta\sin\theta \qquad (1)$$

The applied field also exerts a torque, Γ_H given by

$$\Gamma_H = \mu_0 M_s H\cos\theta \qquad (2)$$

where M_s is the saturation magnetisation within the domain. In equilibrium $\Gamma_K = \Gamma_H$ and hence

$$M = \left\{\frac{\mu_0 M_s^2}{2K}\right\} H = \chi H \qquad (3)$$

and the M-H response is characterised by a constant susceptibility.

1.2 Magnetostrictive response

For a cubic crystal, the spontaneous magnetostrictive strain in a direction defined by the direction cosines $\beta_1\beta_2\beta_3$ when the sample is magnetised in a direction defined by the cosines $\alpha_1\alpha_2\alpha_3$, λ_i, is given by

$$\lambda_i = \frac{3}{2}\lambda_{100}\left(\alpha_1^2\beta_1^2 + \alpha_2^2\beta_2^2 + \alpha_3^2\beta_3^2 - \frac{1}{3}\right) + 3\lambda_{111}(\alpha_1\alpha_2\beta_1\beta_2 + \alpha_2\alpha_3\beta_2\beta_3 + \alpha_3\alpha_1\beta_3\beta_1) \quad (4)$$

For an isotropic material this reduces to

$$\lambda_i = \frac{3}{2}\lambda_s\left(\cos^2\theta - \frac{1}{3}\right) \quad (5)$$

where θ is the angle between the direction in which we require the strain and the initial direction of magnetisation (that in zero field). The maximum value for λ_i is $1.5\lambda_s$.
For devices either open loop or closed loop methods may be used. In the closed loop case, the quadratic dependence of λ on H (or M, given constant χ) is the key factor. For a material of sufficiently low anisotropy constant the response around zero field can be quite strong. In open loop mode, it is the differential response which is important. Again a low anisotropy constant will give increased response. In open loop mode the sample requires a bias field, which may be derived from coils or a permanent magnet, in order to operate at the point of highest differential response.

1.3 ΔE effect

Application of a longitudinal stress, σ, introduces an extra magnetic term into the free energy of the material. This is due to the coupling between strain and the direction of magnetisation arising from the inverse of the magnetostrictive response.

$$K_\sigma = -\frac{3}{2}\lambda_s\sigma\cos^2\theta \quad (6)$$

The total strain on application of a load, ε, is made up of an elastic, ε_{el}, and a magnetostrictive, ε_λ, part. Whence

$$\varepsilon = \frac{\sigma}{E_s} + \frac{3\lambda_s}{2}\left(\frac{\mu_0^2 M_s^2 H^2}{(2K - 3\lambda_s\sigma)^2} - \frac{1}{3}\right) \quad (7)$$

where E_s is the Young's modulus in magnetic saturation. If E is the modulus in a magnetically unsaturated state, then it is possible to show [1] that

$$\frac{E}{E_{s\ min}} = \left(1 + \left(\frac{9\lambda_s^2 E_s}{2K}\right)\right)^{-1} \quad (8)$$

and thus the elastic stiffness may be a function of magnetic field, the magnitude of the effect depending on the ratio of λ_s to K. In materials of low K the minimum ratio can be as small as 0.1.
For the purposes of this discussion it should be remembered that the natural frequency of vibration of a cantilever or bridge is directly related to the value of Young's modulus. If E is field dependent, then so is the resonant frequency of the structure (see section 4).

2. Advanced piezomagnetic materials

2.1 Ferromagnetic amorphous thin films

Amorphous ferromagnets based on the transition elements, Fe, Ni or Co, possess the low K discussed above, and also demonstrate very high χ. Recent effort has concentrated on producing thin films which mimic the ribbon form of these materials which have been known and studied since the early 1970's.

Growth has been by means of rf magnetron sputtering [2], and films have been successfully grown on a range of technically important substrates including Si (001), GaAs (001), Si_xN_y, glass and polyimide. Fig.2 and 3 show typical data for such materials.

Fig.4 shows the hysteresis loop for such films. One data set is taken using an ultra sensitive hysteresis loop plotter, and the other is taken using the magneto-optic Kerr effect (MOKE). Importantly these results show that the films are magnetically homogeneous in that there is no surface or substrate dead layer. Note also that the loops measured in orthogonal directions have different form, indicating the presence of uniaxial anisotropy in the films. Our best results in this area are $\lambda_s = 35$ ppm and coercivity 6 A/m for a film 600 nm thick.

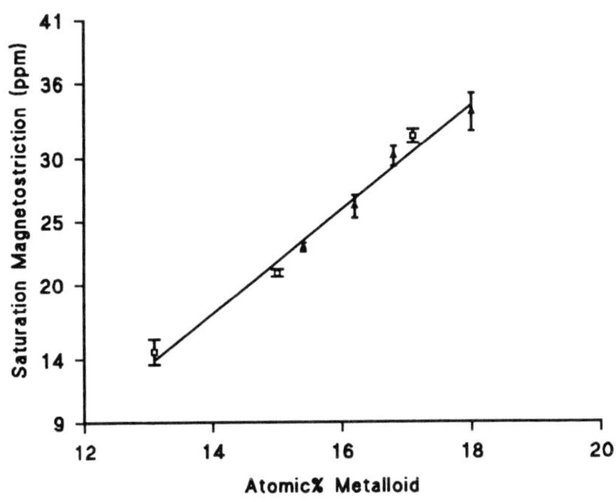

Fig.2
The variation of saturation magnetostriction with metalloid content in an Fe-Si-B-C amorphous thin film.

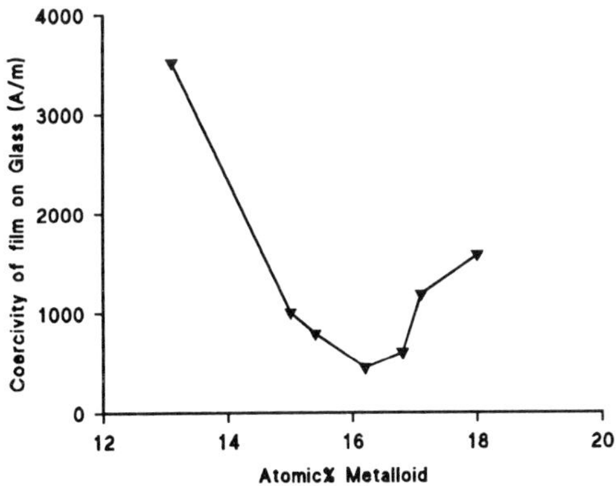

Fig.3
The variation of coercivity with metalloid content in an Fe-Si-B-C amorphous thin film

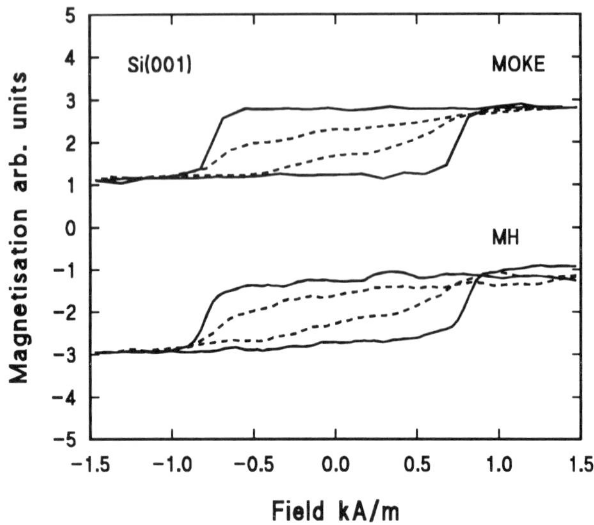

Fig.4
Magnetisation loops for an Fe-B-Si-C amorphous thin film grown on Si.

2.2 Magnetic multilayers

There has been considerable interest in recent years in the physics and applications of magnetic multilayers. They offer a number of key engineered properties such as giant magnetoresistance suitable for read-write heads in the recording industry, and high perpendicular anisotropy for recording media.
The piezomagnetic properties of such materials have received less attention, but there is ample evidence that the net saturation magnetostriction constant is a strong function of the thickness of layers in the structure, the nature of any interface region between layers, and the growth parameters used [3,4,5].
We have, in particular, made a detailed study of the FeCo/Ag multilayer system. Here the FeCo polycrystalline material has high intrinsic saturation magnetostriction constants, and the Ag and FeCo are immiscible. With British Technology Group plc we have patented [6] methods of enhancing the saturation magnetostriction constant in multilayers. The principles rest on the effects at the interfaces in terms of lattice strain of the magnetic component, alloying of the magnetic and non-magnetic components, pseudomorphic or textured growth of the polycrystalline layers.
It is again important to maintain a low anisotropy constant, and low coercivity. The small grain size of sputter deposited polycrystalline films (in our case sub 10 nm) produces such properties.
Fig.5 shows the variation of saturation magnetostriction constant and coercivity in an FeCo/Ag multilayer series as a function of the $Fe_{50}Co_{50}$ layer thickness. The Ag layer was held at 2 nm thickness, and the sputtering power at 250W. The sputtering pressure was found to have the most significant effect on the physical properties, and the data shown in Fig.5 is at a pressure of 1.5 mTorr of Ar. Fig.6 shows the pressure variation.
The best performance here is $\lambda_s = 98 \pm 2$ ppm at $H_c = 250$ A/m. Whilst this is not as magnetically soft as the amorphous ferromagnetic films, the thermal stability is far higher, and there is a factor of two improvement in the saturation magnetostriction constant.
Further systems to be considered are the CoPd/Ag series [5] where the CoPd layer has $\lambda_s = 170$ ppm, but the coercivity remains close to 1000 A/m. Further work is in hand to try to improve the softness by changing the Ag for other materials. Rare earth transition metal thin films offer $\lambda_s > 500$ ppm, but with very high anisotropy constant because of the rare earth element, and high coercivity. Again attempts are being made to reduce K and H_c by varying the properties of an interlayer in a multilayer series.

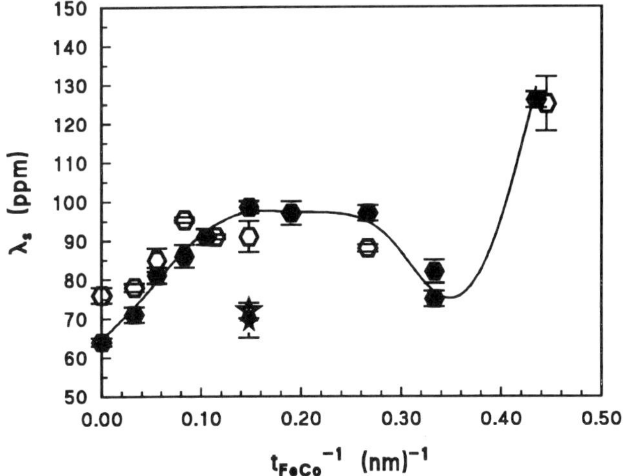

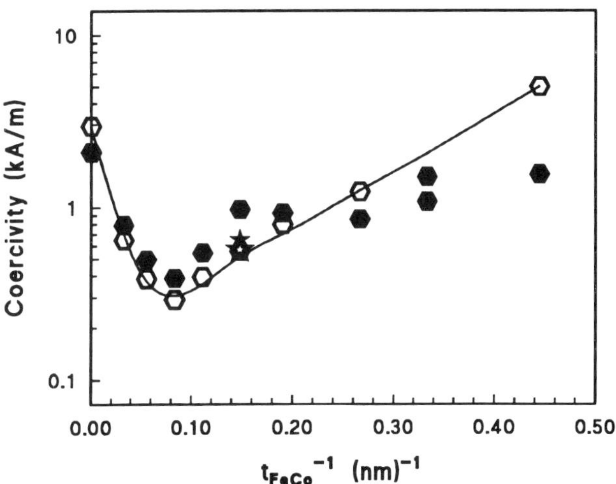

Fig.5
The variation of saturation magnetostriction constant, λ_s, and coercivity, H_c, with the thickness of the FeCo layer in an FeCo/Ag multilayer series

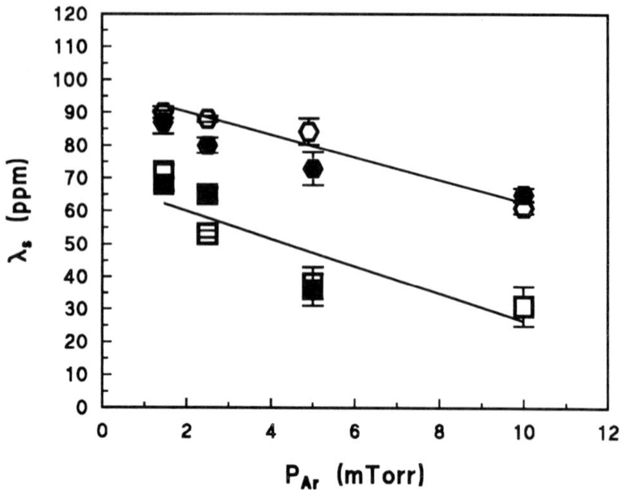

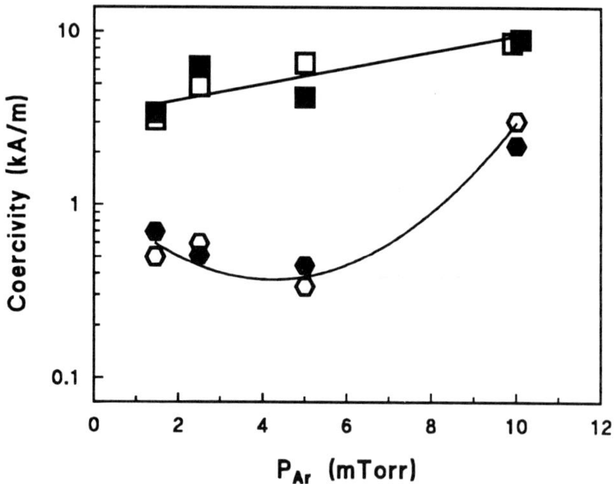

Fig.6
The dependence of saturation magnetostriction and coercivity on the Argon sputtering pressure in an FeCo/Ag multilayer series

3. Advantages offered by piezomagnetic materials

Piezomagnetic films are expected to achieve the goal of higher device integrity through enhancing reliability and providing diagnostics with the inclusion of self-test mechanisms. For example, in accelerometers, one method of self-test is thermal actuation. A magnetostrictively actuated device would have the very important operational advantage in that the actuation mechanism has DC stability. Piezomagnetic materials could overcome demanding packaging requirements, particularly for pressure transducers, which are being required to operate at extremely high pressures (> 1000 bar). Current pressure sensors require electrical connections to the diaphragm which must be lead out through the material of the pressure cell. The ability to sense diaphragm stress in a non-contacting manner would represent a major breakthrough. Another area of advantage could be in allowing the sensing of magnetic fields through a method of robust optical interrogation. This would find application in, for example, gear tooth sensing for aero engine systems.

4. Demonstrator devices

It is important to establish that piezomagnetic films can be patterned. Fig.7 is an atomic force microscope image, taken in Sheffield, of "wires" formed by the etching of an Fe-Si-B-C ferromagnetic amorphous thin film grown on glass. Even in this trial using optical lithography and lift-off techniques there is good definition of the structures.

By way of a demonstration project, and in close collaboration with Lucas Advanced Engineering Centre, we have coated a Si microbridge with the FeCo/Ag multilayer described above. This microbridge formed part of a demonstrator pressure sensor head. In the demonstrator the bridge was driven in to a resonant vibration with pulsed laser light. The frequency of vibration was detected optically. As a DC magnetic field was applied to the device, with the field in the plane of the bridge, the resonant frequency of the bridge could be tuned. The form of the response (see Fig.8) was that of a classic ΔE effect [1]. Whilst the effect was not large, it must be remembered that nothing was done to try to optimise the performance. The importance of this result is twofold. It shows that piezomagnetic films can be integrated in to complete device structures, the device and the film keeping their respective integrities at all stages of fabrication. The result also shows that piezomagnetic films can be grown showing near classic responses, opening up the possibility of a range of applications.

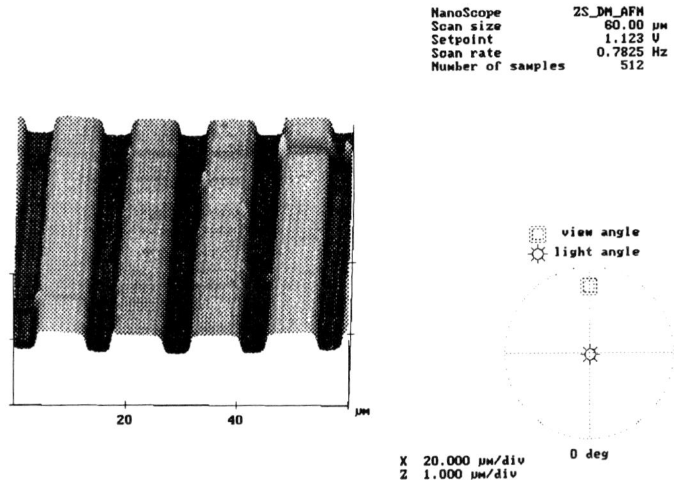

Fig.7
An atomic force microscope image of lithographically defined "wires" formed in an Fe-Si-B-C ferromagnetic thin film on glass

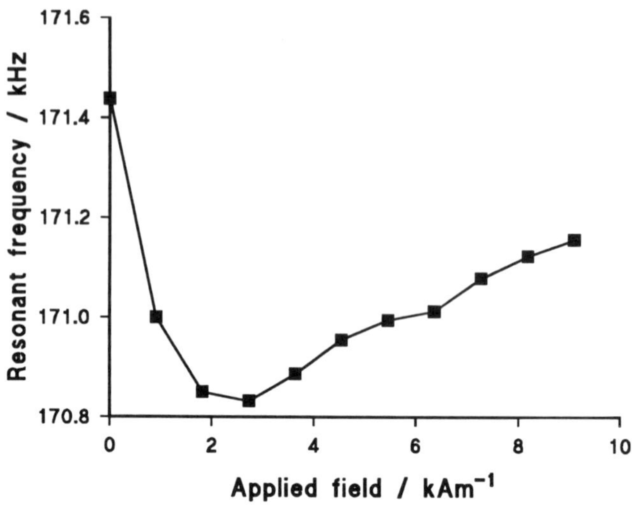

Fig.8
The variation of resonant frequency of a Si microbridge coated with an FeCo/Ag piezomagnetic film as a function of applied DC magnetic field. The form of this curve is a classic ΔE response.

There have also been reports of cantilever deflection using piezomagnetic films [7]. Here a polyimide cantilever was coated with Tb-Fe on one side ($\lambda > 0$) and Sm-Fe ($\lambda < 0$) on the other. A deflection of several microns was observed at the end of the cantilever, but fields of several Tesla were required due to the high anisotropy constant of the rare earth films. With the continued research to reduce the effective K in such systems, it remains to be seen how low a field will ultimately be required to activate such cantilevers.

5. Concluding remarks

The advent of new thin film and multilayer magnetic materials has re-opened the possibility of the use of piezomagnetic properties in devices. The ease of integration of magnetic films directly in to structures based on Si or GaAs gives impetus to research on novel microelectromechanical systems. Whilst basic materials research continues in many laboratories, demonstrator devices have succeeded. The ability to remotely activate or interrogate devices using magnetic materials may provide an answer for sensors and actuators in a range of hostile environments.

6. Acknowledgements

The author particularly wishes to thank Dr.C.Shearwood and Mr.A.D.Mattingley for their contributions to the experimental programme on piezomagnetic films at the University of Sheffield. They are responsible for the data shown in section 2 of this paper. The author is indebted to Mr.R.B.Yates and Prof.C.R.Whitehouse of the University of Sheffield Microelectromechanical Systems Unit, and Prof.S.J.Prosser and Mr.J.Dancaster of the Lucas Advanced Engineering Research Centre for detailed discussions. The work on developing piezomagnetic materials has been supported by the UK Engineering and Physical Sciences Research Council.

7. References

[1] Livingston, J.D.: "Magnetomechanical properties of amorphous metals": Phys.Stat.Sol. (a) 70 (1982) 591-596

[2] Mattingley, A.D., Shearwood, C. And Gibbs, M.R.J.: "Magnetic and magnetoelastic properties of amorphous Fe-Si-B-C films": IEEE Trans.Mag. 30 (1994) 4806-4808

[3] Lafford, T.A., Gibbs, M.R.J. and Shearwood, C.: "Magnetic, magnetostrictive and structural properties of iron-cobalt/silver multilayers": J.Magn.Magn.Mat. 132 (1994) 89-94

[4] Lafford, T.A., Gibbs, M.R.J., Zuberek, R. And Shearwood, C.: "Magnetostriction and magnetic properties of iron-cobalt alloys multilayered with silver": J.Appl.Phys. 76 (1994) 6534-6536

[5] Lafford, T.A., Zuberek, R. And Gibbs, M.R.J., "Magnetic properties of $Co_{32}Pd_{68}/Ag$ multilayers": J.Magn.Magn.Mat. 140-144 (1995) 577-578

[6] UK Patent GB 2 268 191 B

[7] Hayashi, Y., Honda, T., Arai, K.I., Ishiyama, K. And Yamaguchi, M.: "Dependence of magnetostriction of sputtered Tb-Fe films on preparation conditions": IEEE Trans.Mag. 29 (1993) 3129-3131

Current-mode application specific integrated circuits for resistive sensing arrays

P I Neaves and J V Hatfield

Department of Electrical Engineering & Electronics, UMIST, PO Box 88, Manchester M60 1QD

Abstract: There are many sensing systems that rely on resistive sensors; examples include gas sensors, strain gauges and magneto-resistive sensors. This paper describes work on the development of Application Specific Integrated Circuits (ASIC's) employing current-mode techniques as part of an "electronic nose". The ASIC's can be employed in any sensing system utilising resistive sensors.

1. Introduction

A number of sensing solutions rely on changes in conductance or resistance of an array of sensors. The signal processing problem is essentially one of multi-channel resistance measurement. A typical example of such a system, currently under development at UMIST, is a multi-element gas/odour sensing array, often referred to as an "electronic nose".

The UMIST electronic nose is currently a bench mounted instrument. However, a hand-held or portable electronic nose would be most useful; environmental monitoring, for example. This paper describes work towards integrating much of the analogue signal processing circuitry using Application Specific Integrated Circuits (ASIC's) employing current-mode techniques.

2. Current-mode signal processing

Traditionally, analogue circuit design, and in particular integrated circuit design has been concerned with manipulating signal voltages as the desired variable. For example, basic building blocks such as Voltage Operational Amplifiers (VOA), Analogue to Digital Converters (ADC) and Digital to Analogue Converters (DAC) are assembled to operate on signal voltages. An alternative approach is to design circuits that operate on signal currents. Such an approach, often termed the "current-mode" approach has received much attention in recent years[2][3][4][5].

Current-mode configurations offer a number of advantages over their voltage-mode counterparts and include:

(i) Wide bandwidth - voltage swings in current-mode circuits are of the order of hundreds of millivolts and hence the charging and discharging of stray or inherent capacitance is rapid.

(ii) Low supply voltage operation - this is of particular importance in view of the trend towards 3.3V as the standard supply voltage for integrated circuits.

(iii) Low power consumption.

(iv) Flexibility - the majority of analogue functions can be realised with a basic sub-set such as multiplication, subtraction, division and addition [6]. These operations are much easier to achieve with signal currents than with signal voltages.

(v) Signal routing/multiplexing - the multiplexing or routing of signals, in an ASIC solution, is much more readily achieved with currents than voltages. Poor quality analogue switches can be used resulting in a large saving of silicon real-estate.

3. Current-mode ASIC's for an electronic nose

The electronic nose is made up of a multi-element gas sensing array of conducting polymers, signal processing electronics, 16 bit microcontroller and a LCD display[7]. In the UMIST electronic nose each sensor element changes in resistance when exposed to a volatile compound. The degree of response to a given substance depends on the type of polymer element used so that a pattern of resistance changes can be recorded and processed to produce a set of descriptors for that particular substance thus, providing that the pattern of resistance changes can be recognised, the gas can be identified. The signal processing problem is then one of multi-channel resistance measurement. A block diagram of the analogue signal processing section of the electronic nose is shown in Figure 1.

Operation is straightforward. The current multiplexer forces a fixed voltage across one or more sensors. The resulting current, the *basal* current, is then offset by a programmable current source on chip. This is necessary, since in this application we are looking at resistance deviations from the basal resistance of the sensor(s). In order to enhance the signal to noise ratio, a programmable current amplifier provides current gain

The two ASIC's employed, a current amplifier and a current multiplexer will now be described. Note, that the arrangement in Figure 1 can be employed in *any* multi-element sensing engine employing resistive transducers. Typical examples include strain gauges, bio-sensors and magnetoresistive sensing arrays.

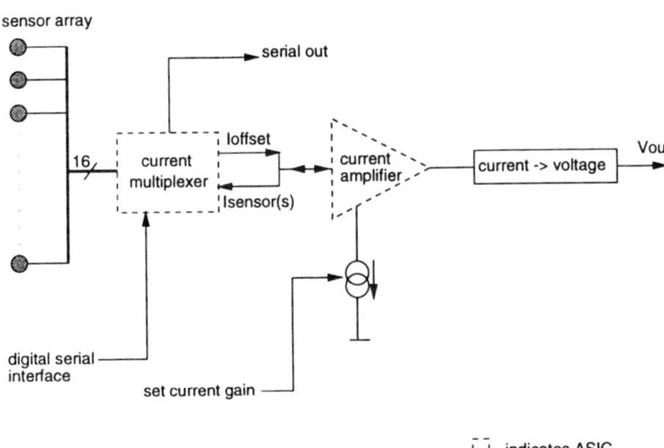

Figure 1
Block diagram of the electronic nose (analogue part only)

3.1. Current amplifier

The current amplifiers are based on the Linear Transconductance Multiplier (LTM), sometimes referred to as the Gilbert multiplier [8][9]. The current amplifier employs complementary LTM's and a block diagram is shown in figure 2b [10]. They can provide high gain and simplicity at the expense of accurate gain settings. This is not a problem in many applications as the gain of the current-amplifier can be calibrated using known and accurate currents.

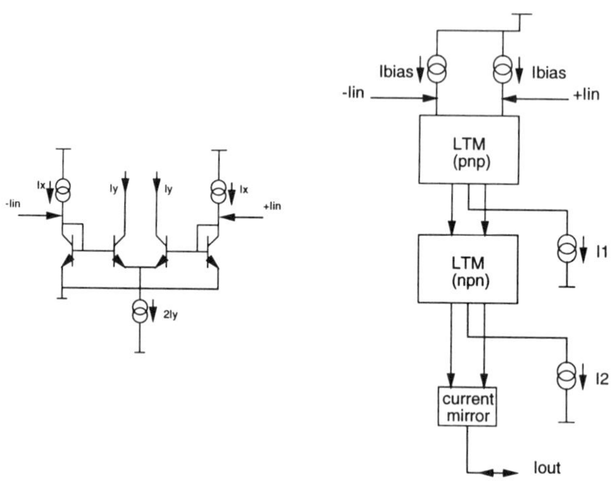

Figure 2
The LTM (2a) employed in the current amplifier (2b)

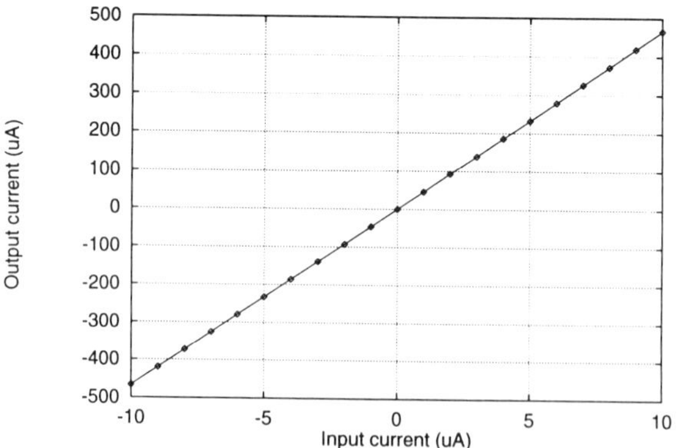

Figure 3
Typical transfer function of the current amplifier

A typical LTM topology is shown in figure 2a. The gain of the cell is approximately:

$$\alpha \frac{I_y}{I_x} \qquad 1$$

where α has its usual meaning when applied to bipolar transistors. The circuit of figure 2a can only support gains up to the beta (β) of the transistors. In practice, gains are restricted to $\beta/10$.

The gain of the current amplifier is only a function of the bias currents, I_{bias} and I_2:

$$\alpha_{npn} \frac{I_1}{I_{bias}} \alpha_{pnp} \frac{I_2}{I_1} = \alpha_{npn} \alpha_{pnp} \frac{I_2}{I_{bias}} \qquad 2$$

The current amplifier has been fabricated on SGS-Thomsons $2\mu m$ BiCMOS process under the EUROCHIP initiative. A typical transfer function of a device is shown in Figure 3. Equation 2 indicates that the current amplifier should be relatively insensitive to transistor beta. Measurements of ten devices with a nominal gain of 48 resulted in a mean gain of 47.3.

4. Current multiplexer

The current multiplexer forces a voltage across a known sensor or sensors and provides a current output. In order to measure *resistance deviations*, the basal current of each sensor can be offset by a digital programmable current source on chip[11].

The current multiplexer enjoys the following features:

- Accomadation of up to 16 sensing elements.
- Ability to cascade many devices for larger sensing arrays with *no* glue logic.
- Precision voltage reference on chip.
- Programmable offset current.
- Accurate summation of up to 16 sensing elements.
- Simple serial digital interface.
- Power down.

A block diagram of the current multiplexer is shown in Figure 4. The multiplexer is made up from 16 "current sense" blocks constructed from 16 current conveyors [12].

Each current-sense unit forces a constant voltage across each sensor. The sensed current is then made available via the collector of a NPN transistor. Such an arrangement allows the output of each conveyer to be wire OR'ed and hence allows accurate summation of different sensor responses. This would be much more difficult to achieve if signal voltages were used and effectively increases the sensitivity of the electronic nose as a number of like sensors can be selected. Control of the chip is via a simple serial digital interface. A "serial out" pin is available allowing many chips to be cascaded without any glue logic.

The current-multiplexer has been fabricated on SGS-Thomson's $2\mu m$ BiCMOS process. A typical transfer function of one of the current-sense units is shown in Figure 5. In this example a voltage of 2 volts was forced across a resistor whose resistance was varied between 10k and 11k.

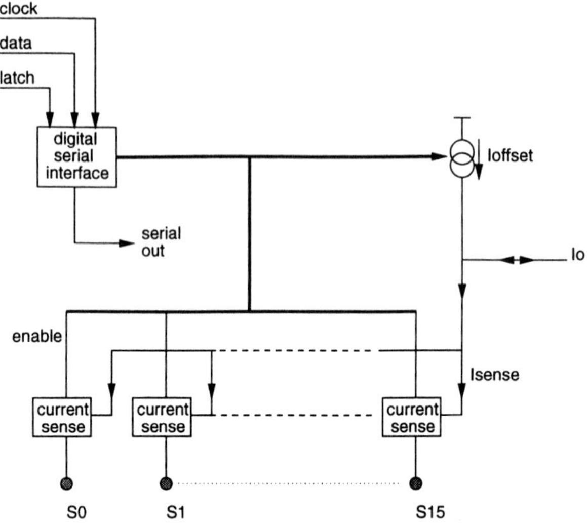

Figure 4
Block diagram of the current multiplexer

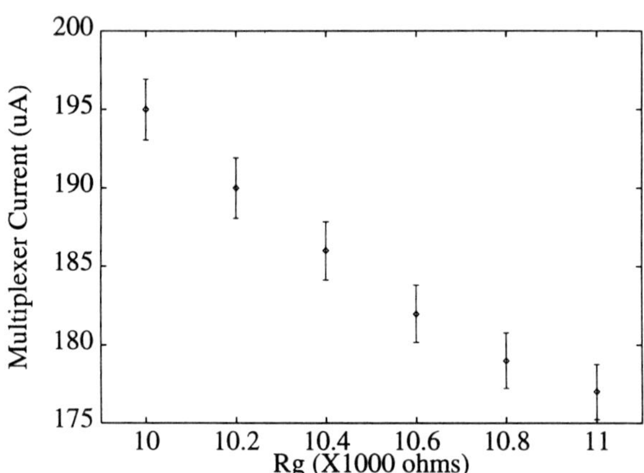

Figure 5
Typical transfer function of the current multiplexer

5. Conclusion

The majority of the analogue signal processing requirements for the UMIST electronic nose has been replaced by ASIC's employing current-mode signal processing techniques. Current-mode signal processing allows much more flexibility in the circuit design and analogue operations are much more readily accomplished when compared to voltage-mode techniques. The two ASIC's can be used in any sensor system that utilises resistive sensing elements.

6. Acknowledgement

The authors gratefully acknowledge Aromascan PLC for their generous support of this project and EUROCHIP for providing micro-fabrication facilities.

References

1. B. Wilson, "Analogue Current-mode Circuits," *International Journal of Electronic Engineering Education*, vol. 26, pp. 206-233, 1989.
2. J. Lidgey and C. T. Toumazou, "Current-Mode Analogue Signal Processing," in *Proceedings of the IEEE Bipolar Circuits and Technology Meeting*, pp. 224-232, Minneapolis, 1991.
3. Z. Wang, "Current-Mode CMOS Integrated Circuits for Analog Computation and Signal Processing," *Analog Integrated Circuits and Signal Processing*, vol. 1, pp. 287-295, 1991.
4. "Special Issue on Current-Mode Analogue Signal Processing Circuits," *IEE Proceedings*, vol. 137, no. Part G, pp. 61-184, 1990.
5. C. T. Toumazou, E F. J. Lidgey, and D. G. Haigh (eds.), in *Analogue IC Design: the current-mode approach*, Peter Peregrinus Ltd., London, 1990.
6. D. L. Grundy and J. Raczkowicz, "Structured Analogue Electronics," *Electronics and Wireless World*, pp. 965-969, Nov 1991.
7. P. I. Neaves and J. V. Hatfield, "A New Generation of Integrated Electronic Noses," *Sensors and Actuators B*, 1995. To be published.
8. B. Gilbert, "A New Wide-Band Amplifier Technique," *IEEE Journal of Solid-State Circuits*, vol. sc-3, no. 4, pp. 353-365, 1968.
9. B. Gilbert, in *Analogue IC Design: the current-mode approach*, ed. C. T. Toumazou, F. J. Lidgey, and D. G. Haigh, pp. 11-92, Peter Peregrinus Ltd., London, 1990.
10. P. I. Neaves and J. V. Hatfield, "An Analogue Current-mode Signal Processing ASIC for Interrogating Resistive Sensor Arrays," *IEEE International Symposium on Circuits and Systems (ISCAS '94)*, vol. 5, pp. 405-408, London, 30 May - 2 June.
11. P. I. Neaves and J. V. Hatfield, "Current-mode multiplexer for interrogating resistive sensor arrays," *Electronics Letters*, vol. 30, no. 12, pp. 942-943, 1994.
12. B. Wilson, "Recent Developments in Current Conveyors and Current-Mode Circuits," *IEE Proceedings G*, vol. 137, no. 2, pp. 63-77, 1990.

Smart silicon Hall devices

Sandra Bellekom, Paul de Vries and Paul Simon

Delft University of Technology, Electrical Engineering Department, Electronic Intrumentation Laboratory, P.O. Box 5031, 2600 GA Delft, The Netherlands. Phone: +31 15 783342, Fax: +31 15 785755.

Abstract. Integrated silicon Hall devices can be used to measure magnetic fields. Unfortunately they can not be used in some applications because they suffer from a large unpredictable offset and sensitivity drift. The offset can be reduced to a few micro Tesla, for temperatures up to about $80°C$, using the spinning-current method. If the sensor can calibrate itself, changes in the sensitivity will not result in errors in the output of the sensor. Both autocalibration and spinning-current offset reduction at a temperature range from room temperature to $110°C$, have been verified experimentally.

1. Introduction

Magnetic sensors are applied in two ways, first to directly measure magnetic field strengths and second to measure mechanical quantities using indirect measurement methods. Examples of the first group of applications are the measurement of geomagnetic fields or data on a magnetic tape. The indirect measurement of mechanical quantities, using a magnetic sensor combined with a small permanent magnet, is often used in the automotive industry for example to determine the throttle angle or the rotational speed of a toothed wheel.

1.1. The Hall effect

Hall devices measure magnetic fields. If a bias current flows through a thin piece of (semi)conductor, a voltage that is proportional to the magnetic field perpendicular to the Hall device is generated (see figure 1). The advantage of Hall devices compared to, for example, magnetoresistive sensors is that they can be realised in standard IC processes such as bipolar or CMOS processes.

Hall devices can be used to measure current without making contact. The magnetic field generated by the current is a measure for the amplitude of the current. To measure small currents, small magnetic fields must be detected. In this application the two major problems of Hall devices are offset and sensitivity drift.

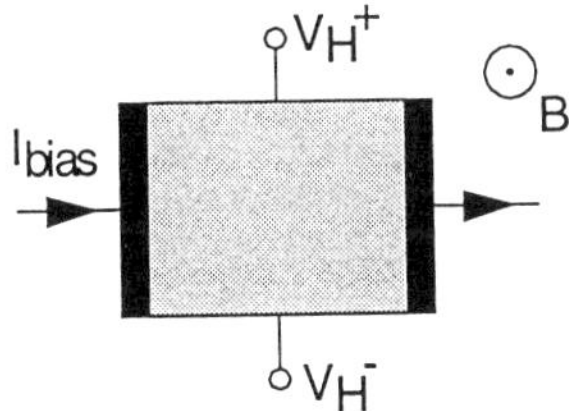

Figure 1. The Hall effect.

1.2. Offset

Unfortunately Hall devices show a large and unpredictable offset (i.e. they generate an output voltage even when no magnetic field is applied). This offset causes an error in the measured magnetic field strength and therefore in the calculated amplitude of the current. One way to reduce the offset is to use spinning-current Hall devices.

1.3. Sensitivity drift

Another disadvantage of integrated silicon Hall devices is the change of sensitivity with time, temperature etc.. This reduces the accuracy and the repeatability of the measurement. The problem can be solved by calibrating the device before each measurement. The calibration should preferably be performed automatically by the sensor itself, without intervention from the user. In this case we speak of autocalibration. To achieve autocalibration a well defined magnetic field is needed.

In this paper the spinning-current Hall devices are described in section 2, in particular the stability and the temperature dependence of the offset are discussed. In section 3 the autocalibration of silicon Hall devices is presented.

2. Spinning-current

As an offset reduction technique the spinning-current method, first described by P. Munter [1] is used. This method requires some electronics which might be integrated on the same chip as the sensor. The measurements presented here have been carried out using a programmable multimeter, a programmable current source and a computer.

2.1. The spinning-current principle

The spinning-current method is a way of separating transduction effects according to their current-orientation dependence. Offset effects can be separated from the desired transduction effect if their radial spatial periodicities are different from the radial spatial periodicity of the desired effect. In a silicon Hall device the Hall effect is current-orientation independent while most offset generating effects show a radial spatial periodicity. Using a symmetrical spinning-current Hall device the bias current is made to spin with small steps around an axis through the centre of the Hall device as shown in figure 2. The output voltage is sampled at the Hall contact pair perpendicular to the bias current. As can be seen in figure 2, the output voltage at zero magnetic field shows a periodic behaviour, so a major proportion of the causes of offset are periodic. By determining the part of the output signal which has a radial spatial periodicity that equals zero, the Hall effect can be separated from radial spatial periodic offset causes. Using this method the offset of the Hall device can be reduced from milli Teslas to micro Teslas.

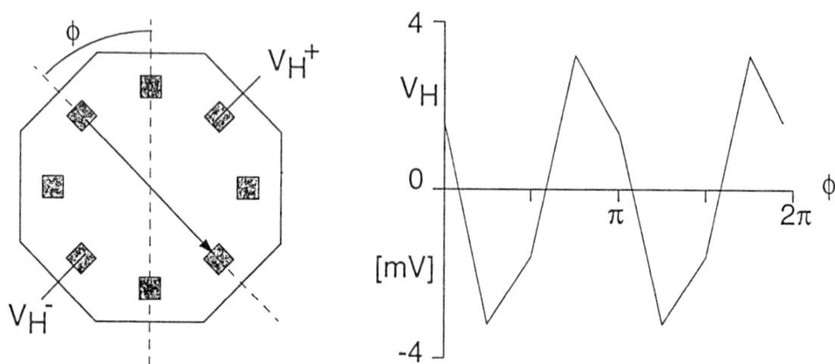

Figure 2. The spinning-current Hall device and its output signal.

2.2. Measurements

The experimental spinning-current Hall devices were realised in a CMOS process. The n-well defines an octagonal Hall device with a diameter of $400\mu m$ and n^+ regions are used as contacts. A bias current of $5mA$ is applied, this current is switched using analogue switches (74HCT4351) which are directed by the computer. The voltage perpendicular to the current is measured using a 2 second integration time, the switching is controlled by a relais card inside the multimeter. The Hall device is directly bonded on a printed circuit board (to reduce the influence of magnetic material in chip housings) and inserted in a zero Gauss chamber. The zero Gauss chamber and the Hall device could be placed in an oven to measure the influence of the temperature on the offset. The stability and the temperature dependence of the offset were measured as described below.

2.2.1. Stability Figure 3 shows the results of 20 hours of measuring. The residual offset field is very low, avarage $-0.66\mu T$ with a standard deviation of $0.8\mu T$.

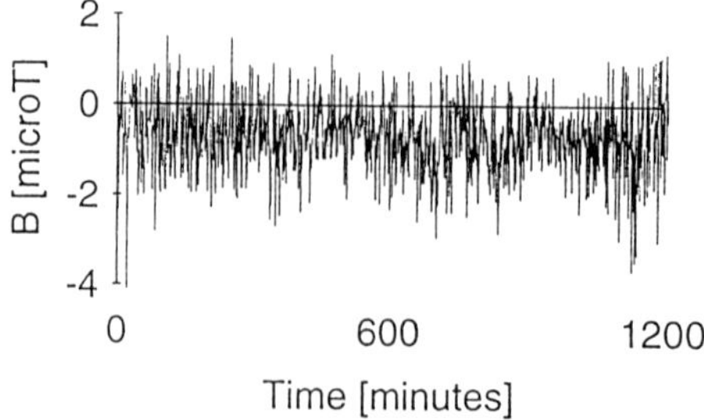

Figure 3. The measured offset versus time.

2.2.2. Temperature dependence of the offset The offset of standard Hall devices changes with temperature, for each of the eight current directions the offset versus temperature is drawn in figure 4. After applying the spinning-current method the offset is reduced to a few micro Tesla as shown in figure 5. The sudden increase in offset at about $80°C$, occuring in most devices, is not yet understood. A simular curve was measured by R. Gottfried-Gottfried [2] on a spinning-current Hall device that uses the channel of a CMOS transistor. For temperatures up to about $80°C$ the offset of the spinning-current Hall device is very low and almost constant.

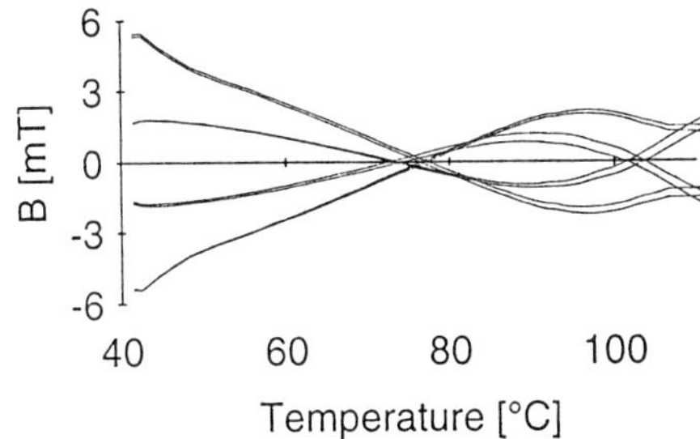

Figure 4. The measured offset in each direction.

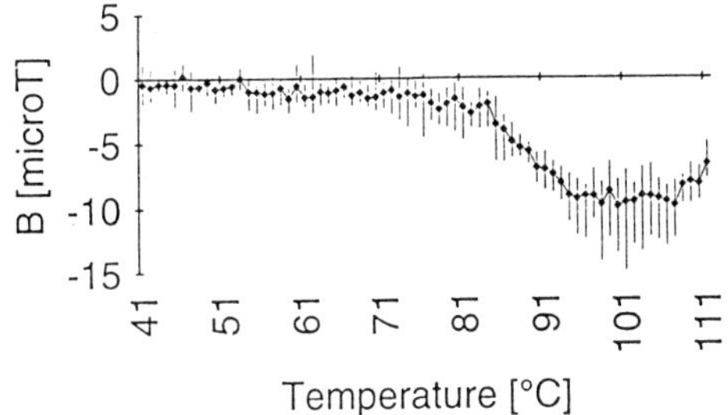

Figure 5. The residual offset after applying the spinning-current method.

3. Autocalibration

Calibrating the sensor without user interference is called autocalibration. An accurate actuator to generate a well-known and stable magnetic field at the location of the Hall device is required. During calibration the actuator is activated two times with different currents to produce two exitation fields. The corresponding output signals of the Hall device are measured. Using this output voltages and the known magnetic fields the sensor can be calibrated.

3.1. The principle of autocalibration

The actuator used for the Hall device is a coil that was integrated together with the Hall device. During the calibration the external magnetic field B_{ext} must be constant. For the determination of the sensitivity S two different fields B_1 and B_2 must be applied:

$$V_1 = S * (B_{ext} + B_1) + Offset \quad (1)$$
$$V_2 = S * (B_{ext} + B_2) + Offset \quad (2)$$
$$V_1 - V_2 = S * (B_1 - B_2) \quad (3)$$
$$S = (V_1 - V_2)/(B_1 - B_2) \quad (4)$$

3.2. Measurements

The Hall device and coil have been realised in DIMES (Delft Institute of MicroElectronics and Submicrontechnology). Figure 6 shows the $100 \times 200\mu m$ Hall device and the coil with 12 windings. Using this coil, fields of up to $0.3mT$ can be obtained without heating up the device.

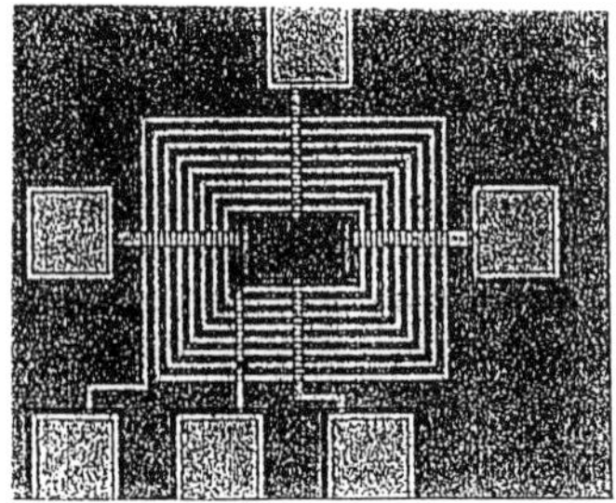

Figure 6. Chip photo of the Hall device with integrated coil.

The measurements were performed in a $11mT$ external magnetic field. By changing the current through the Hall device the sensitivity can be changed. Under normal working conditions the sensitivity of the Hall device will not change to such a great extend, but it is used here to demonstrate the autocalibration. Figure 7 shows the sensitivity versus the measured magnetic field after calibration of the device. It is clear that the measured magnetic field does not change as the sensitivity changes as expected.

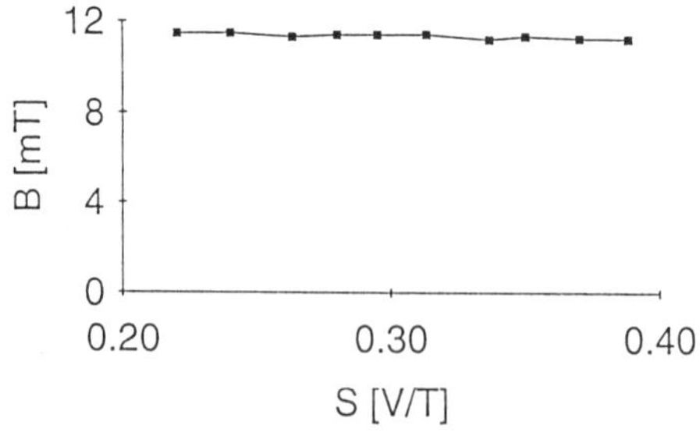

Figure 7. Measured magnetic field versus sensitivity for the autocalibrated Hall device.

4. Conclusion

Integrated silicon Hall devices can be released from offset by using the spinning-current method and the sensitivity drift can be taken care of by autocalibration using an integrated coil and a current source. The offset can be reduced to a few micro Tesla and stays in that range for temperatures up to about $80°C$. Future devices will include both systems and the electronics to create a smart silicon Hall sensor with a low offset and a low sensitivity drift.

References

[1] P. J. A. Munter. A low-offset spinning-current Hall plate. *Sensors and Actuators*, A21–A23:743–746, 1990.

[2] Ralf Gottfried-Gottfried. Thermal behaviour of CMOS Hall sensors for different operating modes. *Sensors and Actuators A*, A41-A42:430–434, 1994.

A high resolution two dimensional electron sensing integrated circuit

J V Hatfield and D G Lomas

Department Electrical Engineering & Electronics, UMIST, PO BOX 88, Manchester M60 1QD

A new kind of two-dimensional electron imaging Integrated Circuit is presented. The charge sensing IC has 65 536, 20μmX20μm, discrete, charge-collecting, aluminium electrodes (pixels) fabricated on the surface of a silicon chip. Although developed to detect the spatially resolved electron image produced by X-ray photoelectron spectrometers the device could be utilised in image intensifiers or astronomical cameras.

1. Introduction

This paper describes a new kind of electron imaging integrated circuit. The chip itself was originally developed to detect electron images in an X-ray photoelectron spectrometer. At the final exit plane of such instruments there exists a spatially (and hence energy) resolved electron image, to be amplified by Micro-Channel Plate electron multipliers (MCPs) and output to the new detector. It could equally find application as an image intensifier or astronomical camera. X-ray Photoelectron Spectroscopy usually referred to as XPS or ESCA (Electron Scattering for Chemical Analysis) gives chemical analysis of surfaces with much reduced damage, [1]. In this technique the sample surface is irradiated by a source of monochromatic X-rays. The X-rays cause photoionisation of atoms in the surface and a chemical image is obtained by measuring the energy spectrum of the emitted photoelectrons. The technique, however, has always suffered from relatively poor spatial resolution and hence a limited imaging capability. The 2-D imaging chip was developed to address some of these problems.

A common feature of most position-sensitive detectors is the use of MCPs, [2]. These consist of an array of channel electron multipliers which preserve the spatial information in the particle image while amplifying it with a gain which is typically 10^6. This enables individual particles incident on the microchannel plates to be detected by the sensor. Various techniques have been used to detect the electrons and these have been reviewed by a number of authors, [3,4]. Currently MCPs are available with 8, 10, 12 or 25μm pore sizes which correspond to centre-to-centre spacings of 10, 12, 15 and 32μm. As each channel operates, essentially, independently of its neighbours such an array can resolve events that are spatially separated by distances of the order of the channel size. For optical imaging MCPs fitted with photo-cathodes can be used.

2. The new 64k pixel two dimensional integrated detector

The charge sensing IC has 256X256 (65 536), 20μmX20μm, discrete, charge-collecting, aluminium electrodes (pixels) fabricated on the surface of an integrated circuit from the second (top) metal layer. The charge detection circuitry is fabricated directly beneath them. The design methodology involves treating the electrode array as a dynamic memory. If each electrode is considered to be

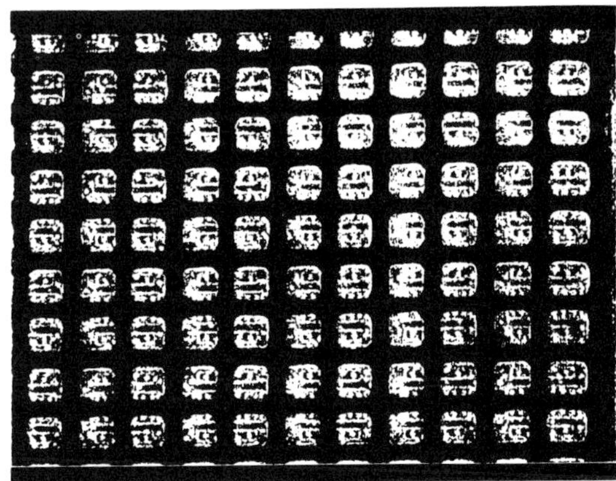

Figure 1 SEM photomicrograph of a section of the electrode array

an elementary charge storage site then the whole array is effectively a DRAM. The chip was fabricated in a two level metal, 1.5μm CMOS process under the EUROCHIP initiative. An SEM photomicrograph of a section of the array is shown in Figure 1. The metal 2 electrodes are seen through windows etched through the chips final overglaze.

A simplified schematic of a portion of the array is shown as Figure 2. In this approach a sensing electrode is directly connected to the gate of a p-channel MOS transistor such that when the electrode potential falls sufficiently, due to the arrival of an electron pulse, the transistor will be held in an "on" state. Provided that the charge site is addressed by the series access transistor before significant charge leakage can occur the pixel will register the arrival of an electron pulse. A detailed schematic of a single cell is shown in Figure 3. The capacitance to ground of the 20μmX20μm charge capture electrode is 20fF, worst case. If this captures 10^N electrons the potential on it will fall by $8x10^{-6}x10^N$ volts. For $N > 6$ such voltage drops could easily damage the gate of the sensing transistor MP1. It is, therefore, necessary to fabricate a clipping diode under the electrode, (Figure 3).

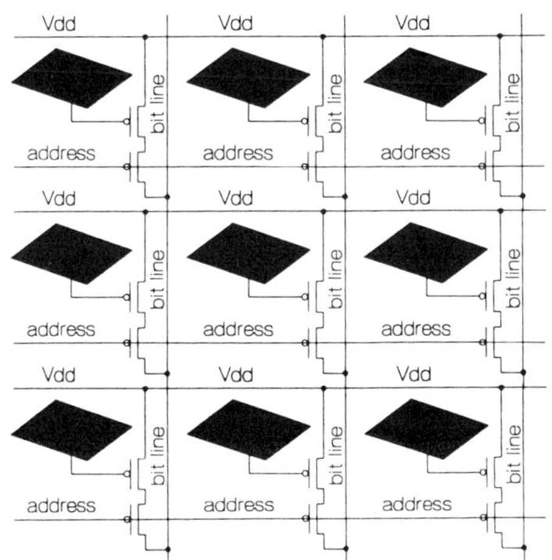

Figure 2 Illustrating sensing and readout mechanism

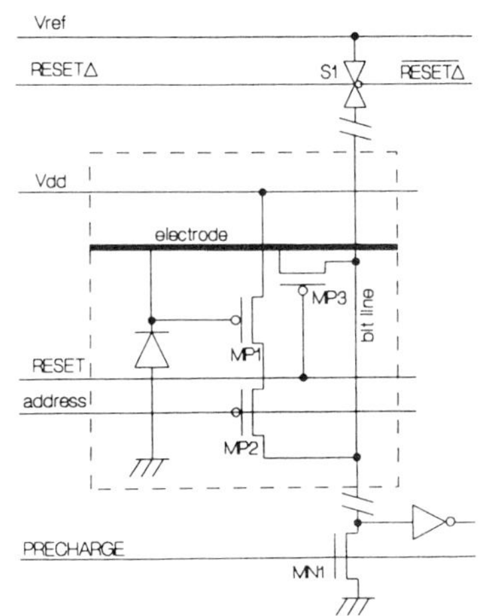

Figure 3 Circuitry of a single pixel site

The bit-line, which has an entire column of 256 pixel sites hooked to it, can be pulled to ground by the nMOS transistor MN1, switched to a voltage Vref by the analogue switch S1 or left high impedance. A readout sequence begins by precharging the bit-line to zero volts by switching on and then off transistor MN1. A row of charge sensors is then accessed by switching on transistors MP2 by driving the address line low. If the electrode potential has fallen below V_{dd}-V_T due to the arrival of an electron pulse (V_T is the threshold voltage of the pMOS transistor and is typically 1V), the potential on the bit-line will rise to V_{dd}, otherwise it remains at ground. Bit-line capacitance thus only affects the rise time to V_{dd}. The readout sequence for a particular row is ended by discharging the electrodes to a potential of $V_{ref}=V_{dd}-V_T+\delta$. This is achieved by gating a RESET pulse to transistors MP3 whilst the row is still addressed and switching the bit-line to V_{ref} by means of switch S1 and the RESETΔ pulse. RESETΔ is a "stretched" RESET pulse which allows MP3 to turn off in advance of switch S1. In this way digital noise fed through to the electrode is reduced somewhat. The incremental voltage δ is large enough to ensure that transistor MP1 remains off. This is experimentally set above noise levels by adjusting V_{ref} with no electrons falling on the detector. Transistors MP1, MP2, MP3 and the clipping diode all fit underneath the charge capture electrode.

2.1 Chip architecture

A block diagram of the chip is shown in Figure 4. It consists of a 256X256 electrode array, a 256-stage address scanning shift register, a 256-bit latch, a 16-bit latch scan register and control logic. Each 256-bit row is sequentially addressed by clocking a single bit along the address shift register. The addressed row is read out in parallel and latched into the 256-bit latch. The latch contents are then shifted off-chip as sixteen, 16-bit words by means of the latch scan register. A design constraint was that the elements of the latch had to be on the same pitch as the sensor pixels, i.e., each latch bit can occupy no more than 22.5μm of silicon. The simple latch circuit of Figure 5 accomplishes this, where the feedback inverter is much "weaker" than the driving inverter, i.e., transistor gate width to length ratios W/L for the feedback inverter are much less than the corresponding ones for the driving inverter. Transistor MN1 of Figure 5 (the same MN1 as in Fig. 3) is turned on to precharge the bit-line to 0V. Currently 1μs is allowed for bit-line rise times and the data latch is clocked out in 16μs. It thus takes just over 4ms to read the entire array, although

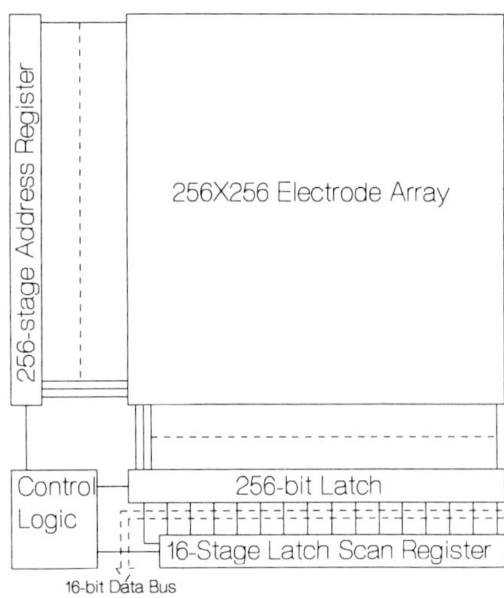

Figure 4 Architecture of charge sensing IC

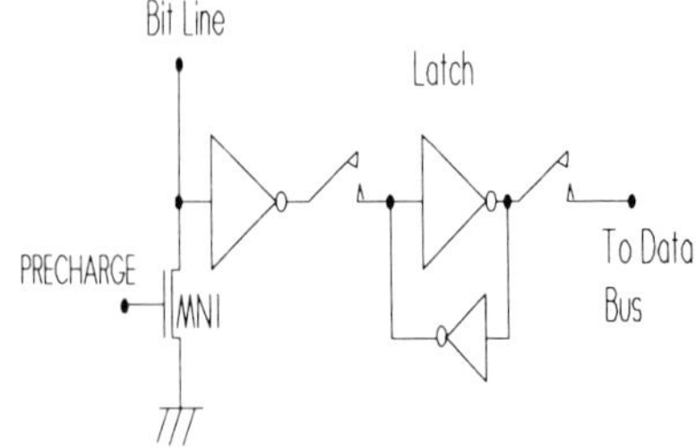

Figure 5 Single bit of 256-bit data latch

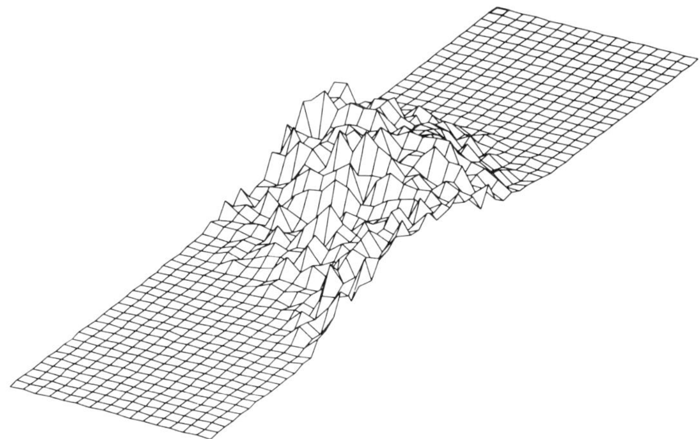

Figure 6 Electron beam incident on array surface. The array was read 255 times.

with a faster data acquisition system this could be significantly speeded up. A number of synchronisation pulses are available for external data acquisition systems.

3. Results

The chip, with 10μm pore chevron channel plates glued directly to its surface has been successfully tested in a vacuum system by scanning an electron beam across it. Figure 6 shows an electron beam incident on a portion of the array. It consists of 255 overlays. It is probable that pulses with less than 10^5 electrons are not being detected. It would also be desirable to use a single channel plate rather than a chevron stack [2] to eliminate pulse spreading. Current work is well advanced towards producing a similar array with enhanced sensitivity. The future work with regards to this sensor is to further reduce the pixel dimensions by moving to a 1μm process. This should produce at least a 40% reduction and thus allow a greater number of detectors to be integrated on a single chip with a considerably higher spatial resolution.

Acknowledgements

The authors acknowledge with gratitude the Science and Engineering Research Council of the UK for providing funding for this project, and EUROCHIP for making commercial micro-fabrication facilities available.

References

1 D Briggs and M. P. Seah, 'Practical Surface Analysis', *John Wiley*, Chichester, 1983.

2 J L Wiza, 'Microchannel plate detectors', *Nucl. Instr. Methods*, 1979, **162**, pp.587-601.

3 Lee J. Richter and W. Ho, 'Position-sensitive detector performance and relevance to time-resolved electron energy loss spectroscopy', *Rev. Sci. Instrum.* **57**, 1469 (1986).

4 O. H. W. Siegmund and R. F. Malina, in *Multichannel Image Detectors, Vol.II,* edited by Y. Talmi, *ACS Symposium Series 236* (ACS, New York, 1983), p.253.

Uniformity In An Ion Counting Detector Array

D P Langstaff

Department of Physics, University of Wales, Aberystwyth, SY23 3BZ

Abstract A novel ion counting array for mass spectrometry applications is described. Results of experiments to determine the uniformity of response across the array are presented. Analysis of the causes of non-uniformity is given together with techniques to minimise its effect. The performance of the array is discussed as well as its implications for mass spectrometry.

Introduction

The performance of a mass spectrometer is largely determined by the detector used. Existing detectors based around a channeltron electron multiplier are able to count single ions but can only analyse a tiny portion of the beam at a time, leading to a low overall sensitivity as the majority of the beam is filtered out

Position Sensitive Detectors have become increasingly important in mass spectrometry and other applications as they allow spatial information to to be recovered as well as information on the number of particles. Most of these detectors consist of a Microchannel plate (MCP)[1], which acts both as the photocathode and also as a charge amplification element preserving position information, an arrangement of anodes which encodes the position information and decoding electronics which process and output this position information.

Different schemes have been devised for encoding the information such as wedge and strip anodes[2], discrete multianodes[3] and resistive anodes[4]. These all suffer from disadvantages, wedge and strip or resistive anode schemes can only handle one ion event at a time, and thus have low bandwidth whereas existing multianode systems suffer from crosstalk between anodes and also require a large amount of electronics to process the signals coming from each anode.

The Ion counting detector array described here overcomes many of these disadvantages by integrating the anodes, sensor amplifiers and counting circuitry for 192 channels of a system onto a single custom silicon microcircuit.

Principle of Operation

The detector has been described previously[5] and only the main points will be reviewed here.

The main elements of a detector module are shown diagrammitically in Figure 1. An incoming ion strikes the front face of a microchannel plate assembly where secondary emission along the tubes in the plates causes a pulse of about 10^6 - 10^7 electrons to be emitted from the rear face of the second plate. This pulse of electrons lands on the detector electrodes patterened on the top metal layer of the custom detector chip where they induce a potential of $\approx$0.15V. The electrodes are 2mm long by 18μm wide on a 25μm pitch, giving an active detedector area of 2mm × 4.8mm

Associated with each collection anode on the detector chip is a sensitive charge amplifier and pulse shaping circuit, some simple gating and an 8-bit counter. Each arriving pulse is registered in the counter during the sampling period of the array. After the sampling period, the accumulated counts are read out from a data port under control of a simple external interface. In the current experimental set-up, the data is collected by a standard PC via an off the shelf interface card.

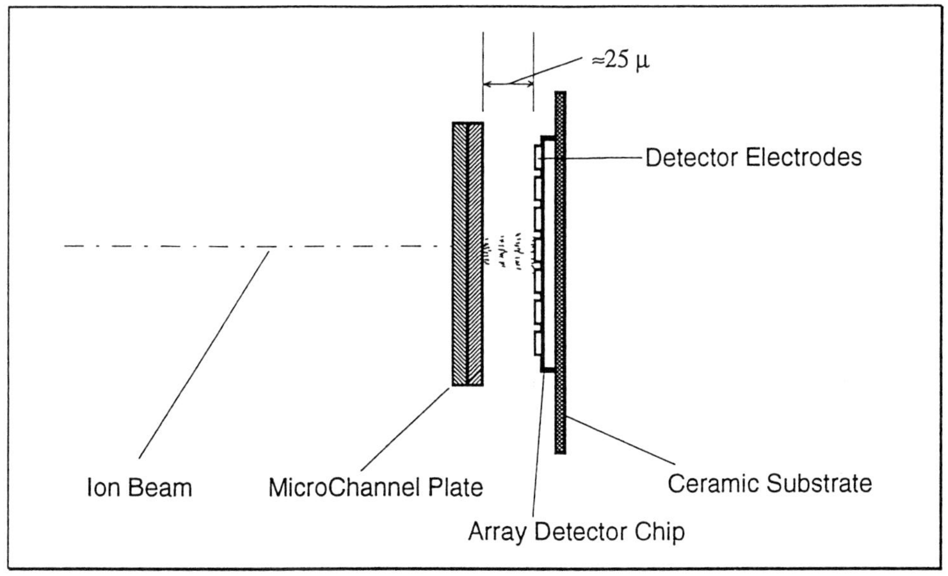

Figure 1. Module Construction

Testing In Vacuum

After wafer test and assembly of working devices onto modules, further testing was carried out at VG Analytical, Wythenshawe. The module was position behind an existing photmultiplier based detector, so that switching between the two detectors could be easily accomplished by turning off the deflection voltage on the photomultiplier.

Initial experiments were conducted with a beam of ions selected at mass 28, this provides a known doublet of peaks comprising N_2^+ (mass 28.013) and CO^+ (mass 28.010) which are normally present in the residual atmosphere in a mass spectrometer.

Gain Variations

Slight differences in sensitivity between the channels, arising from manufacturing tolerances in the IC fabrication process, give rise to differences in gain between the channels.. In order to quantify this effect, an experiment was conducted whereby a double peak was moved across the array by means of an electrostatic deflection system and sampled at a number of points across the array. It was assumed that the ratio of the two peak heights would remain constant over the duration of the experiment, even if slight drift in machine conditions could give rise to changes in absolute amplitude.

Data from the doublet in was accumulated at 16 different positions across the array and is shown in raw form in Figure 2. The ratio of peak heights exhibited a standard deviation of 6.2%.

The variation in peak height comes from two sources, an 'aliasing' error caused by the peak being at different positions relative to the electrodes and a difference in sensitivity caused by variations in MCP gain and detector sensitivity.

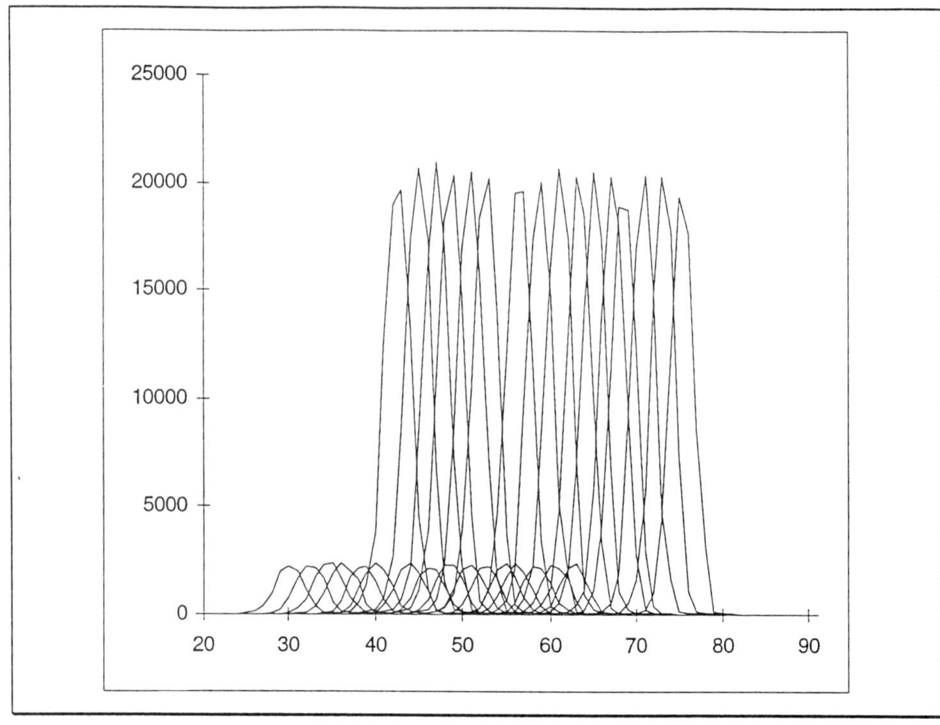

Figure 2. Overlaid traces from doublet swept across array

Interpolation and Correction

In order to realise the full potential of the detector array, techniques to reduce the distortions caused by these two effects were studied.

Correction for differing channel trigger threshold is achieved by first calibrating the array by uniformly illuminating with an ion beam and accumulating counts. From this calibration phase, a series of correction factors are derived which are then applied to the accumulated counts from each channel. Becasuse the correction factors are derived from illuminating the array, they incorporate gain variations caused both by the MCP and also those due to varying sensitivity of detectors on the chip.

Reconstructing the peak shape from the sampled waveform was achieved by the interpolation technique of oversampling. Although this normally finds application in digital singnal processing applications where it is used to increase the data rate of a signal, the technique is equally applicable to spatially separated sampling points. The interpolation algorithm used increases the sampling frequency of the array by a factor of 4. This is done by inserting zeros in the data stream and then putting the resultant through a low-pass filter.

Applying gain correction followed by interpolation to the raw data reduced the non-uniformity in peak height ratios from 6.2% to 3% as shown in Figure 3.

Both of the above techniques were first developed as spreadsheet models and have subsequently been implemented in the control program for the array.

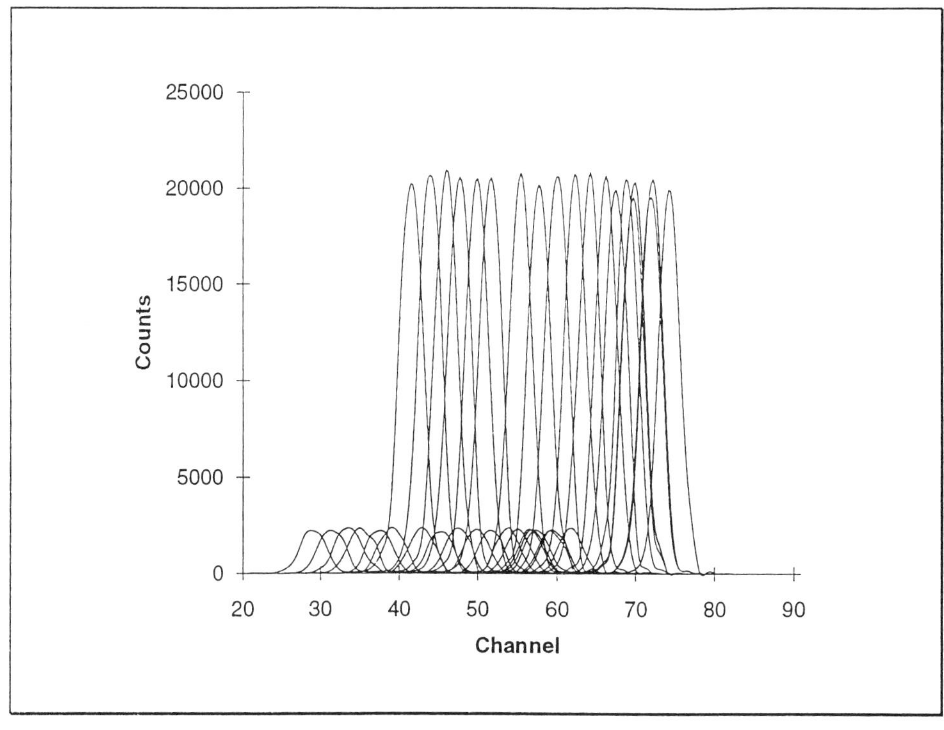

Figure 3. Overlaid traces after corrections

Conclusions

A novel ion counting detector array has been designed and fabricated. Testing has been carried out in a mass spectrometer. The performance of the array has been evaluated and found to surpass other forms of detector in terms of dynamic range and spatial resolution. Non-uniformity in sensitivity across the array was noted and techniques have been developed to correct for this.

References

[1] Wiza J.L. ,"Microchannel Plate Detectors", Nuclear Instruments and Methods, Vol 162, pp587-601, (1979)

[2] Matin C., Jelinsky P., Lampton M., Malina R.F. & Anger H.O. "Wedge-and-Strip anodes for centroid-finding position-sensitive photon and particle detectors", Review of Scientific Instruments, Vol. 52, No. 7, pp 1067-1074, (1981)

[3] Richter L.J & Ho W., "Position-sensitive detector performance and relevance to time-resolved electron energy loss spectroscopy", Review of Scientific Instruments, Vol. 57, No. 8, pp 1469-1482, (1986)

[4] Firmani C., Ruiz E., Carlson C.W., Lampton M. & Paresce F., "High-resolution imaging with a two-dimensional resistive anode photon counter", Review of Scientific Instruments, Vol. 53, No. 5, pp 570-574, (1982)

[5] Langstaff D.P., Lawton M.W., McGinnity T.M., Forbes D.M. & Birkinshaw K., "A new ion detector and digital-signal-processor-based interface", Measurement Science Technology, Vol. 5, pp389-393, (1994).

STRAIN MEASUREMENTS IN MULTI-LAYERS

P.J. French*, C. de Boer+ and P.M. Sarro+

*Lab. for Electronic Instrumentation, Dept. of Electrical Eng., +DIMES
Postbus 5031 2600 GA Delft, Delft University of Technology, The Netherlands.

Abstract: In the development of both micromechanics and microelectronics it is important to be able to characterise the mechanical properties of layers deposited on the silicon surface. It is therefore important to minimise stress levels. In the case of micromechanics the emphasis is often on maintaining low tensile strain to ensure that membranes do not buckle or break when released. An additional problem in predicting the mechanical properties of multi layers is the considerable effects of deposited layers on underlying layers. In this work a detailed study is made on the interaction between mechanical layers in a multi-layer construction. Due to this interaction, knowing the mechanical properties of each layer is not sufficient to predict the properties of the total structure. Furthermore, the order of processing, such as etching and annealing have to be carefully chosen to optimise the properties of the final structure.

This paper examines the importance of the effect of each layer on underlying layers and on the total structure, using polysilicon and nitride as the mechanical layers. The problems of measuring the mechanical strain in multiple layers will also be discussed.

1. Introduction

Intrinsic stress can cause severe detrimental effects in both microelectronics and micromechanics. In the case of microelectronics, deposited films which have high stress can cause leakage currents of p-n junctions in the substrate [1] and it is therefore important to minimise stress levels. In the case of micromechanics the emphasis is often on maintaining low tensile strain [2]. If the films are in compressive stress, they may buckle when released and high tensile stress may result in structural failure.

At Delft University a simple method has been developed for measuring strain using a micromachined pointer [3]. This structure is incorporated into a surface micromachining structure where the pointer is released by removing the underlying sacrificial oxide layer using HF. In many cases multiple layers are used and this technique can also be applied to measuring stress in such multi layer structures, to establish both the total stress and the contribution of each layer. This has been achieved by measuring the mechanical strain in multi layers and also in structures where the upper layer has been removed. For surface micromachining applications the layer structure of main interest for is that of polysilicon/silicon nitride/polysilicon. This combination benefits from the high mechanical strength of the nitride, as applied to the development of a tactile sensor [4], and its low optical absorption, as applied to a micromachined interferometer [5]. This layer structure is not however mechanically symmetrical as often assumed. The resulting stress profile presents severe problems in producing flat structures. A nitride-polysilicon combination has also been applied to bulk micromachining in the fabrication of polysilicon resistors on large nitride membranes [6]. In this case the sequence of anneal and patterning may have a significant effect on the final mechanical characteristics of the device.

The aim of this work is to establish the effect of LPCVD depositions on the underlying layers in the fabrication of multiple layer structures.

2. Processing

The structures tested in this work were based on a surface micromachining process. Firstly a sacrificial phospho-silicate glass (PSG) layer was deposited and patterned. The mechanical layers were then deposited, annealed and patterned. In the case of double layers it was important to establish the characteristics of the underlying layer. Therefore, after all thermal processing the upper layer was completely removed and the underlying layer patterned. The structures were then released by etching the sacrificial layer in HF. In the case of polysilicon both doped and undoped films were investigated, doping being achieved by implantation of phosphorus at a dose of $5 \times 10^{15} cm^{-2}$ at 80keV. The layer combinations used were as follows:

Single layer	-	polysilicon, silicon nitride
Double layer	-	polysilicon/silicon nitride, silicon nitride/polysilicon
Triple layer	-	polysilicon/silicon nitride/polysilicon

Both the silicon nitride and the polysilicon were deposited using LPCVD with processes developed at Delft University for surface micromachining applications [2,7]. As single layer structures these layers have low tensile strain and do not require high temperature strain anneals. However, if these layers are to be incorporated into a given process the layers may be subjected to high temperature process steps and therefore the effects of anneals of 850°C and 1000°C were investigated. The final stage of the process was the sacrificial etching to release the structures and this was achieved by using HF.

3. Strain measurement results

The strain measurements were made in an SEM by measuring the displacement. The first stage of this work was to establish the mechanical strain in these layers both with and without further thermal processing. The results of these investigations are shown in figure 1. The nitride can be seen to be in low strain as deposited but this rises considerably with annealing. Polysilicon, on the other hand, has a strain level of about $900\mu\epsilon$ after an anneal of 850°C, which is used in this process to activate the dopant, and this reduces to zero for an anneal of about 1000°C and is forced into compression for higher temperature anneals. This figure would appear to indicate that an anneal of 900°C should yield a perfectly match double layer. However, as shall be shown below, the interaction between the layers makes the matching of the layers considerably more complicated.

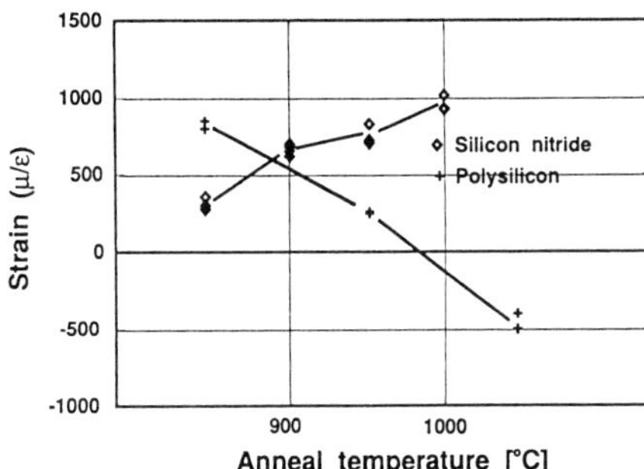

Figure 1 Measured strain as a function of anneal temperature for polysilicon and silicon nitride.

In order to test this effect, double layers were deposited and the top layer was then removed after the completion of all thermal processing. The strain levels in these layers was compared to the strain in films which had not seen the addition deposition. The first layer combination to be investigated was that of polysilicon deposited on silicon nitride. With an anneal of 850°C the polysilicon was found to have little effect on the underlying silicon nitride, since after stripping of the polysilicon, the strain levels in the nitride were similar the single layer values. This seems reasonable since the polysilicon deposition and anneal temperature had not risen above the deposition temperature of the nitride. However, when the sample were subjected to an anneal at 1000°C a clear effect was observed. Figure 2 shows the measured strain in the double layer structure. As expected, when referring to figure 1, the strain has been reduced by the presence of the polysilicon. On samples where the polysilicon was removed after annealing the strain levels remained lower than that of a single nitride layer, as shown in figure 3. This indicates that the presence of polysilicon has effected the mechanical properties of the nitride.

If the layer structure is reversed it could be expected that the effect would be larger due to the high mechanical strength of the silicon nitride and its relatively high deposition temperature (850°C). Once again both doped and undoped polysilicon was used. The results of these experiments are shown in figures 4 and 5.

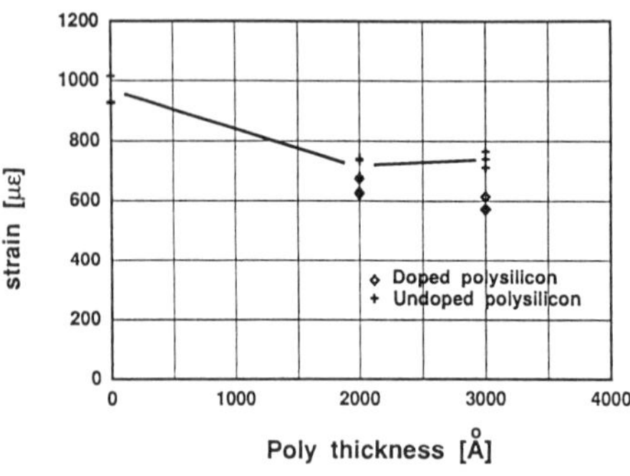

Figure 2 Measured strain for a polysilicon on silicon nitride structure after an anneal of 1000°C for 30 mins.

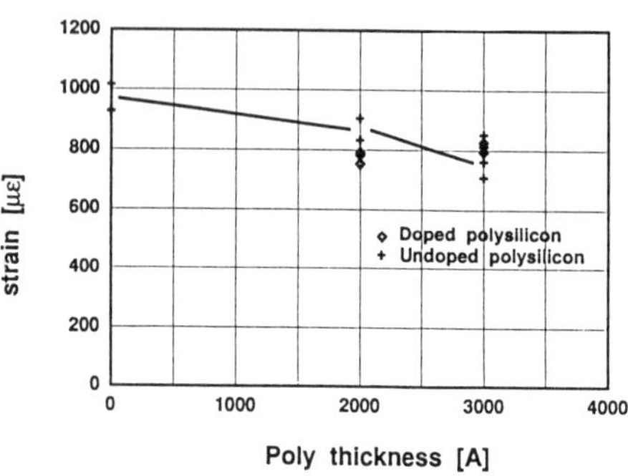

Figure 3 Measured strain of silicon nitride after the removal of the upper polysilicon layer as a function of the removed layer

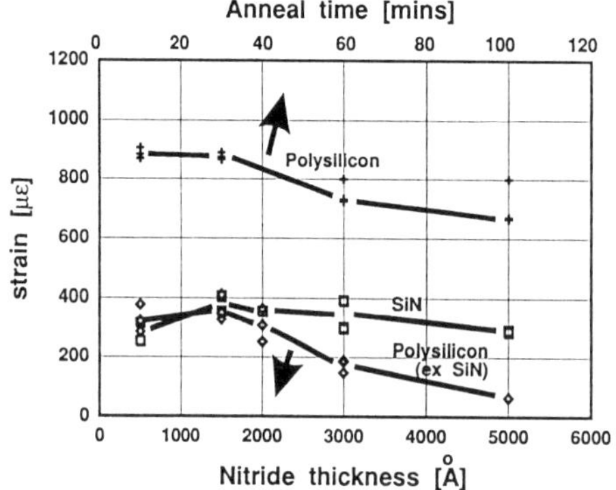

Figure 4 Measured strain for single polysilicon and nitride layers and polysilicon after the removal of a covering silicon nitride layer, for undoped polysilicon.

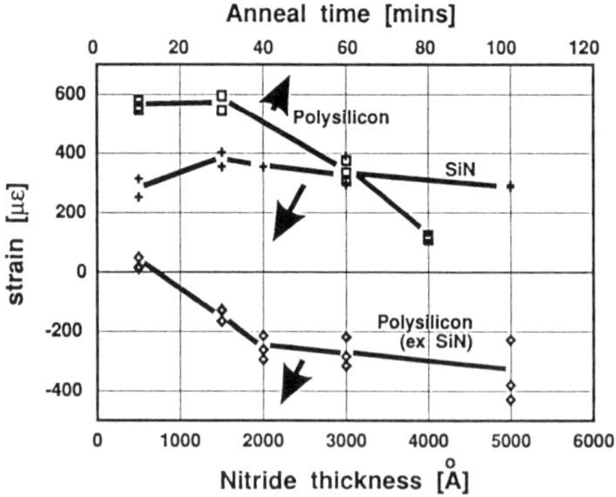

Figure 5 Measured strain for polysilicon, silicon nitride and polysilicon after the removal of a covering silicon nitride layer, for doped polysilicon.

Due to the relatively high deposition temperature of silicon nitride (corresponding to the normal anneal temperature for polysilicon) single polysilicon layers were annealed at times corresponding to the deposition time of the silicon nitride. This was to separate the effect of the additional thermal budget from the deposition. These experiments show a considerable effect of the silicon nitride on the polysilicon. The polysilicon films which were subjected to an anneal without nitride deposition displayed some lowering in strain levels for increased anneal times. The sharp drop in stress, however, shows that the main effect is the deposition process itself and the thermal budget is reduced to a second order effect.

The measurements have shown the considerable effect on the nitride on the underlying polysilicon. If an additional polysilicon layer is grown over the nitride the structure cannot be symmetrical. In the example shown in figure 6 a triple layer was deposited using 500Å polysilicon

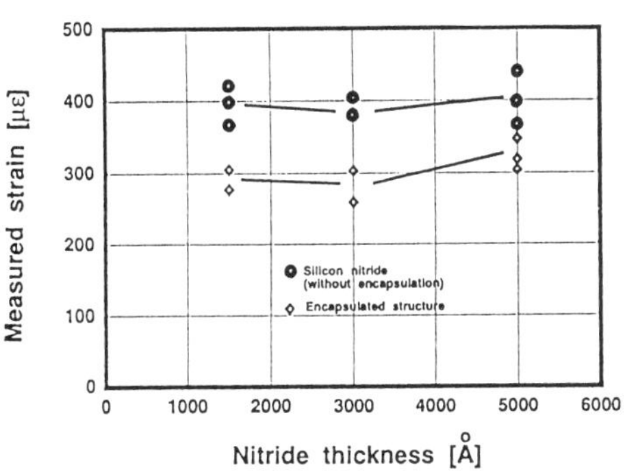

Figure 6 Measured strain of triple polysilicon-nitride-polysilicon structure as a function of n itride thickness.

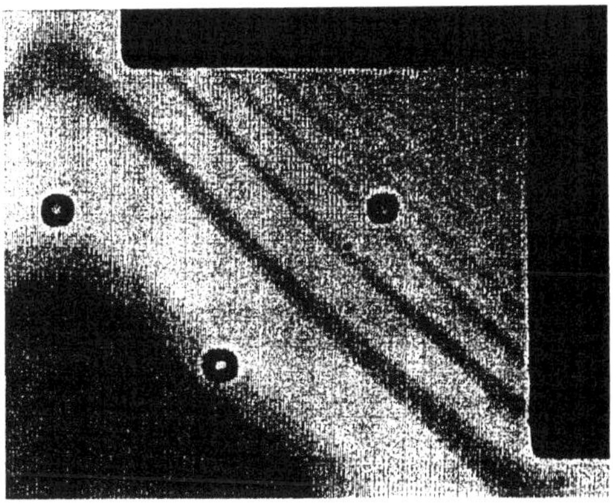

Figure 7 Effect of a stress profile on the corner of a surface micromachined membrane.

layers encapsulating a single nitride layer. Although a single polysilicon layer of the same thickness would have a strain levels higher than those of silicon nitride, the total strain was lower. As the nitride thickness is increased the total value approaches that of a single nitride layer, as expected. The low strain levels in themselves do not create a problems, provided the final strain is tensile. However, the resulting stress profile can create bending problems as shown in figure 7. In this case the corner of a membrane has been bend upwards by the stress profile in the film.

4. Conclusions

These studies have shown that knowledge of each layer in a membrane is not sufficient to predict the mechanical properties of the whole membrane. Structures which appear to be symmetrical can have a strong stress profile resulting in buckles structures. A nitride layer deposited on polysilicon has a considerable effect on the polysilicon. Polysilicon which has seen a nitride deposition has far lower tensile stress and in some case compressive stress. The additional thermal budget associated with the nitride deposition had only a second order effect and the main effect was the deposition itself. Polysilicon deposited on nitride has little effect on the nitride if the anneal temperature is 850°C. After stripping on the polysilicon there was only a slight difference in the nitride stress levels. If a polysilicon on nitride structure is annealed at 1000°C then the effect of the polysilicon is increased. If the polysilicon is not stripped the strain levels were in the order of 600-800$\mu\epsilon$, compared to 1000$\mu\epsilon$ for a single nitride layer annealed at the same temperature. If the polysilicon is removed the strain of the remaining nitride layer was 800-900$\mu\epsilon$. This is still lower than the single nitride layer which has not seen a polysilicon deposition.

Considerable care must therefore be taken when using multiple layers for micromachining applications. In building multi layers it is impossible to ensure that all layer see the same processing. Therefore for a given layer combination adjustments must be made to the processing of each layer to minimise the effect. If this is not done, the stress profiles may cause severe problems in fabricating microstructures. Further investigations will include materials studies of the polysilicon to establish the effects on the electrical and structural properties.

5. References

[1] L. Nanver, P.J. French, E.J.G Goudena, H.W. van Zeijl, "Low stress nitride as surface isolation in bipolar transistors", To be published, Journal Material Science and Technology

[2] P.J. French, B.P. van Drieënhuizen, D. Poenar, J.F.L. Goosen, P.M. Sarro and R.F. Wolffenbuttel, "Low stress polysilicon process compatible with standard device processing", Sensors VI Technology, systems and applications, IOP Publishing London, pp 129-133.

[3] B.P. van Drieënhuizen, J.F.L. Goosen, P.J. French and R.F. Wolffenbuttel, "Comparison of techniques for measuring both compressive and tensile stress n thin films", Sensors & Actuators, A37, 1993, pp. 756-765.

[4] M.R. Wolffenbuttel, P.J. French and P.P.L. Regtien, "Integrated capacitive tactile image sensor", Proceedings Eurosensors VIII, Toulouse, France 26-29 September 1994, p 33.

[5] K. Aratani, P.J. French, P.M. Sarro, D. Poenar, R.F. Wolffenbuttel and S. Middelhoek, "Surface micromachined tuneable interferometer array", Sensors and Actuators, A43, (1994), pp 17-23.

[6] P.M. Sarro, A.W. van Heerwaarden and W. van der Vlist, "A silicon-silicon-nitride membrane fabrication process for smart thermal sensors", Sensors and Actuators, A41-42, (1994), pp 666-671.

[7] P.J. French, P.M. Sarro, R. Mallée and R.F. Wolffenbuttel, "Optimisation of a low stress silicon nitride process for surface micromachining applications, Proceedings Eurosensors VIII, Toulouse, France 26-29 September 1994, p 205

Section C

Acoustic and Ultrasonic Sensors

Leak detection under overwhelming ambient noise conditions

R Ohba*, T Tanaka*, Y Tamanoi, M Matsumoto**, T Shinagawa**, T Ohtsuka** and Y Noguchi*****

* Department of Applied Physics, Faculty of Engineering, Hokkaido University, N-13 W-8, Sapporo 060, Japan

** Division of Equipment Maintenance, Marifu Oil Refinery, Koa Oil Co. Ltd., Iwakuni 740, Japan

*** Department of Aeronautical and Mechanical Engineering, University of Salford, Salford M5 4WT, England

Abstract. A new method based on an acoustic signal processing is proposed for detection of a possible leak under overwhelming ambient noise conditions. A set of residual spectra are estimated from inverse filtered signals and a possible leak is decided by a statistic hypothesis-testing on measurements of difference between maximum and minimum power of the residual spectrum. Experiments on leak detection are successfully performed in an oil refinery applying a hand held data processor with a specially designed sound collecting microphone. The results of the experiments are also presented as well as the idea and principle of the present detection method.

1. Introduction

Any leak in industries working on inflammable gases or materials should cause a quite serious even fatal damage to the industry. Quite a lot of manpower of maintenance staff is devoted to detect possible leak as soon as possible in real plants in those industries. Recently in Japan, those who want to be maintenance engineers are decreasing rapidly, so that divisions for maintenance in factories are keen to develop automated maintenance/diagnosis methods. The authors had successfully worked on developing acoustic methods to detect possible defects of bearing in practical rotary machines in real plants[1,2]. A development work has started on an acoustic method for leak detection applying the method developed for machinery diagnosis.

It is well established that a leak of gaseous fluid from a small pin hole consists of a subsonic/sonic gas jet and that choked gas flow in the jet stream generates a broad band noise[3-5]. A recent review on the subject is given by Lighthill[6]. It may be possible, then, in a quiet circumstance such as in a laboratory room to detect any leak by using acoustic energy flow in appropriate frequency bands. There are several instruments which can detect a leak based on sound pressure level in certain ultrasonic frequency ranges. It is, however, not possible to utilize them in real plants since the ambient noises in ultrasonic frequency range due to continuous fluid flow in piping or intermittent air/steam purges such as steam-trap, control valve or hydraulic instrument/controller are at considerable sound pressure level.

In real plants, the sound pressure level of ambient noises may be as high as 90 dB and it is almost impossible to detect a leak by human auditory sense. Therefore any leak in real plants is likely to be of serious one possibly leading to a fatal disaster of the plant. It is one of the most important and urgent themes to develop a method and instrument to detect a leak as early as possible in order to prevent any disastrous consequences. In the present paper, experiments on leak detection performed on a real plant of an oil refinery are described. The data processing and sensors used in the experiment are also described together with the idea and principle of the present detection method.

2. Background of the problem

2.1. Inverse filtering of signal

Operation of plants generates noises and each of which has its own source such as vibration of mechanical parts, rotation of fans, steam purge from traps etc. Even a constant flow of fluid in piping may be one of the sources.

In a linear predictive modeling theory, it is possible to model any colored signal by using either an appropriate ARMA or AR model. Most of noises originated by above mentioned sources are, however, well modeled by a simpler AR model. If an AR model is determined for a noise, then it is possible to remove the noise using the model. This operation is called as the inverse filtering and the filter used for this operation is called as the inverse filter.

When a signal is inverse filtered, then output of the inverse filter is a white noise and which is actually the residue of the inverse filtering operation and is the virtual input to the model. If the signal is of a different model, the residual signal of the inverse filtering is never a white noise and has a considerable amount of power at some frequency bands. It is then possible to determine the model used to design the inverse filter is valid or not by the power spectrum of the residual signal (residual spectrum). When the power of the residual signal of a noisy signal processed by an inverse filter is less than a certain value the signal has the same model that is used to design the inverse filter. If the value is higher it must have a different model.

2.2. Sound generated by leak

Leaks of gas/fluid from piping or pressure vessels of industrial plants treating inflammable materials such as in an oil refinery can cause a catastrophic damage to the plant. It is one of the most important tasks of field men in the plant to detect any leak as early as possible to prevent the leak from developing to a dangerous level. Most of early leaks from piping or pressure vessels in industrial plants are a high pressure liquid or gas escaping to the air through a small pin hole or crack. It is well known in the fluid dynamics[3)] that such a leak, above a certain pressure ratio between inside the pipe/vessel and outside, composes a sonic jet in which flow of the fluid is choked. The jet generates a broad band acoustic noise. It is well studied especially screech tones at the audible frequency range in the aeronautical field relating to the effects of jet noise pollution[3-5)]. Though mechanism of the screech tone generation is studied in the audible frequency region, there seems little literature treating beyond the audible frequency.

Experiments are carried out in an ordinary laboratory room to clarify basic characteristics of the sound signals from a leak at choked conditions. Simulated leak sources are used for the experiments. The air jet from a pin hole or slit simulates a leak. One of caps (made of cast iron with a hole of 0.3, 0.5, 1.0 and 2.0 mm diameter and a steel cap with a hole of 0.5 mm diameter) or flanges (steel with a slit of 10 mm long and either 0.1 or 1.3 mm wide) is fixed at the end of a pipe, which is connected to a reservoir tank and is supplied with a constant pressure of air via a small pressure regulator. The sounds from the leaks are collected with varying combinations of cap/flange, supplied air pressures and distances between the leak and microphone. The air supply pressures tested are 0.1, 0.2, 0.3 and 0.4 MPa gauge and the distances are 1, 2, 4 and 5 meter. The sound signals are picked up by a special sound collecting microphone[7)] and are band-pass filtered from 20 kHz to 100 kHz. The special sound collecting microphone is composed of a precision quarter inch condenser microphone whose frequency response is flat from dc to 100 kHz and a parabolic reflector made of glass fibre strengthened epoxy whose focal length and length from the bottom to mouth are 7.5 mm and 400 mm, respectively. Then the signal is sampled by 1,024 points at 250 kHz and stored into the memory of a computer after a 12 bit analog to digital conversion.

The example spectra of a set of successively sampled ten sound data, which were taken at a point 2 m distant from a leak with the cast iron cap of 0.3 mm hole and 0.4 MPa pressure, are shown in Fig. 1. The spectra are estimated using the above mentioned data by an AR modeling with 23 degrees, which is determined by the AIC criterion. It can be seen from the figure that the power spectrum of the sound produced by a leak is decreasing inversely proportional to the frequency over the range though it has several small peaks over the whole frequency range. When the pressure of the air supply is increased, the spectra of sound signals remain

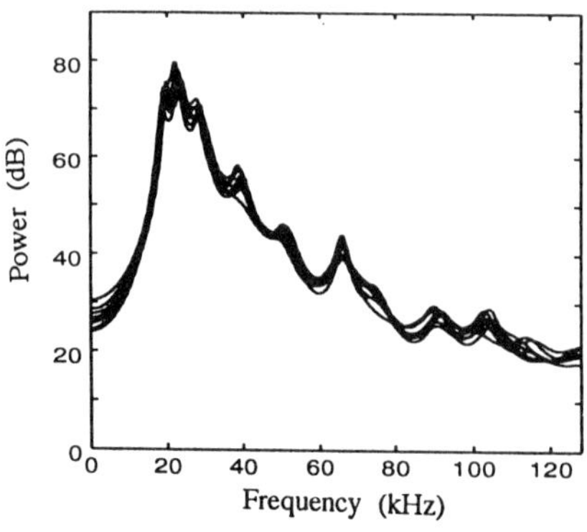

Fig. 1 Power spectra of subsonic air-jet noises.

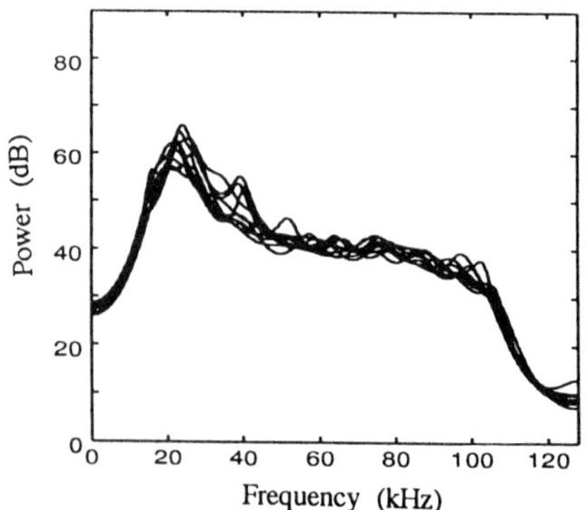

Fig. 2 Power spectra of ambient noises.

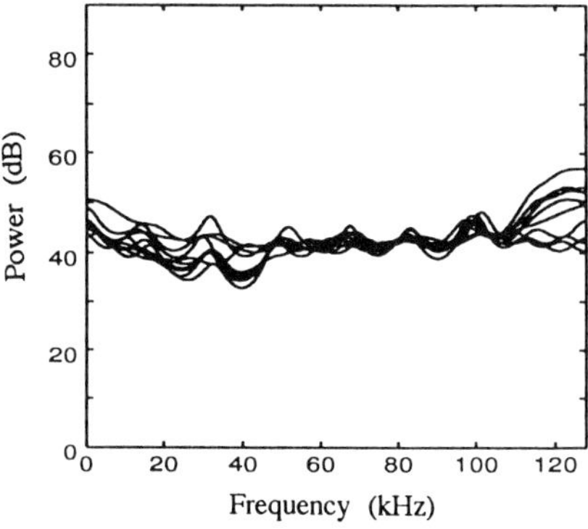

Fig. 3 Residual spectra of ambient noises.

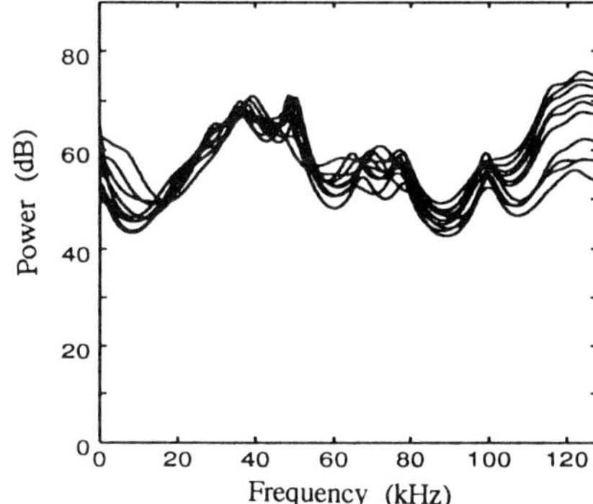

Fig. 4 Residual spectra of leak sounds under ambient noise conditions.

almost identical though a higher sound pressure level is observed. It could easily detect any leaks based on the spectra provided there is no ambient noise. However ambient noises as strong as nearly 88 dB (A) in real plants make the detection of leak quite difficult.

2.3. Ambient noise

Figure 2 shows some example power spectra of ambient noises in a real plant at an oil refinery. They are also estimated from a set of ten data sampled and processed by the same procedure as the laboratory experiment. It can be seen from the figure that though the spectra of noises are rather flat and also of wide band, they concentrated in higher frequency range compared with that of a leak. It can also be seen from the figure that the power distributions are essentially flat in the higher frequency range though they increase in the lower frequency range, say under 50 kHz.

An inverse filter with 23 degrees is designed based on a sample of the ambient noise data. This filter is used to filter the other samples of the ambient noise data whose spectra are shown in Fig. 2. Figure 3 shows the residual spectra of the data, that is the spectra of the inverse filtered noises. It can be seen from the figure that the estimated residual spectra are rather flat over the frequency range from 50 kHz to 100 kHz though they are considerably different in the frequency range of less than 50 kHz. The differences appeared in the residual spectra over the frequency range less than 50 kHz may due to operational sound of plants near the point of experiment. The leak sound data taken under the ambient noises using the simulated leak is also treated by the inverse filter. Figure 4 shows the example residual spectra of ten sound data taken successively at the same point where the ambient noise data used to design the inverse filter was obtained. It can be seen from the figure that they have peaks and valleys over the frequency range of 50 kHz - 100 kHz where the ambient noises have rather flat residual spectra.

3. Detection of leak

3.1. Principle

Figure 5 shows histograms of difference between maximum and minimum power of estimated residual spectrum for the ambient noise (hatched) and for the leak under the noise conditions (clear), respectively. The frequency range is between 50 kHz and 100 kHz and the microphone is 2.0 m away from the leak source at site III using a steel cap with a 0.5 mm hole and 0.2 MPa supply air pressure. The total sample numbers are 80 for the former and 200 for the latter, respectively. It can be seen from the two histograms that the distribution of the difference for the ambient noises is narrower than that for the leak sounds and that the average value of the distribution of the former is smaller than that of the latter. Sample mean m and sample variance V for the ambient noise are 5.62 dB and 1.14 dB^2, respectively. Whereas, they are 18.03 dB and 7.93 dB^2, respectively, for the leak noise. Then it may be possible to detect a leak using this statistic characteristics.

3.2. Statistical testing

A hypothesis-testing in the statistics enables to determine whether a set of statistics belong to a certain population or not, using small number of samples from the set. The testing can be applied to determine whether it is a mere ambient noise or leak in an acoustic field at a point using several samples of the power differences described before. It can be seen from Fig. 5 that difference between the two distributions is quite apparent with variance and/or mean. The hypothesis "variance V_i and mean m_i of the statistic $\{x_i\}$ of difference between maximum and minimum power over the frequency range from 50 kHz to 100 kHz of inverse filtered sound data are identical to those of ambient noise" is tested using both the F-test and t-test, respectively.

Assume that a pair of sample sets of statistic, $\{x_{1i}\}$ and $\{x_{2i}\}$ are given, and that the sizes of the sample are n and k, respectively. The problem is to determine whether the set $\{x_{2i}\}$ is from the identical population of the set $\{x_{1i}\}$ or not. If we assume the hypothesis "both variances are not identical," the statistical theory tells the statistic F_0 satisfies Eq. (1).

$$F_0 = V_1/V_2 > F(n-1, k-1 ; \theta), \quad (1)$$
$$V_1 = \Sigma(x_{1i} - m_1)^2/(n-1), \quad (2)$$
$$V_2 = \Sigma(x_{2i} - m_2)^2/(k-1), \quad (3)$$

where, m_1 and m_2 are sample mean of $\{x_{1i}\}$ and $\{x_{2i}\}$, respectively. θ is a risk threshold of the test, that is, result of the test is valid with a risk of θ x 100 %. F(a, b ; c) is the value of the F-distribution[8] with degrees of freedom a and b with risk c. The failure of the test indicates the both variances are identical. It is then possible to test the hypothesis "both of the means are identical" using the following statistic t_0 (t-test).

$$t_0 = (m_2 - m_1)/\{(V_1(n-1) + V_2(k-1))(1/n+1/k)/(n+k-2)\}^{1/2} > t(n+k-2 ; \theta), \quad (4)$$

where t(a ; b) is the value of the t-distribution[9] with degree of freedom a and risk b. If the relation is valid, then the hypothesis "both of the means are identical" is rejected.

The values of V_1 and m_1 are calculated in advance for pure ambient noise sample with size n = 10. At first the hypothesis-testing for variance is carried out by F-test using ten samples, that is k = 10. If Eq. (1) is satisfied then the hypothesis "variance of the data is identical to that of the ambient noise" is rejected, which indicates that they are different with a risk θ and the test is finished after giving an alarm for a possible leak. When the hypothesis is accepted, then the other test is carried out for the hypothesis "mean is identical to that of ambient noise" using t-test Eq. (4). If the hypothesis is rejected, the test is finished also giving an alarm for a possible leak. Otherwise, the test is finished without any alarm. Figure 6 shows the flowchart of the hypothesis-testing.

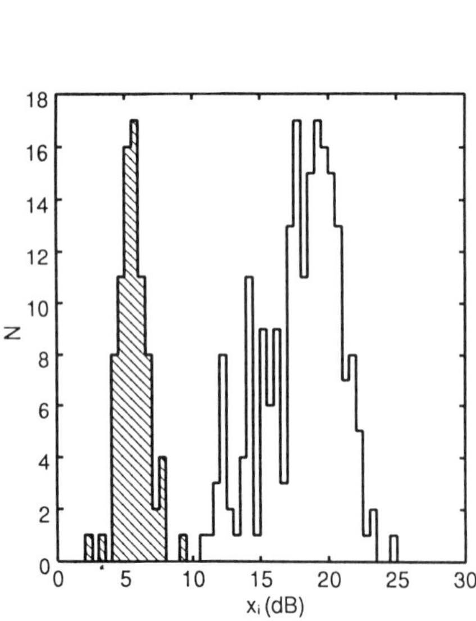

Fig. 5 Histograms of $\{x_i\}$ for ambient noise (hatched) and for leak sound under the noise condition (clear).

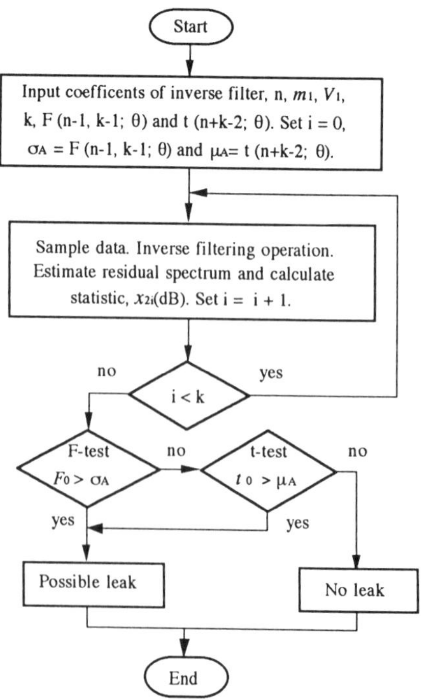

Fig. 6 Flowchart of hypothesis-testing.

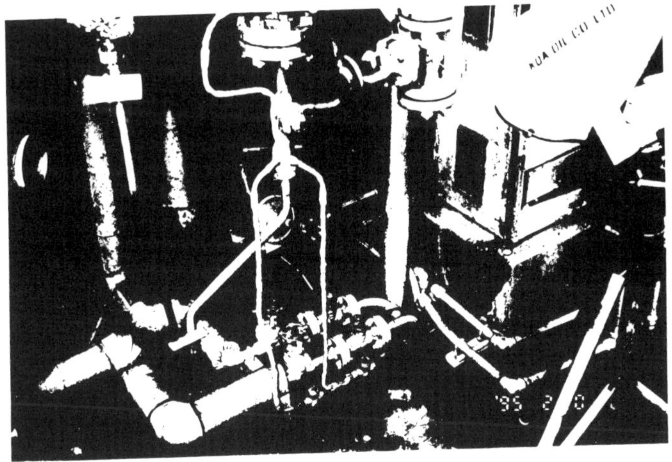

Fig. 7 Overview of the experimental setup on site III.

Tab. 1 Conditions by which data were acquired.

Source/	Cast Iron Cap																Flange								Steel Cap	Ambient
Size (mm)	0.3				0.5				1.0				2.0				0.1				1.3				0.5	
Pressure (x 0.1 MPa)	1	2	3	4	1	2	3	4	1	2	3	4	1	2	3	4	1	2	3	4	1	2	3	4	2	
Distance (m) 1	II	II	II	II	II	II	II	II	II	II	II	II	-	-	-	-	III	III	III	I,III	-	-	-	-	-	I, II, III
2	-	-	-	-	I	I	-	I	I	I	-	I	-	-	-	-	-	-	-	-	-	-	-	-	I	I
4	-	-	-	-	I	-	-	-	I	I	I	I	I	I	I	I	-	-	-	-	-	-	-	-	-	I
5	-	-	-	-	-	-	-	-	I	I	I	I	I	I	I	I	I	I	I	I	I	I	I	I	-	I

I, II and III: site number.
- : data were not available.

3.3. *Experiments by real plant*

Experiments are carried out using the sound signal samples of both ambient noises and simulated leak sounds obtained at three points (site I, II and III) on a real oil refinery. Sites I, II and III are a pump station, a drain hole with a vent of a steam trap and another drain hole with vents of two steam traps, respectively. When one of the traps purges drainage, it generates a single puff of choked steam jet and makes extremely powerful intermittent ambient noise which contains a large portion of ultrasonic components. The experiments on a real plant are performed by using the hand held data processor, Ono Sokki: CF-1200[2] with a renewed software and the special sound collecting microphone mentioned in section 2.2. Figure 7 shows the overview of the experimental setup on site III. Two vertical thin pipes directing to vents via stream-traps can be seen at the center of the figure. Behind the pipes, the simulated leak source with a flange can be seen together with the air tank. Upper right is the special sound collecting microphone set on a tripod.

At first, the ambient noise data are collected containing both the operational sound of normal plant and the steam-jet noise due to operating steam-traps at several distant points from the drain hole by setting the special sound collector. Coefficients of the inverse filter, sample variance V_1 and sample mean m_1 of the statistic x_1, difference between maximum and minimum power of the residual spectrum over 50 kHz - 100 kHz are calculated based on the data. Various leak conditions are tested with the simulated leak source just beside a drain hole. The statistic x_2 is successively measured ten times by sampling data, inverse filtering with the inverse filter and calculation. Finally, both the sample variance V_2 and sample mean m_2 are calculated on the ten measurements. The hypothesis "the statistic $\{x_{2i}\}$ is from the identical population of ambient noise" is tested following the flowchart of Fig. 6 with $\theta = 0.05$.

Experiments are carried out on 75 sets of data, which were acquired with changing in sites, sources and pressures. Table 1 shows all of conditions used in the present experiments. 74 sets out of the 75 sets of data except one were correctly diagnosed as a possible leak or a mere ambient noise. Only one set of leak data from

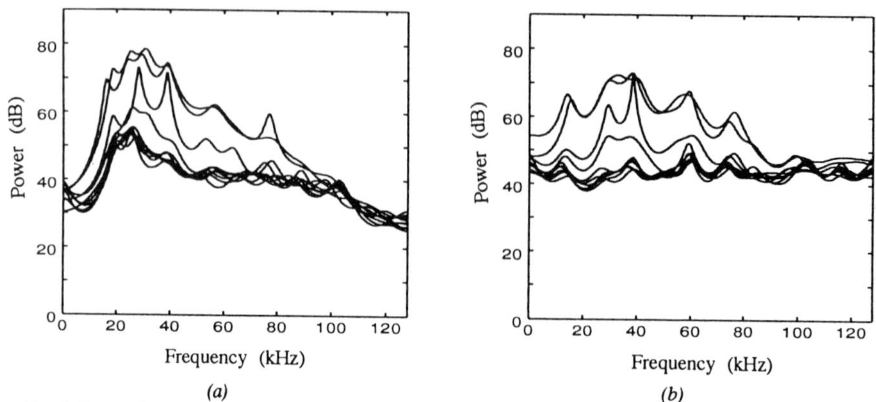

Fig. 8 Original- (a) and residual-spectra (b) of ambient noises at site III when one of steam-traps operates.

the 0.1 mm slit flange with 0.1 MPa pressure and distance of 1 m at site III, which is by a drain hole, is not correctly diagnosed. Fig. 8 shows example power spectra of the ambient noise at site III. Figs. 8(a) and 8(b) show the continuously sampled ten original and residual spectra. In spite of the largely deviating four cases which were due to the steam-jet from an operating steam-trap, it is possible to decide correctly that they are from the ambient noise.

4. Conclusions

A method to diagnose possible leaks is proposed by applying the inverse filtering technique and statistic hypothesis-testing. A specially designed sound collecting microphone shows excellent performances to separate target acoustic signal from ambient noise. It can detect a possible leak under 80 to 90 dB surrounding noise environment. Although the procedures used in the hypothesis-testing are not optimized and require further studies, results of the experiments suggest a highly possible applicability of the present method in real plants. It is necessary to confirm the applicability of the method and further experiments are carried out to test a pilot model by a real oil refinery.

References

1) Ohba, R. and Tamanoi, Y.: Method and apparatus for machine diagnosis, Jpn. Pat. File No. H4-138681 (1992)

2) Ohba, R. et al: Machine Diagnosis by Acoustic Signal Processing, *Sensors VI* (K T V Grattan and A T Auguousti ed.), IOP Publishing (Bristol), pp. 193-200(1993)

3) Lighthill, M. J.: On Sound Generated Aerodynamically, Part I, General Theory, Proc. Roy. Soc. (London), **A221**, pp. 564-587(1952)

4) Mawardi, O. K. and Dyer, I.: On Noise of Aerodynamic Origin, Jour. Acoust. Soc. Am., **25**, 3, pp. 389-395 (1953)

5) Westley, R. and Woolley, J. H.: An Investigation of the Near Noise Fields of a Choked Axi-Symmetric Air Jet, Proc. of AFROS-UTIAS Symposium, Univ. of Toronto Press, pp. 147-167(1968)

6) Lighthill, M. J.: The Inaugural Theodorsen Lecture -- Some Aspects of the Aeroacoustics of High-Speed Jets --, Theoretical and Computational Fluid Dynamics, **6**, pp. 261-280(1994)

7) Ohba, R. and Tamanoi, Y.: Sound collector, Jpn. Pat. File No. H6-21346 (1994)

8) Snedecor, G. W and Cochran, W. G. : *Statistical Methods* (7th ed.), The Iowa Univ. Press, Ames., pp. 221-222 (1980)

9) Fisher, R. A. : *Statistical Methods for Research Workers* (14th ed.), Oliver and Boyd, Edinburgh, pp. 358 (1970)

Optical fibre sensing of acoustic emission in fibre reinforced composites

P W R Baillie, K F Hale[1], G F Fernando[2] and B E Jones

The Brunel Centre for Manufacturing Metrology and [2]The Department of Materials Technology, Brunel University, Uxbridge, Middlesex, UB8 3PH, UK
[1]To whom correspondence should be addressed.

Abstract: Over the last decade there has been a rapid growth in the use of advanced composite materials offering enormous potential for use in a wide number of engineering applications. Therefore, information about the integrity of the composite component during its service life is an ongoing concern. In many industrial applications the composite component being assessed is usually taken out of service and tested under conditions alien to the material. The development of a real-time, in-situ damage detection system would ensure that material inspection could be conducted on site under normal service conditions without withdrawing the component from service. This paper describes the development of an acoustic emission (AE) detection system, where the monitoring of AE activity emitted from within a carbon/epoxy composite material is achieved using an all-fibre Mach-Zehnder interferometric sensor. This single mode optical fibre sensor has been embedded in the material to give an in-situ capability.

1. Introduction

In recent years there has been an increasing use of advanced composite materials for a variety of applications, for example, aircraft structures, sports equipment and automotive components. These materials offer high specific strength and stiffness properties, excellent fatigue and corrosion resistance and can be used to manufacture complex component shapes. However, the structural integrity of the composite material may deteriorate under service conditions due to manufacturing defects such as voids, impact damage, fatigue loading, and environmental effects such as ingression of moisture.

Once initiated, damage propagation can develop and is not readily evident from visual inspection or other non-destructive techniques such as x-ray, ultrasonics, and edge replication. Therefore, the development of a built-in damage detection system based on optical fibres embedded within the composite structure at the time of manufacture represents a very attractive proposition for the non-destructive evaluation (NDE) of composite structures. This clearly has both safety and economic advantages and therefore could lead to a greater confidence in the use of advanced composite materials, possibly even permit current design criteria to be reviewed.

The onset of micro-damage in composites is accompanied by a sudden release of energy within the material. Some of this energy is dissipated in the form of elastic waves, known as Acoustic Emission (AE). Numerous mechanisms have been proposed and confirmed as sources of AE, for example, fracture of fibres and matrix, fibre/matrix debonding, initiation and propagation of intra-laminar and inter-laminar cracks, and debonding between lamina (delamination).

Therefore, the goal of the AE measuring system is to detect the acoustic event and provide suitable signal processing to characterise it and determine its significance. The most critical component of any AE measurement system is the transducer. The conventional NDE technique for monitoring material degradation through AE is based on the piezoelectric effect. Piezoelectric (PZ) transducers are sensitive over the frequency range of interest (100kHz-2MHz), easy to use, relatively cheap and are an established technology. However, these type of transducers are contacted externally to the material and are not suitable for monitoring AE in-situ within the material. This paper reports the use of an all optical fibre Mach-Zehnder interferometric sensor to monitor AE. The optical fibre sensor (OFS) is embedded into the composite at the time of manufacture. Any AE activity due to internal composite damage will modulate the dimensions and refractive index of this embedded optical fibre, and result in the modulation of the phase in the sensing arm of the interferometer.

2. Optical fibre sensing of AE in composite materials

In 1950 Kaiser showed that the micro-damage present in many materials can be evaluated by the detection of internally generated acoustic stress waves, or AE. Since then AE has received growing attention due to the relative ease of detection and in-situ and real-time mode of operation. It is possible to obtain information about the defect (type, geometry, and possibly location) through the detection of acoustic events using the conventional PZ transducer[1]. However, there are many advantages of using optical fibres over their electrical counterparts, such as immunity from electromagnetic interference, high electrical isolation and corrosion and fatigue resistance[2]. An optical sensor is a device in which an optical signal can be modulated by an external stimulus, such as temperature, pressure, and strain. There has been some published work in the open literature covering embedded optical fibre sensors for measuring AE[5-8], and sufficient work has been done to show that such a sensing system is feasible.

In 1989 a breakthrough in the development of a localised embedded AE detection system for composite damage monitoring based on fibre optic Michelson interferometry was reported by Liu et al [3,4]. At this time, little had been done to develop a fibre optic sensor for AE detection that could sense damage in real time. Their research included the development and testing of a system employing an active homodyne detection scheme (used to maintain linearity and maximum sensitivity). The system provided single-ended sensing with real-time monitoring capabilities and a certain degree of localisation. The optical fibre AE sensors were embedded into Kevlar/epoxy composite specimens which were subject to tensile loading. The resulting AE signals were detected and found to have a broad-band response of 100kHz to 1MHz. This OFS detected acoustic signals associated with the formation of threshold regions of delaminations[5, 6].

The research work reported to date has shown that the high sensitivity of an interferometric OFS allows it to detect damage within a composite material. The all-fibre Mach-Zehnder interferometer reported here is being used to monitor AE emanating from damaged carbon fibre reinforced plastic (CFRP) specimen under tensile stress. The Mach-Zehnder interferometer has the advantage of not requiring mirrors at the end of the fibres

3. Experimental techniques

3.1. The fibre optic AE detection system

The design of the optical-fibre phase-modulated sensor was a two-stage process. The first stage involved the sensing element, where the mechanical interactions between the measurand and the optical fibre produces a phase shift in light transmitted. In the second stage the interferometer detects these phase shifts by modulating the output intensity. In the all-fibre Mach-Zehnder interferometer[7] (Figure 1) the light is coupled from a coherent source (for example, a polarised He-Ne laser or a laser diode) into a monomode optical fibre, and the optical signal is divided into two paths by means of a 2x2 directional coupler (DC1).

In the signal arm of the interferometer, polarisation controllers (PC1 and PC2) enable the recombining beams to be mutually coherent and have identical polarisation states. In the reference arm, several turns of fibre are wrapped around a PZ drum, which is driven by a high voltage feedback to produce a controllable phase shift. In the sensing arm, a length of single-mode optical fibre is embedded in the material under consideration. A section of the sensing arm, equal to the embedded and pigtail lengths of the test piece, has been cut out and replaced with a demountable composite test section.

The signal and reference beams are recombined at a second directional coupler (DC2), giving two optical outputs at the photodiode detectors (PD1 and PD2). The signal beam is phase modulated by the external stimulus (that is, the AE activity), whereas the phase of the reference beam remains constant. The reference is set at the quadrature point of the system by implementing an active homodyne feedback technique[8]. This technique locks out any environmental phase perturbations. After passing through a high-pass filter (100kHz cut-off frequency) the output signal becomes the optical phase change induced by high frequency AE waves. The output from the fibre optic AE detection system has been monitored using an AE signal analyser (AET5500). This collects the output AE events which cross its preset threshold level and determines in real-time a series of AE parameters which will help distinguish the level of damage.

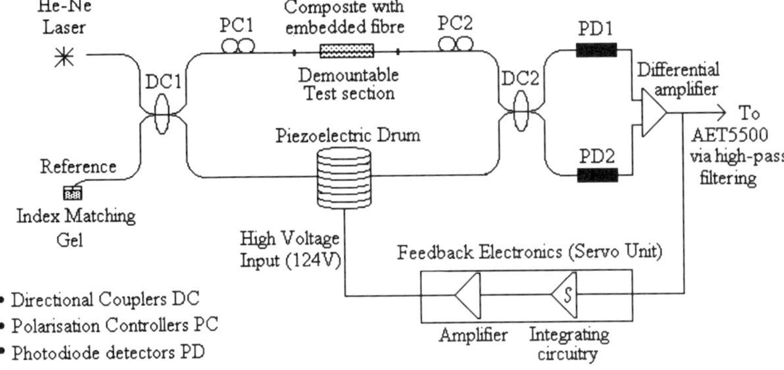

Figure 1. Schematic diagram of the all optical fibre Mach-Zehnder interferometric sensor.

3.2. Material and specimens

The tensile specimens were manufactured by hand laying 16 plies of pre-impregnated carbon fibre (Ciba Geigy T300/920) prepreg using a $[0, 90, 90, 0, 0, 90, 0, 90]_S$ sequence. The prepregs were cut into strips and placed in individual specimen moulds. The optical fibre's acrylate layer was stripped and the fibre was placed along the length of the composite specimen at the 8/9 ply interface. The samples were cured at 0.62MPa for 1 hour at 125°C. A schematic illustration of the test specimen is shown in Figure 2. In order to distinguish the AE activity of damaged and undamaged composite materials, samples were pre-damaged by drilling two 2mm holes either side of the optical fibre in the centre of the composite specimens.

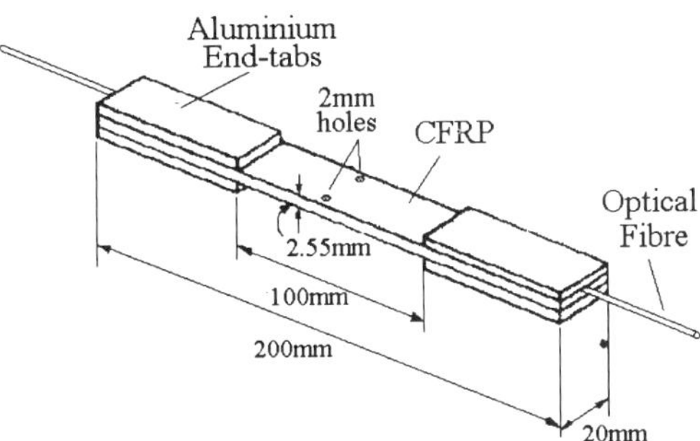

Figure 2. The dimensions of the CFRP tensile specimens. The optical fibre is protected from damage at the fibre/composite interface by embedded PTFE tubing.

4. Experimental results and discussion

It is essential to characterise the AE detection system before it can be used to progressive damage. Several tests have been carried out to characterise the system for known acoustic events. These tests were as follows: active acoustic excitation and simulated AE event detection. Once this was achieved the composite specimen was tensile tested using a Hounsfield tensometer. The AE activity was monitored using the embedded optical fibre sensor and the PZ transducer so that a comparison between the two sensors could be made.

4.1. *Active acoustic excitation*

The first set of characterisation experiments involved the acoustic excitation of the composite specimen using a separate PZ transducer, driven sinusoidally at approximately 160kHz. This source and the broadband PZ detector were placed approximately 50mm apart. The responses of the optical sensor and the PZ transducer are shown in Figure 3. The interferometric sensor results compare favourably with its electrical counterpart. The difference in the two signals was due to the fact that the OFS gain was limited by the processing unit.

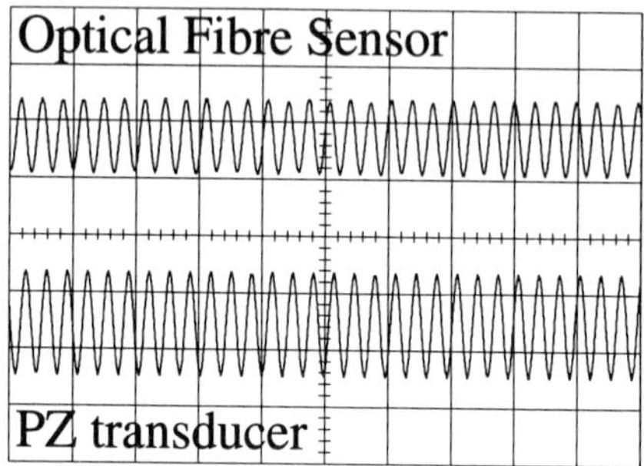

Figure 3. A comparison between optically and electrically sensed signals due to an applied AE signal at a frequency of 160kHz. The amplitude scale for the OFS and PZ transducer was 1V and 5V per division respectively whereas the time base for both traces was 20μs per division.

4.2. *Simulated acoustic emission*

The most common method of simulating broadband AE signals in a material involves the fracture of a 0.5mm 2H graphite pencil lead (Nielson source) while the lead is in contact with the surface of the material. The pencil-break tests were performed using a Teflon jig which maintained a constant angle between the pencil lead and the material surface. This pencil-break technique was used to evaluate the response of the fibre optic system prior to collecting damage-induced AE data from materials. Once again, the response of the optical fibre sensor was compared to a PZ transducer subjected the same excitation (Figure 4). The results yielded a reproducible high frequency AE signal, which took the form of an exponentially decaying sinusoidal oscillation.

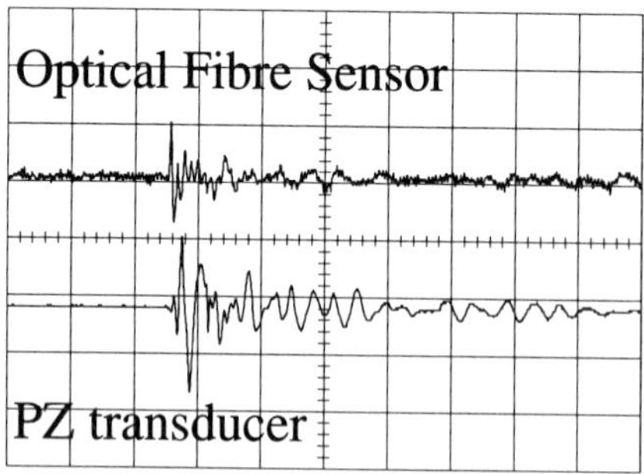

Figure 4. A comparison between optically and electrically sensed AE activity caused by a pencil break on a composite sample. The amplitude scale for the OFS and PZ transducer was 1V and 5V per division respectively whereas the time base for both traces was 20μs per division.

4.3. *AE from a damaged composite*

The pre-damaged composite sample with the embedded optical fibre was placed in the grips of the Hounsfield tensometer. The emphasis of the test was placed on counting the number of AE signals generated as damage progressed through the specimen. On the application of the load there was a small amount of AE activity, indicating the onset of micro-damage. As the amount of damage in the composite increased, there was a point where the rate of AE activity dramatically increased. In the cumulative AE event count graph this can be seen by the 'knee' in the curve (Figure 5). The dramatic increases in AE could be due to significant matrix cracks, delaminations, and ultimately, fibre fractures.

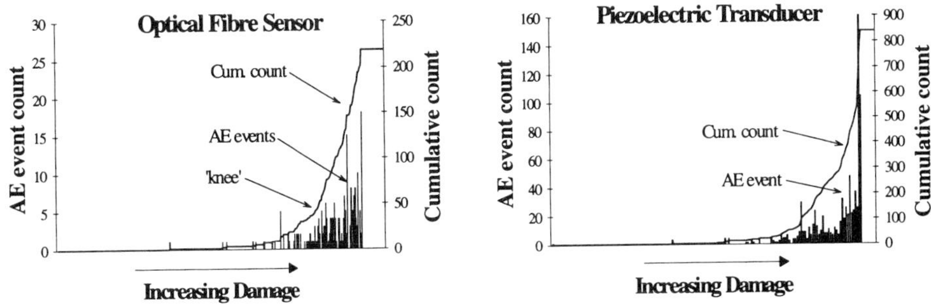

Figure 5. Acoustic Emission from a CFRP specimen loaded in tension to failure

The tensile loading of the composite samples produced AE 'events v time' graphs which was consistent from test to test. The cumulative AE event count graphs were also all similar to the exponentially increasing trend found using PZ transducers[9,10] by other authors for CFRP samples loaded in tension. The AE from the composite sample in tension was also monitored using a PZ transducer mounted on the composite surface. The combined event and cumulative count plots for both the OFS and PZ transducer are shown in figure 5. A comparison of the cumulative count plots for the OFS and the PZ transducer show that the embedded OFS was the less sensitive (Figure 6).

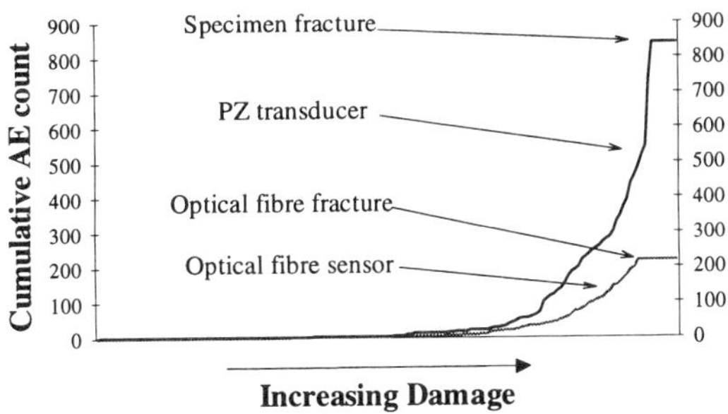

Figure 6. A comparison between optically and electrically sensed cumulative AE count plots.

The overall gain of the OFS was 10dB less than the electrical system and therefore fewer signals crossed the threshold voltage of the AET5500. The gain of the OFS was limited by the processing electronics. The number of events captured by the OFS was also limited by the fact that the embedded optical fibre fractures just before specimen failure. This was due to the high crosshead speed used for the tensile test.

5. Conclusions and future work

In these experiments it has been shown that AE activity can be observed using an embedded optical fibre and that the results compared favourably with those obtained from the standard PZ transducer under the same conditions. A Mach-Zehnder all-fibre interferometric OFS has been applied to monitoring the AE resulting from progressive damage that occurs during loading of a CFRP composite. The sensitivity of the interferometer is dependent on the optical path difference (OPD) between the two arms. The introduction of fibre collimators into the sensing arm would improve the OPD towards the ideal zero. As the gain of the OFS is governed by its processing unit, different AET5500 threshold levels will need to be investigated to ensure that the system is sensitive to smaller amplitude AE signals. Further characterisation of the AE signals generated by progressive composite damage will be performed to determine the most effective operation of the sensor. Frequency and time domain analysis of the AE will enable characterisation of the severity of the accumulated damage.

Acknowledgements

The authors gratefully acknowledge financial support from UK Engineering and Physical Sciences Research Council.

References

(1) Zimcik DG, Proulx D, Roy C, Mashouli A.
"Real time monitoring of carbon epoxy composites using acoustic emission "
NDE, SAMPE Quarterly, January, pp5-11, (1988)
(2) Jackson DA.
"Monomode optical fibre interferometers for precision measurement"
J.Phys.E: Sci.Instrum.,18 , pp981-1001, (1985)
(3) Liu K, Ferguson SM. and Measures RM.
"Damage detection in composites with embedded fiber optic interferometric sensors"
Fibre optic smart structures and skins II , pp205-210, (1989)
(4) Liu K, Ferguson SM, Measures RM.
"Fiber optic interferometric sensor for the detection of acoustic emission within composite materials"
Optics letters, 15 No.22, pp1255-1257, (1990)
(5) Measures RM, Valis T, Liu K, Hogg D, Ferguson SM, Tapanes E.
"Interferometric fiber optic sensors for use with composite materials"
Optical testing and metrology III, SPIE 1332, pp421-430, (1990)
(6) Liu K, Ferguson SM, McEwan K, Tapanes E, Measures RM.
"Acoustic emission detection for composite damage assessment using embedded ordinary single mode fiber optic interferometric sensors"
Fiber optic smart structures and skins III, SPIE 1370, pp316-323, (1990)
(7) Zheng SX, McBride R, Barton JS, Jones JDC, Hale KF, Jones BE.
"Intrinsic optical fibre sensor for monitoring acoustic emission"
Sensors and Actuators A, 31, pp110-114, (1992)
(8) Udd E.
"Fiber Optic Sensors- An introduction for engineers and scientists"
Ch.10, (John Wiley & Sons, Inc.), (1991)
(9) Fuwa M, Bunsell AR, and Harris B.
"Tensile failure mechanisms in carbon fibre reinforced plastics"
J.Material Science, 10, pp2062-2070, (1975)
(10)Miller RK. and McIntire P. (Editors)
"Nondestructive Testing Handbook: Volume 5- Acoustic Emission Testing (2nd Ed.)"
Sect.12, Part 3, (Amer. Soc. NDT), (1987)

A NOVEL INTEGRATED OPTICAL MICROPHONE UTILIZING THE THERMOOPTIC EFFECT IN POLYMERS

K H Cazzini, F R Akkari* and W Blau

Department of Physics
Trinity College
Dublin 2
Ireland

Tel: +353-1-7022404
Fax: +353-1-6711759

Direct sound modulation of a guided optical beam was demonstrated using thermooptic effect in thin film polymeric waveguides. Sound waves impinging on a polymeric thin film single mode waveguide biased thermally near extinction drives the waveguide back into guiding mode. The amplitude of the propagating guided beam is modulated accordingly. This was realized in polyurethane thin film structures for frequencies below 1 kHz.

Classical microphones have played a vital role in present communication and sound reproducing systems. Similarly optical microphones and/or acoustic sensors can play a major role in future all-optical communication systems or signal processing systems. Basically an optical microphone is a transducer that transforms variations in sound intensity into amplitude modulation of a guided optical beam directly without going through a middle stage electrical transducer. To achieve this in an integrated optical form adds to it the advantages of thin film guided wave integrated optics. Namely small size, immunity from environmental effects, freedom from mechanically vibrating components and the fact that the guided optical beam is the medium of detection and transmission at the same time [1]. To our knowledge this is the first time that such a device has been

* on leave from EE Department, Al-Fateh University

demonstrated in this form. To transform variation in sound waves amplitudes into an optical form, thermooptical and elastooptical effects were utilized. It is well known that thermooptic effect in polymers is dominated by volume expansion [2] where the increase in volume will lead to a decrease in the material refractive index. Similarly when the polymer is compressed under the effect of an increase in pressure its refractive index will increase. In this microphone we used a thermooptic mode extinction modulator [3] biased thermally near extinction as the sensing element. When the polymer film is electrically heated via a strip heater it will become elastic and therefore sensitive to variations in pressure, sound waves or gas flow. Any increase in pressure will revive the guided mode with its amplitude directly proportional to its increase. The microphone was fabricated on a BK-7 glass substrate. A weakly guiding polyurethane thin film was deposited on the substrate. A buffer layer, PMMA thin film, was spin coated on the polyurethane film and finally a metallic strip heater was sputtered on top. The polyurethane film has a thickness of 3.4 μm and a refractive index of 1.522 determined by M-line technique at room temperature at 633 nm. The buffer layer has a thickness of 0.3 μm and a refractive index of 1.488. The strip heater was cold sputtered using a mechanical mask. The refractive index of the BK-7 glass substrate is 1.515. Conventional prism coupling was used to couple light (HeNe λ = 0.6328 μm) in and out of the polyurethane thin film waveguide. The strip heater was used to thermally bias the waveguide near mode extinction. Sound waves generated by a small loudspeaker were directed to the interaction region of the device. The output of this microphone is focused on a photodetector connected to a storage oscilloscope to detect any variations in the guided optical beam intensity. A sine audio frequency wave generated by the loudspeaker was used to test the response of the microphone at different frequencies and sound levels.

Figure 1 demonstrates the correspondence between the electrical signal applied to the loudspeaker input and modulated guided optical beam at 200 Hz. We were able to verify modulation of the guided beam directly by sound waves at several audio frequencies below 1 KHz. The measurement indicated also an increase in the optical microphone response with increasing the sound level. In addition we were able to detect gas flow levels in the sub-millilitre per second range when N_2 gas was passed over the device. Further investigations are in progress to optimize this device and determine its response in different applications.

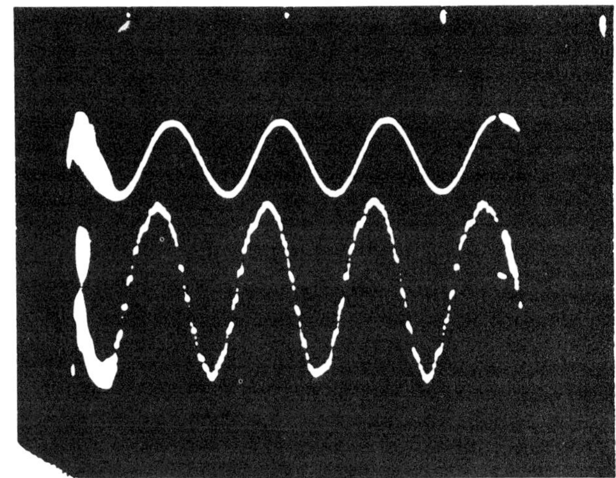

Figure 1: Comparison of the detected microphone output with the input to the loudspeaker at 200 Hz.

In conclusion we have demonstrated an integrated optical microphone/acoustic sensor. This device can play a major role in all-optical systems.

References

1. R E Kunz, "Totally Integrated Optical Measuring Sensors", SPIE Vol 1587, 98 (1991).

2. J M Cariou, J Dugas, L Martin and P Michel, "Refractive-index Variations with Temperature of PMMA and Polycarbonate", Appl Opt Vol 25, 3, 334 (1986).

3. F R Akkari, K H Cazzini and W Blau, "Thermooptic Mode Extinction Modulation in Polymeric Waveguide Structures", accepted for publication in Journal of Non-Crystalline Solids (1994).

Application of Ferroelectric Ceramic/Polymer Composites for Detection of Acoustic Emission in Composite Plate Structures

M.P. Wenger, P. Blanas†, R.J. Shuford† and D.K. Das-Gupta

School of Electronic Engineering Science, University of Wales, Dean Street, Bangor, Gwynedd, LL57 1UT, UK

†U.S. Army Research Laboratory, Materials Directorate, Arsenal Street, Watertown, MA. 02172-0001, USA

Abstract

Most commercially available acoustic emission sensors employ the piezoelectric properties of a ferroelectric ceramic e.g.. PZT, the majority of these being resonant transducers with resonant frequencies of a few kHz. Acoustic emission frequencies are predominantly in the range of a few kHz to 1MHz. Ultrasonic transducers with resonant frequencies of a few MHz are wideband sensors when used in this range. All these sensors measure the displacement of the surface of the structure under test. Work has been done to compare sensors constructed from piezoelectric composites consisting of ceramic particles embedded in a polymer matrix. These materials have been fabricated into surface mounted acoustic emission transducer and directly compared to the commercially available transducers. The obvious advantage of a piezoelectric element consisting of a ceramic/polymer composite is that it can be made into very thin films (<50μm) which can then be embedded into laminate structures for *in situ* sensors. Preliminary investigations into the viability of embedded composite sensors have been performed.

Introduction

Acoustic emission (AE) is the elastic energy that is spontaneously released by materials when they undergo deformation. An acoustic emission sensor is used to detect the dynamic motion resulting from the AE events and to convert the detected motion into an electrical signal. Most commercially available AE sensors are used to detect the motion of the surface of the material at a point some distance from the source of the emission. The signal at this point will have been altered in some form due to the geometry and properties of the structure concerned. When considering structures with plate like geometry, i.e. two dimensions much larger than the third, the propagation of the elastic waves will be governed by Lamb's Homogeneous equations ([1], and references therein). When the wavelength of the elastic wave is much larger than the thickness of the plate the set of governing equations are greatly simplified. In the 'thin plate' region equations derived from the classical plate theory can be used to understand the propagation of acoustic waves.

Plate waves in aluminium plates have been investigated previously [1-3]. It was found that the acoustic waves in the plate propagated in two distinct modes, the extensional and the flexural mode. The extensional modes have higher velocities than the flexural mode and frequencies predominantly above 300kHz as compared to those of the flexural mode with frequencies predominantly less than 300kHz. The higher frequencies of the extensional mode suffer from greater attenuation than the flexural mode. AE measurements typically concentrate on threshold measurement for triggering of the detection electronics and location detection of source. When plate waves are present the mismatch in velocities will cause the extensional mode to arrive earlier than the flexural mode. This is fine if all the transducers are triggered by this mode. But due to the greater attenuation of the higher frequencies present, this may not be the case. The flexural mode, while having a greater amplitude perpendicular to the plane of the plate, is a dispersive mode. Therefore inherent errors will occur if this mode is used to detect the location of the source.

Detection of the whole of the AE signal without corruption by the transducer will yield a signal containing the separated modes and their relevant frequencies. The use of all the information contained within the signal could be used to accurately locate the source. In this work we have fabricated surface mounted AE transducers and in situ embedded transducers from a piezoelectric ceramic and polymer composite. The ceramic, calcium modified lead titanate grains of ~10μm, have been dispersed into a polymer matrix to form a so called 0-3 connectivity [4]. The spontaneous polarisation of the ceramic/(polymer) can be aligned in a DC field to produce a net polarisation. The resulting ferroelectric material can posses piezoelectric figures of merit of 2.01pPa^{-1} and electromechanical coupling coefficients of 0.24 [5].

Experimental

The frequency response of the commercial transducers and the ceramic/polymer transducers were evaluated using a face-to-face method where two AE transducers are coupled with their sensing faces together. Silicone vacuum grease was used to ensure the transducers were acoustically coupled. One of the transducers was used as a driving transducer while the other was used as a sensing transducer and the output of the second transducer was recorded on a storage oscilloscope at a sampling rate of 2MHz.

Assuming that the attenuation due to the coupling interface is negligible the output of the sensing transducer (TO) will be a convolution ($\otimes$) of the electrical input signal (EI), the response of the driving transducer (DR), the response of the sensing transducer (TR) and the response of the detection electronics (DE) such that

$$TO = EI \otimes DR \otimes TR \otimes DE \qquad \text{...(1)}$$

In the frequency domain the convolution of the responses becomes a straight multiplication so

$$TO(f) = EI(f) \times DR(f) \times TR(f) \times DE(f) \qquad \text{...(2)}$$

Assumptions can be made that $DE(f)$ is flat and equal to unity over the range of frequencies we are concerned with and that the transducer responses are reversible, i.e. input response is equal to the output response. If the two transducers are of the same type then the driving response (DR) can be evaluated from

$$TO(f) = EI(f) \times [DR(f)]^2 \qquad \text{...(3)}$$

$$\therefore DR(f) = \sqrt{\frac{TO(f)}{EI(f)}} \qquad \text{...(4)}$$

Once DR has been determined then the response of another transducer can be found from

$$TR(f) = \frac{TO(f)}{EI(f) \times DR(f)} \qquad \text{...(5)}$$

In this work the electrical drive signal applied to the driving transducer was produced by a 150V spike generator [Par Scientific Instruments SPIKE 150 PR] which produced a spike of rise time 90ns and duration 280ns. Through Fast Fourier Transform (FFT) analysis the frequency components over the range 0 - 1MHz was calculated.

Two identical transducers [Dunegan/Endveco S 140 B/HS] were used to evaluate their responses. One of these was then used to provide a driving transducer with known response to evaluate the ceramic/polymer transducers compared to two widely used commercially available transducers [Dunegan/Endveco S 140 B/HS, Physical Acoustics Corp. R15]. It is noted here that the characterisation of the transducers are not an absolute characterisation but a comparative one.

The responses of a commercial transducers and an average response of a surface mounted ceramic/polymer transducer, both normalised to the response of the driving transducer, are shown in figures 1 & 2.

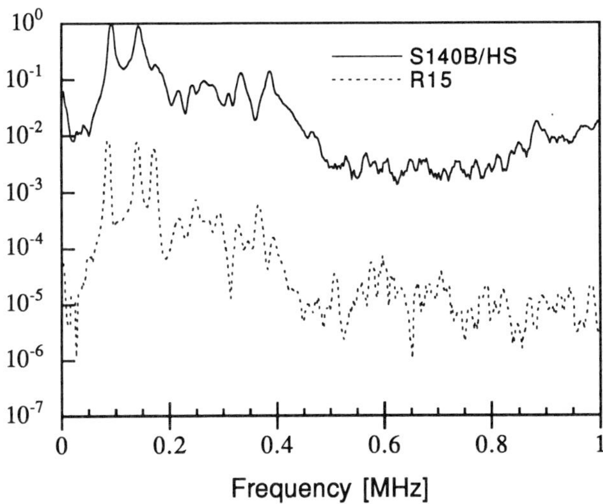

Figure 1: Normalised response of the driving transducer [Dunegan/Endveco S 140 B/HS] and the normalised response of another commercially available resonant transducer [PAC R15].

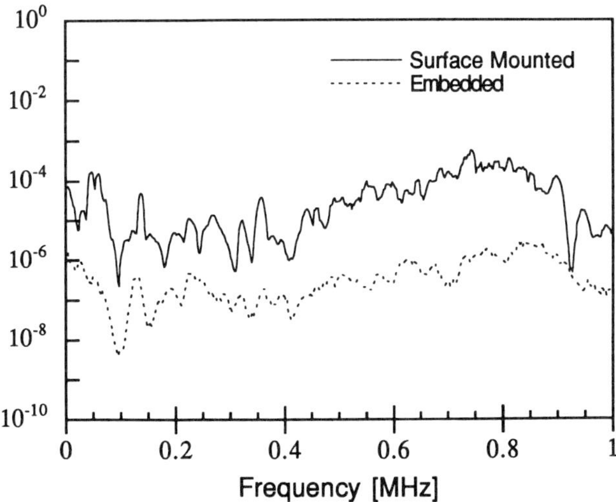

Figure 2: Normalised responses of ceramic/polymer surface mounted and embedded transducers. Greater attenuation of the signal from the embedded transducer is present due to the nature of the epoxy plate.

It can be seen from these figures that although the ceramic/polymer transducer is considerably less sensitive than the commercially produced ceramic transducers it possesses a flat response over the frequency range 0 to 1MHz, showing it to be viable for the detection of acoustic waves with these frequencies.

A thin sensor film [~100μm] was embedded into a test piece of glass fibre reinforced epoxy plate with electrical connections to the edge of the plate. The driving transducer was then placed on the surface of the plate and excited in the same manner as the face-to-face method. The output of the embedded transducer was then digitised and analysed as before. Figure 2 shows the response of the embedded sensor.

Again the embedded sensor is less sensitive than the commercially available sensors. A wide band flat response is noted. It should also be noted that this preliminary investigation of an embedded sensor is neither an absolute measurement nor a relative one. Due to the nature of the epoxy plate greater attenuation of the signal, as compared to the face-to-face method, is present showing the embedded transducer to be less sensitive than it appears to be in practice.

The suitability of the ceramic/polymer sensor for detecting plate waves in aluminium plates was investigated and compared to a commercially available transducer. The sensors were mounted on the surface of an aluminium plate using silicone vacuum grease to ensure an acoustic coupling between the plate and the transducer. An acoustic emission source was simulated by breaking of a pencil lead on the surface of the plate a distance of 127mm from the sensing transducer. The output of the sensing transducer was directly recorded, without amplification, on a digital oscilloscope at a sampling rate of 20MHz. Care was taken not to record signals corrupted by reflections from the edges of the plates.

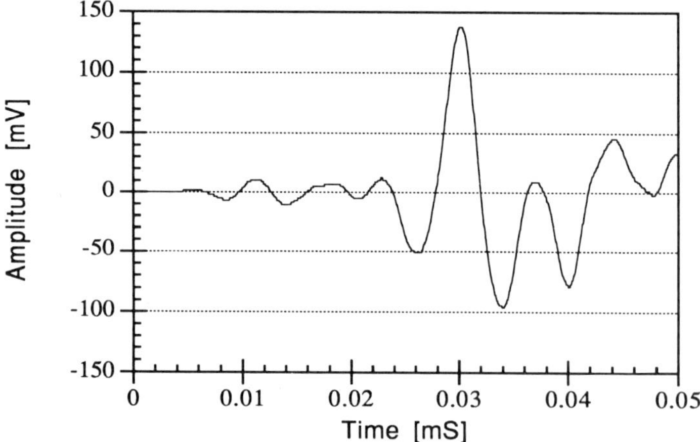

Figure 3: Response of a PAC R15 AE sensor to a lead break 127mm away on an aluminium thin plate.

Figures 3 & 4 show the plate waves detected by the commercial sensor and the ceramic/polymer sensor respectively. The extensional mode and the flexural mode of the waves can be seen clearly from both of these figures. The resonant frequency of the commercial transducer of 150kHz is present in both the extensional and the flexural mode of the transducer output while the signal produced by the ceramic/polymer transducer showed a distinct difference in modal frequencies. This ability to discern the difference in frequency from mode to mode exhibits the wideband nature of the transducing element.

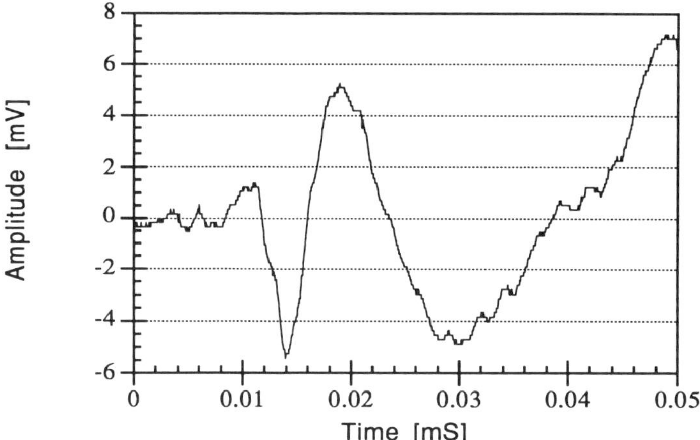

Figure 4: Response of a ceramic/polymer composite surface mounted transducer to a lead break 127mm away on an aluminium thin plate.

Figure 5 shows the response of an embedded transducer to a lead break on the surface of the epoxy plate. It should be noted the difference in time axis from the previous figures. It is believed that the signal from the embedded sensor is a signal generated from an averaging of the acoustic waves present in the epoxy plate. This is due to the large size of the transducer [3cm x 4cm].

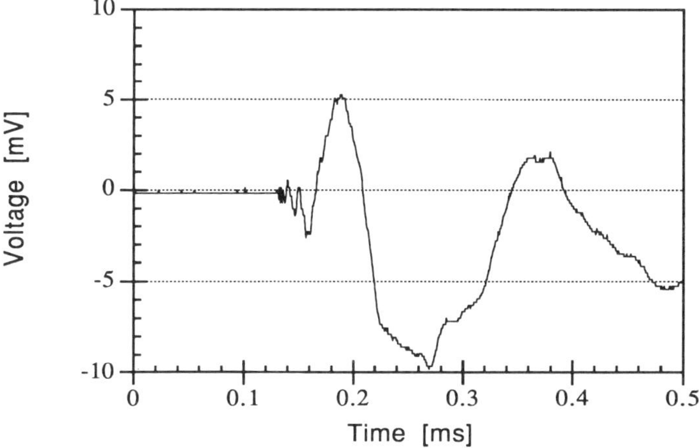

Figure 5: Response of an embedded ceramic/polymer composite transducer to a lead break on the surface of the epoxy plate a distance of 8.5cm from the transducer.

Conclusion

In this work we have considered the detection of acoustic emission particularly in plate like structures using relatively new materials of ferroelectric ceramic and polymer composites. Their flat band response over the 0-1MHz range make them suitable for detection of acoustic waves in solids of frequencies within this range. Their ability to be embedded within laminate composite structures make them appealing for in situ transducers. Further work is being conducted to improve the response of the embedded sensor and further characterise its behaviour within the laminate composite.

References

1. Gorman, M.R., *Plate Wave Acoustic Emission.* J. Acoust. Soc. Am., 1991. **90**(1): p. 358 - 364.

2. Prosser, W.H., *The Propagation Characteristics of the Plate Modes of Acoustic Emission Waves in Thin Aluminium Plates and Thin Graphite/Epoxy Composite Plates and Tubes*.1991, NASA:

3. Gorman, M.R. and W.H. Prosser, *AE Source Orientation by Plate Wave Analysis.* Journal of Acoustic Emission, 1990. **9**(4): p. 283 - 288.

4. Dias, C., D.K Das-Gupta, Y. Hinton and R.J. Shuford, *Polymer/Ceramic Composites for Piezoelectric Sensors*. Sensors and Actuators A, 1993. **37-38**: p. 343-347.

5. Dias, C.J.M.M., *Ferroelectric Composites for Pyro- and Piezoelectric Applications 1994*, University of Wales, Bangor:

ULTRASONIC SENSOR SYSTEM FOR DAIRY INDUSTRY

Alf Püttmer[a], Peter Hauptmann[a) b)], Bernd Henning[b], Ralf Lucklum[a]

a) Institut für Prozeßmeßtechnik und Elektronik, Otto-von-Guericke-Universität Magdeburg, Universitätsplatz, PF 4120, 39016 Magdeburg, Germany, Tel: +49 (391) 5592-2570, FAX: +49 (391) 561 6358

b) Institut für Automation und Kommunikation e.V. Magdeburg, Steinfeldstr. (IGZ), 39179 Barleben, Germany, Tel: +49 (39203) 81030, FAX: +49 (39203) 81100

Abstract: The described ultrasonic measurement system offers the possibility of in-line quality and process control in the production of food emulsions and suspensions, e.g. in the dairy industry. The principle and features will be described. Results of laboratory measurements with the measurement system and possible applications in the process of dairy industry will be shown.

1. Introduction

Various applications of ultrasonic sensors for the characterization of liquid systems in the chemical or food industry and in biotechnology have been known. Due to the continuous measuring principle ultrasonic sensors allow in-line process monitoring. On the other hand these measuring principle works without direct contact to the investigated substance mixture, which means it can be applied even with aggressive liquids like acids. If the acoustic behaviour of the investigated liquid is well known, acoustic sensors permit the direct and continuou analysis of material properties like density or compressibility of the liquid, the concentration of selected substances in multicomponent systems, or technological quantities, e.g. turnover or speed of chemical reactions. The paper reports about the ultrasonic sensor system for simultaneous measurement of sound velocity and damping.

2. Theory

The propagation of ultrasound waves in liquids can be described with help of the acoustic pressure function field $p = p(x,t)$. For harmonic waves equation (1) is valid

$$p(x,t) = p_0 e^{j\omega(t-x/c)} e^{-\alpha x} \qquad (1)$$

with time t, distance x, frequency ω, sound pressure p_0 at $x = 0$, phase velocity c and absorption coefficient α.

For ideal liquids the sound velocity is determined by the bulk module K and the density ρ

$$c^2 = K / \rho . \qquad (2)$$

The determination of the sound velocity is usually done with a discontinuous method. An acoustic pulse or burst is transmitted into the liquid to be received and measured after propagating through a well defined distance. Actually the transmission time of the acoustic signal is finally a value for the sound velocity.

The absorption coefficient α of a liquid is dependent on the material properties in a more complex manner. For simple homogeneous liquids α is determined by the shear viscosity η_S and the bulk viscosity η_V

$$\alpha = \frac{\omega^2}{2\rho c^3}\left(\eta_v + \frac{4}{3}\eta_s\right). \qquad (3)$$

In contrast to the sound velocity the absorption coefficient shows a quadratic dependence on the frequency.
Attenuation of acoustic waves in liquids is caused by absorption (equation 3), relaxation losses and losses due to scattering effects from particles. Acoustic attenuation is usually measured with a pair of transducers by changing the distance between them. The amplitude of the received signal as a function of the distance is a value for the acoustic attenuation. Due to the necessity of changing the distance this measuring principle is not applicable in the process environment.
For industrial applications a pair of two transducers with a fixed distance is used. The relation of the received signal to the transmitted signal gives you a rough estimation of the attenuation of the medium. The damping of the acoustic signal within the measuring distance is influenced by different factors of the investigated liquid: scattering, acoustic impedance, attenuation. The attenuation of acoustic waves in liquids is caused by different absorption processes, viscous losses, thermal losses, relaxation losses, and losses due to scattering effects from particles. Shape, structure and arrangement of the transducers as well as the transmitted signal wave form have a significant effect on the form and amplitude of the received signal, too. Because we are not interested in the acoustic parameters themselves our concept divides into material (liquid) dependent and arrangement dependent correlations. The latter determine the optimal transducer design and arrangement for the measurement of sound velocity and attenuation or acoustic impedance of the liquid investigated.

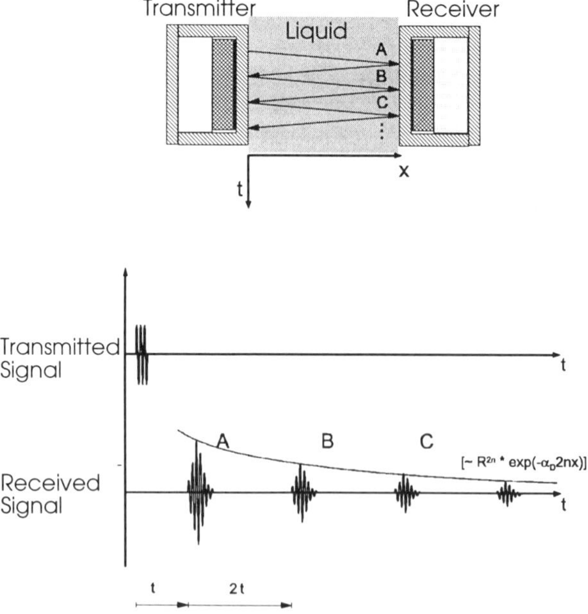

Fig. 1: Measuring principle

3. Ultrasonic sensor system

Ultrasonic sensors for sound velocity are well known. The characteristic of this sensors is a high accuracy of measurement. A high reproducibility of the correlations between the acoustic quantity and the process parameter is the advantage of measuring sound velocity. In an application in the dairy industry process parameters of special interest are the concentration of fat or solid-non-fat (SNF) in milk.
The velocity does not often supply sufficient information about the liquid mixture. Due to the dependency of sound velocity on temperature and pressure the acquisition of these quantities is necessary. Additional quantities would provide more information about the investigated liquid mixture. This quantities can be acoustic quantities (damping calculated from the signal amplitudes) or non-acoustic quantities like, pH, conductivity. Together with such sensors the sensor system can characterize complex liquid systems like milk or other multicomponent mixtures.
New developed high precision analogue circuits as well as high speed complex digital circuits like LCAs and micro controllers made it possible to measure the values and calculate concentrations of substances in high accuracy under industrial conditions.
Figure 2 shows the scheme of our sensor system for in-line process monitoring [1]. The heart of the system, the micro controller unit with a 80C196, provides functions like self adjustment, self diagnosis and data transfer to a PC or process control computer. The micro controller unit is responsible for the calculation of the interesting value (e.g. concentration) from the measurement values and the stored characteristic of the investigated liquid. With the signal multiplexer one measuring system can use different pairs of transducers to measure the sound velocity on different locations in the process.

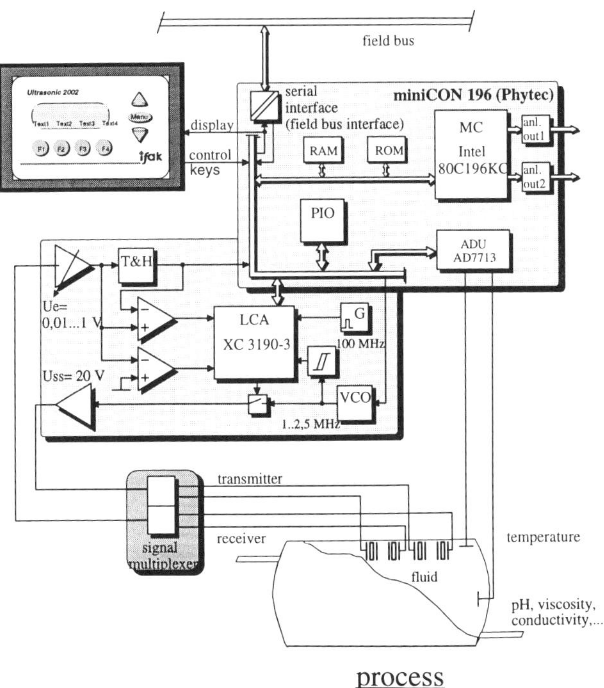

Fig. 1: Principle of ultrasonic sensor system

4. Application and results in milk

Major application fields of the ultrasonic measurement system are the food industry as well as the biotechnological and chemical industry. The main advantage of ultrasonic sensors is the possibility of in-line process monitoring. That means the continuous determination of concentration of substances inside a pipe or a reactor. Interesting is the application of ultrasound is the dairy industry.

Milk can be considered as a fluid consisting of the three components water, fat and SNF. Investigations about ultrasound velocity and attenuation in milk have been done since the 60s [2][3][4]. This studies show a correlation between the concentrations of milk fat and SNF in milk and the propagation constants ultrasonic velocity and attenuation.

Our ultrasonic measurements were carried out on samples of pasteurized milk consisting on skimmed, semi-skimmed, whole milk and cream. The measurements were made at a frequency of 1 MHz with the above presented measurement system.

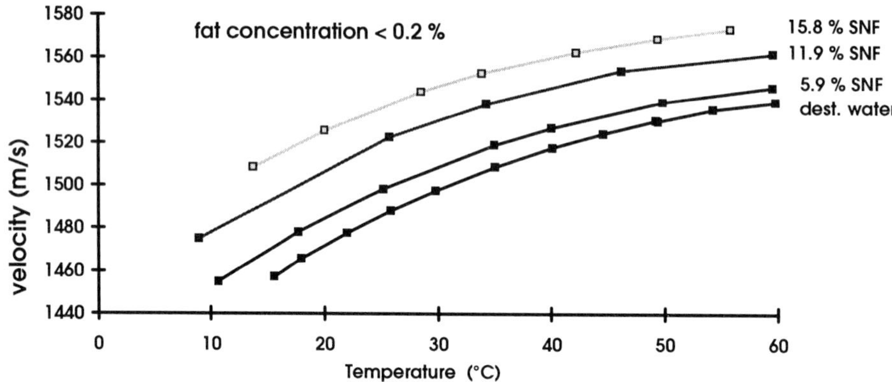

Fig. 3: The dependency of ultrasound velocity on temperature and SNF concentration

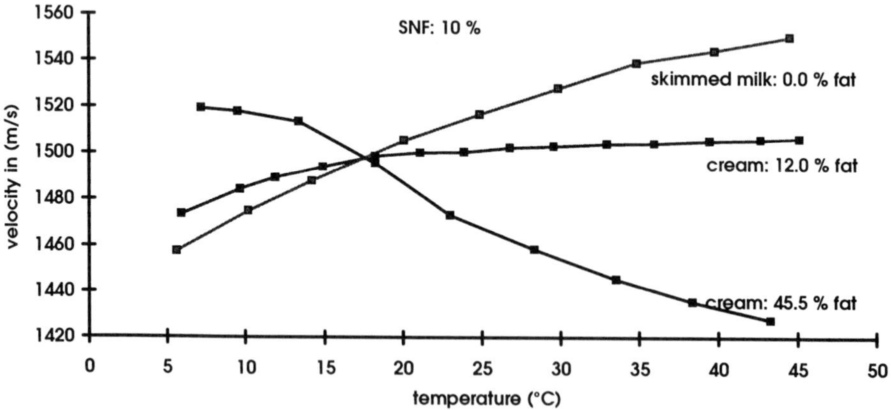

Fig. 4: The dependency of ultrasound velocity on fat concentration for constant SNF concentration over the temperature

The dependency of velocity of skimmed milk with different SNF concentration on the temperature is plotted in figure 3. As you can see the SNF concentration increases the propagation velocity of ultrasound. The ultrasound velocity shows a linear dependency on SNF concentration. An application of SNF measurement is the quality control during the production of milk powder.
Figure 4 shows the dependency of ultrasound velocity on the fat concentration. For a constant SNF concentration the slope of the curve is influenced. Referring to figures 3 and 4 measuring the propagation velocity of ultrasound on two temperatures permits the determination of SNF and fat concentration in milk. One application of continuous fat measurement is the process of combination of skimmed milk with the desired amount of fat. The in-line measurement system can also be used to distinguish diary product and waste water during the cleaning process. In this case our measurement system can help to avoid waste water.

5. Prospects

There is a high potential for application of ultrasound in the dairy industry. Our measurement system is usable for monitoring fat and SNF concentration in dairy products as well as for process control. Today the fat must be completely separated from the milk in the process. Semi skimmed milk and whole milk are produced by combining skimmed milk and fat in certain relations. With in-line measurement of fat it would be possible to separate only the desired amount of fat from the milk. This would increase the quality of milk because some sensitive ingredients e.g. vitamins are damaged during the separation process.

References

1 Henning, B., Lucklum, R., Kupfernagel, B., Hauptmann, P., Ultrasonic sensor system for characterization of liquid systems. Sensors and Actuators A, 41-42 (1994) 476-480.
2 Miles, C.A., Shore, D., Langley, K. R., Attenuation of ultrasound in milks and creams, Ultrasonics (1990) 28 394-400.
3 Fitzgerald, J. W., Winder, C., Ringo, G. R., Litovitz, T. A., Shadyside, M., An ultrasonic method for measurement of solids-not-fat and milk fat in fluid milk. I. Acoustic properties J Dairy Sci (1961) 44 1165.
4 Graefe, V., Studies of milk with the aid of ultrasonics. Kiel Milchwirtsch (1965) 17 175-205

Section D

Optical Sensors

Novel signal processing scheme for enhancing the central fringe in a white light interferometer

Y. N. Ning, Q. Wang, A. W. Palmer, K. T. V. Grattan, B. T. Meggitt, and K. Weir

Department of Electrical , Electronic and Information Engineering, City University, London EC1V OHB, U.K.

Abstract A simple, and fast signal processing scheme using a low-cost optical filter and a multi-stage squaring signal processing circuit to enhance the peak value of the central fringe in the output fringe pattern of a white light interferometer is described. The SNR_{min} required to identify the central fringe can be reduced considerably from ~50 dB to ~17 dB, as show in a practical application using the method.

1. Introduction

In recent years, various optical-fibre schemes based on the white light interferometer (WLI) have been actively studied and developed for optical sensing and measurement [1, 2]. In order to obtain measurements with high accuracy, it is important correctly to identify the central fringe in the zero-order fringe packet of the output intensity of the interferometer, and precisely determine the peak position. This may be achieved with, for example, the use of a two- or three- wavelength combination source [3, 4] to generate a fringe beating pattern, and thus enhance the relative peak value of the central fringe and reduce the value of the signal-to-noise ratio (SNR) required in its identification. Another approach [5] is to use an electronic device to switch on/off two low coherence sources sequentially to generate two autocorrelation functions respectively, then use a computer to multiply the two functions and obtain an output signal with the peak value of the central fringe thus being increased. In this work, a simple, fast and low-cost approach based on contrast enhancement is reported. With the use of this technique, two autocorrelation functions, which correspond to the same value of the change of optical path difference (OPD), are measured simultaneously and then processed directly using electronic circuits. Hence, the accuracy of the identification of the central fringe from the output of a white light interferometer can be enhanced and the associated signal-to-noise ratio (SNR) required can be reduced.

In this scheme, two multimode laser diodes with different central wavelengths (670 nm and 780 nm respectively) are used to illuminate an interferometer via a multimode fibre, as shown in Fig. 1. The output light from the interferometer is then divided by an optical filter, which, in fact, is a "hot mirror" (part No. 16 BH 16, supplied by Comer Instruments [6]) that allows the beam with a wavelength of 670nm to be transmitted and that of 780 nm to be reflected. This mirror is not optimized for use at other than normal incidence, but it provides sufficient wavelength separation to prove the principle discused in the paper. These two divided beams are then detected with two photodiodes to give two autocorrelation functions simultaneously. Following this, the two resulting electronic signals are fed into a three-stage squaring circuit which first multiples the two signals and then squares the output signal of the multiplier twice. This gives a group of fringes with the peak value of the central fringe being significantly increased with respect to the adjacent fringes, to aid in the identification of the peak, especially in situations where there is a high level of noise present.

2. Theoretical considerations

For the experimental arrangement shown in Fig. 1, the output intensities of the two beams, and $I_1(\Delta L)$ and $I_2(\Delta L)$, can be described by the autocorrelation functions of the two light sources, respectively given by

$$I_1(\Delta L) = (1/2)[1+\exp[-(2\Delta L/L_{c1})^2]\cos(2\pi\Delta L/\lambda_1)] \quad (1)$$

and

$$I_2(\Delta L) = (1/2)[1+\exp[-(2\Delta L/L_{c2})^2]\cos(2\pi\Delta L/\lambda_2)] \quad (2)$$

where λ_1, λ_2 are the central wavelengths of the two sources respectively, ΔL is the optical path difference (OPD) introduced by the interferometer, and L_{c1} and L_{c2} are the source coherence lengths. When the detected output signals are multiplied, the output signal, $I_n(\Delta L)$, can be then written as

$$I_n(\Delta L)=[I_1(\Delta L)*I_2(\Delta L)]^n \quad (3)$$

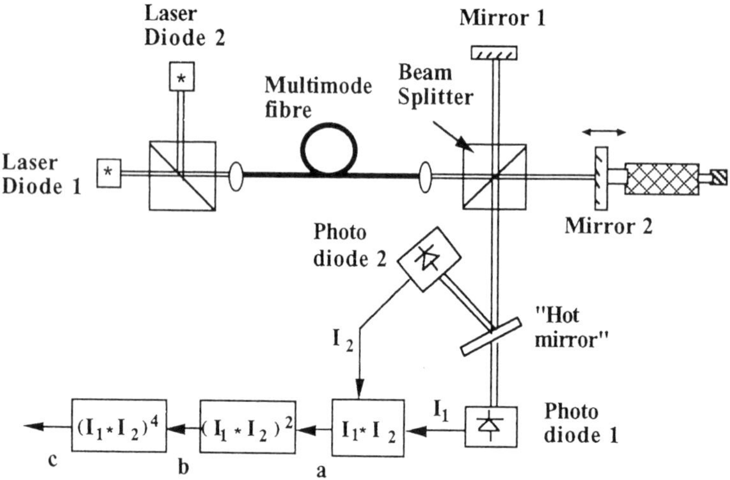

Figure 1. *Schematic diagram of the experimental arrangement used in this work. (the experimental results were recorded at a, b, and c respectively).*

where n =1, 2, 4, $I_n(\Delta L)$ represents the output of the multiplier, and the first and a second squaring operation respectively. With the use of a simple computer program, the output intensity as a function of the OPD may be calculated for the case where the coherence length is 25 mm, the fringe visibility is 1, and the number of the power is 1, 2, and 4 respectively. The results of a simulation based on the above are shown in Fig. 2.

Clearly, if a system has a noise level which is equal to or greater than the amplitude difference, ΔI, between the central fringe and the second largest fringe within the zero-order fringe packet, the central fringe cannot be identified directly through a simple inspection of its amplitude. In order to determine the minimum value of the signal-to-noise ratio (SNR_{min}) required to identify the central fringe, ΔI is used to represent the maximum noise level and this can be written as

$$\Delta I= I_n(0) - I_n' \quad (4)$$

where I_n' is the peak value of the second largest fringe. If the normalized peak value of the central fringe is defined as a unit signal, a minimum signal-to-noise ratio, SNR_{min}, required to identify the central fringe may be given by

$$\mathrm{SNR_{min}(dB)} = -20\log(\Delta I/2) \quad (5)$$

Thus it can be seen that circumstances can arise where the use of such a central fringe enhancement can mean the difference between successful identification and failure due to the prevailing SNR.

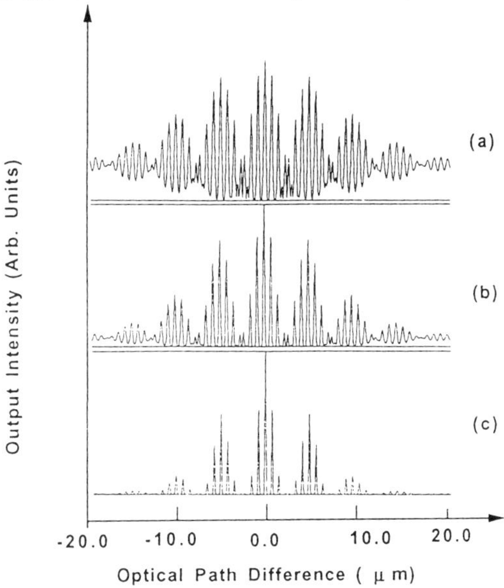

Figure 2. *Simulated results showing the fringe intensity as a function of the optical path difference for the cases of (a) n = 1, (b) n=2, and (c) n=4 respectively.*

3. Experimental results and discussions

In order to verify the principle of this signal processing scheme, the experimental arrangement shown in Fig. 1 was used in this work, where light from two multimode laser diodes (LPM3.670 and LT023MDO, with central wavelengths of 670 nm and 780 nm respectively) was injected into a multimode fibre (with a core diameter of 200 mm and a length of 4 metres) via an 10X objective lens. The collimated beam was modulated by a Michelson interferometer and then divided by the filter, and finally detected with two photodiodes. The $\mathrm{SNR_{min}}$ required to identify the central fringe for either sources is typically >50 dB [7].

Since the two autocorrelation functions of the light sources are measured simultaneously, they correspond to the same change of the OPD and are completely in phase one with the other. The output signals after the multiplying operation (for the case of n=1) at each stage of the squaring operation are recorded, and the results are shown in Fig. 3. It can be seen that the relative peak value of the central fringe has been increased considerably in comparison to either of the two autocorrelation functions shown in Fig. 2, and it also is gradually enhanced as the number of the squaring operation increases. The values of the SNR for each of the cases are measured and plotted in Fig. 4, in which the theoretical values of the $\mathrm{SNR_{min}}$, for the case that the coherence length is 25 mm, and the fringe visibility is 1, are also shown. The enhancement effect will, of course, also occur for a wide region of contrast of the original autocorrelation functions.

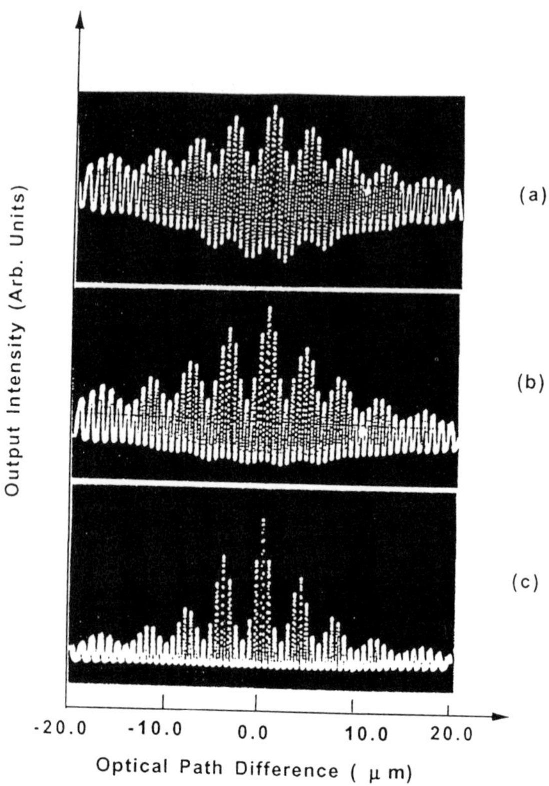

Figure 3. *Experimental results showing the fringe intensity as a function of the optical path difference for the cases of (a) n = 1, (b) n=2, and (c) n=4 respectively.*

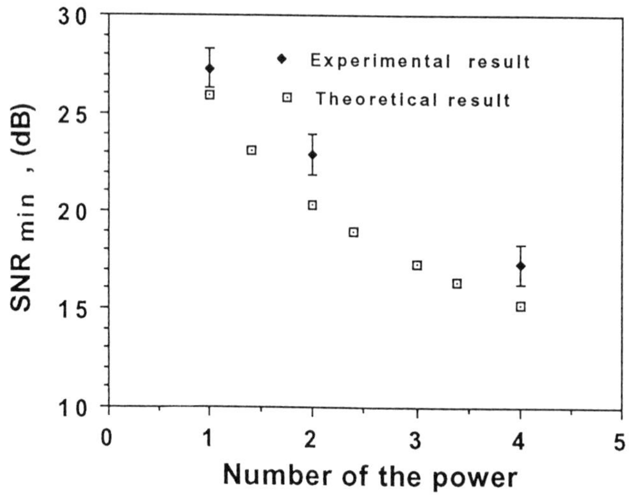

Figure 4. *Experimental and theoretical results showing that the value of the SNR_{min} is reduced with the increase of the number of the numerical power applied.*

By comparing Figures 2 and 3, it can be seen qualitatively that the results obtained from the experiment are in good agreement with those obtained theoretically, the figures showing the same structure in both cases. It is also notable that as the number of the squaring operation increases, the relative peak value of the central fringe becomes larger, and as a result, identification becomes easier by the use of a conventional electronic preset threshold device.

In Fig. 4, a similar trend in which the value of the SNR_{min} is reduced with the increase of the number of the operation power can be seen. This is because the maximum "noise" defined here is, in fact, the difference between the peak values of the two fringes and it can be directly increased by a squaring operation, after which the identification of the central fringe becomes easier, and the value of the SNR_{min} required is lower. It also can be seen from Fig. 4 that there is a difference (of about ~3 dB on average) between the results obtained from the theoretical simulation and the experiment. This may be partially explained by the following, in that the enhancement of the difference between the values of the zero-order and the adjacent fringes is a function of the fringe visibility. Thus the value of the SNR_{min} is also dependent of the fringe visibility, and the greater the fringe visibility, the lower the value of the SNR_{min} that may be obtained. The relationship between them can be seen in Fig. 6, where the theoretical results obtained after a two-step squaring operation are shown for a region of fringe visibility up to and including unity. Hence if the fringe visibility obtained in the experiment is less than one, all other conditions being the same, the value of the SNR_{min} after the squaring operations will be slightly higher than that shown in Fig. 5.

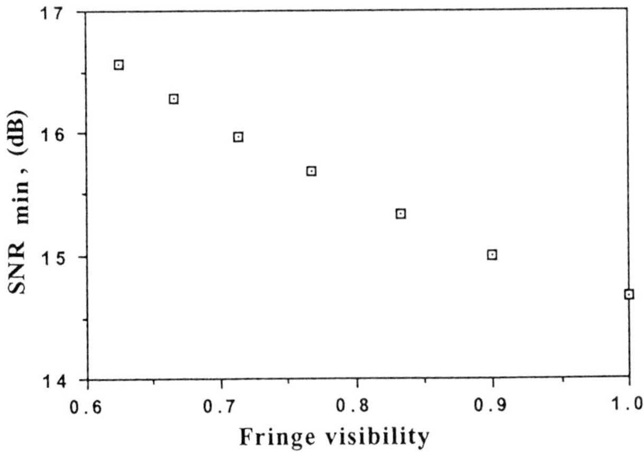

Figure 5. *Theoretical results showing that the value of the SNR_{min} is reduced with the increase of the fringe visibility.*

It is important to point out that the squaring operation, as would be expected, also increases the noise level induced by other sources of noise present such as light source intensity noise, detection shot noise etc., and thus the actual SNR of the detected signal cannot simply be improved by this operation. Thus the familiar electronic "contrast enhancement" that results derives from a different set of criteria, as discussed below, and not from an operation which preferentially increases the signal while leaving the noise level unaffected. The real value of the system is that the amplitudes of the central fringe and the second largest fringe are much larger than those of the noise, and the amplification of the amplitude difference (i.e. the "noise" used to calculate the SNR_{min} required) between these two fringes is larger than that of the overall noise in the squaring operation. This results in an enhancement of the central fringe and a suppression of the peak values of the other fringes and thus represents the essential value of the operation of this relatively simple process on the data, to yield valuable results, especially for the case where there is real difficulty in central fringe identification from the raw data. It is true that some additional noise may derive from the operation of the squaring circuits. However, according to the manufacturer, the typical noise induced by such a squaring circuit (MPY634, supplied by Burr-Brown International Ltd) is about 0.8 mV/$\sqrt{Hz}$, whilst the amplitude of the central fringe in the output signal is about 0.2 V, hence the effect of such a noise source is so small as that it may be ignored in this case. This is an important result for the broader use of the technique.

4. Conclusion

In conclusion, with the use of a low-cost optical filter to separate the sensing beam generated by a synthesized source into two individual beams divided according to their wavelengths, and the operation of the multi-stage squaring signal processing circuit to enhance the peak value of the central fringe in the output fringe pattern, the SNR_{min} required to identify the central fringe can be considerably reduced, as illustrated, to a value of about 17 dB, therefore making fringe identification more easily achieved with the use of simple, conventional analogue devices. This represents an advance on simple contrast enhancement methods in an important application of white-light interferometry.

The authors are pleased to acknowledge the support from UK Engineering and Physical Sciences Research Council.

5. References:

[1] S. Chen, A. W. Palmer, K. T. V. Grattan and B. T. Meggitt, "Digital signal-processing techniques for electronically scanned optical-fibre white light interferometry", *Appl. Opt.* **31**, pp 6003-6010, (1992).

[2] R. Dandliker, E. Zimmermann and G. Frosio, "Noise-resistant processing for electronically scanned white light interferometry", *Proceedings of the 8th Optical Fibre Sensors Conference*, pp 53-56, Monterey, CA, USA, (1992).

[3] Y. J. Rao, Y. N. Ning and D. A. Jackson, "Synthesized source for white-light sensing systems",*Opt. Lett.*, **18(6)**, pp462-464, (1993).

[4] D. N. Ning, Y. N. Ning, K. T. V. Grattan, A. W. Palmer and K. Weir, "Three-wavelength combination source for white light interferometry", *IEEE Photon. Technol. Lett.* **5(11)**, pp1350-1352, (1993).

[5] Y. J. Rao, D. A. Jackson, "Improved synthesized source for white light interferometry", *Electron. Lett.*, **30(17)**, pp1440-1441, (1994).

[6] Comar Instruments, 70 Hartington Grove, Cambridge, UK, *Manufacturer's Data*, Section 7.7, (1993).

[7] S. Chen, K. T. V. Grattan, B. T. Meggitt, and A. W. Palmer, "Instantaneous fringe-order identification using dual broadband sources with widely spaced wavelength", *Electron. Lett.*, **29**, pp. 334-335, (1993).

Wavelength Stabilisation of Laser Diodes Using Optical Feedback

R C Addy, A W Palmer, & K T V Grattan

Department of Electrical, Electronic, and Information Engineering, City University, Northampton Square, London, EC1V 0HB, UK.

Abstract. Wavelength stability is an important factor in the potential accuracy of a laser diode based interferometric sensor. Two aspects of the use of optical feedback for wavelength stabilisation are discussed; the use of weak feedback for longitudinal mode stabilisation and the use of strong feedback to reduce the temperature dependence of the longitudinal mode wavelength. Experimental results are presented to demonstrate the latter effect, and confirm its value in optical measurement techniques.

1. Introduction

The effects of self-mixing interference in laser diodes [1] have recently been employed for various sensor applications including velocimetry and the measurement of displacement [2,3]. Self-mixing interference occurs when light emitted by a laser diode is reflected from an external surface (which, in practice, may be a mirror, a grating, or a scattering surface) and re-enters the laser cavity. Coupling between the cavity photon density and the injected charge carrier density results in a modulation of, in turn, the cavity refractive index, the threshold gain, and the intensity of the laser output. With the correct level of feedback a self-mixing interference-based system produces a modulated output with one fringe per half-wavelength of movement of the external reflector, similar to the output of a conventional interferometric system. Self-mixing interference- based systems, however, are of particular interest due to their potential advantages in terms of the size of the equipment, simplicity and cost.

2. Brief Description of Feedback

In order to illustrate both the effects and the use of such a feedback system, Fig. 1 shows the compound optical system formed by a laser diode with an external reflector. The laser diode has a cavity length, l, and facets with amplitude reflection coefficients r_1 and r_2. The external reflector has an amplitude reflection coefficient, r_3, and forms an external cavity of length, L.

The amount of light re-entering the cavity is characterised by the coupling coefficient, K, where

$$K = (1 - | r_2 |^2) r_3 / r_2 \ . \qquad (2.1)$$

The strength of the feedback in terms of its effects depends not only on the coupling strength but also on the length of the external cavity and is quantified by the feedback coefficient, C, where

$$C = K (1 + \alpha^2)^{1/2} L / nl \ , \qquad (2.2)$$

where n is the laser cavity refractive index and α is the linewidth enhancement factor and is a measure of the coupling between the real and imaginary parts of the refractive index [4].

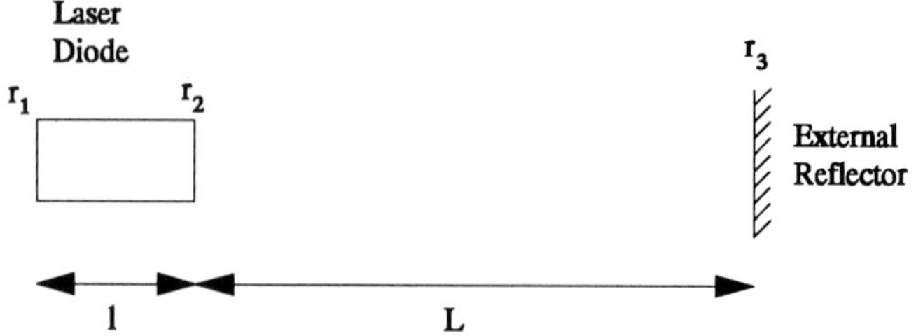

Fig. 1 Schematic representation of a laser diode with an external reflector.

3. Wavelength Stabilisation of Laser Diodes

There are a number of major considerations in the use of such systems in practical optical sensor devices. One important factor in the accuracy of a laser diode-based interferometric sensor is the stability of the wavelength of the light emitted by the laser diode, which typically varies with the laser diode drive current and with temperature. In practice the sensitivity to temperature is the more significant and that will considered here; the effect of the laser drive current, above threshold, is mainly due to its effect on the temperature of the device and is qualitatively similar to the effect of any other externally caused temperature change.

There are two major temperature-dependent factors which affect the wavelength: the gain peak wavelength shift of typically about 0.3 nm / K and the change of the refractive index and hence the optical length of the cavity, which shifts the longitudinal mode wavelength by about 0.07 nm / K [5]. The combined result of these effects is that there are discrete regions with a lower sensitivity in-between "hops" between longitudinal modes, as shown in Fig. 2.

Two possible methods of improving the wavelength-temperature characteristics by using either weak or strong feedback are discussed below, with the aim of offering a system of satisfactory stability for practical applications.

3.1 Weak Feedback

A technique already in commercial use (e.g. the Sharp LTO80MD laser diode [6]) uses an external cavity which is shorter than the laser cavity, in which case the feedback coefficient is low ($C < 1$) and the feedback can be described as *weak.* The periodic variation of threshold gain with feedback phase modulates the wavelength gain distribution and determines the position of the gain peak and therefore the longitudinal

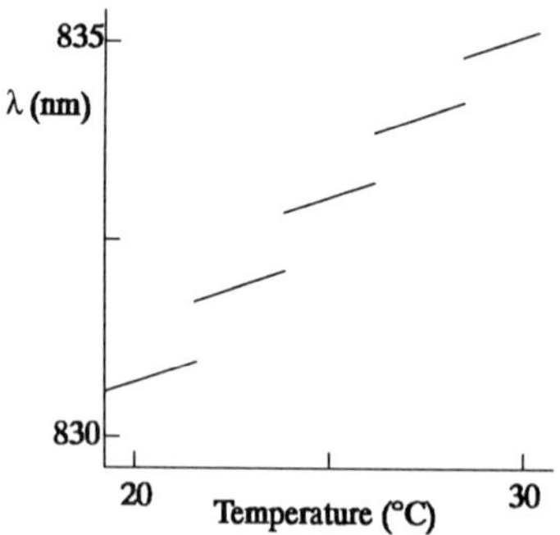

Fig. 2 Typical variation of laser diode wavelength, λ, with temperature.

mode in which the laser diode operates. If the variation of the feedback phase with temperature can be controlled so that the gain peak wavelength has the same temperature dependence as the longitudinal mode wavelength, then the wavelength interval between mode hops can be greatly increased.

Whilst this technique may eliminate potential problems caused by longitudinal mode hopping, it does not affect the temperature sensitivity of the wavelength within the longitudinal mode, and thus is limited in its applicability.

3.2 Strong Feedback

For very high accuracy applications, or for the use of a self-mixing interference system with a long external cavity, the wavelength stability criteria may need to be very strict. For example in a self-mixing interference system with an external cavity of length, L, and a laser diode wavelength, λ, the fringe number, n, is given by

$$n = 2L / \lambda \tag{3.1}$$

and the variation of n with wavelength is thus:

$$\partial n / \partial \lambda = - 2L / \lambda^2 . \tag{3.2}$$

For an external cavity length of 1m with a laser diode operating at 830 nm, the wavelength tolerance for single fringe accuracy is $\Delta\lambda < 3 \times 10^{-4}$ nm, and with a temperature sensitivity of 0.07 nm / K, the corresponding temperature tolerance required is $\Delta T < 4$ mK. A typical laser diode temperature controller using a thermistor and a Peltier element can achieve a long term stability of $\Delta T \approx 10$ mK [7], and so it is desirable to be able to reduce the temperature sensitivity within a single longitudinal mode.

Previously published theoretical studies of feedback in laser diodes [8,9] suggest that this is possible when the feedback coefficient is high ($C > 1$), which can be achieved by a combination of high coupling and the use of an external cavity longer than the diode cavity, producing *strong* feedback. Under these conditions, external cavity longitudinal modes exist whose spacing is less than the spacing of the intrinsic laser diode modes and whose temperature sensitivity is also less.

4. Experiment

4.1 Method

The wavelength stability characteristics of a laser diode have been measured with respect to drive current as this offers greater resolution and accuracy, but the qualitative features of the results may equally well be applied to temperature stability.

The experimental arrangement used to measure the wavelength change in a laser diode with feedback is shown in Fig. 3. Light was fed back into the laser diode, LD1, from a 1800 line / mm holographic diffraction grating, DG, which formed an external cavity of length, L. Part of the beam from LD1 was diverted by the beam-splitter, BS1 and mixed with the beam from a second laser diode, LD2, by the beam-splitter, BS2. The combined beam was analysed by the Michelson interferometer consisting of the beam-splitter, BS3, the mirrors, M1 and M2, and the photodiode, PD. The neutral density filters, ND1 and ND2, were used to minimise feedback from the interferometer into the two lasers. The difference in wavelength between the outputs of the two lasers was found by measuring the separation of the minima in the visibility pattern observed with the Michelson interferometer. Both lasers were temperature stabilised to about 10 mK and their drive currents were stable to about 0.02 mA. These were the main factors affecting the uncertainty in the wavelength difference measurements, which was estimated in this work to be $\pm 5 \times 10^{-4}$ nm.

4.2 Results

Fig. 4 shows the measured variation of wavelength difference, δλ, with the drive current of the laser diode with feedback. For this measurement, the optical length of the external cavity was 195 ± 5 mm.

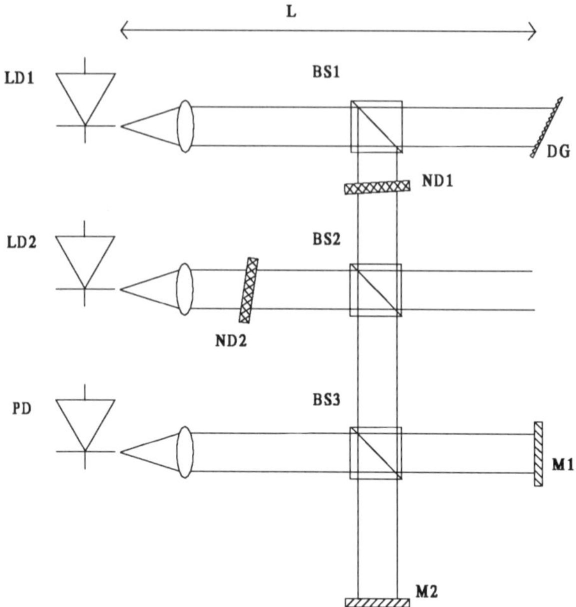

Fig. 3 Experimental arrangement; LD1, LD2: Laser diodes; DG: Diffraction grating; PD: Photodiode; BS1, BS2, BS3: Beam-splitters; ND1, ND2: Neutral density filters; M1, M2: Mirrors.

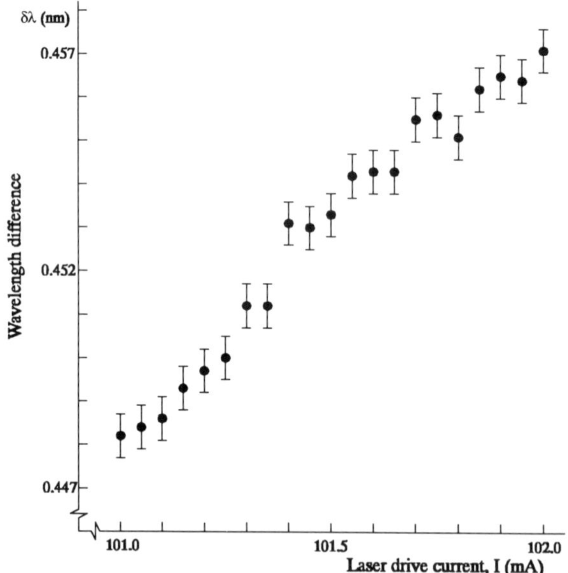

Fig. 4 Measured variation of wavelength with drive current for a laser diode with strong optical feedback.

The average gradient of the graph of wavelength difference versus drive current is 8.9 ± 0.6 nm / mA but it can be seen that there are groups of two or three points in which the gradient is significantly less than the large-scale average and between which there are sudden jumps, corresponding to hops between external cavity longitudinal modes.

The mean wavelength separation of adjacent groups is $1.3 \pm 0.2 \times 10^{-3}$ nm, which is in reasonable agreement with the calculated external cavity mode spacing of $1.75 \pm 0.05 \times 10^{-3}$ nm.

5. Conclusion

Two different aspects of the use of optical feedback for wavelength stabilisation have been discussed. The first, using weak feedback to reduce the occurrence of longitudinal mode hops, is already in current commercial use. This work has emphasised an approach which builds upon this to enable the use of a laser diode-based system in optical measurement applications by developing a method of reducing the temperature dependence of the wavelength within a given longitudinal mode.

Further work is required, to achieve a greater experimental resolution, in order to determine the practical limitations of this technique in optical measurement situations. Specific areas for future investigation include the process of transition between the external cavity modes and the possibility of bistability around the region of transition.

6. Acknowledgements

The authors are pleased to acknowledge the support of the Engineering and Physical Sciences Research Council in this work.

7. References

1. W M Wang, W J O Boyle, K T V Grattan, and A W Palmer, "Self-mixing interference in a diode laser: experimental observations and theoretical analysis", *Appl. Opt.*, **32**, 1551-1558 (1993).
2. M H Koelink, M Slot, F F M de Mul, J Greve, R Graaff, A C M Dassel, and J G Aarnoudse, "Laser doppler velocimeter based on the self-mixing effect in a fibre-coupled semiconductor laser: theory", *Appl. Opt.*, **31**, 3401-3408 (1992).
3. S Donati, Guido Giuliani, and S Merlo, "Laser diode feedback interferometer for measurement of displacement without ambiguity", *IEEE J. Quantum Electron.*, **QE-31**, 113-119 (1995).
4. K Petermann, *Laser Diode Modulation and Noise*, Kluwer Academic Publishers, Dordrecht, The Netherlands (1991).
5. C E Wieman and L Holberg, "Using diode lasers for atomic physics", *Rev. Sci. Instrum.*, **62**, 1-20 (1991).
6. *Laser Diode Users Manual,* Sharp Corporation, Japan (1988).
7. P Saunders and D M Kane , "A driver for stable frequency operation of laser diodes", *Rev. Sci. Instrum.*, **63**, 2141-2145 (1992).
8. S Saito, O Nilsson, and Y Yamamoto, "Oscillation centre frequency tuning, quantum FM noise, and direct frequency modulation characteristics in external grating loaded semiconductor lasers", *IEEE J. Quantum Electron.*, **QE-18**, 961-970 (1986).
9. G P Agrawal, "Longitudinal mode stabilisation in semiconductor lasers with wavelength selective feedback", *J. Appl. Phys.*, **59**, 3958-3961 (1986).

Two dimensional ranging using multidetection

S L Hill, K M Booth*, B Bury and J O Gray

Telford Institute of Cybernetics, Dept. Electronic and Electrical Engineering, University of Salford, Salford, Manchester M5 4WT, UK

* Science Institute, Dept. of Physics, University of Salford, Salford, Manchester M5 4WT, UK

Abstract. The development of a new, robust and compact system capable of producing two dimensional range information of a target scene for use in short range robotic applications is described. The system is based upon the high speed gating capability of laser diode illumination, and a micro-channel based image intensified CCD camera. The technique differs significantly from conventional time of flight range gated systems, since the range information is determined purely from a measurement of the relative phase and intensity of a reflected light pulse. The range and resolution of the system are currently limited to 2m and 5cm respectively due to dynamic range limitations. However a number of improvements are being implemented which should improve these figures.

1. Introduction

The determination of the range to an object is a desirable sensory attribute for many applications in robotics. Knowledge of object and barrier positions in the physical environment allow a robot to construct a model of the world and plan navigation and manipulation tasks accordingly[1]. Indeed to this end many successful ranging systems have been developed based on both acoustic and optical methodologies. However as the tasks performed by these autonomous devices increase in complexity so the need for more extensive range data about the environment in which they operate is required. This has led to the development of rangefinding sensors able to construct a range image of the surroundings providing sufficient information for complex and detailed task planning.

Optically based two dimensional ranging systems normally employ some form of scanning laser to acquire the range image. Various methods are used to determine the range of the target object from the sensor:Time of flight [2], Amplitude modulation [3], Frequency modulation [4], and triangulation [5] being the most common methods used. However, the beam scanning systems are often slow and bulky mechanical mechanisms utilising reflection techniques or electro/acousto-optic devices which have their own set of problems[6]. With present day small high speed robots there is a clear and definite need for a light weight, compact and robust high speed two dimensional ranging sensor.

One approach which avoids the use of a beam scanning system and also reduce the acquisition time of the sensor system is to use multiple detectors to view the entire image area simultaneously. In the case of many sensing strategies it would be prohibitively expensive to provide a large number of individual sensors [7]. However, optical imaging sensors are available in arrays which are able to detect the light intensity at each element of the array and output this signal as a video data stream (CCD's). Hence the problem of producing the desired object range sensor is now reduced to developing a methodology of deriving range information from the intensity and timing of the reflected light signal. This is achieved through a measurement of the phase shift induced into a target reflected light pulse relative to that of the initial light pulse.

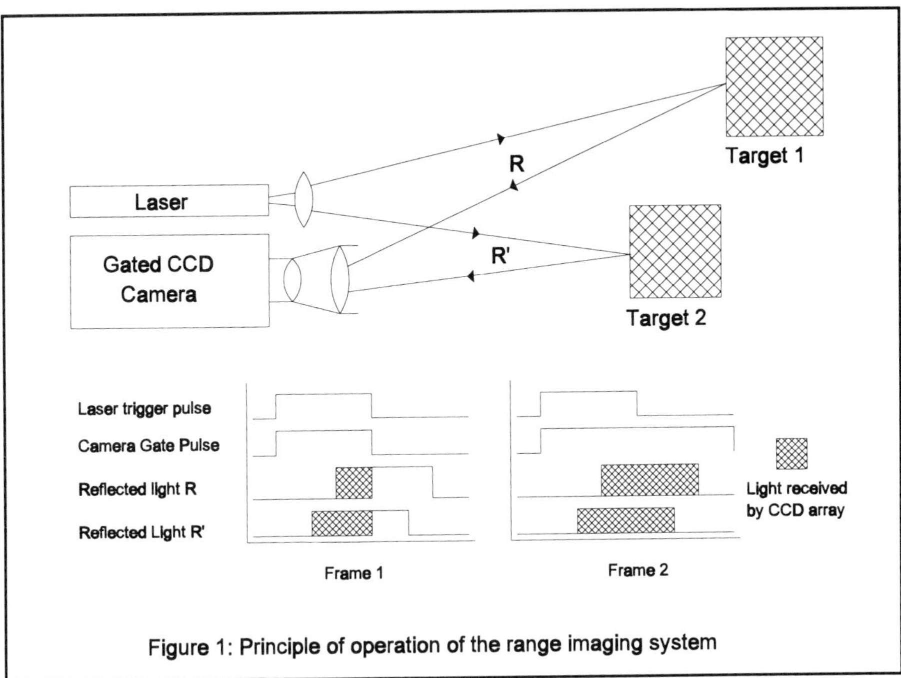

Figure 1: Principle of operation of the range imaging system

2. Ranging technique

The basic principle operation on which this two dimensional multidetecting range finding sensor system is based requires the capture of two distinct image frames. The first frame is captured withthe application of range gating to the image so that the total illumination received by each pixel of the camera is proportional to the distance to the object that the pixel is imaging. This basic principle of range gating is illustrated on the pulse timing diagram for frame 1 in figure 1. However it is not possible to obtain the range directly from this gated image unless the environment is extremely structured. Hence a secondframe is then acquired without the use of camera gating as illustrated in the pulse timing diagram forframe 2 in figure 1. The information in this frame is used to normalise this measurement taking into account the reflective properties of the target objects in the image scene, non-linearities in illumination, detector non-linearities, etc.. i.e. all of the potential unknown variables in the environment.

To be more specific with reference to the timing diagram for frame 1 of the range gated image in figure 1; it is apparent that the light pulse reflected from an object and arriving at the camera will be time shifted with respect to both the illumination pulse and camera gate. This shift is proportional to the laser-object-camera path length. The charge accumulated on an individual CCD pixel is therefore also proportional to its individual laser-object-pixel path length. However, the amplitude ofthe signal from this CCD pixel is also dependent upon the reflectivity of the target object and a number of other factors mentioned above. It is therefore necessary to correct for these. With reference to the timing diagram for frame 2 in figure 1; it is apparent that for the second image frame the entire reflected light pulse is captured by each individual CCD pixel. Hence in this case, the intensity of the captured light is wholly dependent upon these factors and is independent of the target range. The ratio of the intensity in these two image frames is therefore proportional to the round trip path from the laser to the target object and back to the camera. Clearly this principal is easily extendable to any number of pixel elements where each object point in an image scene is brought to focus at a unique and specific CCD pixel. Twodimensional ranging information can therefore be obtained by suitable digitisation and mathematical treatment of the pixels in each of the two image frames.

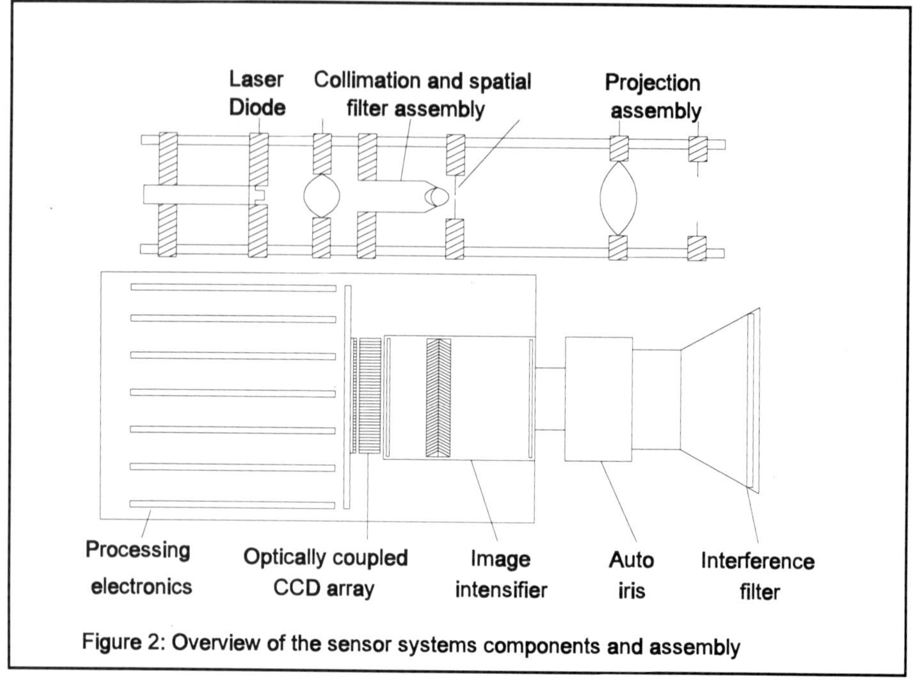

Figure 2: Overview of the sensor systems components and assembly

3. Experimental setup

An overview of the sensor system that has been constructed to produce the two dimensional range image based on the above principle is shown in figure 2. A suitable light pulse is generated using a 40mW laser diode operating at 780nm. A spatial filter and projection lens is used to provide uniform illumination over the target area. Light reflected from the target area is recorded by a camera system which basically consists of an gated image intensifier tube fibre optically coupled to a CCD array. An image intensifier was chosen as the gating element in order to allow flexibility in the detection of low light levels created by the safety restrictions of using laser light in open environments. The camera uses an auto iris zoom lens to image the target area. Overall scene intensity can be controlled by this iris with the iris closing down if saturation of the CCD array is detected. This also serves to protect the delicate photocathode of the image intensifier tube from excessive light. An interference filter with a bandpass FWHM of 10nm centred on the laser line wavelength of 780nm is also mounted on this lens eliminating ambient light at wavelengths other than the laser line. Range gating of the returning light pulse is achieved by using the image intensifier as a light valve. In normal operation the image intensifier has a negative potential applied to its photocathode which causes electrons generated from the photocathode to be accelerated towards the microchannel plates where amplification takes place to produce the intensified image. If a positive potential is applied to the photocathode electrons are no longer repelled and intensifier tube effectively acts as a shutter to light. This gating facility is operated using switching voltages of +/-15V in order to minimise the switching speed delays. Although this is less than the recommended switching voltages of around +/- 100V it still produces a shutter ratio of $1:10^5$ and has been shown to have minimal effect on the focusing properties of the tube [8].

A central pulse generator provides a pulse train with pulses of suitable duration depending on the maximum range required (10ns for every 1.5m of range required) and with a suitable gap between pulses to prevent the camera collecting light from a previous pulsereflecting off a distant target (a mark:space ratio of 1:1 or greater). These pulses are routed to both the laser diode and the photocathode gate of the image intensifier tube via adjustable delay lines to allow for any timing errors introduced by the system setup. The CCD array collects the returning gated light pulses for the duration of theCCD frame averaging the results of several thousand pulses in one frame. Once this first frame has been digitised and stored the accumulation of the second frame is initiated. In this case and identical pulse train is routed to the laser producing a light output identical to the first frame. However the pulses sent to the intensifier gate are now of sufficient duration to capture the entire returning light pulse regardless of the target range. Once again the pulses are averaged over the duration of the CCD frame, digitised and stored. Once both frames have been collected a computer hosted image processing board takes the ratio of the two frames to produce the data from which the ranges canbe determined. Images collected using the complete camera system are digitised and two frames of data processed on a computer hosted image processing board to yield the final range image.

4. Theory

In the following theory, the system is assumed to behave ideally so that the illumination light pulse and camera gate may be considered to be perfectly square in the time domain. Furthermore, the non linear characteristics of the image intensifiers screen phosphor are neglected, and ambient light is assumed to be blockd from the system. The light intensity captured by the gated camera in the first image frame is given by.

$$L_1 = \frac{kRI_0T}{r^2} \tag{1}$$

where k is a constant taking into account the efficiency of the light collecting optics and the quantum efficiency of the system, R is the target objects reflectivity coefficient, I_0 is the illuminating light intensity (which is also function of r), r is the object-camera distance, and T is the effective gate width given by:

$$T = T_0 - \delta t \tag{2}$$

where T_0 is the width of the illuminating light pulse, and δt is the time delay introduced into the reflected light pulse by the object-camera separation, such that $\delta t = 2r/c$, where c is the speed of light (refer to figure 2). The light intensity captured by the gated camera in the second image frame is given by:

$$L_2 = \frac{kRI_0T_0}{r^2} \tag{3}$$

where all the terms are as previously defined. Hence the ratio of the two image frames is:

$$\frac{L_1}{L_2} = \frac{T_0 - \delta t}{T_0} \tag{4}$$

which after substitution for δt, may be rearranged to give the object range as:

$$r = \frac{cT_0(L_2 - L_1)}{2L_2} \tag{5}$$

Hence the pixel by pixel object to camera distance may be obtained from the measurement of the intensity in the two image frames. Furthermore, the object range is directly proportional to the incident pulse width.

However, equation (5) strictly represents an over simplification of the situation since it is also necessary to consider the response characteristics of the various system elements. These principaly include the rise and decay characteristics of the laser diode illumination, the photocathode gate, and the output of the image intensifiers phosphor screen. The response envelope for the laser diode and photocathode havean equal effect on both image frames, and can be taken in to the constant k in equations (1) and (3). The most significant source of error is the time response of the phosphor screen. The rise and decay characteristics of the phosphor are inherently non linear. The P20 phosphor used forthe output screen of the image intensifier, has a characteristic rise time, but has a decay time that is highly dependent on the width of the input excitation pulse [9-11]. However it can be shown that these nonlinearities of the phosphor cancel out.

5. Errors And Limitations

At present the major limitation on range depth resolution is due to the limited dynamic range of the image intensifier tube. The camera is limited by the addition of the front end image intensifier to a maximum signal to noise ratio of approximately 32dB [EEV] at 25°C. This is only sufficient to give a maximum frame digitisation is 40 grey levels, hence a resolution of about 15cm with a single 40nS pulse covering an approximate 10m range. A number of methodologies to increase this dynamic range are being implemented.

The first method of increasing the dynamic range is to directly improve the signal to noise ratio of theintensifier tube by cooling. A Peltier driven cooling unit is being constructed to surround and cool the intensifier tube. This will have the effect of doubling the signal to noise ratio for every 8°C of temperature drop. Hence using this method it should be possible to increase the signal to noise ratio by a factor of four with a cooling unitthat can achieve a 16°C temperature drop. Insulation restrictions and problems with condensation and frosting would prohibit the use of a greater temperature drop in the current camera system.

Another method that would help reduce the dynamic range of the target scene being viewed when the camera is operated in gating mode would be to adjust the relative timings ofthe laser pulse and the camera gate pulse. At present the amount of light getting through the gate is proportionally less as the target distance increases. However a distant target also reflects proportionably less light back to the camera compounding thedynamic range problem. If the timing of the laser and camera gate pulses is adjusted to be out of phase then proportionally more light will get through the gate from targets at greater distances. Hencethe dynamic range of the scene as received by the camera is effectively compressed from both ends of the scale by using this technique. The range may still be determined simply by interpreting the ratiometric division of the two scenes from the opposite end of the range scale.

Additional dynamic range improvement may be achieved by changing the intensity of illumination through the duration of the laser pulse by ramping the laser current. This method is not particularly desirable since it is not easy to implement and will introduce other non-linearities that will introduce complexities into the range calculation. This method will therefore only be used as a last resort if other methods of reducing the dynamic range fail.

The image intensifier tube also presents an additionalproblem due to some non-linearity in gain of the intensifier as a function of scene brightness [12]. Since portions of the scene areinherently less bright when the camera is gated this will have the effect ofproducing a range error which is a function of the objects reflectivity. However this function is easily determined and should be easy to remove with some additional processing of the image.

A further source of error not so far discussed, arises due to ambient light entering into the system and therefore into the captured images. To included this effect, the captured light intensity can be considered to be the sum of a signal and ambient component, hence $L_n = S_n + A_n$. The ambient light will have an equal spatial distribution in both image frames, but different levels of intensity as the camera gate is open for different time intervals. Consequently, equation (5) which describes the objects range becomes,

$$r = \frac{cT_0(S_2+A_2-S_1+A_1)}{2(S_2+A_2)} \qquad \textbf{(6)}$$

where $A_1 = K.A_2$, and K is a constant. The inclusion of an ambient light level in the captured image frames will therefore introduce a non linear error into the measurement of the objects range. This would also require the system to be calibrated and then recalibrated for different levels of ambient light. An interference filter can eliminate much of the ambient light. Increasing the laser power in the pulse and reducing the camera aperture or using a neutral density filter will also reducethe relative amount of ambient light entering the system, however this would require the pulse mark:space ratio to be increased to meet safety criteria. For its complete elimination, the measured ambient light level must be subtracted from each image frame during image processing. This will however further reduce the overall dynamic range and hence resolution of the system.

A final limitation of this rangefinding system is its ability tohandle fast moving objects. Any object that moves any significant distance between the acquisition of the two frames of data required to produce the range image will produce false results. However it is expect that these results will produce distinct features in the final range image which could be detected and eliminated by the application of suitable image processing techniques. This is however beyond the scope of this project at the current time.

6. Image Ranging Results

Due to the problems with the dynamic range of the system the sensor system is currently only capable of deteting an objects range between 1m and 2m to an accuracy of 5cm. The objects being ranged must also meet certain reflectivity criteria to prevent saturation of the system.

7. Conclusions

A new compact and robust solid state multidetecting range imaging sensor system has been designed and constructed. This system is able to acquire the data required to produce a range image in two video frames taking less than 100ms. The number of range image points is determined by the size of the CCD array and is currently over 250,000. The maximum range of the system is in theory 10m butdue to dynamic range limitations is only operable to 2m. The range resolution is at present only 5cm. Work is proceeding to correct the dynamic range problems and other problems affecting the range resolution. A range resolution of 1cm could be ultimately expected if these problems can be completely rectified. If achieved this would make this sensor system an ideal unit for use in mobile robots or other applications with similar near range environmental range sensing requirements

References

[1] Kak A C (1986) Depth Perception for Robotic Vision *Handbook of Industrial Robotics* (Ed. S Nof.) Wiley, New York p272-319

[2] Heikkinen T Ahola R Manninen M Mylltia R (1986) Recent Results of the Performance Analysis of a 3D Sensor Based on Time of Flight *Proc.*

[3] Sampson R E (1987) 3D Range Sensor Via Phase Shift Detection *IEEE Computer* **20** p23-24

[4] Beheim G Fritsch K (1987) Range finding using a frequency modulated Laser Diode *Appl. Opt.* **25,9** p1439-1442

[5] Faugeras O D Hebert M (1986) The representation, recognition and locating of 3D Objects *Intern. J. Robotics Res.* **5,3** p27-52

[6] Karim M A (1992) Electro-optical Devices and Systems *PSW-KENT,* Boston ISBN 0-534-91630-9

[7] Cathey W T Davis W C (1987) Imaging System with Range to Each Pixel *J. Opt. Soc. Amer A..***3,9** p1537-1542

[8] Hirsch K Kochel M Salzmann H (1987) Shutter Ratio of a Gated ITT F4128 Microchannel Plate Photomultiplier *Rev.Sci.Instr.* **58** p2339

[9] Flynt W E. 1990 Phosphor Behavior in High Speed Gated Image Tubes *SPIE.* **1243** 96

[10] Flynt W E 1989 Characterisation of Some Common CRT Phosphors *SPIE.* **1155**

[11] Majumdar Serdyuchenko Y Platonov V Phosphor Non Linerities in Picosecond Streak Cameras *SPIE.* **1155** 375

[12] Image Intensifiers *Hamamatsu Product Guide and Technical Information*

AN OPEN LOOP FIBRE OPTIC GYROSCOPE FOR ROBOT NAVIGATION SYSTEMS

Brian Bury and Julian C. Hope

Dept. Electronic & Electrical Engineering
University of Salford
Salford Lancs. M5 4WT. UK.

Abstract: Automated manufacturing systems must increasingly be seen as a 'society' of manufacturing 'experts' and 'craftsmen' where co-operative task achievement is the key to success. Research at the University of Salford involves the design of relatively simple multiple mobile robots that are able to communicate and co-operate with each other to complete a given task. Part of the experimental programme includes the development of cost effective sensory systems for object detection, obstacle avoidance and global navigation. This paper will describe the design and development of an open loop fibre optic gyroscope for autonomous mobile robot navigation.

1. BACKGROUND

Five years ago, a major new research initiative in the area of co-operant mobile robots was established by members of the Control and Instrumentation Group at the University of Salford (Barnes, 1989a). This work was funded partly by the EPSRC ACME Directorate and Industrial collaboration and its aim was to investigate the use of co-operant multi-agent mobile robots for advanced manufacturing and material handling applications (Barnes, 1989b; Barnes, et al., 1991). Part of this project was the development of cost effective sensory systems for obstacle avoidance and global navigation.

The following sections will describe the design and development of a low-cost Fibre Optic Gyroscope (FOG) and the integration of this with a Digital Compass to produce a global navigation system for an autonomous mobile robot agent.

2. GYROSCOPE OPERATING PRINCIPLES

The fundamental principle of operation of a FOG is based on the well known Sagnac effect (Sagnac, 1913). Figure 1 shows a simplified block diagram of an open loop gyroscope. This basically consists of a single light source, a detector, piezo-electric modulator, 3dB couplers and a long length of single-mode fibre with which to create an optical path for the light. As light from the source passes through coupler 2 it is divided into two parts. Each part, travelling in opposite directions around the fibre loop, describes the same optical path. The returning light waves recombine and are partially coupled into the detector at coupler 1. While the fibre loop is stationary the phase delays acquired by each beam of light match exactly, producing a maximum output (due to constructive interference) at the detector. However, if the fibre loop is rotated, one of the paths of light appears longer with respect to the other path of light and (because the speed of light is constant) a net phase difference results. This phase difference, $2\phi_s$ is proportional to the rate of rotation, Ω of the fibre loop as described by Equation 1.

$$2\phi_s = \frac{4\pi N R^2 \omega}{c^2}\Omega \tag{1}$$

where **R** is the loop radius, ω is the source frequency, **N** is the number of turns and **c** is the speed of light.
The purpose of the phase modulator in the loop is to provide a nonreciprocal phase shift to bias the gyroscope to it's maximum sensitivity point. Since the normal output of the interferometer is cosinusoidal, with it's maximum at 0 and 2π, the gyroscope has to be biased by π/2. This is achieved by asymmetrically positioning the phase modulator toward one end of the fibre loop. The detected signal is then demodulated using a lock-in analyzer. This produces the characteristic sinusoidal output described by Equation 2.

$$V = E_1 \sin(O_1\Omega + O_2) + E_2 \tag{2}$$

where E_1=Electrical scale factor, E_2=electrical offset
O_1=Optical scale factor, O_2=Optical offset.

Further signal processing electronics are necessary to produce a linear relationship between rotation rate, Ω and output voltage, V. This may then be integrated to obtain angular displacement, Θ.

A number of FOG systems have already been reported (Bergh, et al., 1981; Lefevre, et al., 1982), but these have mainly been high accuracy devices (with sensitivities in the region of 0.01 deg/hr) for use in the aerospace and sub-sea industries. As a result, these were considered to be too expensive (typically tens of thousands of pounds) for the multi-agent robot system outlined above. Therefore, there was a need to investigate and subsequently develop a low-cost FOG for this application.

3. GYROSCOPE DEVELOPMENT

To reduce the cost of the gyroscope, standard commercially available components have been used throughout the design. This has been achieved using a 1300nm LED source, PIN diode detector and 1300nm single-mode optical fibre. These components are traditionally used in optical communication systems. Standard 3dB optical couplers have also been used in preference to the higher priced integrated optic couplers. The design does not utilize an optical polariser, which also gives a considerable cost saving. The choice of standard components, however, does reduce the sensitivity and increases the output drift of the gyroscope, but these are within acceptable limits for the chosen application area. The following parameters have been used in the design.

Loop Radius, R:	0.17m
No. of turns, N:	300
Fibre Cladding/Core Dia:	125/9μm
Source Wavelength, :	1293.6nm
Source Power, P_o:	4.9μW
Source 3dB Bandwidth, δB:	59.75nm
Detector Responsivity, r:	0.77μA/W
Insertion Loss, I_L:	15dB

The phase modulator was implemented using a piezo-electric cylinder which provided phase modulation of 1.91 radians at a frequency of 50KHz. This frequency was derived from a crystal oscillator operating at 4MHz. A constant current supply was designed to stabilize the output power of the LED source which was also maintained at a fixed temperature of 26°C +/-0.025°C. Signal demodulation and processing electronics were performed using commercially available instrumentation packages.

3.1 Analysis/development software

In order to aid the design and development of open loop fibre optic gyroscopes a computer programme has been written by the authors. This provides a database to store the choosen design parameters and a variety of utilities to determine the output response and characteristics of the sensor. The programme consists of three sections: Data storage and retrieval, Analysis & development and Simulation as shown in Figure 2. Features available include: Output characteristics, Temp/drift measurement, Minimum rotation rate measurement, Temperature stability, Effects of optical feedback, Degree of coherence etc. This programme has proven indispensible in evaluating and developing the gyroscope for this application. It is believed that it will also be a useful tool in the development of other open loop gyroscopes for future projects. Further details of this software package are available from the authors.

3.2 Results

The gyroscope system has been extensively bench tested and the following results have been obtained. The output characteristic of the open loop FOG is described in Figure 3. This diagram specifies the electrical scale factor, E_1, 7116.9mV, optical scale factor, O_1, 1.652 and electrical offset, E_2, 72.6mV. The optical offset, O_2, was negligible. The maximum detectable rotation rate, Ω_{max}, which corresponds to a net phase shift of 90 degrees, is equal to 54.5 degs/sec. Figure 4 shows the sensitivity of the system at low rotation rates. (Note: the oscillatory behaviour of the graph at +/- 78°/ was due to a small non-uniform motion in the rotational system used.) The resultant sensitivity of the gyroscope was 1.5°/hr. The output drift of the gyroscope is shown in Figure 5. A large percentage of this was produced by temperature variations in the fibre coil and external lock-in analyzer. These factors will be reduced in the near future resulting in a drift of better than 10°/hr.

4. ROBOT NAVIGATION

To date, a number of navigation systems have been implemented on single autonomous mobile robots (Thompson, 1979 Harmon, 1987; Giralt, et al., 1977), some of which have included inertial sensors (Kuritsky and Goldstein, 1983). The FOG described above, however, has been developed for use in a hierarchical navigation system for co-operating multi-agent mobile robots (Bury, et al., 1992; Hope, 1992). This system provides for a number of navigational levels which include: Global, Local, Object and Docking.

In the early days of the mobile robotics project at Salford, the global navigational level was controlled by inputs from a single flux gate (FGM) digital compass (Azimuth, 1990) sensor. However, because the FGM was sometimes influenced by large metal objects and underground power lines within the workspace, their was a need to include an inertial sensor for additional control. Hence, it was for this reason that the FOG described above was developed. The Global navigation level has subsequently been re-implemented using multi-sensor inputs consisting of the FOG , Digital compass and a rotational optical encoder as shown in Figure 6.

4.1 Multi-sensor analysis/fusion

The global navigation system provides sub-procedures for the calibration of the output drift of the FOG and integration of the rotation rate to obtain angular displacement. This is fused (Luo and Kay, 1989) with the output from the digital compass and optical encoder using an adaptive Kalman filter (Maybeck, 1972) to provide positional control of the robot. The filter also allows compensation for electromagnetic interference and positional errors due to wheel slippage. The effect of these anomalies has been simulated using a software package designed by the authors and the results used to develop adaptive control algorithms for the filter.

This multi-sensor global navigation system is presently being evaluated using a Cybermation K2A mobile robot (Cybermation, 1988) with the eventual aim of installing it on the departments co-operating mobile robots 'Fred' and 'Ginger' (Barnes, et al., 1991).

Although this application has been directed towards the control of autonomous mobile robot agents, the basic gyroscope configuration described in this paper could also be used in manipulator robots where inertial control was also required (Weisbin, et al., 1990; Janocha and Schmidt 1990). Current research at Salford is presently concentrating on the development of a 3-axis fibre gyroscope system for station position stabilisation of remote observation vehicles (ROVs) and end-effector stabilisation of very long manipulators in marine and nuclear applications.

ACKNOWLEDGEMENT

The authors would like to express their appreciation for the support of this work by the Engineering and Physical Sciences Research Council (EPSRC/ACME), ref: GR/F 71454, and the Department of Electronic and Electrical Engineering, University of Salford.

REFERENCES

Azimuth., (1990). 314 Digital Compass, *KVH Industries, Inc. 850 Aquidneck Avenue, Middletown, Rhode Island, 02840 U.S.A.*

Barnes D.P., Bury B., Gray, J,O., (1989a). 'Co-operant Mobile Robots' in 'Robotics and Manufacturing: *Recent Trends in Research, Education and Applications'*, **Vol.3**, Eds. Jamshidi, M and Saif, M, ASME Press, USA, pp. 899-904.

Barnes D.P., Bury B., Gray J.O., (1989b). 'Co-operant Mobile Automata for Advanced Manufacturing and Material Handling Applications', *SERC ref: GR/F 71454.*

Barnes D.P, Bury B., Gray J.O., Hill S.L., and Eustace D., (1991). 'Research Platforms for Investigating Mobile Robot Co-operancy', *Int. Conference on Advanced Robotics, Pisa, Italy, June.*

Bergh R.A., Lefevre H.C., Shaw H.J., (1981). *'All Single-mode Fibre Optic Gyroscope with Long Term Stability', Opt. Lett.*, **Vol 6**, pp. 502-504.

Bury B., Hill S.L., Hope J.C., (1992). 'Hierarchical Navigation System for Autonomous Mobile Robots', *Inst M.C. Symposium on Sensory Systems for Robot Control,* London, Oct.

Cybermation., (1988). K2A Mobile Robot, *Cybermation, 5357 Aerospace Rd, Roanoak, Virginia 24014, U.S.A.*

Giralt G., Sobek R., Chatila R., (1977). 'A Multi-Level Planning and Navigation System for a Mobile Robot: A First Approach to HILARE', *Proc. IJCAI-6.*

Harmon S.Y., (1987). 'The Ground Surveillance Robot (GSR): An Autonomous Vehicle Designed to Transit Unknown Terrain', *IEEE Journal of Robotics and Automation,* **RA-3,** 3, pp. 266-279.

Hope J.C., (1992). 'Global Navigation for Autonomous Mobile Robots', *MSc Thesis, Dept of Electronic & Electrical Engineering, University of Salford.*

Janocha H and Schmidt D., (1990). 'Requirements for Inertial Sensor Systems for Measuring Robot Positions', *Robotica,* **Vol 8**, pp. 145-150.

Kuritsky M.N and Goldstein M.S., (1983). 'Inertial Navigation', *Proc. of the IEEE,* 71, No 10, pp. 1156-1174.

Lefevre H.C., Bergh R.A., Shaw H.J., (1982). 'All Fibre Gyroscope with Inertial Navigation Short Term Sensitivity', *Opt. Lett.,* **Vol 7**, pp. 454-456.

Luo R.C and Kay M.G., (1989). 'Multisensor Integration and Fusion in Intelligent Systems', *IEEE Trans. Systems, Man and Cybernetics,* **Vol.19**, No 5, Sep/Oct, pp. 901-931.

Maybeck P.S., (1972). TM-72-3, *Air Force Flight Dynamics Laboratory, Write-Patterson AFB, Ohio.*

Sagnac G., (1913). 'L'ether lumineux demontre par l'effet du vent relatif d'ether dans un interferometre en rotation uniforme', *C.R. Acad. Sci,* **Vol 95**, pp. 708-710.

Thompson A.M., (1979). 'The Navigation System of the JPL Robot', *Proc. of 5th Int. Conf. AI, Tokyo,* pp 335-337.

Weisbin C.R., etal., (1990). 'HERMIES-III: A step toward autonomous mobility, manipulation and perception', *Robotica,* **Vol 8**, pp 7-12.

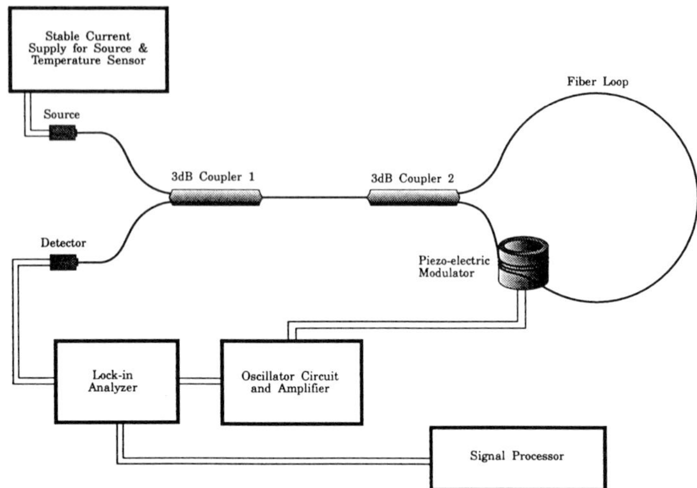

Figure 1: Simplified block diagram of a Fibre Optic Gyroscope.

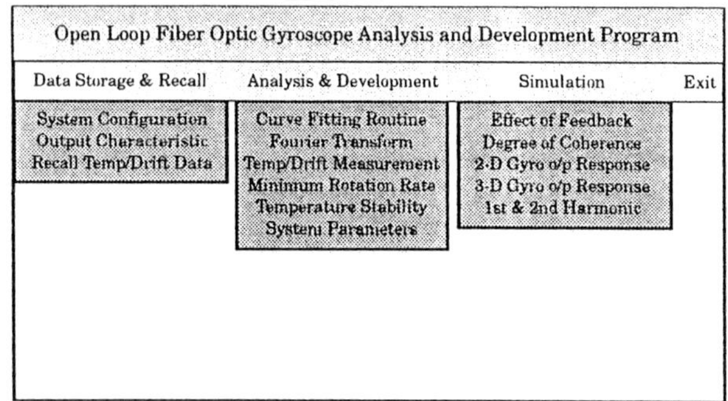

Figure 2: Gyroscope Analysis/Development Programme.

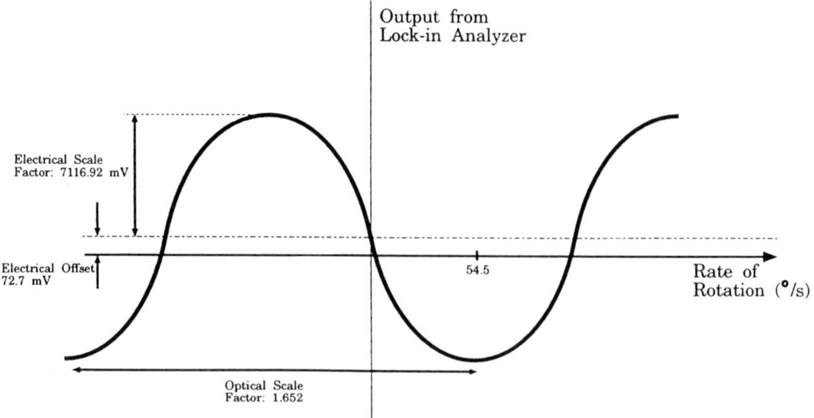

Figure 3: Output characteristic of fibre optic gyroscope.

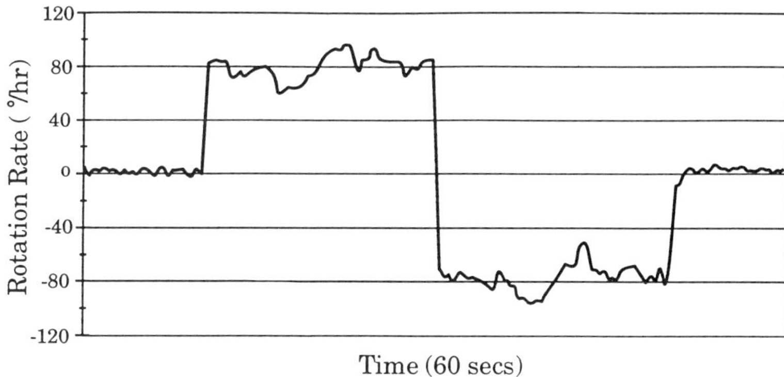

Figure 4: Short term sensitivity.

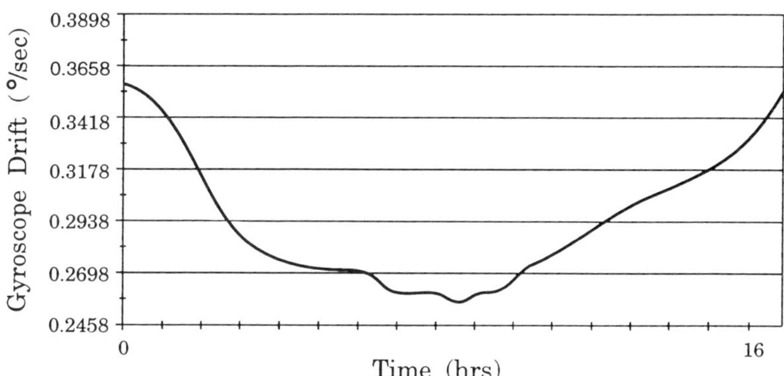

Figure 5: Long term drift.

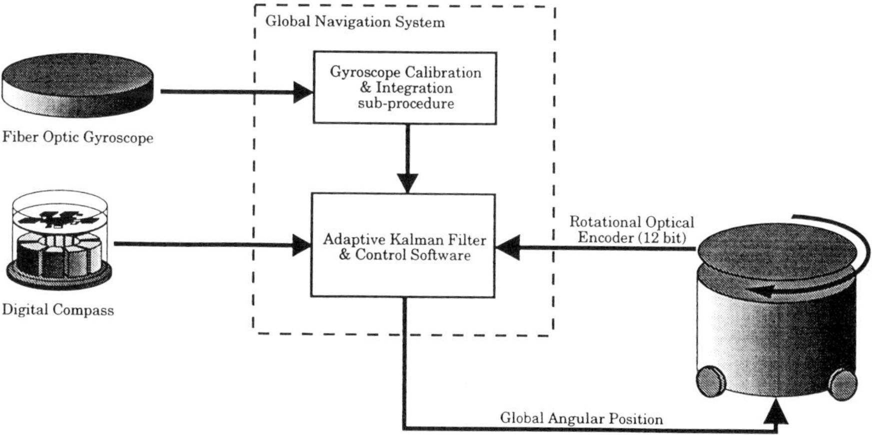

Figure 6: Multi-Sensor Global Navigation System.

Fibre optic Sagnac interferometer for non contact structural monitoring in power plant applications

T.A.Carolan, R.L.Reuben, J.S.Barton*, R.McBride* and J.D.C.Jones*

Department of Mechanical and Chemical Engineering, Heriot-Watt University, Riccarton, Edinburgh, EH14 4AS, United Kingdom.
*Department of Physics, Heriot-Watt University.

Abstract

This paper details the design of a non-contact fibre optic sensor for the detection of acoustic emission (AE) for structural integrity monitoring in power plant applications. The sensor is based on a Sagnac interferometer and produces an output proportional to target velocity, without the need for active phase stabilisation. It is inherently insensitive to low frequency perturbations of the instrument or the target and incorporates an environmentally insensitive downlead which may be of arbitrary length. It is shown that the sensor is capable of meeting the specifications for structural integrity monitoring of high temperature power plant components based on acoustic emission detection and has a velocity resolution of 50 $nms^{-1}Hz^{-\frac{1}{2}}$.

1. Introduction

On-line structural integrity monitoring has an essential role in power plant operations to avoid unplanned shutdowns, reduce maintenance costs and improve safety. Such technology is also important in other industries, such as chemical, oil and aerospace, where components are highly stressed and may operate at high temperatures.

Monitoring of metal-borne stress waves in the 0.1 to 1 MHz range, known as acoustic emission (AE), is considered a very promising method for long term on-line structural integrity monitoring, due to its potential for real time non-local detection and localisation of propagating flaws in stressed structures [1]. These AE monitoring techniques generally use contacting piezoceramic transducers which, when monitoring critical high temperature components present in a power plant (e.g. boiler membrane panels, steam headers and steam lines in fossil fuel stations which typically have operating temperatures in the range 350-650°C), require to be mounted onto metallic waveguides welded to the structure [2]. This disruption of the structure by welding waveguides is inevitably undesirable and a non-contact detection method would be preferred.

The objective of this paper is to demonstrate that a new type of optical fibre interferometer, based on the Sagnac configuration, is capable of non-contact detection of acoustic emission events. It is further shown that the performance of the interferometer can meet the specifications required for condition monitoring in power plant components.

2 Design considerations

Detection is required over the frequency range of approximately 0.1 to 1.0 MHz. The stress waves generate equivalent out-of-plane vibrations of the surface, and for sensitivity comparable with a contacting piezoelectric transducer it is necessary to be able to detect vibration amplitudes in the 0.1

nm range. In the power plant environment, high levels of ambient vibration generate out-of-plane motion with amplitudes of up to 1 mm, and at frequencies up to 10 kHz. High surface temperatures and an aggressive operating environment demand the use of a robust passive measurement probe, with the active parts of the instrument located remotely in a control room perhaps 100 m from the measurement volume.

Optical interferometry offers the necessary non-contact measurement resolution, with two basic design options: *displacement*-measuring and *velocity*-measuring interferometers. In displacement-measuring interferometers, the test-surface effectively forms the mirror in one arm of the interferometer. In velocity-measuring interferometers, an unbalanced interferometer is used as a frequency discriminator to measure directly the Doppler shift in light scattered from the test-surface. In both cases, fibre-optics offer the possibility of practical systems for use in difficult practical environments, with the active components located in the control room connected to the passive probe by a long fibre downlead.

A basic problem in the use of a displacement interferometer is that if the active components used for demodulating the phase output (by frequency shifting in heterodyne techniques, or phase control in homodyne techniques) are located in the control room, then the downlead forms part of the interferometer and suffers excessive environmentally-induced phase modulation; they cannot be located in the probe, which must remain passive. Phase control by wavelength modulation of the optical source demands a large pathlength imbalance in the interferometer, and leads to excessive source-induced phase noise, thus preventing adequate measurement resolution.

Velocity interferometers, such as the unbalanced Michelson [3] and Fabry-Pérot [4], are as sensitive to source frequency fluctuations as to the Doppler shift, and without sophisticated stabilisation (impractical in the present application) have inadequate insensitivity for our needs.

Our design is thus based on an adaptation of the fibre Sagnac interferometer. We have previously shown that by breaking the Sagnac loop near to one end, and introducing a bulk-optic probe, it is possible to derive an optical signal proportional to the velocity of a test surface illuminated by the probe, without sensitivity to the source wavelength [5].

In this paper, we describe a development of the technique, allowing the Sagnac interferometer to be used with a downlead of arbitrary length, separating the interferometer from its probe. We show that the downlead is bend insensitive and immune to environmentally-induced phase perturbations, and that the interferometer has the sensitivity required for the intended application.

3 Theory

The interferometer is shown in figure 1 and its basic mode of operation is as follows. Light from a laser source is coupled into a single-mode optical fibre and amplitude divided at a 50:50 directional coupler to follow clockwise and anti-clockwise paths in the Sagnac loop. The loop, of length typically 200 m, is interrupted near to one end and is interfaced via a polarising beamsplitter such that the two beams are coupled as the orthogonal polarisation eigenmodes in the highly birefringent fibre downlead. The downlead terminates in the probe, where the beams are conditioned by a lens to produce a beam waist on the test surface. Because the counter propagating beams traverse different optical path lengths from the source to the test-surface, they 'see' the test-surface at times separated by the loop delay, τ. Also contained in the probe optics is a quarter wave plate which causes the signals to return through the downlead in the orthogonal polarisation eigenmode to the out going beams and thus they continue to propagate around the loop until they reach the detectors where they interfere, giving an intensity

$$I_i = \tfrac{1}{2} I_0 (1 - (-1)^i V_i \cos\phi) \qquad (1)$$

$i=(1,2)$

where I_0 is a constant intensity, V_i is the fringe visibility and ϕ is the phase difference between the two beams which will be proportional to the change in position of the target during the loop delay, τ, and hence to the target velocity averaged over τ.

It can be shown [5] that the phase difference has a frequency response to the velocity of a target vibrating at a radian frequency ω and a peak velocity v is given by

$$\phi = 2kv\,\tau \mathrm{sinc}\left(\frac{\omega\tau}{2}\right)\cos\omega\left(t-\frac{\tau}{2}\right) \qquad (2)$$

where $k = 2\pi/\lambda$, where λ is the wavelength of the source, and t is time.

A roll-off in this response occurs at vibration frequencies around $1/2\tau$ with a null at $1/\tau$, thus determining a maximum loop delay for the required system bandwidth, indicating the use of a 200m fibre delay loop in our application.

As both beams travel around the same path around the loop the interferometer is intrinsically balanced and is thus wavelength insensitive

By controlling the birefringence of the fibre loop, via the fibre polarisation controller BC, it is possible to generate a passive phase offset between the clockwise and anti-clockwise beams [6]. To achieve this the polarisation states of the counter propagating beams travelling through the birefringence controller must be different to enable control of their relative phase. As the polarisation planes of the counter propagating beams are forced to be identical at the polarising beam splitter (PBS1) of the downlead interface, a non reciprocal Faraday rotator (FR) element is required between the downlead interface and the birefringence controller. To maximise sensitivity, the phase bias is set to $\pi/2$ radians, known as the quadrature condition, such that equation (1) becomes

$$I_i = \tfrac{1}{2}I_0(1-(-1)^i V_i \sin\phi) \qquad (3)$$

The input polarisation to the Sagnac loop is also controlled with a fibre polarisation controller (PC1) which, along with the birefringence controller, allows the polarisation states of the counter propagating beams to be aligned with the polarisation axes the polarising beam splitter (PBS1).

Bend insensitivity in the probe downlead in this interferometer is achieved by using highly birefringent (hi-bi) fibre for the downlead. Such fibre has a high degree of intrinsic birefringence due to its structure, producing two orthogonal polarisation modes in the fibre (which have distinct propagation constants), so that when light is launched into one of these modes its polarisation state will be maintained. Only the probe downlead is fabricated of hi-bi fibre as this is the only part of the system exposed to dynamic bending. Additional phase insensitivity in the downlead is achieved using a quarter waveplate at the probe which causes light reflected from the target to be reflected back down the other orthogonal polarisation mode of the downlead. As both counter propagating beams travel out to the probe and back to the loop in both of the polarisation modes only phase perturbations that are comparable to the loop delay (i.e. ~ 0.1-1 MHz frequency range for which the fibre downlead can be efficiently shielded) can produce any effect on the interferometer operation. This property also means that the fibre downlead may be of arbitrary length.

As the counter propagating beams leave the loop with orthogonal polarisation states, due to the Faraday rotator, beam recombination is achieved by setting the orthogonal beams at 45° to the transmission plane of a second polarising beam splitter (PBS2) using another fibre polarisation controller, PC2. This gives the interferometer two antiphase (I_1 and I_2) outputs which by subtraction at quadrature operation produce an interferometer output of

$$I = I_0 V \sin\phi \qquad (4)$$

for $V_1 = V_2 = V$ and $I_{01} = I_{02} = \tfrac{1}{2}I_0$

The subtraction of antiphase outputs is essential for compensation of intensity noise from both the source and the target.

Now for small signals around phase quadrature which the AE vibrations would produce, i.e. $\phi << 1$ radian, equation (4) becomes

$$I = I_0 V \phi \qquad (5)$$

So the interferometer produces an intensity which is linear with target velocity provided ϕ remains significantly less than 1 radian (see figure 2 for a plot of the required vibration amplitude against frequency to produce phase changes of 1 radian).

4 Practical instrument

The source used to illuminate the interferometer was a pigtailed 30 mW 780 nm laser diode. Due to the interferometer configuration being intrinsically balanced laser frequency noise due to path imbalance was minimised. Parasitic reflections from interfaces and Rayleigh back scatter in the fibre with imbalances within the coherence length of the laser diode source will, however, generate noise in conjunction with laser frequency noise. To minimise this effect a high frequency modulation, above the cut off of the photodetectors (~ 4 MHz), was applied to the laser diode to reduce its coherence length.

A piezoelectric cylinder fibre phase modulator [7] was located beside the downlead interface to provide calibration signals.

The focusing element of the probe consisted of a 0.23 pitch graded index lens which was fixed, using high temperature epoxy, at a controlled distance from the fibre end inside a protective aluminium housing. The probe beam N.A. was set to 0.01, yielding a 40μm spot size with a depth of focus of ~1 mm, suitable for the level of low frequency vibrations encountered in plant operations. The downlead fibre was protected using Kevlar reinforced fibre cable (Seicor fan-out tubing) which had previously been demonstrated to provide efficient shielding of the downlead from AE pick-up [8].

5. Performance Tests

With the interferometer set at quadrature its output was monitored over the 0.1-1 MHz frequency range. Over this range it displayed a phase resolution of 0.5 mrads which at 0.5 MHz corresponds to a velocity resolution of around 50 μms^{-1} (50 $nms^{-1}Hz^{-\frac{1}{2}}$) or, in terms of displacement for simple harmonic motion, ~ 0.08 nm (80 $fmHz^{-\frac{1}{2}}$).

For comparison the Sagnac interferometer was tested in conjunction with an active homodyne balanced Michelson interferometer designed to measure AE displacement in metal cutting operations [8,9], with both instruments probing a mirror attached to a piezoelectric element driven by a 0.4 MHz sinusoidal signal. The result shown in figure 3 clearly displays the velocity measurement given by the Sagnac which is $\pi/2$ out of phase with the displacement signal given by the Michelson interferometer. Also, assuming simple harmonic motion, the calculation of the peak target velocity from the maximum target displacement given by the Michelson interferometer shows both instruments to measure the same maximum velocity of ~ 1.6 mms^{-1}

6 Conclusions

The suitability of the fibre optic Sagnac interferometer for the non-contact detection of acoustic emission has been demonstrated. The developed interferometer gives an output proportional to target velocity, without the need for active phase stabilisation and is inherently insensitive to low frequency perturbations of the instrument or the target. It provides antiphase outputs which are necessary for compensation of intensity noise from both the source and the target with the intrinsic balance of the Sagnac interferometer minimising the effects of source frequency noise. This fibre sensor also incorporates an environmentally insensitive downlead of arbitrary length. We have shown that the sensor is capable of meeting the specifications for structural integrity monitoring of power plant components based on acoustic emission detection and has a velocity resolution of 50 $nms^{-1}Hz^{-\frac{1}{2}}$, but the sensor is suitable for more general application.

7 Acknowledgements

The authors gratefully acknowledge the support of the European Community BRITE programme, project No. BE 6056.

8. References

[1] M.Mocchetti, E.Fontana, D.Bozzetti, F.Cattaneo and S.Ghia, "Application of AE technology to header monitoring in thermal power plants.", 10th International Symp., Sendia, Japan, October 1990, pp 528-536.

[2] J.E.Coulter et al, "Acoustic emission monitoring of fossil fuel power plants.", Materials Evaluation, **46**, Feb 1988, pp 230-237.

[3] G.Smeets and A.George, "Michelson spectrometer for instantaneous Doppler velocity measurements.", Journal of Physics E: Sci. Instruments, **14**, 1981, pp 838

[4] M.D.Paul and D.A.Jackson, "Rapid velocity sensor using a static confocal Fabry-Perot and a single frequency argon laser.", Journal Sci. Instruments, **4**, 1971, pp 170.

[5] D.Harvey, R.McBride, J.S.Barton and J.D.C.Jones, "A velocimeter based on the fibre optic Sagnac interferometer." Journal Measurement Sci. & Technology, **3**, 1992, pp 1077-1083.

[6] R.McBride and J.D.C.Jones, "A passive phase recovery technique for Sagnac interferometers based on controlled loop birefringence.", Journal Mod. Optics, **39**, 1992, pp 1309.

[7] D.E.N.Davis and S.Kingsley, "Method of phase-modulating signals in optical fibres: application to telemetry systems.", Electronic Letters, **19**, 1974, pp 21.

[8] T.A.Carolan, "Acoustic emission detection by fibre optic interferometry.", PhD thesis, Heriot-Watt University, Edinburgh, May 1994.

[9] R.McBride, T.A.Carolan, J.S.Barton, J.S.Wilcox, W.K.D.Borthwick and J.D.C.Jones, "Detection of acoustic emission in cutting processes by optical fibre interferometry.", Journal Measurement Sci. & Technology, **4**, 1993, pp 1122-1128.

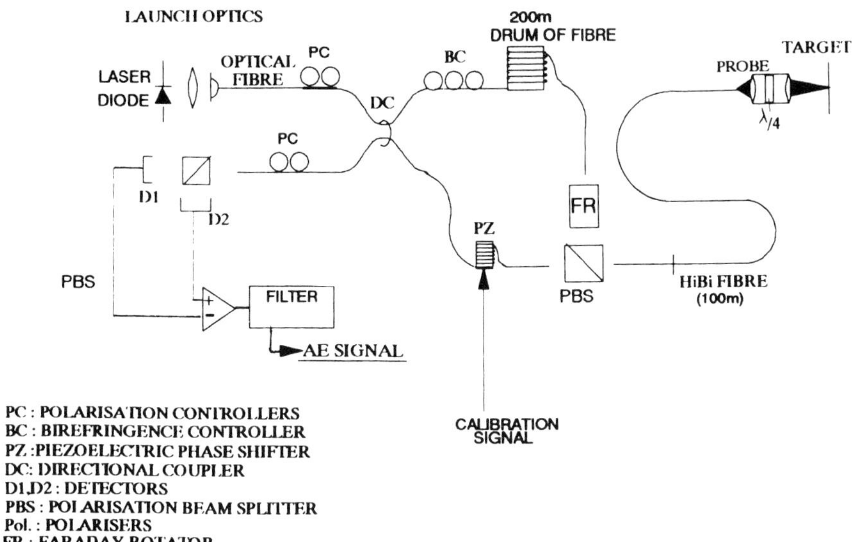

Figure 1: Sagnac interferometer for non-contact AE detection in power plant monitoring.

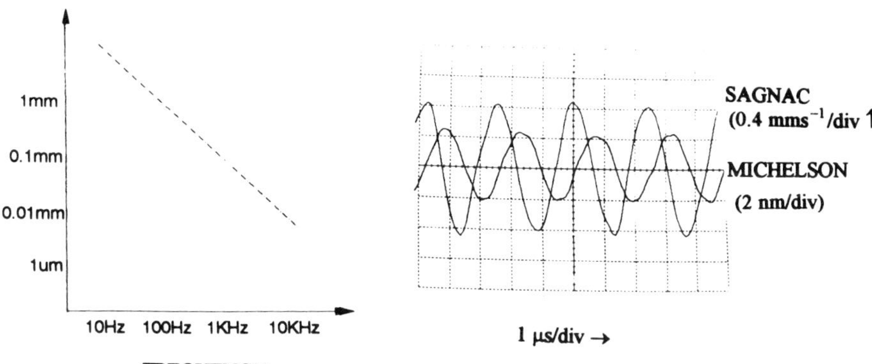

Figure 2: Plot of vibration amplitude against frequency required to produce 1 radian phase change in Sagnac AE interferometer.

Figure 3: Comparison of the velocity output of the Sagnac interferometer with a displacement measuring Michelson interferometer when probing an oscillating PZT target.

An interferometric optical fibre sensor system suitable for large scale structure monitoring

S. Chen, Y. Hu

School of Electronic, Electrical and Information Engineering,
South Bank University, 103 Borough Road, London SE1 0AA, UK

Abstract. An interferometric type optical fibre sensor system employing a novel digital spatial multiplexing technique is presented. It makes use of digital image processing technology to greatly simplify the network configuration. With a minimum number of optical devices in the system, the cost of building, installation and maintenance of the sensor array can be greatly reduced. Strain measurement experiments have been undertaken to assess the feasibility and performance of the system.

1. Introduction

The optical fibre is an one-dimensional, low-cost, dielectric, flexible, light-weight and yet very strong media with a very small cross-section and it is able to carry light wave over a very long distance with minimum loss. These distinctive features have attracted intensive research interest in the subject area of optical fibre sensing. After nearly two decades of research work, optical fibre sensors have established their importance in several specific application areas, including where fire and explosion hazards[1] or strong electromagnetic interference[2] restricting electronic based sensors, and where light weight or miniature size are crucial [3]. Particularly, interferometric optical fibre sensors offer very high resolution and very large dynamic range[4]. However, the interference output of such a sensor is rather difficult to interrogate. In normal interferometric sensors, this leads to some special system requirements such as quardrature position holding, phase modulation etc. Because of these requirements, more optical devices, such as couplers, modulators etc. have to be connected to the fibre network. Contrast to the optical fibre itself, optical fibre devices are bulky, rigid, heavy, fragile and expensive. More devices in the system not only complicate the system configuration, increase building, installation and maintenance cost but also prevents applications such as structure monitoring to enjoy the full benefits of the distinct features of the optical fibre. Multiplexing techniques developed so far, which can be classified into time[5], frequency[6], wavelength[7] and coherence[8] domain multiplexing techniques, have not been successful in reducing the number of devices per sensor in an integrated interferometric sensor array. More often, the system complexity is increased to the extend that it limits the maximum number of sensors in the array. In another word, the capacity of existing multiplexing technique is far from satisfactory, especially for applications such as monitoring large or complex structures.

On the other hand, systems with "down-lead sensitivity" have long been considered unacceptable in interferometric sensors by the academic community because it can significantly diminish the high resolution advantage of this type of sensors. However, for most of applications in the industry, the resolution of interferometric sensors is more than enough. It is the cost associated with such systems that is deterring the industry from accepting the technology. In applications such as structure monitoring, an extra length of fibre which interacts with the measurand can be arranged so that the measurand sensitivity can be still very high even with a modest phase resolution.

Chen et al have, for the first time, studied the effect of a multimode down-lead to an interferometric optical fibre sensor system[9]. Prior to that work, he also presented a single interferometric sensor with reduced number of couplers and modulators using an "electronically scanning" method and Young's interferometer topology[10]. These two separate research work have led to the development of the "digital spatial domain multiplexing" technique presented here.

The technique features a very simple system architecture and a minimum number of required optical fibre devices. These lead to a lower system building and maintenance cost, a potentially much higher multiplexing capacity and a better utilisation of the unique characteristics of optical fibres for sensing applications. The system demonstrated here using this technique is especially suitable for monitoring vibration characteristics of a large mechanical structure.

2. Operating principles

2.1 System description

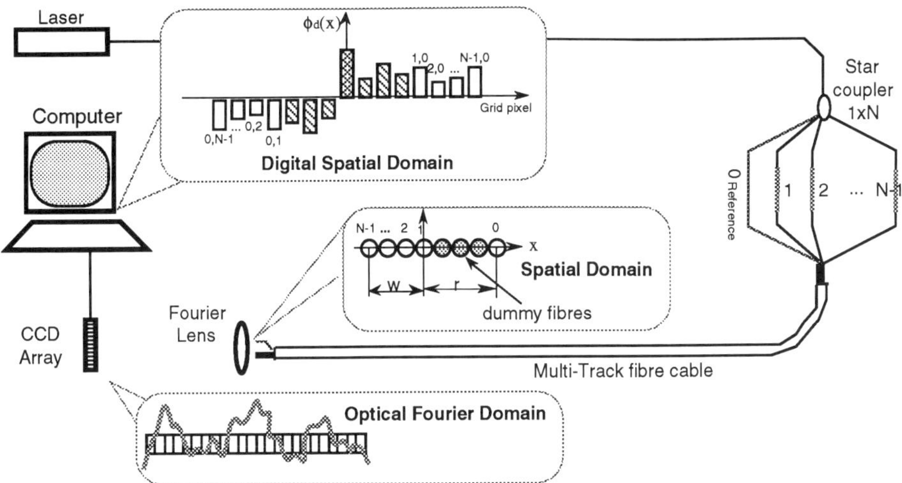

Figure 1 The schematic of optical fibre sensor system using the digital spatial multiplexing technique

The interferometric optical sensing system using the electronic scanning spatial multiplexing technique is shown schematically in Figure 1. The light from the laser is coupled into a single-mode 1xN tree-coupler via a lead fibre. Sensing fibres, connected to the coupler, carry the light passing the sensing field, then are bundled together and led to the processing unit in a multi-track cable. On the plane where all fibres terminate, the end-faces of the fibres are arranged in a line parallel to a linear CCD array. As shown in Figure 1, the sensing fibres are evenly positioned close to each other, while a reference fibre is at some distance away. The fringe pattern, which is formed by the mutual interference of the fields from all fibres, is sampled by the CCD and sent to a computer for processing. When the number of optical fibres, N, equals two, the system reduces to the optical fibre Young's topology investigated before. Here, where N>>2, the fringe pattern on the CCD is much more complicated but the phase information in each sensing fibre is still encoded, or multiplexed, "spatially" in the fringe pattern and can be retrieved by the computer using an appropriated digital processing algorithm.

2.2 Theoretical bases

If a Fourier transform (FT) lens is placed close to the fibre end-face array with the CCD in its rear focal plane, the light amplitude field on the CCD, F(u), is the Fourier transform of the field on the fibre end-face plane f(x) and the signal recorded by the CCD is the intensity or the power spectrum of f(x), $F(u)F^*(u)$, where * stands for complex conjugate. Without the FT lens, the above relation exists in approximation if the distance between the fibre and the CCD is large enough.

The task to recover the complete optical information of an object, including phase and amplitude, from its power spectrum by digital image processing has been referred as the "phase retrieval" problem[11]. However, conventional algorithms have been unsuccessful in recovering coherent objects, as the one in this system. In this paper, we present a non-iterative, fast phase retrieval method for the described electronic scanning spatial multiplexing technique.

Because of the small core size of the single mode fibre, light from each fibre can be treated as from a point source on the object plane. Referring to Figure 1 and without losing generality, the light field at the end-face of the reference fibre can be expressed by a real Dirac function, δ(x-r). Then the light field on the fibre end-face plane can then be expressed as

$$f(x) = f_s(x) + \delta(x-r) \qquad (1)$$

where $f_S(x)$ represents the light field from the sensing fibre group. It is a complex function with values only at discrete points where the fibre cores are located. The phase of each complex value is the phase of the light in the fibre. Hence the phase element of $f_S(x)$ is one to be detected. In the proposed system, a digital Fourier transform algorithm such as Fast Fourier Transform (FFT) is applied to the digitized fringe pattern, $F(u)F^*(u)$, in the computer. According to the Fourier transform theory, the output $\Gamma(x)$ can be expressed as

$$\Gamma(x) = \mathbf{FFT}\{ F(u)F^*(u)\}= f(x) \otimes f(x) \tag{2}$$

where $\otimes$ stands for correlation and * for complex conjugate. Bring Equation 1 into 2, we get

$$\Gamma(x) = \delta(x) + f_S(x) \otimes f_S(x) + f_S(r-x) + f_S{}^*(x+r) \tag{3}$$

From Figure 1, if

$$r > w \tag{4}$$

we get

$$\Gamma(x) = f_S(r-x) \qquad \text{when} \quad r+w \geq x > r \tag{5}$$

The FFT can produce a phase and an intensity array of $\Gamma(x)$ simultaneously. From Equation 5, a range of values in the phase array directly provides digital values of the relative phases in the sensing fibre array. If the time taken to retrieve all the relative phases in each sensing fibre in the bundle is defined as a period of one frame, phases in each fibre can be traced in a very wide range, which is only limited by the coherence length of the light source, as long as they do not change more than π within such a period. This digital processing based technique only requires the satisfaction of sampling rule, i.e. there should be at least two CCD pixels to sample the narrowest fringe formed by the pair of fibres farthest apart. This means a minimum of 4N pixels are required to interprete a N fibre array. No other mechanism is required to hold the interferometer array at quadrature. This unique feature greatly simplifies the system architecture.

In summary, with careful arrangement of fibre end-face array in the spatial domain, the phase signals in the sensing arms are encoded in the optical Fourier domain in the form of interference fringes and finally retrieved by a computer using FFT based algorithms in the digital spatial domain. Hence we have named this technology.

3. Experiments

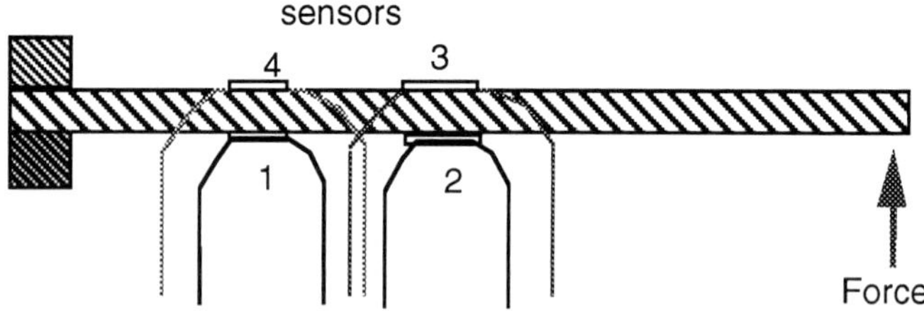

Figure 2 An optical fibre sensor array attached to a suspension beam used in the experiment

Preliminary experiments were carried to evaluate the phase retrieval and tracing performance of the proposed system depicted in Figure 1, where a laser diode with a wavelength of 830 nm and output power of 3mw was used as the source. A single-mode 1x8 tree coupler was connected in the system but only five of the eight output fibre were used in the preliminary experiment. The distance from the CCD array to the fibre end-face plane was 215mm and the FT lens was not used. A line of 512 pixels was read out in each frame and the inverse FFT was performed by a 486 DX2 66 PC. A industry standard "real-time" frame rate of 24 frames per second (fps) was achieved using the system. Along each of the four sensing fibres, short section of about 20mm in length were bound to a suspension beam at four different positions using super glue, as shown in Figure 2. A short, concentrated pulse force was applied at the free end of the beam and the result phase variations over time in the sensing fibres were retrieved and traced as shown in Figure 3. A strain sensitivity of 2 microstrain was achieved. The behaviour of this simple mechanical structure under such an excitation can be predicted precisely in the theory and the measured results is consistent with the theoretical calculation in terms of the levels and frequencies of the strain variations at different positions..

If the state of polarisation (SOP) of the light is taken into consideration in the proposed system, the amplitude of light field in a sensing fibre, f_s, can be separated into two vector components: $f_{s||}$, which is parallel to the SOP of the light in the reference fibre and f_{sT}, which is perpendicular to it. $f_{s||}$ is termed as "effective amplitude" because it is this part of the light amplitude in the sensing fibre that contributes to the formation of interference fringes. It is well known that light varies its SOP in a normal single mode optical fibre. So the normalised effective amplitude (NEA) of a particular sensing fibre can vary between 0, when the SOP of the light in this sensing fibre is perpendicular to that in the reference fibre and 1, when these two are in parallel. Since the effective amplitude only determines the visibility, not the positioning, of the fringe set related to any two fibres, the variation of SOP in the sensing fibre will not directly change the retrieved phase value. However, the accuracy of the phase retrieval is affected. This aspect of system performance is assessed using computer simulation.

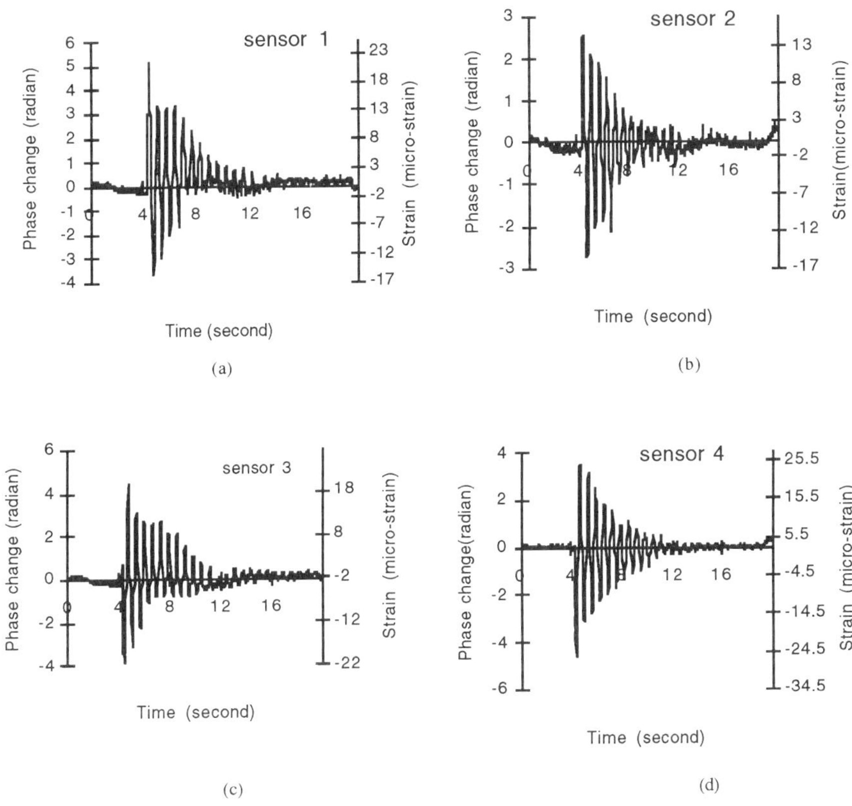

Figure 3. Output of the optical fibre strain sensor array when the suspension beam was subject to a pulse excitation

A worst case in a proposed system with N=16 was investigated where the NEA of an arbitrary sensing fibre was pre-set at various levels while the NEAs of the light in the rest all equal to 1. Random numbers were added to the computer generated fringe pattern to simulate different levels of intensity noise, which is defined as the average amplitude of the noise divided by that of the fringe. The result is shown in Figure 4, where it can be seen that the system can achieve a phase retrieval accuracy of one 10th of a fringe with the NEA as low as 0.2 in a 10% intensity noise environment. Therefore, although the system is not immune to the SOP variation, it is not sensitive to it.

The particular system evaluated in the experiment is designed to sense the dynamic strain response of a structure under a pulse excitation. The typical signal duration is within 20 seconds. The SOP variations in sensing fibres were much slower and were found to have no noticeable effect to the system functionality in this particular application. In fact, a large force was applied to the beam in a slow load and un-load cycle and the phase changes in the sensing fibres during the process were successfully traced as shown in Figure 5. This experiment reveals that the technique is able to measure the phase change to a very wide range only limited by the coherence length of the laser source.

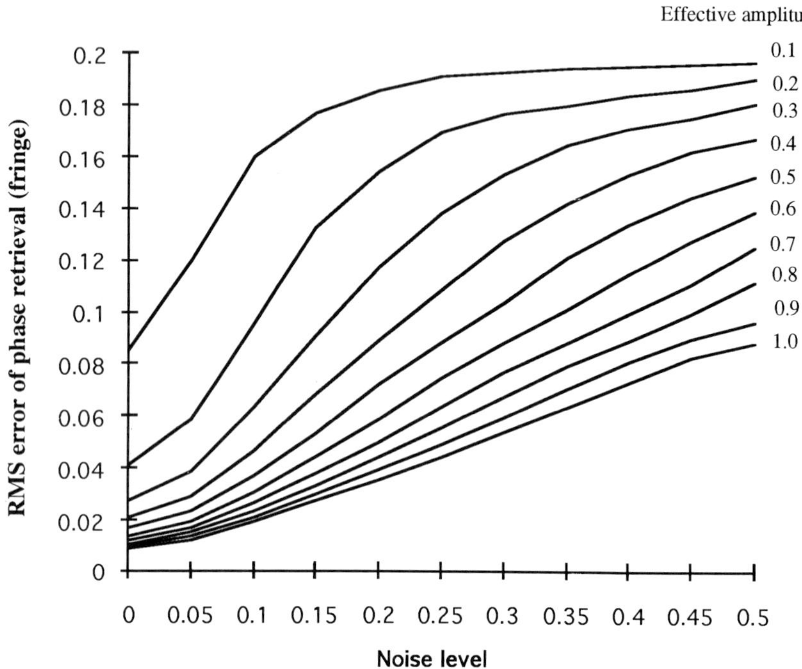

Figure 4 The computer simulation results of the RMS errors of phase retrieval versus intensity noise levels under different normalized effective amplitude in the sensing fibre

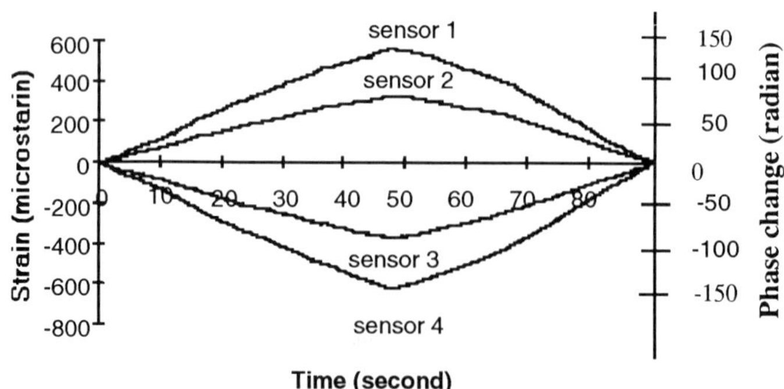

Figure 5. Output of the fibre sensor array when the beam was subject to a slow load/unload cycle

4. Conclusion

The theoretical study and experiment demonstration shown that phase information in an optical fibre sensor array can be encoded spatially in far-field fringe pattern and later retrieved using digital image processing techniques with a non-iterative algorithm. The electronic scanning spatial multiplexing technique is feasible. In a system using this technique, there are no modulation and tracking devices, no mechanical moving parts. Its simple and robust architecture allows the sensor array to take a full advantage of all the benefits by the unique characteristics of the optical fibre. This makes the array suitable to be attached onto or even embedded into a structure without affecting its functionality, thus has a huge potential in application areas such as Smart Structures and Materials It is obvious that the fibre arrangement and the CCD array in Figure 1 can be extended to 2-dimension allowing a very large number of sensing units to be multiplexed in this system. Further research is going on to improve the performance of the system.

5. Acknowledgements

Mr. Y. Hu is supported by the Overseas Research Scholarship.

6. Reference

1. M. Aizawa et al, "30Km long remote sensing of methane gas using a 1.65μm DFB LD and single mode fibre", Proceedings of 9th International Optical Fibre Sensor Conference (OFS'9), Firenza, Italy, pp285-288, (1993)
2. M. Martinelli, "Applications of optical fiber sensors in electrical power plants (Invited presentation)", Proceedings of 9th International Optical Fibre Sensor Conference (OFS'9), Firenza, Italy, pp403-408, (1993)
3. G. A. Sanders et al, "Progress in interferometer and resonator fiber optic gyros (Invited presentation)", Proceedings of OFS'8, Monterey, USA, pp26-29, (1992)
4. D. A. Jackson, "Monomode optical fibre interferometers for precision measurements", Current Advance in Sensors, IOP Publishing Ltd., ISBN 0-85274-509-5, pp75-95, (1987)
5. Brooks J. L., Tur M., Kim B. Y., Fesler K. A. and Shaw H. J., Fibre-optic interferometric sensor arrays with freedom from source phase-induced noise, Opt. Lett., **11,** pp473-475, (1986).
6. Collins S. F. , Meggitt B. T., Palmer A. W., and Grattan K. T. V., A multiplexing scheme for optical fibre interferometric sensors using an FMCW generated carrier, Proceedings of 8th optical fibre sensors conference, pp209-212, (1992).
7. Duratians R., Anglaret G., Hugues C. J. And Fehrebach G. W. :"Specific design of optical fibre sensor systems for wavelength division multiplexed networks", Springer Proceedings in Physics, Optical fibre sensors, Spring-Verlag Berlin Heidelberg, **44**, pp504-512, (1989).
8. Santos J. L. and Leite A. P., Multiplexing of polarimetric sensors addressed in coherence, Proceedings of 9th optical fibre sensors conference, pp59-51, (1993).
9. S.Chen, A.W.Palmer, K.T.V.Grattan and B.T.Meggitt, "An extrinsic optical-fibre interferometric sensor that uses multimode optical fibre: system and sensing head design for low noise operation", Opt. Lett., Vol.17, No.10, (1992).
10. Chen S., Rogers A. J., and Meggitt B. T. :"Electronically scanned optical-fibre Youngs white-light interferometer", Opt. Lett.. , **16**, pp761-763, (1991).
11. Fienup J. R.: "Phase retrieval algorithm: a comparison", Appl. Opt., **21**, pp2758-2769, (1985).

POTENTIAL OF FIBRE OPTIC POINT AND DISTRIBUTED FIBRE OPTIC SENSORS FOR STRUCTURAL MONITORING.

K. Kalli, Y. Rao, D.J. Webb and D.A. Jackson

Applied Optics Group, Physics Laboratory, The University, Canterbury, Kent, CT2 7NR, U.K.

Abstract. This paper examines the potential applications of fibre optic point sensors and distributed fibre optic sensors for structural monitoring. We present results from two systems developed at Kent: one is able to address a network of over 32 point sensors that may be interferometers or fibre Bragg gratings; the other is a truly distributed sensor using Brillouin scattering that is able to measure temperature and strain over a range of 50km.

1. Introduction

It is now recognised that fibre optic sensors offer significant advantages when compared with conventional sensors for a number of 'niche' applications. These new sensing devices are particularly suited for applications where it is necessary to make precise, remote and electrically passive measurements. Another major advantage of fibre optic sensors is that their cross section can be relatively small ranging from 100-500 μm thus it is possible to incorporate them into buildings or composites without being unsightly in the case of the building or without compromising the structural integrity in the case of the composite. For monitoring of such large structures it is necessary to have a large number of sensing points, thus the design of the sensors must be such that they can be readily and reliably multiplexed.

Clearly absolute strain is the key parameter to be determined in structural monitoring, however, in many cases such as bridges which are subject to large daily temperature changes it will generally be necessary to determine the temperature of the location where the strain is to be measured, particularly when long term structural changes are to be followed. In applications where only the effects of dynamic loading are important it may not be necessary to measure the temperature.

We have developed two systems which are ideally suited for structural monitoring

1) A multiplexed network for fibre optic point sensors capable of supporting >32 sensors where these sensors can be miniature fibre optic interferometric sensors (FOIS) or fibre optic Bragg grating sensors or any combination of the two ;

and 2) A distributed sensor based on Brillouin scattering capable of simultaneously measuring temperature and strain over a 50 km range.

2. Point Sensors

2.1 Fibre Optics Interferometric Sensors

Although fibre optic interferometric sensors offer micro-strain resolution (or better) the requirement to make long term absolute measurements limits the interrogation techniques which can be used. We have introduced a fibre optic sensor network based on low coherence (or White Light) interferometry which can multiplex and demultiplex up to 32 miniature Fizeau/Fabry Perot (F/FP) fibre linked point sensors.

The basic network is shown in Figure 1 [1]. The output beams from 2 super luminescent pigtailed sources at ~800 nm (λ_1,λ_2) are coupled into a transmission Michelson interferometer (TMI) with a selectable path imbalance of x microns, which is greater than the coherence length of the sources. The F/FP sensors are designed to have the same optical path differences of ~ x microns. The output signal from each sensor is taken back to a separate detector. The scanning range of the TMI is $\lambda_1/2$, which can be scanned at rates up to 500Hz. The wavelength separation of the sources is chosen to be ~7 nm giving an effective wavelength $\frac{\lambda_1\lambda_2}{\lambda_1 - \lambda_2}$ of ~ 50 μm. The TMI has a capacitive sensor which enables its OPD to be determined absolutely with a resolution of ~10^{-10} m. The absolute measurement range of the system is 50μm with a resolution of 1 nm. The system was initially designed for hybrid pressure sensors [2] where the optical cavity is formed between the (uncoated) fibre tip and a pressure sensitive surface (typically a diaphragm). The cavity length being about 500 μm to minimise the temperature sensitivity. F/FP temperature sensors with similar cavity lengths have also been developed [2].

High resolution strain sensors could be readily implemented as shown in Figure 2(a). With a gauge length of 1 cm, and an OPD of 1mm, an absolute displacement of 10 nm is equivalent to a microstrain. Thus high resolution would be readily obtainable with the present multiplexing system. This sensor will exhibit a temperature sensitivity of 0.1 μ strain 1°C; hence temperature compensation is necessary. A compatible temperature probe is shown in 2b where the sensing element is a short length of fibre also with an OPD of 1mm giving an absolute temperature resolution of $\sim 2 \times 10^{-2}$k. The dynamic range of the system is $\sim 5 \times 10^5$; exceeding the possible operating of the strain gauge by an order of magnitude.

2.2 Bragg Grating Sensors

Bragg grating sensors (BGS) are attracting considerable attention as they offer high strain sensitivity and are ideally suited for embedding in composite structures. As discussed above for effective structure monitoring it is necessary to deploy a large number of sensors. Hence it is essential to be able to multiplex a large number of BGS. As BGS are also temperature sensitive a strategy to allow temperature compensation must be developed. Various techniques have been proposed based on techniques such as collocated gratings [3] and dual wavelength illumination which exploit the differential sensitivity of the fibre strain and temperature coefficients [4]

Various techniques have also been proposed to multiplex BGS [5][6] however the numbers have been relatively small < 5. The most sensitive technique reported to date for interrogation of BGS is to use a pseudo frequency tuned source based upon a broad band source coupled into a scanning interferometer [5] The combination of one of the sources shown in Figure 1 with the TMI is effectively a (pseudo) frequency tuned source, thus provided the free spectral range (FSR) of the TMI is compatible with the working range of a BGS it can directly replace a FP/P. Figure 3 shows the output signals from BGS substituted for one of the F/FP sensors shown in Figure 1when subject to strain or temperature. The strain resolution is ~ 300 nano strain, here the TMI cavity length is ~ 500 μm. The simplest method to compensate for thermal cross sensitivity of the BGS is to collocate one of the F/FP temperature sensors. Without any modifications the multiplexing system could be used to interrogate 16 BGS and 16 temperature sensors . In principle the number of BGS which can be simultaneously interrogated could be significantly greater if wavelength and time division were incorporated into the network, also the resolution can be increased by increasing the FSR of the TMI.

3. Long Range Distributed Sensor

Both techniques described above are limited by the number of points sensor which could be practically deployed. For very large structures such as dams, tunnels etc, a fully distributed fibre optic sensor offers a potential solution. Distributed strain measurements have been demonstrated using a combination of microbending loss in conjunction with OTDR techniques [7]. However, as these systems use multimode fibre their range is somewhat limited. A distributed sensor based upon Brillouin loss/gain [8] in a monomode fibre offers the greatest range of all the reported distributed sensors with the advantage that simultaneous measurement of temperature and strain is possible. A typical experimental configuration for this sensor is shown in Figure 4. In this technique light from a CW laser is launched into one end of the monomode sensing fibre while short pulses of light from a pulsed laser probe are injected into the other end of the fibre. When the probe laser frequency is less than that of the CW laser by the Brillouin frequency shift (in the fibre) the probe beam will be amplified, experiencing Brillouin gain at the expense of the CW beam. The Brillouin frequency shift depends on several parameters such as the temperature and the Bulk modulus of the fibre, which is dependent of the local fibre strain. Hence when the fibre is subject to non uniform temperature/strain, the probe will experience gain for a given laser frequency difference only in those parts of the fibre that are at a specific temperature/strain.

If the intensity of the CW beam emerging from the fibre is monitored following the launch of a pump pulse, a decrease in intensity will be observed whenever the Brillouin interaction occurs. The time delays between launch of the pump pulse and these decreases in the CW power signal correspond to round-trip times for light travelling to and from the regions of gain. These times provide positional information. If the laser frequency difference is adjusted, then the probe will experience gain in parts of the fibre at a different temperature/strain. Therefore by slowly scanning one of the laser frequencies the temperature/strain distribution along the fibre length can be mapped.

As strain is the main parameter of interest and as it is virtually impossible to deploy the fibre such that it is not sensitive to temperature variations, the sensing fibre must be deployed such that both the temperature and the strain can be uniquely determined at every point along the fibre. In principle, it might be possible to separate the two measurands by using the dispersion in the Brillouin temperature and strain coefficients by illuminating the sensing fibre at 2 different wavelengths eg 1.3 μm and 1.55 μm, however, this experiment has yet to be performed. An approach which we have adopted is to mount the fibre such that on the 'outgoing leg' the fibre is sensitive to both temperature and strain whilst on the 'return leg' the fibre is mounted such that strain is temperature independent.

The results of an experiment demonstrating this concept have been reported [9]. The process of recovery the strain signal involves a 'two step' measurement procedure. The fibre is first calibrated by measuring the Brillouin shift as a function of strain at a specific temperature. The frequency shift corresponding to a range of temperatures is also determined. As the slope of the frequency verses gain curve is virtually independent of temperature, as shown in Figure 5 the strain can be recovered by first measuring the Brillouin frequency shift corresponding to the maximum Brillouin loss at a specific location in the fibre section sensitive only to temperature changes. This gives the temperature of the location. The maximum Brillouin loss frequency is now determined for the collocated fibre section which is sensitive to both measurands and the strain recovered by subtracting the Brillouin frequency corresponding to the temperature of region. Using this approach a strain resolution of 20 micro strain with a 2° C temperature resolution combined with a spatial resolution of 5 m has been obtained. The modification to the system shown in Figure 4 to enable simultaneous measurements of temperature strain is shown in Figure 6.

4. Conclusion

Two fibre optic based systems with potential for structural monitoring have been presented. One system is capable of multiplexing and demultiplexing up to 32 point temperature or strain sensors which can be Fizeau/Fabry Perot interferometers or Bragg grating sensors or any combination of the two. The second system is a distributed sensor based on Brillouin gain/loss which offers unprecedented sensing range with the capacity of simultaneously measuring temperature and strain.

5. Acknowledgements

Dr X Bao made a major contribution to the development of the Brillouin distributed sensor. We also gratefully acknowledge Professor I Bannion of Aston University who supplied us with the BG at 800 nm. The work was supported by EPSRC and DTI by the Nanotechnology LINK scheme.

6. References

[1] Y J Rao and D A Jackson "Multiplexed sensor network exploiting low coherence interrogation", in preparation.

[2] Y J Rao, D A Jackson, R Jones and C Shannon, Journal of Lightwave Technology, Vol. 12, No.9, September 1994, "Development of Prototype fibre-optic based Fizeau pressure sensors with temperature compensation and signal recovery by coherence reading".

[3] M G Xu et al, Elec. Lett. **30** No. 13, 1085 (1994), " Discrimination between strain and temperature effects using dual-wavelength fibre grating sensors".

[4] K Kalli, G Brady, D J Webb, L Reekie, J L Archambault and D A Jackson, "Possible Approach for the Simulataneous Measurement of Temperature and Strain via First and Second Order Diffraction from Bragg Grating Sensors", Post deadline paper OFS 10 Glascow 1994.

[5] A D Kersey, SPIE proceedings, Vol. 2071, 30, 1993 Boston, "Interrogation and Multiplexing Techniques for Fiber Bragg Grating Strain Sensors".

[6] D A Jackson, A B L Ribeiro, L Reekie and J L Archambault, Optics Letters, Vol. **18**(14), p.1192-1194, July 1993, "Simple Multiplexing Scheme for a fibre-optic grating sensor network".

[7] K H Wanser, M Hazelhun and M Lafond, Ed. F. Ansari, "High Temperature Distributed Strain and Temperature Sensing using OTDR", p.193 1993, Applications of fibre-optic sensors in Engineering Mechanics, Am.Soc. of Civil Engineers.

[8] X Bao, D J Webb and D A Jackson, Optics Letters, Vol. **18**, p.1561-1563, 1993, "32 km Distributed Temperature Sensor using Brillouin Loss in an Optical Fibre"

[9] X Bao, D J Webb and D A Jackson, Optics Letters, Vol. 19(2), Jan. 1994, p.141-143, "Combined distributed temperature and strain sensor using Brillouin loss in an optical fibre".

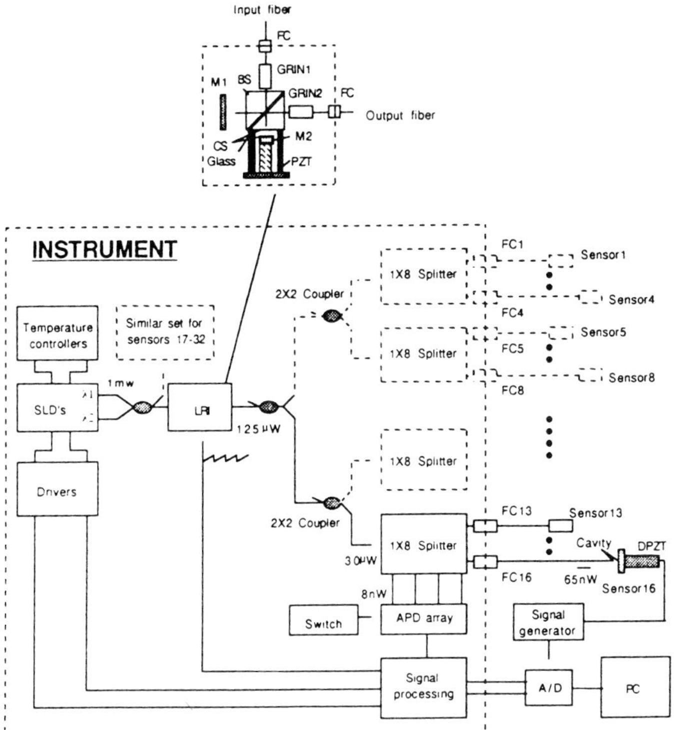

Fig. 1. Schematic diagram of the multiplexing system

SLD's: Superluminescent Diodes, LRI: Local Receiving Interferometer; FC1-16: Optical Fibre Connectors.

—— **implemented** ----- **not implemented**

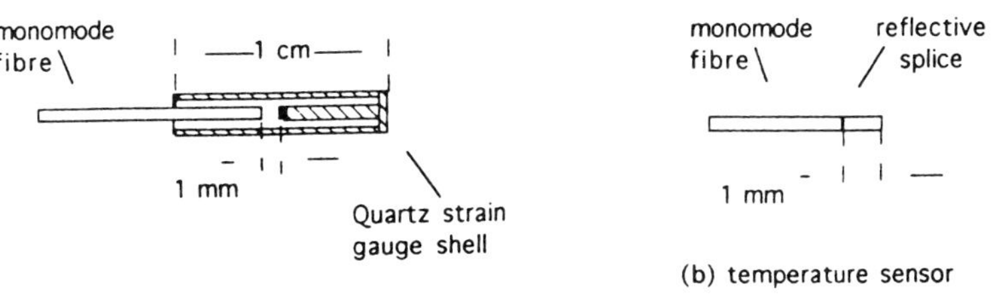

FIG 2 F/FP Strain & Temperature Sensors

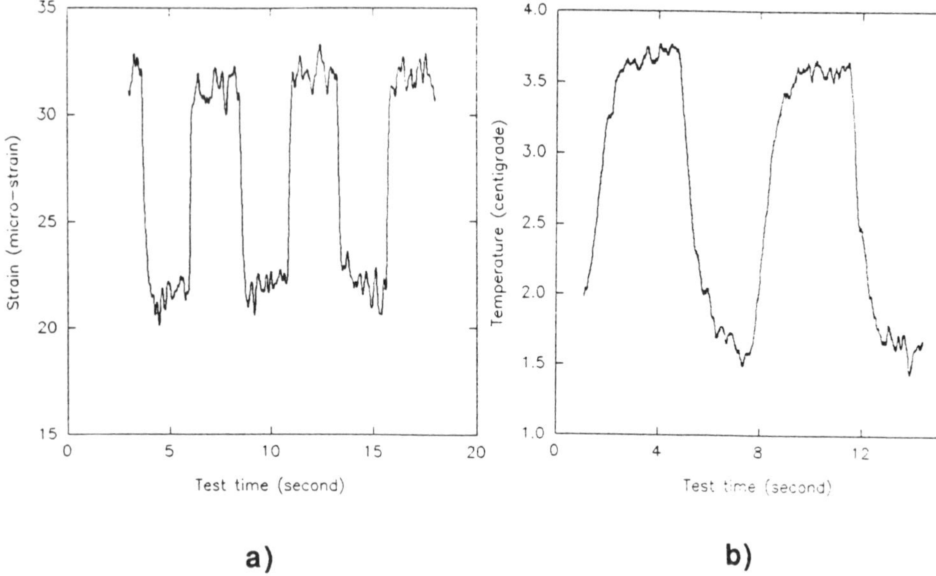

Fig. 3. Output of BGS as a function of a) strain and b) temperature recovered via the multiplexing system shown in Fig. 1.

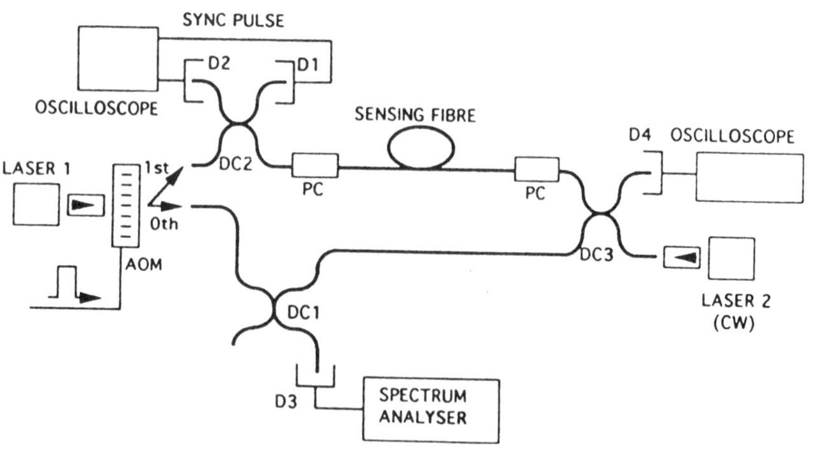

Fig. 4. Distributed sensor based upon Brillouin gain/loss.

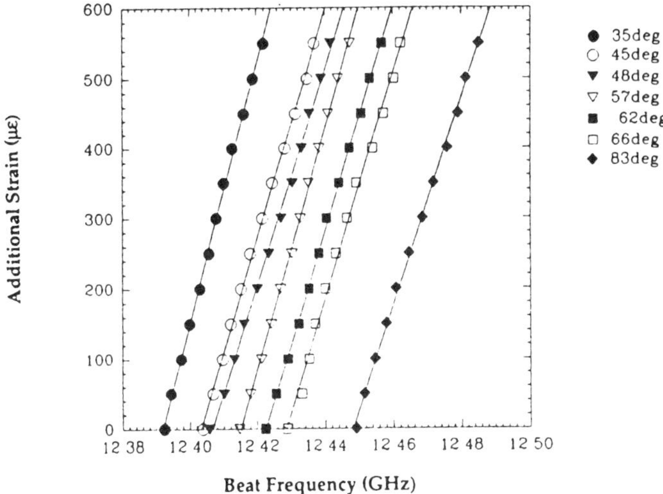

Fig. 5. Variation of the Brillouin peak frequency as a function of strain at various fixed temperatures.

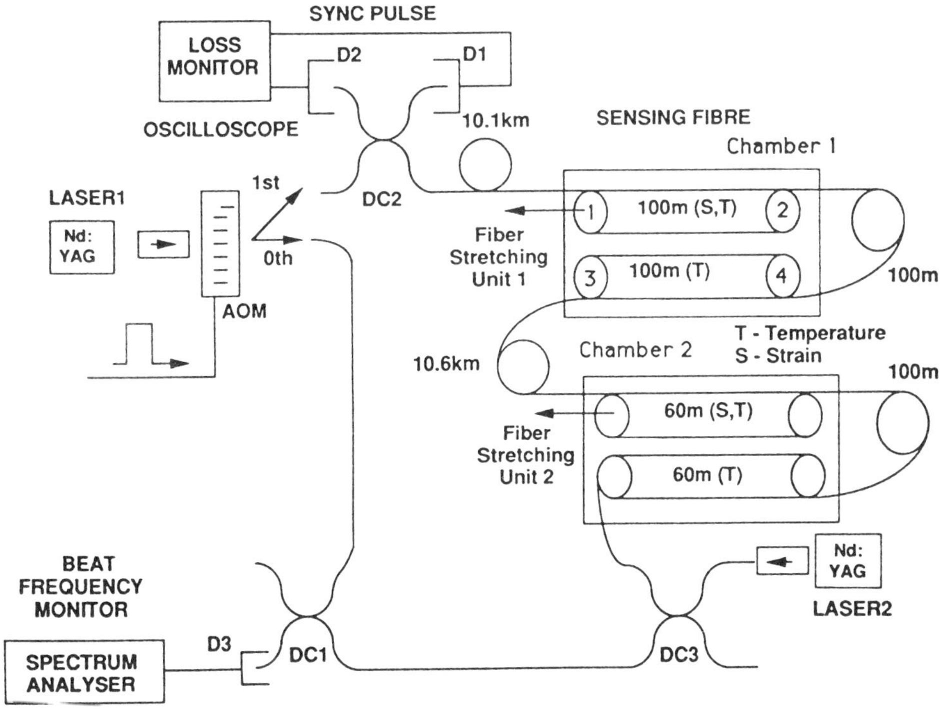

Fig. 6. Modification of the system shown in Fig. 4. to enable the simultaneous measurement of strain and temperature.

A high sensitivity Faraday current sensor using a novel multi-optical-loop sensing element

Z. P. Wang, Y. N. Ning, A. W. Palmer and K. T. V. Grattan

Department of Electrical , Electronic and Information Engineering
City University, London EC1V OHB, U.K.

Abstract. A novel Faraday current sensor with a sensitivity of 6.57×10^{-5} rad/A and a minimum detectable current of 27 mA/$\sqrt{Hz}$, using the so-called "pseudo dual-quadrature" reflection scheme, is reported.

1. Introduction

Research and development into optical Faraday current measurement systems using bulk glass as their sensing elements have been attractive subjects in the field of current measurement and monitoring in recent years. As a result of a series of intensive studies, a number of new topologies for the sensing elements involved have been introduced and investigated [1-5]. The basic schemes used to design these sensing elements are the so-called "dual-quadrature" reflection [1, 2] and "critical angle" reflection approaches [3 - 5]. By using these arrangements, the reflection-induced phase difference in linearly polarized light may be effectively cancelled or avoided, so as to ensure the state of polarization (SOP) of the light can be preserved along the optical path, within the sensing element.

In this work, the design and performance of an optical Faraday current sensing system employing a novel sensing element is reported, in which a linearly polarized beam is constrained to encircle the current carrying conductor three times and thereby increase the sensitivity also by a factor of three, compared to that reported by Chu et al [4]. The new sensor topology used in this work shows some similarities to the sensor developed by Westinghouse [1], but it uses a more compact arrangement of a multi-loop optical path. This is achieved by allowing the input beam to have a carefully calculated and implemented offset angle, resulting a large number of "pseudo dual-quadrature" reflections occurring with the sensor.

2. Theoretical background

Optical current measurement systems can make use of the Faraday effect or the magneto-optic effect to exploit the interaction between an optical beam and the magnetic field associated with the current, **I**. When a linearly polarized beam propagating inside a dielectric material, its polarization azimuth being Φ_F, is rotated under the influence of a magnetic field, the component along the direction of the light propagation, **H**, generated by the electrical current to be measured is given by the following relationship

$$\Phi_F = \int_l V \, \mathbf{H} \cdot dl \qquad (1)$$

where V is the Verdet constant of the dielectric material and l is the interaction length. For a closed optical path, this parameter, Φ_F, is proportional to the current to be measured, as described by Ampere's circular law. Therefore, in order to measure the current flowing in a conductor, the light beam must encircle the conductor, without altering its SOP. Under this condition

$$\Phi_F = VN\,I \qquad (2)$$

where I is the current in the conductor and N is the number of times that the light encircles the current.

3. Configuration of the sensing element

According to equation 2, it can be seen that if the factor N is increased, the sensitivity of the system increases by the same amount. The newly introduced 3D topology (with three optical loops) of the sensing element is shown in Figure 1. This optical element itself is made of SF6 Schott glass (the refractive index is 1.798), of size 100 X 100 X 60 mm, and has a 35 mm diameter central aperture through which the current-carrying conductor passes.

A linearly polarized beam from the light source is coupled into the sensing element through a small input prism. When the beam is injected at specific angles, the light can undergo as many as 22 internal reflections, at angles which are slightly offset from 45° (by about 0.90° with $\pm$ 0.09° of tolerance), forming 3 optical loops around the current carrying conductor, and finally emerging at the other output prism, as shown in figure 1. In this new topology, each optical loop needs 8 interreflections (both the first and the last loops need 7 reflections respectively) where the number of the loops is determined by the offset angle and the size of the sensing element. Since each inter-reflection is not at 45°, the reflection-induced optical phase difference will not be cancelled to zero by each pair of "pseudo-quadrature" reflections, and as a result, the overall measurement sensitivity (or the scale factor of the system) for each higher order of the loop will slightly reduce. However, the sensitivity, or the measured Verdet constant of the sensing element remains unchanged and as the number of the loop increases, the overall sensitivity increases accordingly.

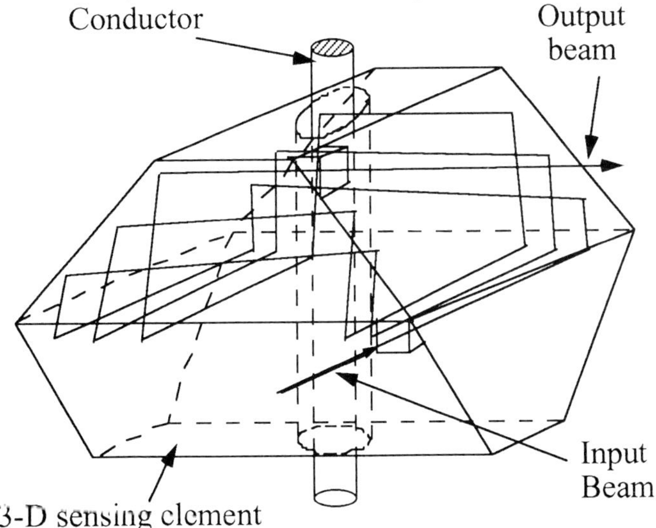

Figure 1. *Schematic of the 3-D sensing element.*

The SOP of the output beam may be predicted by employing Fresnel's equations and Jones calculus [6], and the details of the computer simulation involved will be discussed subsequently. Although a maximum number of three loops of the beam can be achieved using this sensing element, one and two loops may also obtained by altering the incident angle appropriately, and hence the sensitivities of the device with one and two loops can be measured respectively, to verify the basic concept of the design.

4. Experimental arrangement and results

The experimental arrangement used to test the sensing element is shown in Fig. 2, where a 5 mW Helium Neon laser operating at a wavelength of 633 nm is employed as the light source. The light beam is first passed through a quarter waveplate which is used to generate a circular polarized beam, and this beam is then injected in the sensing element via a Glan-Thompson polarizer in order to obtain a highly linearly polarized state. The principal axis of the Glan-Thompson polarizer is aligned such that the azimuth of the output beam is at 45° to the axis of the Wollaston prism which converts this beam into its two orthogonal states of equal amplitude. Two identical photodiodes are then used to convert two output beams into two electronic signals, I_1 and I_2, respectively. A signal processing unit is then used to give an output proportional to the ratio of the difference and sum of I_1 and I_2. Hence the resulting signal which is directly proportional to the Faraday rotation (provided the value of the rotation is small) is then virtually independent of variation in the intensity of the source. In order to generate a relatively large current, a coil of 300 turns of insulated copper wire capable of carrying a maximum current of 6 A is wrapped through the centre aperture of the sensing element. A 10A DC power supply (DCV80-10A, Sorensen power supplies) and a variable transformer are used to generate DC and AC currents respectively.

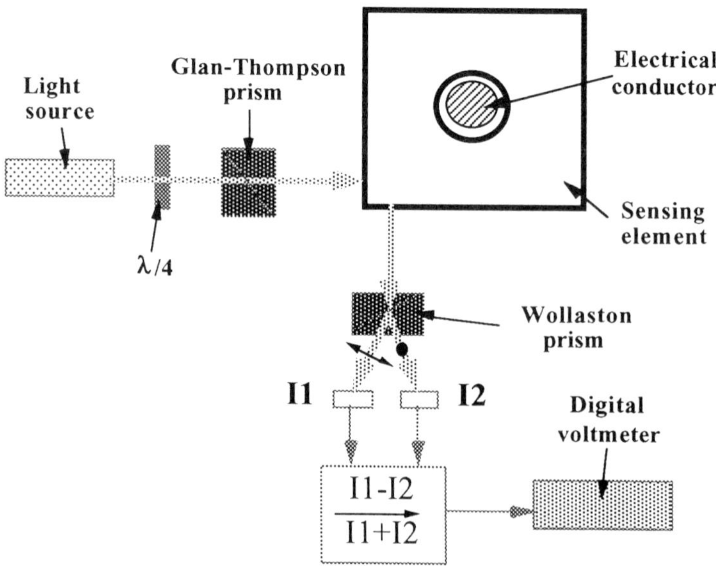

Figure 2. *Experimental arrangement used in this work.*

Figure 3 shows the experimental results obtained when a DC current is measured. It can be seen that the Faraday rotation angle is a linear function of the input current for an optical loop number of 1, 2 and 3 respectively. The measured sensitivities for these three cases are 2.19 $X10^{-5}$, $4.35X10^{-5}$, and $6.57X10^{-5}$ rad/A (± 1% in each case) respectively. Their ratio is 1: 1.99 : 3 which agrees well with the designed specifications. The measured Verdet constant in each case is also in good agreement with the manufacturer's data of 2.23×10^{-5} rad/A, at a wavelength of 633 nm [7].

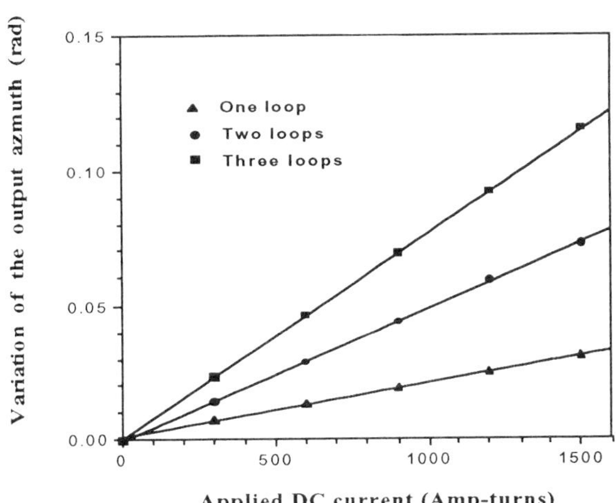

Figure 3. *Variation of the Faraday rotation as a function of the applied DC current for the cases where one, two and three optical loops are formed inside the sensing element respectively.*

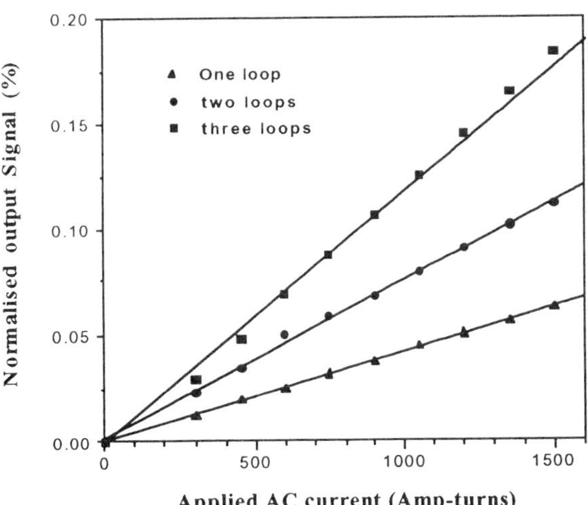

Figure 4. *Variation of the output signal as a function of the applied AC current for the cases where one, two and three optical loops are formed inside the sensing element respectively.*

Figure 4 shows similar measurement results when an AC current supply is used. The outputs of the system vary linearly as a function of the input current and the normalized scale factors for one, two and three loops are 0.042, 0.078 and 0.126 mV/A (± 1% in each case) respectively, giving a ratio of 1: 1.86 : 3.

5. Discussion

The main advantage of this sensing element is its larger incident angle tolerance by comparison to the ring-shaped device which was designed and based upon the scheme of critical angle reflections [3]. According to the experimental results, the sensitivity of the device may be kept unchanged within $\pm 0.16^{\circ}$ of the designed incident angle, and this is a vital aspect which makes it easier for the sensor to be used in practice. Since the light beam inside the sensing element experiences a large number of pairs of "pseudo-quadrature" reflections, this will introduce a small phase difference between the two orthogonal states after each pair of reflections, resulting in an elliptically polarized state of the output. This effect may reduce the scale factor of the system: however, after the normalization of the measured data, the relative scale factors still show a correct relationship in these three cases, as illustrated in Figure 4.

It should be pointed out that the phase difference induced by the reflections in this sensing element is different from that induced by the birefringence inside an optical fibre, because this phase difference is dependent of the incident angle. Once this angle is fixed, the value of the phase difference will be fixed accordingly, forming a known part of system error which can be removed in the calibration. However, the phase difference induced by the birefringence keeps changing with the variation of the temperature, as the birefringence itself can vary with the temperature.

Since the Verdet constant is also a function of temperature, a multi-loop design will increase this effect by the same factor. Hence, for the condition in which the temperature constantly varies over a larger excursion, a temperature sensor may be needed, and its output used to modify the measured data with reference to a "look-up" table.

6. Conclusion

In conclusion, a novel miniature optical current sensing element with a multi-optical-loop configuration based upon the Faraday effect has been demonstrated. This new sensing element not only increases the overall device sensitivity, but also overcomes the launch difficulty encountered in the ring-shaped sensing element [3]. In order to address both the basic science and the technologies of developing a stable and effective sensor system, recognizing both the potential of the overall system and the need for careful, precision engineering to enable the scheme to be optimized in a practical sensor to achieve the expected sensitivity and performance, a detailed study of the characteristics of the "pseudo-quadrature" reflections and the performance of the sensor using this new scheme is currently being carried out.

One author (ZPW) would like to acknowledge the support from British Council by way of a Technical Co-operation Training Award.

7. References

[1] T. W. Cease, J. G. Driggans and S. J. Weikel, *IEEE Trans.on Pwr. Del.*, **6**, 1374, 1991.

[2] A. H. Rose, M. N. Deeter and G. W. Day, *Proc. of 8th OFS Conf.*, 394, 1992.

[3] Y. N. Ning, B. C. B.Chu, and D. A. Jackson, *Opt. Lett.* **16**, 1996, 1991.

[4] B. C. B. Chu, Y. N. Ning, and D .A. Jackson, *Opt. Lett.* **17**, 1167, 1992.

[5] Y. N. Ning and D. A. Jackson, *Opt. Lett.,* **18**(10), 835, 1993.

[6] R. C. Jones, *J. of Opt. Soc. of America,* **31**, 488, 1941.

[7] Schott Optical Glass Company, Technical Information, Optical Glass, Schott Glaswerke, Postfach 2480, 1985.

A MULTIPROBE OPTICAL FIBRE IN-SITU DETECTOR FOR MONITORING FLUORESCENT DYE TRACER PARTICLES

R.G.Milne, D.McStay, P.Pollard
School of Applied Sciences, The Robert Gordon University, St Andrews St, Aberdeen, AB1 1HG

Abstract
The use of fluorescent dye tracer particles in sediment flow and sewage effluent discharge have been investigated using an optical fibre fluorescence detection system comprising of an array of twelve probes. Characterisation of the fluorescent dye tracer spectral properties as well as particle size distribution, were also reported. Results from a test at Aberdeen harbour involving the monitoring of sediment flow within the harbour and the river Dee estuary using one dye tracer and the in-situ detector are presented. Results from a second test was carried out at Esholt sewage treatment plant, where the monitoring of effluent discharge through various stages of the plant was investigated using two dye tracers and the in-situ detector are also presented. The results obtained using the in-situ detector show good agreement with results obtained from water samples collected during the tests and analysed later in the laboratory using standard analytical techniques.

Introduction
Pollution of our oceans, rivers, estuaries and harbours is a serious and immediate problem, which involves every living creature on Earth. The discharging of waste due to increased industrialisation near our coasts and harbours along with increasing sewage waste disposal into our oceans, due to increasing populations, means that monitoring and detection of water pollutants has become both more important and difficult using the present instrumentation and techniques.[1,2].
The complex chemical nature and close proximity of other discharges often make it impossible to assess the distribution and thus the environmental impact of material from a given source. In such cases it is standard practice to employ a tracer which can be injected into the discharge and subsequently monitored. Traditional tracing methods such as bacterial spores, radioactive tracers, fluorescent dyes (eg fluorescein and rhodamine), chemical tracers (eg lithium chloride and sodium bromide) give limited information, and are becoming environmentally unacceptable due to their adverse effects on marine organisms and the food chain [3]. Additionally they do not always accurately mimic the motion of the target species. Enviromentally many of the problems encountered using conventional tracers can be overcome by using a benign tracer. If the size, shape, duration of detectability and surface charge of the particle are all variable the tracer particle can be manufactured to accurately mimic the motion of the target species thus providing a more reliable result.
Traditional fluorescence methods of detection with or without tracers, have meant that water samples have to be collected for later analysis in the laboratory, either using a standard fluorometer or an analytical flow cytometer (AFC). The instrumentation used for in-situ measurements are basically adaptations of traditional spectroscopic instruments. The main disadvantages in the former methods are that water sampling is very time consuming and contamination of samples can occur before reaching the laboratory. In the latter method the instruments are often bulky and unsuitable for applications involving remote or inaccessible locations or shallow water, such as in rivers, estuaries or sewage treatment plants. Additionally most do not provide continuous real-time insitu measurements [4,5,6].
These problems can be overcome by employing an optical fibre based fluorescence sensor as reported previously [7]. In this device an array of optical fibre sensors are deployed in the water column while the expensive and bulky optoelectronics package is kept safely out of the water. This system has the added advantage of providing continous real time data thus giving a clearer

picture of the particle dispersion and allowing this to be followed. In this paper we report the results obtained from these initial fluorescent dye tracer particle studies using the multiprobe optical fibre in-situ detector (MOFID).

Dye Tracer Characteristics

Initial laboratory analysis was carried out on the fluorescent dye tracer particles, supplied for the field tests by Environmental Tracing Systems (ETS). Unlike Fluorescein and Rhodamine dyes the dye tracer is particulate with many dye molecules encapsuled in a polymer shell. The physical size and density of these particles can range from (0.01 to 1000μm), and (0.9 to $3gcm^{-3}$) respectively, the shape, duration of detectability and surface charge can also be altered. Therefore the physical properties of the dye tracer can be altered to mimic the target particle under investigation. The optical properties of the dye tracer were characterised using a fluorescence spectrophotometer (Perkin Elmer MPF-3). The measured emission characteristics for varying excitation wavelengths is shown in figure 1. It can be seen from figure 1 that the dye tracer has an excitation wavelength around 470nm, with a broadband fluorescence around 540-595nm with a distinct fluorescence peak at 570nm.

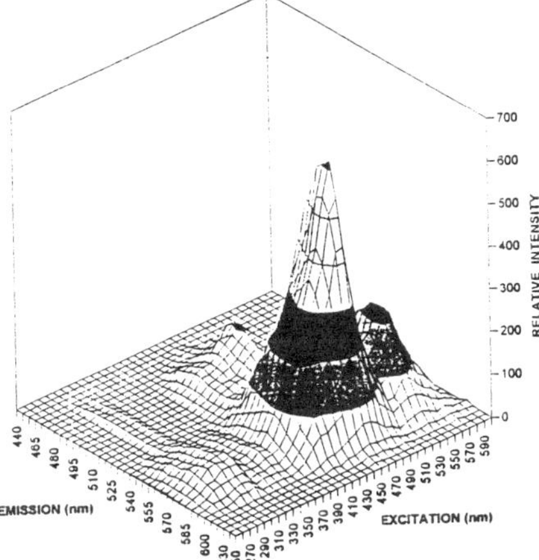

Figure 1. The emission characteristics for varying excitation wavelengths for a fluorescent dye tracer.

In monitoring the flow characteristics it is very important that the tracer particles have a size distribution and buoyancy which is similar to the target species. The particle size distribution of the dye tracer was therefore compared with a water sample from the river Dee estuary using the Malvern SB.OD particle sizer. Both samples were stirred and unstirred for comparison. The stirred samples were representative of a typical turbulent environment as the large particulates may remain in suspension. It can be seen from figure 2 that the river water sample has a peak distribution centred around 250μm with about 70% of particles between 100-550μm, however, large particulates are also present shown by the total obscuration of light above 550μm. The dye tracer distribution peak is around 120μm with about 60% of particles between 40-230μm, but again a significant number of larger particulates are present. These results show that the dye tracer although not an ideal match for the water particle distribution provides a reasonable match for the current experiments.

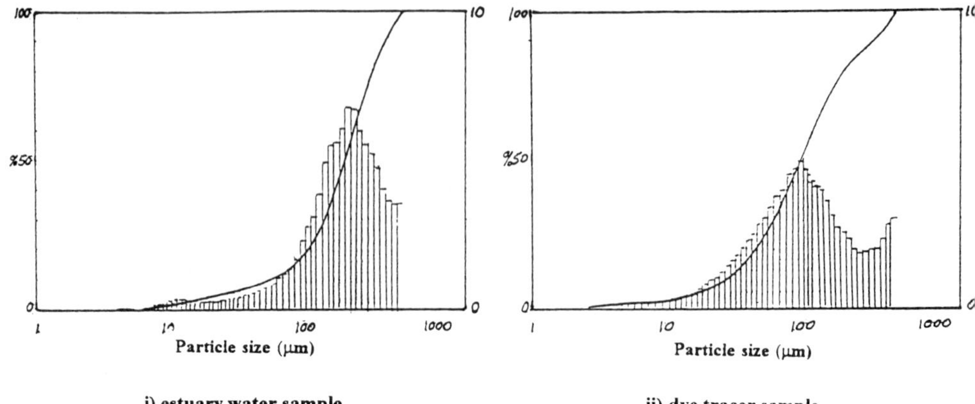

Figure 2. The particle size distributions for i) estuary water sample, ii) dye tracer sample containing estuary water.

Multiprobe Optical Fibre In-situ Detector

The in-situ detector system reported here incorporates the previously reported Optical Fibre Marine Fluorosensor [7] and its application to dye tracer studies. Amplitude modulated light (0.9kHz) from the excitation source is launched into the twelve fibres equidistantly spaced using a central spacer, at an equal radius, mounted on a purpose built holder. This arrangement ensures that the power launched into each fibre is equal. A similar arrangement in front of the photodiode detector is used along with a filter wheel which allows several different bandpass filters to be used while the system is in operation to reject the background light. In this way the average signal for all the probes is obtained. The output of the detector is connected to a lock-in and the resultant signal logged on a computer. Calibration of the MOFID was carried out using a range of concentrations of the dye tracer from a stock of 0.5g/l, containing about 5×10^9 particles per gram. As can be seen from figure 3 the system has a linear response.

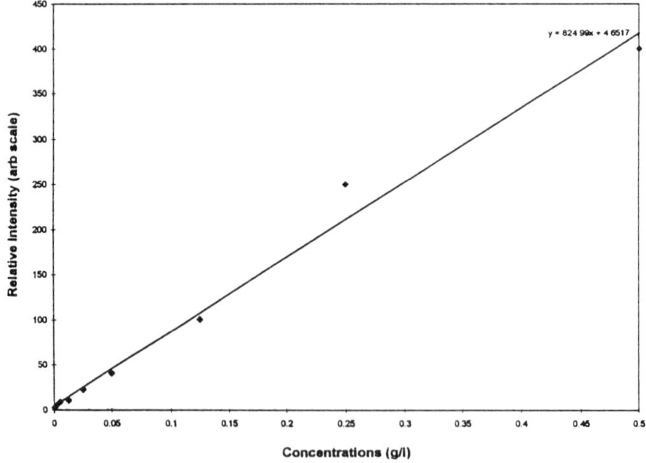

Figure 3. The measured response of the MOFID to known concentrations of the dye tracer.

Applications

1.Sediment movement

The movement of sediment caused by construction and dredging activity in harbours and estuarys can cause the accumulation or loss of sediment which can restrict shipping and result in serious financial loss. Also the flow dynamics of the water system can in turn effect the natural balance of that system resulting in environmental consequences. It is therefore important to conduct reliable measurements of the sediment movement in harbours in order to minimise such problems. Conventionally this is performed using tracer particles and by taking water samples. The combination of dye tracer particles and an in-situ detector were first tested as an improved sediment monitoring technique in Aberdeen harbour. The MOFID was used to monitor the movement of a plume of dye tracer from the river Dee estuary into Aberdeen harbour. The twelve probes of the MOFID were attached to a metal pole deployed over the side of the boat at a fixed distance and depth of around (1-2m). A fixed volume (bullet) of dye tracer was injected upstream from the mouth of the harbour, at Victoria bridge. Typical measurements obtained using the MOFID deployed from the Harbour office boat Sea Echo while traversing the river a little downstream from the initial injection point of the dye tracer are shown in figure 4. Patchiness of the dye tracer was observed, which was to be expected because of the presence of many eddy currents in the river and the non-uniformity of the release. After several hours small patches of dye tracer were still detectable in the harbour itself but not in the river Dee as figure 5 shows, as most of the dye tracer had dispersed in the river Dee estuary. These results are in good agreement with those from water samples collected during the test and subsequently analysed later in the laboratory.

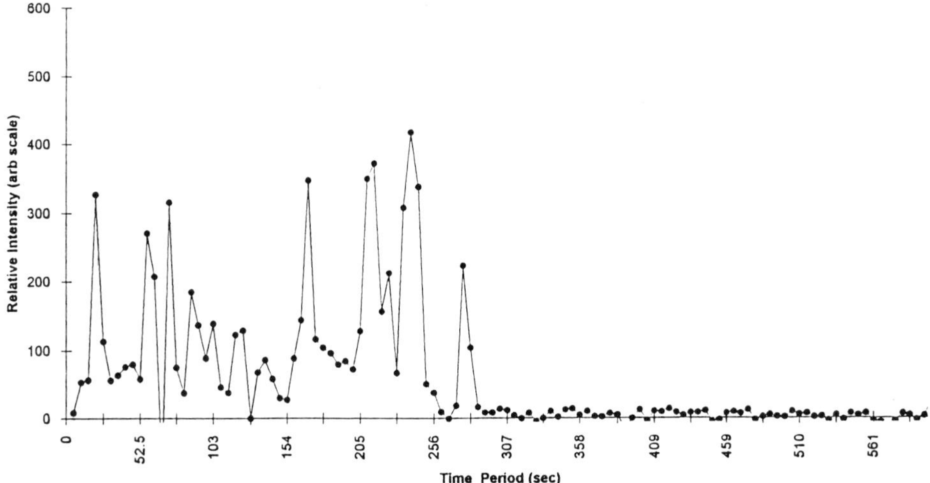

Figure 4. Measurements obtained using the MOFID deployed from the harbour boat down stream from the initial release point shortly after dye tracer release.

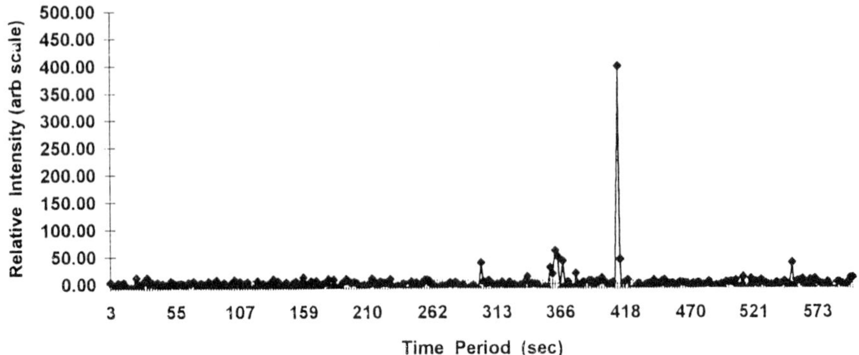

Figure 5. Measurements obtained several hours later show patches of dye tracer still detectable in the harbour.

2.Esholt sewage plant

Deployment of the MOFID to measure the dynamics and efficiency of effluent flow discharge within the various stages in a sewage plant was undertaken using two different fluorescent dye tracers. A Yellow dye tracer was used to monitor each stage within the treatment plant and then a Red dye tracer was used to monitor all of the stages continuously. As previously mentioned each dye tracer has a specific emission spectra, therefore by changing the bandpass filters at the detector simultaneous measurements are obtainable. The configuration of the twelve probes was different at each stage so that the depth and velocity of the effluent flow could be included in the results. The results from water samples taken using conventional water sampling methods and later analysed in the laboratory using an AFC were compared with real-time insitu measurements using the MOFID, figure 6 shows a typical result from the final effluent stage. As can be seen from figure 6 the results obtained from both techniques follow similar trends and are generally in good agreement, giving confidence in the valadity of the MOFID measurements

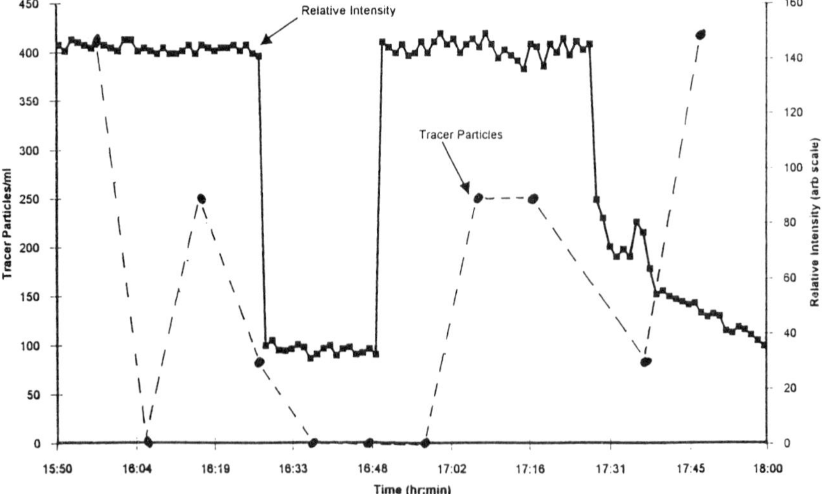

Figure 6. Comparison of the real time measurements obtained from the MOFID with that from conventional water sampling techniques at the final effluent stage.

Conclusions

It has been shown during two different field tests that dye tracer particles have several advantages over traditional dyes in that they can be altered to mimic the target particle, are environmentally benign and are detectable over long periods of time. The MOFID has demonstrated that it is capable of detecting fluorescent dye tracer particles within sediment movement or sewage effluent discharge, using different dye tracers. The results obtained from both tests using different techniques follow similar trends and are generally in good agreement, giving confidence in the valadity of the MOFID measurements. A range of environmental applications using dye tracers of different spectral characteristics and the MOFID system are under investigation.

References

1) Mellanby K.(1975) Studies in Biology No.38, The Biology of Pollution (Edward Arnold), pp20-29

2) Mellanby K.(1975) Studies in Biology No.38, The Biology of Pollution (Edward Arnold), pp40-43

3) Barnes R.S.K.(1976) Studies in Biology No.49, Estuarine Biology, pp1-11

4) Chudyk W. and Pohlig K.(1988) SPIE Vol.990, Chemical, Biochemical and Environmental Applications of Fibers, pp45-48

5)Vodacek A. and Philpot W.D.(1987) Remote Sensing of Environment, Vol.21, pp83-95

6) Althausen J.D. and Kjerfve B.(1992) Estuarine, Coastal and Shelf Science, Vol.35, pp517-531

7) McStay D. and Milne R.(1993) Sensors VI, Technology, Systems and Applications, Inst. Physics Publishing. ISBN 0 7503 03166, pp209-214

A Self-compensated optical fibre displacement sensor

Yicheng Lu, James McKenzie, Clive Butler
BCMM, Brunel University, Uxbridge, Middlesex UB8 3PH, UK

Abstract

A new optical displacement sensor, which utilizes 3 fibres working in a reflecting mode, has been developed. An aperture serves to modulate signals in a differential manner so as to compensate intensity variations and increase sensitivity. The sensor structure is very simple, rugged and easy to set-up. It has a resolution of 0.2μm and can be used to measure vibrations, pressure or other measurands involving small displacement.

1. Introduction

Optical fibre sensors have received considerable attention in recent years because they have distinct advantages over conventional sensors for some applications. Among these, intensity modulated sensors have been successfully used in industry because of their simplicity, reliability and low cost. However, the sensor has the main draw back that some referencing or compensation technique is required for good accuracy and stability. A large number of techniques have been developed[1-5]. Wavelength referencing is one of the frequently used methods, for which a second light source with a different wavelength or the same source with a filter is used. This not only adds to the complexity and introduces more noise, but also assumes that the transmission and receiving optics and electronics behave identically at the two wave lengths considered"[4]. Other techniques,[1,5] which depend on a 'GRIN' rod lens to produce a parallel beam and adopt a time delay loop to separate two signals, were also used. However, the ideal parallel beam is very difficult to achieve practically. Moreover, a coupler device is required, which increases the sensor cost and introduces instabilities when the ambient temperature changes. As a result, the main advantages of the sensor in terms of simplicity and low cost have been diminished. The main effort of this work was to investigate compensation techniques without increasing complexity and cost.

2. Measuring principle and results

The sensor is similar to the "7 fibre sensor"[6] but only three of the fibres are used in this case(see figure 1). The central fibre is for light input, and the other two fibres symmetrically reside at each side of the central fibre and serve to receive displacement modulated signals. Light from a light source (e.g. a white light incandescent source) is transmitted via the central fibre and is projected onto the reflecting surface, and then received by the two receiving fibres. Because there is an aperture in the reflecting surface, all the light projected onto this area is not reflected back and permanently lost. Therefore, the aperture will act as a "black hole". Obviously, the horizontal position of the aperture to the sensor head will decide the amount of light reflected back into the receiving fibres. In other words, the output from the two receiving fibres measures the horizontal position of the aperture to the sensor head.

For simplicity of discussion, the reflective version of the sensor (figure 1(a)) is converted into a transmission version (figure 1(b)-(d)). The aperture now becomes a opaque disk. This disk produces a shaded area at the receiving projected plane and the light intensity increases with the distance from the centre of the sensor head (figure 1(b)). Figure 1(b) shows the situation when the disk is located in the central axis of the sensor head. In this position, the opaque disk has exactly same effects on both receiving fibres. That is, both fibres receive the same amount of light. If the disk is moving in one direction (left or right), the shaded area is moves in the same direction with an apparent doubling in distance. Figure 1(c) shows the situation when the right edge of the disk (point B in the figure) moves to the central line of the sensor head. The output from the right receiving fibre will reach its maximum and the left fibre reaches its minimum. Similarly, when point B moves to position C (the central line of the right receiving fibre), the right fibre reaches its minimum and the left reaches its maximum. This indicates that the maximum measuring range is the distance between point B and point C, or the diameter of the fibres if the fibres are packed tightly together. The above analysis is based on the supposition that the opaque disk has the same dimensions as the fibres. Actually, the hole size (over quite a large range) does not affect the maximum measuring range significantly. The hole size which is smaller than the fibre diameter is taken as an example (figure 1(d)). When the centre of the hole is located at the middle of the central fibre and the one of the receiving fibre, the fibre reaches its minimum. Due to the symmetry of the sensor, the distance between the minimum from the left fibre and the minimum from the right fibre is the thickness of the fibre, and is the maximum measuring range. Although the maximum measuring range mainly depends on the fibre thickness, the shape of the sensor output curve with displacement will exhibit some changes, especially the linearity of the sensor output. Because of the existence of low sensitivity at the limits of the usable measuring range, is smaller than the maximum range, about 0.6 to 0.85 of the maximum range.

The sensor was tested on a computer controlled profile projector measuring system which has a resolution of 0.5 micron. The fibre was a glass type with a core diameter of 400μm and 424μm including the cladding. The hole diameter was 0.403mm (drilled hole). Figure 2 shows an example of the results. The dashed lines with small markers are from the each channel and the solid black line is the sensor output calculated by the following expression:

$$Output = \frac{(O_1 - O_2)}{(O_1 + O_2)}$$

Based on this expression and the measuring principle, it can be seen that all common mode signals to the two receiving fibres will be cancelled. This means that light intensity changes or drifts, intensity loss due to input fibre connection or microbending and reflectivity changes etc. are totally compensated.

Results in figure 3 were obtained when the hole centre had some offset to the central line of the sensor. This was a simplified test on the effects of hole diameter. The offset distances relative to the first position (figure 3(a)) are respectively 0.1015 (b), 0.15 (c) and 0.204mm (d). The results proved that the measuring range exhibits little change though the output from each channel changes substantially. The figure also shows that the linearity and the sensitivity is related to the hole diameters. According to the experiments, the hole diameter should be selected in the range of 0.65 to 0.9 of the fibre diameter so as to obtain good linearity with a good sensitivity.

Figure 4 shows the compensation ability to light intensity change by manually adjusting the driving voltage of the light source. The graph shows the output from one channel and the calculated output. When the voltage changed from 4.0V to 6.12V, the intensity changed by 3.41 times. But the maximum change of the calculated output only varied 1.83%. In fact, the variation was smaller (only 0.58%) in the central half of the measuring range, and the largest variation occurred at the end of the full measuring range. This means that the maximum variation of the calculated output is only 0.5 percent or less of the light intensity variation.

Figure 5 shows the results from three experiments. The coincidence of the results proved that the sensor has very good repeatability.

3. Discussion

As discussed in the above, the measuring range is dominated by the fibre thickness (or more accurately, the distance between two receiving fibres), it is easy to design sensors to meet different measuring ranges by changing fibre thickness, or changing the distance between the two receiving fibres. The ratio of the resolution to the measuring ranges mainly depends on the A/D resolution, 12 bits in this case. Therefore, high resolution is quite easily realized by reducing its measuring range. For example, if 200μm core fibre is used, a resolution of 0.1μm (or better) can be achieved without introducing much difficulty in making the sensor.

Because the compensation is achieved by two differential sensing channels, it does not introduce any referencing channel which is more commonly used as a compensation technique. In addition, this compensation technique produces higher sensitivity due to the nature of differential effects on the two channels compared with sensors using one sensing channel and one referencing channel.

The sensor made full use of the advantages of intensity modulated sensors. It is very simple to construct, only three fibre and a reflecting surface with an aperture. The sensor is easy to set-up and robust enough for using in an industrial environment. Lenses, couplers, beam-splitters etc. have been eliminated. The sensor itself is not critically affected by the components used such as the aperture size. Furthermore, the sensor allows quite large tolerances in setting up the sensor. Figure 6 is an example which shows that the sensor output is not sensitive to the stand off distance (D_s, changed by more than one millimetre). In the figure, signal levels from both channels decreased greatly with the standoff distance, but the sensor output only changed by 1.5 percent. Similarly, the sensor has some tolerance to the offset distance (see figure 3). In fact, the offset effects can be eliminated by replacing the circular aperture with a rectangular slot. Furthermore, the tilt of the reflecting surface around the measuring axis produces little effect on the sensor output because the tilt generates the same effects on the two receiving fibres. All these features significantly reduce the precision of components required in the sensor design, assembly, set-up and re-calibration. This transducer form is attractive for manufacturers requiring reliable and cheap displacement sensing.

If evenly spaced slots (similar to a grating) are used, the measuring range can be greatly extended. Due to the large tolerance of the stand off distance and the large pitch of the "grating", the measuring range can be extended to any practical length without too much difficulty. Figure 7 shows the results tested on a simple laser cut piece (of quite poor cutting quality) with a coupler connected to the central fibre to collect the signal from the central fibre. It can be seen from the figure, that when outputs from the outside fibres were close to their maxima/minima (the low sensitivity area), the output from the central fibre reached its high sensitivity region, using outside signal as references in this case. Therefore, measurement with a similar sensitivity over its full range can be realized. Similarly, the sensor can use peak values to locate between slots and use the peak sequence of the three fibres to decide moving directions. Better results could be obtained if a chemical etched sample is used.

The sensor has the potential to be extended for two dimensional measurement because of the structural symmetry of the sensor used in conjunction with a 7 fibre sensor head. It may be possible to extend the range of applications to cover 3 dimensional measurement by using the central fibre signal via a coupler. A comprehensive investigation of 3 dimensional measurement using these techniques is currently under way.

4. Conclusions

A new optical fibre displacement sensor has been developed which can reliably compensate most of the variations commonly encountered in intensity modulated sensors. The sensor structure is very simple, rugged and easy to set-up. The sensor which has been built and tested has a resolution of

0.2μm, and will find application in the measurement of vibrations, pressure or others involving small displacement.

Acknowledgement

This is part of the work required in the development of an optical fibre vibration sensor in a Brite-Euram project (OPTIMA) funded by the EC.

References

[1] Beheim, G.: "**Loss-compensation technique for fiber-optic sensors and its application to displacement measurements**", *Applied Optics*, Vol. 26, No.3, pp452-455

[2] Bois, E. et al.: "**Loss compensated fiber-optic displacement sensor including a lens**", *Applied Optics*, Vol.28, No. 3, 1989, pp419-420

[3] Adamovsky, G.: "**Fiber-optic displacement sensor with temporally separated signal and refrence channels**", *Applied Optics*, Vol.27, No. 7 pp1313-1315

[4] Culshaw, B.: "**Optical systems and sensors for measurement and control**", *J. Phys. E: Sci. Instrum.*, Vol. 16, 1983

[5] Davies, D.E.N. et al.: "**Displacement sensor using a compensated fibre link**", *Proceedings of the Society Photo-Optical Instrumentation Engineers*, 1984

[6] Butler, C. & Gregoriou, G.: "**Novel non-contact surface topography measurements using fibre optics**", *SPIE Vol. 1584 Fiber Optics and Laser Sensors IX(1991)*, pp282-293

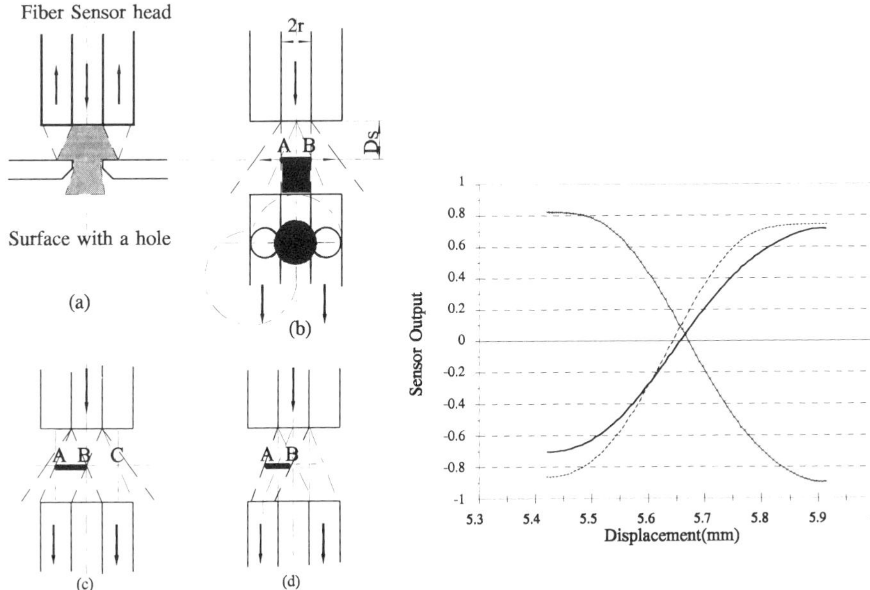

Figure 1 Sensor structure and measuring principle

Figure 2 Typical results

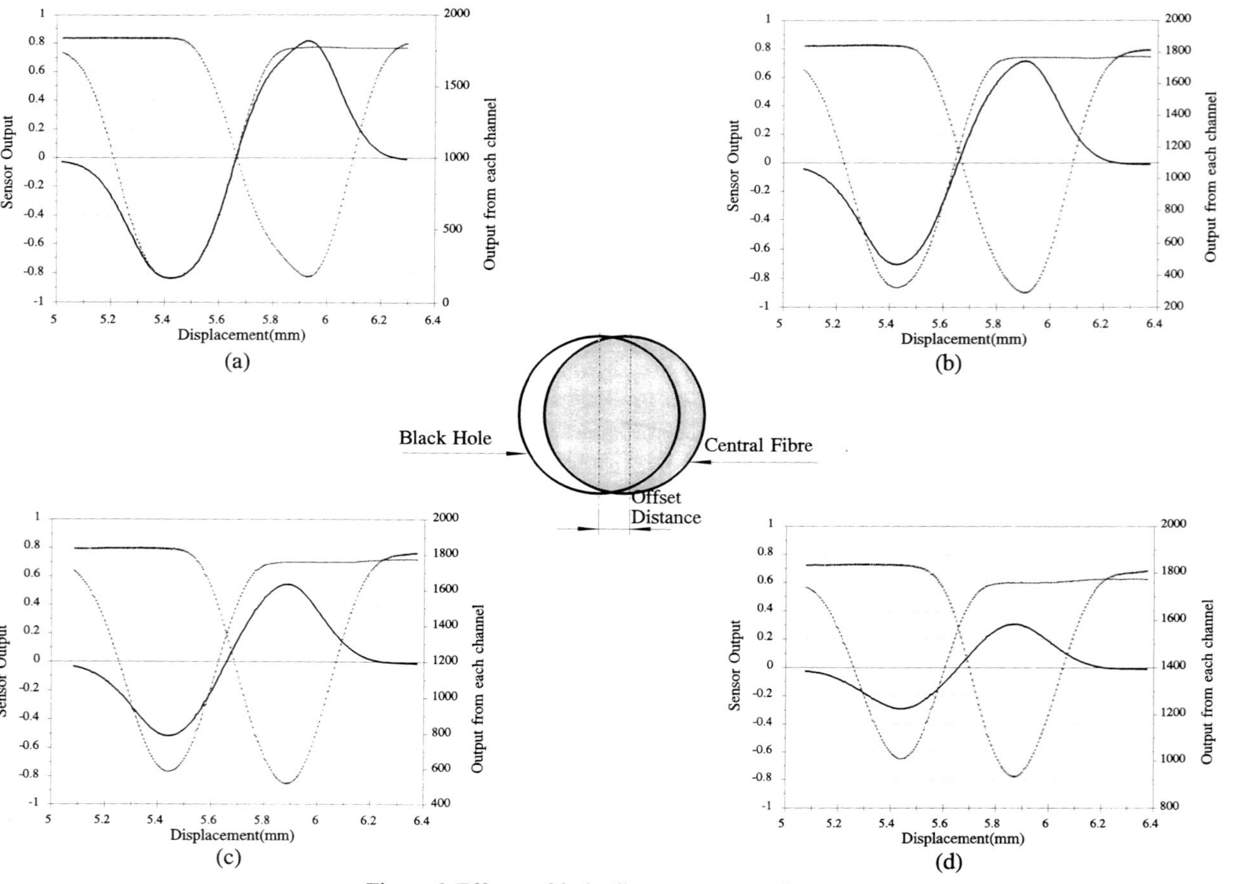

Figure 3 Effects of hole diameter on results

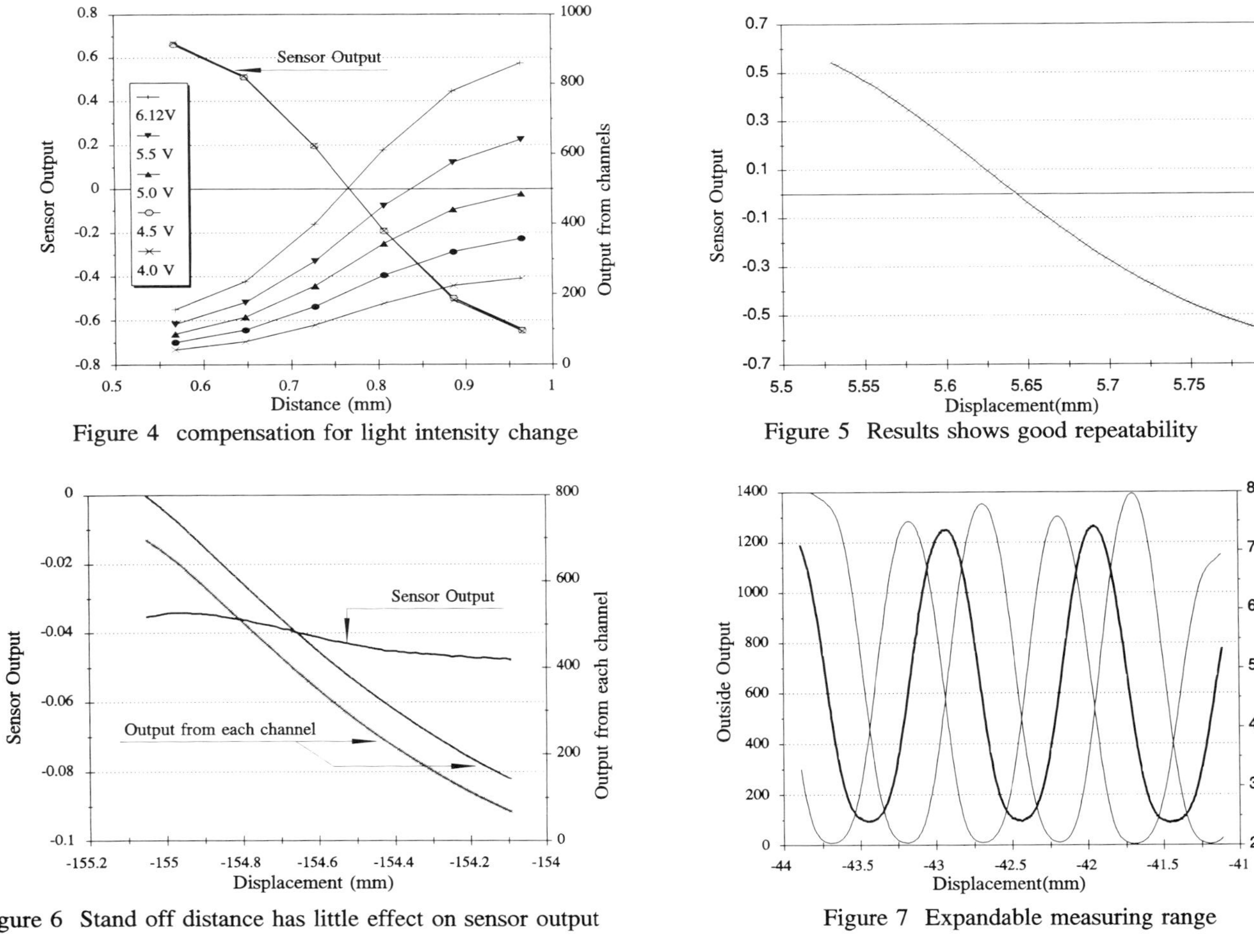

Figure 4 compensation for light intensity change

Figure 5 Results shows good repeatability

Figure 6 Stand off distance has little effect on sensor output

Figure 7 Expandable measuring range

Towards a distributed optical fibre chemical sensor

P. A. Wallace, Y. Yang and M. Campbell
Department of Physical Sciences, Glasgow Caledonian University, Glasgow G4 0BA, UK

Abstract. Progress towards a distributed optical fibre fluorosensor for pH is reported. The operation of the sensor is based on the pH dependent quenching of fluorescein dye immobilised in the porous cladding of a PCS optical fibre which has been stripped and reclad in a sol-gel coating. The analyte distribution is recovered from the OTDR response of the system to a short excitation pulse.

1. Introduction

The problem of detecting chemical species in real time and *in situ* is one which is receiving increasing attention because of its importance in industrial process control, industrial safety and in the protection of the environment. Such problems necessitate the development of robust, inexpensive chemical sensor systems capable of remote deployment and, ideally, capable of making distributed measurements where the presence of the analyte is mapped over a domain extending from metres to kilometres depending on the application. Optical fibre techniques promise to provide solutions to many of these problems [1]. The sensor heads are compact, inexpensive and, because of their all-dielectric nature, suitable for use in hazardous environments and largely immune to electromagnetic interference. The present state of affairs in fibre sensing is that a number of single-point chemical sensors have been reported and that distributed sensors for physical measurands such as strain and temperature are in the commercial domain. Distributed chemical sensors, while being universally regarded as an attractive concept, are in a much earlier stage of development. This work describes a system which brings together a number of individually well proven elements in a novel distributed sensor configuration.

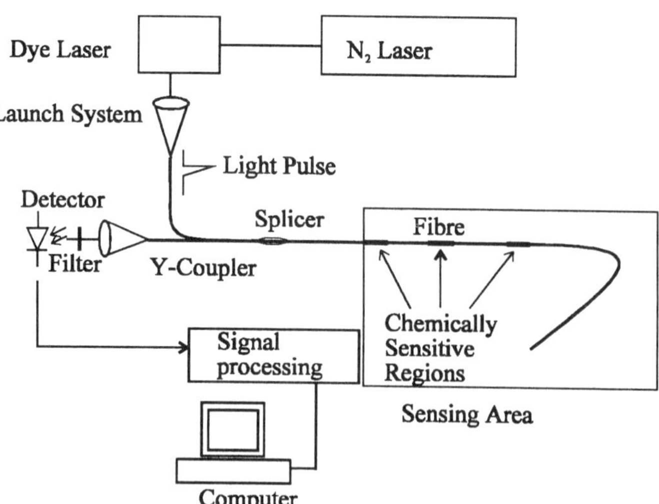

Figure 1. The experimental configuration for the distributed system.

Fibre optic sensors may be described as either intrinsic or extrinsic according to whether the fibre itself forms part of the sensing medium or merely acts as a waveguide. One considerable attraction of the intrinsic type is the possibility of development of a spatially extended, position sensitive sensor system. Such a system is conventionally described as distributed if the position sensitivity extends over its entire active length and quasidistributed if it consists of a series of sensitive regions spaced over the length of the fibre. One method of deriving position information from the sensor is to analyse the temporally extended reflected light pulse originating from a short excitation pulse launched into the fibre. This technique is known as optical time domain reflectometry (OTDR). In this work we propose to investigate a variation upon the theme whereby the excitation pulse produces fluorescence in the sensor sections. The advantage of using a fluorescence signal is that it is separated in wavelength from the excitation. The concomitant drawback is that the time resolution of the system, upon which depends directly the positional information, is limited by the lifetime of the fluorescence deexcitation. Given a lower limit of 10 ns on the lifetime, and a propagation time of 5 ns per metre in the fibre, the lower limit on position resolution of the system would be ~ 1 metre. This would, however, be adequate for many environmental and hazard monitoring purposes.

2. Instrumentation

The model system we have chosen to work with is a pH sensitive system based on the quenched fluorescence of fluorescein dye immobilised in the porous cladding of a modified optical fibre. The experimental configuration is shown in figure 1. The light source is a compact dye laser operated at 440 nm pumped by a N_2 laser producing 50 μJ pulses of duration 1 ns. (ILEE model NN100). The light is launched into a PCS optical fibre (Spectran HCP-M0200T, 200 μm core, NA = 0.37) via a directional coupler. The excitation radiation couples to the dye molecules via the evanescent wave which exists within the fibre cladding close to the core and produces luminescence in the region of 500 - 600 nm, some fraction of which will be launched back into the core to be detected at the end of the fibre after the appropriate propagation delay. The fluorescence light is passed through a 500-580 nm band pass filter and is detected using an EMI 9125 photomultiplier whose output is viewed on a 100 MHz Hewlett Packard storage oscilloscope. The reflectometry response of a 10 metre length of fibre is shown in figure 2.

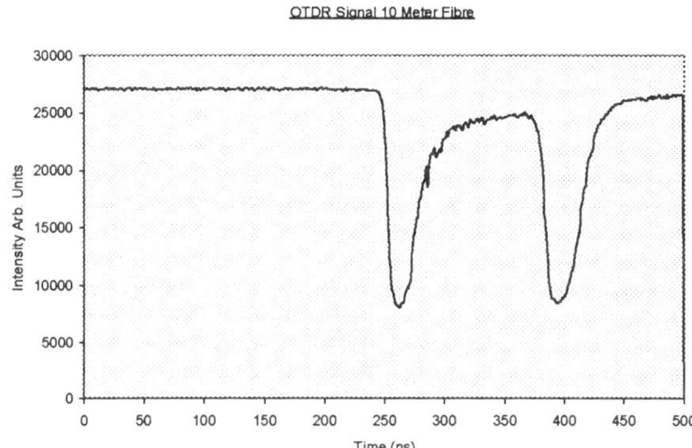

Figure 2. OTDR signal from 10 metres of fibre terminated in a cuvette of fluorescein solution.

The transduction mechanism is based upon the modification of the fluorescence yield of the indicator dye depending on the pH of the surrounding medium. This quenching phenomenon is illustrated in the context of an aqueous solution in figure 3.

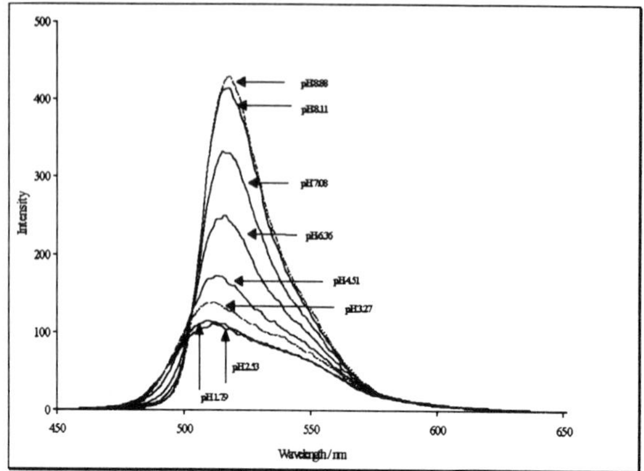

Figure 3. Variation of luminescence intensity of fluorescein over the region pH 2 - pH 8.

3. The sol-gel cladding

One of the most technically challenging aspects of the work is the fabrication of the doped porous fibre cladding. Following other work [2,3] we have elected to utilise a sol-gel process to produce a glass matrix within which to immobilise the indicator dye. The sol-gel process involves the hydrolysis of a metal alkoxide solution to produce a gel which may then be further cured to produce a hard glass material. In this work, tetraethyl orthosilicate $Si(OC_2H_5)_4$ was used to make the solution under acid catalysis. The $Si(OC_2H_5)_4$ was mixed with H_2O and C_2H_5OH with the aid of ultrasonic agitation in a hot water bath. The HCl was added slowly, then the fluorescein dye was added in. The resulting solution was then baked in an oven for about 5 hours at 75 degrees to achieve a suitable viscosity. The fibres used were 200 μm core plastic clad silica (PCS) fibres from which had been removed short lengths of the plastic cladding.

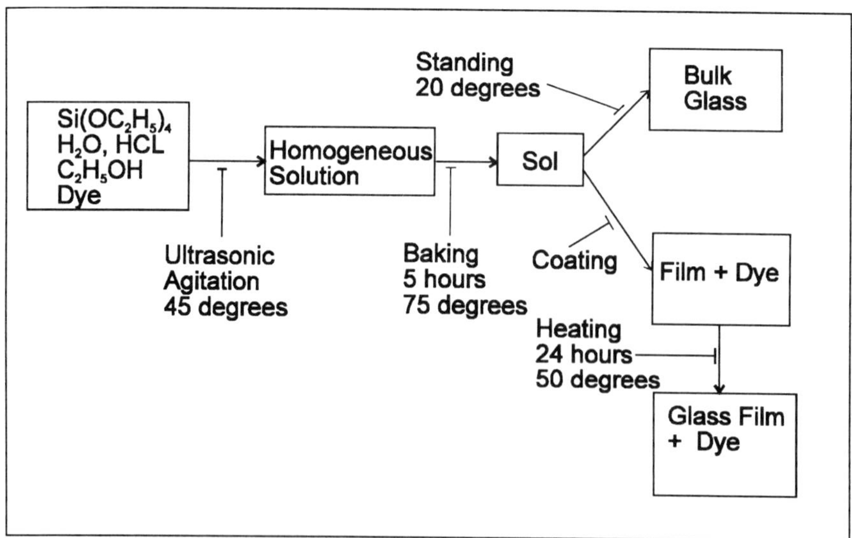

Figure 4. Block diagram of the sol-gel process used to create the fibre cladding

Attempts were made to coat stripped sections both at the end of a length of fibre and in the middle of a length. The end section coating was performed by simply dipping the unclad fibre end into the solution for about 10 minutes and the drawing it out at a controlled speed. The middle section coating was made by immersing the unclad section into a shallow pool of solution lying in a watch glass and then drawing it through the solution at a controlled speed. The coatings made in this way were then heated in an oven for 24 hours in order to cure them. In this process, the solution baking time and the drawing speed are both important for control of the thickness of the coating. The longer baking time results in higher solution viscosity and therefore greater thickness. From the experiment results, higher drawing speeds result in greater thickness. A speed of 0.13 mm/s was found to give a thickness of approximately 1 μm which is in the most suitable range for the present application.

4. Acknowledgments

One of us, Y.Y., wishes to thank Glasgow Caledonian University for support in the form of a research studentship.

5. References

[1]. B. Culshaw, Optical Fibre Sensing and Signal Processing, 1984, Peter Peregrinus
[2]. C. Jeffrey Brinker, George W. Scherer, Sol-Gel Science, 1990, Academic Press
[3]. B. D. MacCraith, V. Ruddy, et al, Optical Waveguide Sensor Using Evanescent Wave Excitation of Fluorescent Dye in Sol-Gel Glass, Electronics Letters, 4th July 1991, Vol. 27, No. 14

A fibre optic based ellipsometer for thin film based sensor systems

R Chitaree, K Weir[†], V Murphy*, A W Palmer, K T V Grattan and B D MacCraith*

Measurement and Instrumentation Centre, Department of Electrical, Electronic and Information Engineering, City University, Northampton Square, London, EC1V OHB, England.

[†]The Blackett Laboratory, Imperial College of Science Technology and Medicine, Prince Consort Road, London SW7 2BZ England

*School of Physical Science, Dublin City University, Glasnevin, Dublin 9, Ireland.

Abstract The performance of a highly birefringent fibre ellipsometer system, based on a polarization modulation technique, as an element in an optical sensor system is described. The system is used to monitor the response of three sol-gel films, representing the sensor elements, in response to temperature and pressure. The extension of the system to chemical sensor systems is discussed.

1. Introduction

Ellipsometry is a well known optical technique for characterization of optical surfaces. The information gained may be related to the optical properties of the sample and calculation provides optical constants such as the index of refraction and thickness of a thin film under study. The main advantages of ellipsometry being that the measurement is both non-contact and non-destructive. Such features are very important and, in principle, such an instrument forms the basis of a highly important measurement tool. However, the high cost and complexity of the instrument has meant that it is a large, laboratory based instrument, mainly finding use in specialised area of measurement.

Recently, a new scheme for an ellipsometric measurement system using a highly birefringent fibre polarization modulation technique was proposed [1]. The technique produces a linear output polarization, the azimuth of which is continuously rotating. This is a major development in ellipsometry as no mechanically adjustment is required during the measurement procedure and the sample may be continually monitored on a time scale associated with the polarization rotation rate. The resulting ellipsometric system is more compact, flexible, non-mechanical and has potential for high speed operation. The performance of the instrument has also been thoroughly investigated and tested in the ellipsometric characterization of a simple system consisting of the interface of two semi-infinite media [2].

Here, the potential of this fibre polarization modulated ellipsometer for application in thin film sensor systems is demonstrated. The thin film represents the sensor element, the optical properties of which respond to some external factor. The reduced cost and simplicity of the fibre based ellipsometer makes the application of this system to such a measurement regime an attractive proposition.

A suitable range of thin films, chosen for this preliminary study, was sol-gel films. Sol-gel thin films are used as antireflection coatings and to form optical waveguides, so detailed information on their refractive index and film thickness is often required [3]. Further their use in immobilising chemically sensitive species is important in the construction of optically based chemical sensors [4,5].

In this work, results are presented illustrating the response of three sol-gel thin films to variation in temperature and humidity. These results clearly demonstrate the potential of the fibre based ellipsometer system in thin film based sensor systems. The application of the fibre ellipsometer to chemical sensors based on sol-gel films is also discussed.

2. Theoretical Background

An ellipsometer is used to detect changes in the polarization state of polarized light reflected from a sample. This change is described in terms of the ellipsometric angles ψ and Δ. ψ is defined as the inverse tangent of the ratio of the amplitude Fresnel reflection coefficients and Δ is simply the phase difference introduced between these two principal directions in the reflected light i.e.

$$\psi = \tan^{-1}\left(\frac{|r_p|}{|r_s|}\right) \qquad (1)$$

$$\Delta=\delta_p-\delta_s \qquad (2)$$

where r_p and r_s are the (complex) Fresnel amplitude reflection coefficients for light polarized parallel (p) and perpendicular (s) to the plane of incidence respectively; δ_p and δ_s are the corresponding phase changes on reflection.

This technique is easily applied to bulk materials [2] and may also be applied to film/substrate systems. ψ and Δ are defined in exactly the same way, but in this case the Fresnel reflection coefficients are complicated by multiple reflections within the thin film. The overall complex Fresnel reflection coefficients for light polarized in the *p* and *s* directions are given by [3]:

$$r_v = \frac{r_{01v} + r_{12v}e^{-i2\beta}}{1+r_{01v}r_{12v}e^{-i2\beta}} \qquad (3)$$

where r_{01v} and r_{12v} are reflection coefficients for the ambient-film (01) and film-substrate (12) interfaces, respectively and v represents either the p or s polarization direction. β is the film phase thickness which is given by:

$$\beta = 2\pi\left(\frac{d_1}{\lambda}\right)\left(N_1^2 - N_0^2\sin\phi_0\right)^{\frac{1}{2}} \qquad (4)$$

where λ is the freespace wavelength, d_1 is the film thickness, N_1 is the refractive index of the film (medium 1), N_0 is the refractive index of the ambient medium (incident medium) and φ_0 is an angle of incidence.

Numerical methods are available to transform the ellipsometric parameters to values of thin film thickness and refractive index but here, for simplicity, only the ellipsometric angles are calculated. Clearly, ψ and Δ depend on the optical properties of the thin film (Eqs. (1) to (4)). Thus, if there is any process that changes any of the optical variables, ψ and Δ will also change, and the ellipsometer may be utilized as part of as a sensing scheme.

3. Experiment

3.1. Apparatus

Fig. 1 illustrates the highly birefringent fibre polarization modulation scheme as used in this ellipsometer. The details of the polarization modulation technique have been given elsewhere [1], [2]. The operation is summarized briefly as follows. Light from a HeNe laser (L) is linearly polarized (P1) at 45° relative to the eigenaxes of the highly birefringent fibre and launched into the fibre (F) by way of x10 objective lens (Ls1). This gives rise to equal population of both eigenmodes of the fibre. The fibre modulation unit (FMU), driven by a sawtooth signal (frequency ≈1Hz) stretches the fibre longitudinally to modulate the birefringence. Thus, the phase between the two orthogonally polarized beams emerging from the eigenmodes of the fibre is similarly modulated. At the output end, these beams are coupled out of the fibre by a x20 objective lens (Ls2) and they propagate through a quartz quarter-wave plate (QWP) orientated at 45° to the eigenaxes of the fibre. The function of the wave plate is to transform the output beams into two orthogonal circularly polarized beams with opposite handedness. This is equivalent to a linear polarization state with an azimuth determined by the phase difference between the two circularly polarized beams. As this phase difference is modulated, the azimuth of polarization is modulated. The frequency of modulation of the azimuth depends on the frequency and amplitude of the sawtooth modulation. For the purposes of this initial investigation the frequency of polarization modulation was limited to 1Hz, though this does not represent the limitation of the system.

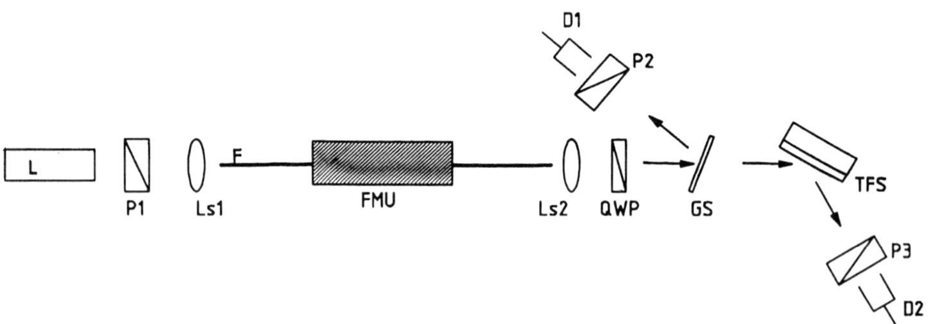

Figure 1. A Schematic of the optical fibre polarization modulation ellipsometer. L HeNe laser; P1,2,3 polarizing prisms; Ls1,2 lenses; F highly birefringent optical fibre; FMU fibre modulation unit; QWP quartz quarter-wave plate; GS glass slide; TFS thin film sample and D1,2 photodetectors.

A fraction of the output is sampled by a glass slide (GS), adjusted to have an angle of incidence of ≈3°. This fraction passes through a polarizing prism (P2) orientated at a pre-determined direction and onto a photodetector (D1) to form a reference signal. From this signal the azimuth of polarization may be determined at any time. The main light beam is incident onto the thin film sample sensor element (TFS) at an appropriate angle of incidence. The reflected light then propagates through a polarizing prism (P3) with its transmission axis orientated at 45° to the plane of incidence, and onto a photodetector (D2) to provide the output signal. Both signals (output and reference) are then transferred to an on-line computer for further analysis.

With the final polarizer (P3) orientated at an angle of 45° to the plane of incidence the detected intensity, I_f, is found to be [1],

$$I_f = \frac{I_0}{2}\left\{\left(\frac{R_p + R_s}{2}\right) + \left(\frac{R_p - R_s}{2}\right)\cos(\varphi(t)) + \sqrt{R_p R_s}\,\sin(\varphi(t))\cos\Delta\right\} \tag{5}$$

where I_0 is a constant, R_p and R_s are the intensity Fresnel reflection coefficients and $\varphi(t)$ is the azimuth of polarization of the incident light.

The reference signal allows for any particular orientation of azimuth of the input rotating polarized light to be identified. As a result, the reflected intensities, corresponding to known input polarization states, can be identified in the final signal. This allows ψ and Δ to be determined through Eq. (5), thus enabling the characterization of the reflection sample without mechanical movement of any optical components.

3.2. Thin film preparation

The sol-gel thin films were prepared on a silicon substrate using the dip-coating method using mixtures of SiO_2 and TiO_2 'Liquicoat' solutions supplied by Merck. The procedure followed was similar to that used in producing sol-gel films for chemical sensors [5]. Three similar films were produced and their refractive index and thickness recorded immediately after curing using a conventional commercial Rudolf AutoEl-III ellipsometer. Each sample had a refractive index of 1.52 but differing thin film thickness as follows; sample 1: 145nm, sample 2 :157nm and sample 3: 179nm. Each sample was then placed in the new fibre ellipsometer system, at an appropriate angle of incidence, for characterization as a sensor element.

4. Results and Discussion

To test the effect of humidity, a source of steam was place in the vicinity of the sample under investigation. Care was taken to avoid significant change in temperature and condensation around the sample. The temperature was continually monitored using a thermocouple.

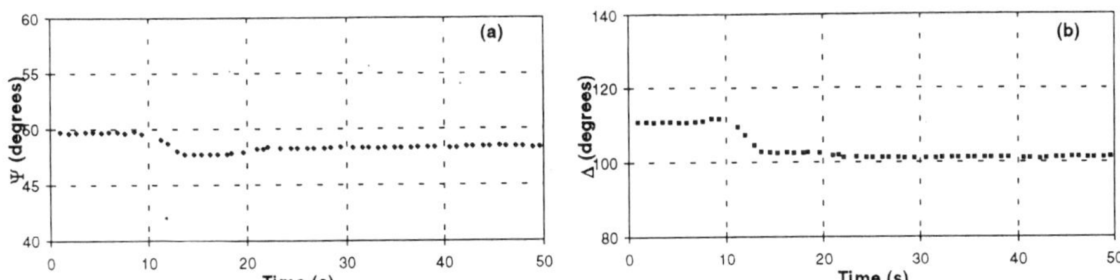

Figure 2. Variation in ellipsometric parameters; (a) ψ and (b) Δ for thin film sample 1 over a 50s period after exposure to humidity. The first 10s provides a reference for the isolated sample. Angle of incidence 60°.

Fig. 2 shows the variation in ψ and Δ for sample 1 on exposure to a humid atmosphere (with an angle of incidence of 60°). The sample was shielded for the first ten seconds to provide a reference value for the unperturbed sample. A clear variation in both ellipsometric parameters may be seen, although it is more pronounced in Δ. The temperature of the sample was monitored to be 35°C. The change in both ψ and Δ was observed to be irreversible under normal laboratory conditions. One possible source of this effect may be the water vapour entering the pores of the film and causing a permanent swelling Fig. 3 shows the results for sample 3 on exposure to a humid atmosphere (angle of incidence 70°) and no discernible change is observed. One possible explanation is that the sample response is highly dependent on the detail of the film fabrication. This is the subject of a continuing investigation of these phenomena.

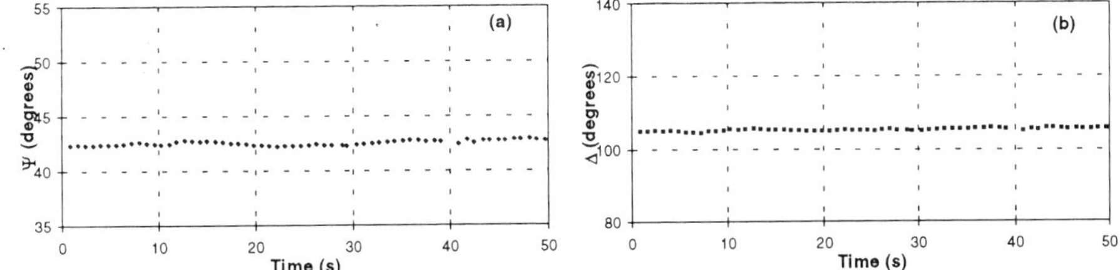

Figure 3. Variation in ellipsometric parameters; (a) ψ and (b) Δ for thin film sample 3 over a 50s period after exposure to humidity. The first 10s provides a reference for the isolated sample. Angle of incidence 70°.

For temperature measurements, a source of hot air was used to produce a temperature rise in the sol-gel sample. The heat generated by the hot air stream could raise the sample temperature to 90°C (monitored by the thermocouple).

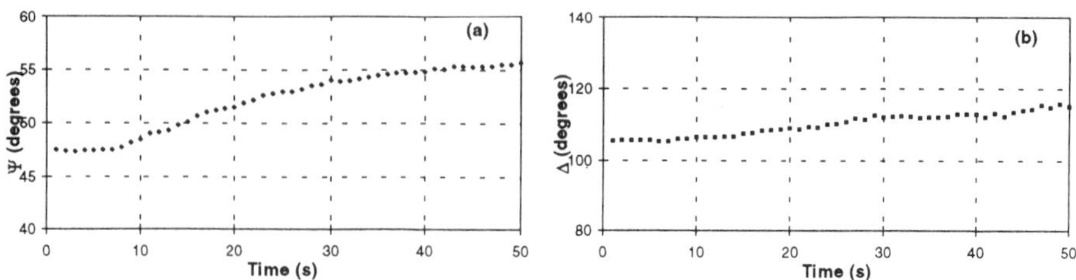

Figure 4. Variation in ellipsometric parameters; (a) ψ and (b) Δ for thin film sample 2 over a 50s period after exposure to increasing temperature. The first 5s provides a reference for the isolated sample. Angle of incidence 60°.

Figs. 4 and 5 show the variation in ψ and Δ for samples 2 and 3 respectively on exposure to an increase in temperature (angle of incidence 60° and 70° respectively). In both cases the sample was shielded for the first 5 seconds to provide a reference value for the unperturbed samples. Both samples show an increase in both ψ and Δ over the time period 5 to 30s. The response is then steady as the samples reach the maximum attainable temperature of 90°C. In both cases a larger response is observed in the variation of ψ. The changes in the samples were again observed to be irreversible.

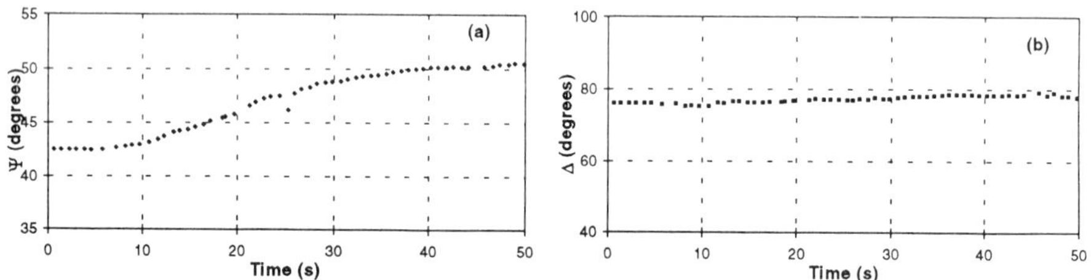

Figure 5. Variation in ellipsometric parameters; (a) ψ and (b) Δ for thin film sample 3 over a 50s period after exposure to increasing temperature. The first 5s provides a reference for the isolated sample. Angle of incidence 70°.

5. Conclusion

A novel reflection ellipsometry employing a highly birefringent fibre polarization modulation technique has been implemented in a sensor system, based on sol-gel thin films. The results obtained from the three samples studied, recorded the change in the ellipsometric parameters ψ and Δ of the thin films in response to humidity and temperature over a period of 50s. The changes were observed to be irreversible under normal conditions and so may not represent useful sensors for temperature and humidity (although sensors may operate using disposable sensor elements).

The results are highly encouraging in that they clearly demonstrate the potential of the fibre ellipsometer as an integral element in sensor systems utilizing thin film sensors. Production of suitable sensor elements, based on sol-gel films, for a range of chemical sensors is a realistic prospect and the subject of continuing work.

6. Acknowledgements

The authors are pleased to acknowledge support from the Engineering and Physical Science Research Council (EPSRC). R C acknowledges support from the Thai Government by way of a studentship.

7. References

[1] R Chitaree, K Weir, A W Palmer and K T V Grattan, *A HiBi Fibre Polarisation Modulation Scheme for Ellipsometric Measurements*, in 'Sensors VI : Technology, Systems & Applications' K T V Grattan and A T Augousti (Eds.), Published by Institute of Physics Bristol, pp 275-280 (1993).
[2] R Chitaree, K Weir, A W Palmer and K T V Grattan, *A Highly Birefringent Fibre Polarisation Modulation Scheme for Ellipsometry: System Analysis and Performance*, Measurement Science and Technology **5** pp1226-1232 (1994).
[3] R M A Azzam and N M Bashara, *Ellipsometry and Polarised light*, North Holland, Amsterdam (1977).
[4] B D MacCraith, *Enhanced evanescent wave sensors based on sol-gel derived porous glass coatings*, Sensors and Actuators B, **11**, pp 29-34 (1993).
[5] B D MacCraith, V Ruddy, C Potter, B O'Kelly and J F McGilip, *Optical waveguide sensor using evanescent wave excitation of fluorescent dye in sol-gel glass,* Electronics Letters **27** pp1247-8 (1991).

Millimetre-wave quasi-optical transmissometry for measuring leaf water content

S Hadjiloucas, L S Karatzas, J W Bowen, D A Keating, and M J Usher

Department of Cybernetics, University of Reading, PO Box 225, Whiteknights, Reading RG6 2AY, Berks., UK

Abstract. The monitoring of water uptake in plants is becoming increasingly important. The performance of a novel transmissometer operating at 94 GHz is evaluated for measuring leaf water content. The instrument is suitable for non-invasive, continuous monitoring of water uptake and translocation from individual leaves at periods of water stress.

1. Introduction

The development of plants is much affected by adverse environmental parameters, collectively known as water stress. The quantitative study of both moderate and severe water stress may provide information that can help to increase plant growth under field conditions.

Moderate stresses have the potential of inducing reversible growth inhibition, whereas more severe stresses may produce irreversible cell injury. Living cells need to be more or less saturated with water to function normally. The water content is usually expressed as relative to that at full saturation, i.e. the relative water content (RWC) or water saturation deficit. The energy status of the water in a leaf, is usually expressed as the total leaf water potential. These two parameters are linked through the moisture release curve or water potential isotherm, which is almost the same for individual leaves of the same plant.

At present, most measuring devices and associated techniques are either bulky or invasive, making continuous monitoring of the water status of the plants difficult and time consuming. Since 1950, thermocouple psychrometers [1] and pressure chambers [2] have been employed to measure total leaf water potential directly. Unfortunately, however, these techniques are destructive and therefore preclude repetitive measurements in a given tissue.

Methods for non-invasive measurements include the use of leaf clamps and LVDTs to measure leaf deformation, correlating the latter to changes in RWC and leaf water potential [3]-[6]. On the other hand, the employment of optical techniques [7] and in particular of optical fibres [8] in measuring leaf water content using a reflectance type, amplitude modulated displacement transducer has shown considerable advantages over conventional LVDTs, in that the former do not introduce errors by applying pressure on the leaf during the measurement, and they do not require temperature compensation for the leaf temperature. Furthermore, the optimisation of the emitting and receiving angles of the fibres [9] has shown two orders of magnitude improvement in responsivity and improved resolution over conventional optical fibre measurement techniques. A disadvantage of this method is the sensitivity of measurements to leaf angular misalignment, (compensated by triangulation and feedback) and the limited surface area of the measurand (a function of the numerical aperture of the fibres).

An alternative approach to measuring leaf water content, using a transmissometer at 94 GHz, is presented. Millimetre wavelength electromagnetic radiation is strongly absorbed by water, a fact that is used in the measurement of atmospheric water vapour profiles using meteorological satellite radiometers. In the system presented here, we measure the millimetre-wave transmission of leaves, this being dependent on the absorption which is, in turn, dependent on the leaf water content.

This technique has a number of advantages: firstly, the transmission directly depends on the water content; the method is non-destructive and non-contact; minor angular misalignments should have a fairly limited effect; and the millimetre-wave beam can have a fairly large cross-section, thereby averaging over irregularities in the leaf structure.

2. Experimental arrangement

Figure 1 is a schematic diagram of the transmissometer layout. The system is quasi-optical, i.e. it makes use of lenses (made from high-density polyethylene) to control the beam but, as the lens apertures are kept to only a few tens of wavelengths in diameter in order to maintain compactness, the beams suffer diffractive spreading as they propagate. It is consequently necessary to have a train of lenses through the system, each lens converging the beam to a beam-waist from which it will diverge as it propagates to be picked up and reconverged by the next lens. In this manner it is possible to transmit a millimetre wave beam with low loss over an appreciable distance. The design of quasi-optical systems is dealt with in [10] and [11].

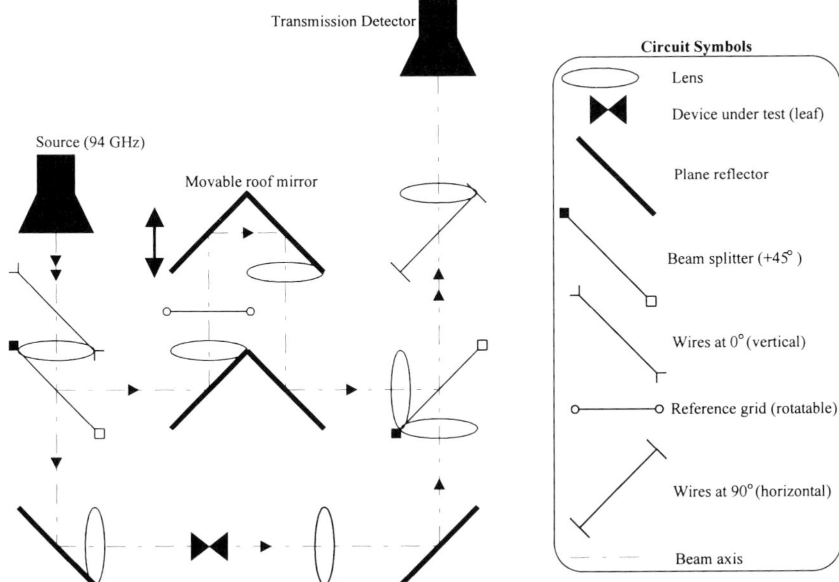

Figure 1. Quasi-optical transmissometer setup.

The system also makes use of wire grid polarising beam splitters made from free-standing 10 μm diameter tungsten wires wound onto frames at a centre-to-centre spacing of 25 μm. A grid resolves an input beam into two orthogonal linearly polarised components: that with its E-vector parallel to the wires being reflected, whereas the perpendicular component is transmitted. A grid may also be used in a time-reversed manner to combine two orthogonal linearly polarised beams to produce an emergent elliptically polarised beam. The description above remains valid for any angle of incidence.

The transmissometer is based on a hybrid Mach-Zehnder/Martin-Puplett interferometer. The beam launched from a 94 GHz IMPATT oscillator (10 mW) via a corrugated feed horn, is passed through a wire grid to ensure polarisation purity. The beam is then split into two orthogonal linearly polarised components by a 45° grid. Each component takes a different path through the instrument: one is transmitted through the sample to be measured, while the second, the reference beam, travels through an adjustable grid and reflector arrangement. The beams are recombined at a second 45° grid, the resultant travelling to the analyser grid where it is again split into two orthogonally polarised components, one of which is transmitted to the detector while the second is dumped.

The adjustable grid in the reference beam may be rotated about its optic axis so that the amplitude of the transmitted beam can be adjusted: this depending on the angle between the grid wires and the incident polarisation. The path length of the reference beam can be modified by movement of the roof mirror. If the amplitude and phase of the reference beam are exactly matched to those of the beam transmitted through the sample the resultant beam, on recombination, will have the same (linear) polarisation as the input beam prior to splitting. The analyser grid is oriented to reflect all of this polarisation and, in this balanced state, a null is recorded at the detector. If the amplitude and phase of the two beams are not matched the recombined beam will have a component which will be transmitted through the analyser grid to the detector, registering an imbalance. Jones matrices [11] may be used to relate the angle between the orientations of the reference beam grid that result in nulls to the amplitude transmission coefficient of the sample, T:

$$T = \cos^2 \theta \tag{1}$$

where θ is half the angle between adjacent null orientations measured about the reference beam grid orientation that transmits all of the incident polarisation. The phase change introduced by the sample may be quantified by noting the distance that the roof mirror has to be moved through to achieve a null.

The use of a null-balancing technique means that the measurement precision depends on the precision with which the null angles can be measured and is independent of source and detector fluctuations (which affect both the sample and reference beams in equal proportion).

The complete measuring system is illustrated in Figure 2. The IMPATT oscillator is square-wave modulated at 1.5 kHz and the output from the detector, a Flann Microwave point-contact crystal detector fitted with a corrugated feed horn, is fed into a lock-in amplifier which drives a chart recorder. The beam at the sample location has a Gaussian transverse amplitude distribution with a $1/e$ amplitude half-width of 10 mm. Quasi-optical components are available from [12].

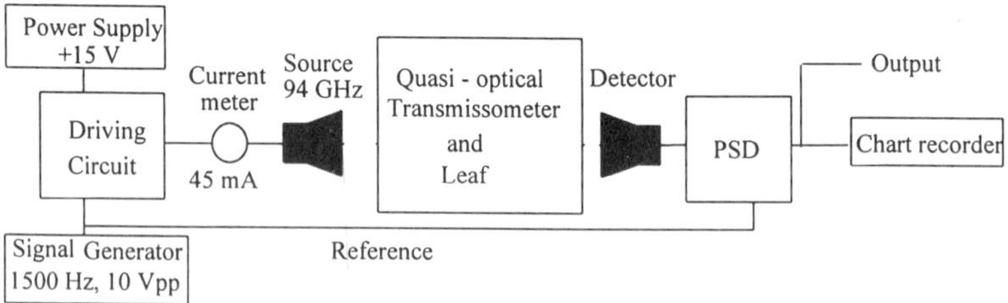

Figure 2. Block diagram of experimental setup.

The practical implementation (using half-cubes) of the quasi-optical transmissometer can be seen in Figure 3. Optical elements (lenses, polarizers, etc) may be attached to the faces of a half-cube. A transmissometer (or any other millimetre-wave optical circuit) can then be assembled by mounting an appropriate arrangement of half-cubes on a flat baseplate.

Figure 3. Transmissometer assembly using half-cubes.

3. Results

Several transmission tests were performed. Initially, a circular cuvette of beam-width size, accommodating a liquid path length of 300 μm, was made of highly-transmissive PTFE and filled with polyethylene glycol solutions (PEG), of molecular weight = 20000, and of different concentrations corresponding to different water potentials. The corresponding transmittances of the PEG solutions were determined using (1) by ratioing the transmittance of the filled cuvette to that of the empty cuvette. The results are shown in Figure 4.

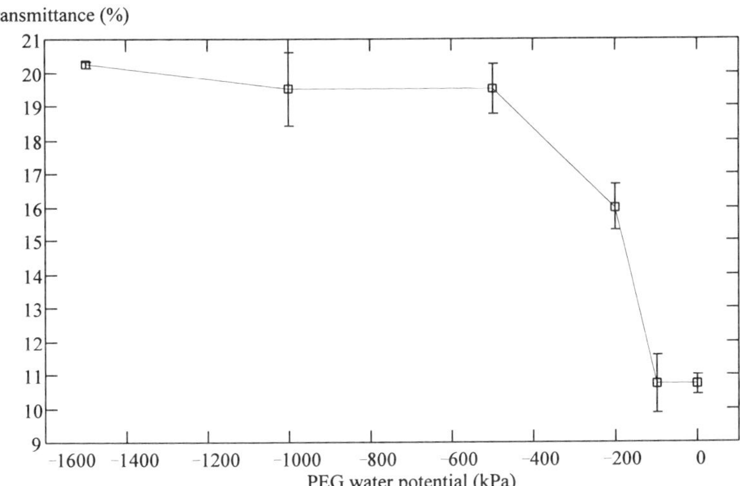

Figure 4. Graph of transmittance against PEG water potential.

Once the relation between water potential and transmission was established the cuvette was replaced by plant leaves (*Fatsia japonica* (Thunb.) Decne. & Planchon). Replicate leaves were immersed in PEG solutions of different concentrations for 24 hours. After equilibrating the samples at different water potentials, they were placed in the transmissometer and measured. The results are shown in Figure 5. In an attempt to quantify and reduce the effect of leaf thickness variations each point is the result of measuring the transmission at three different positions through the same leaf.

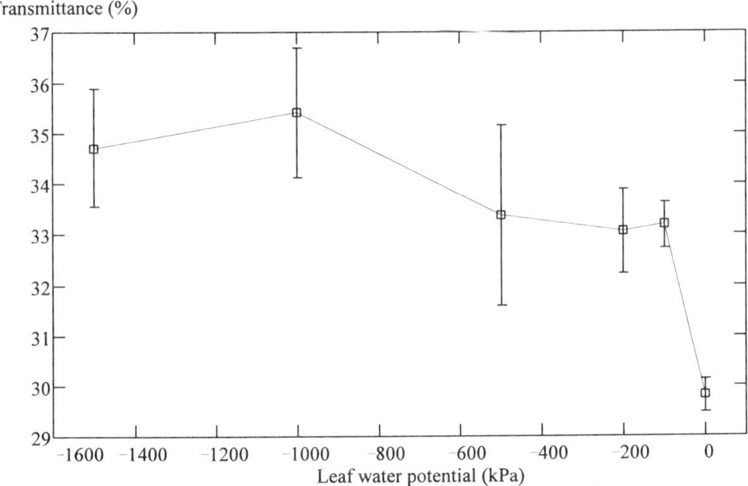

Figure 5. Graph of transmittance against leaf water potential.

Water potentials in plants are usually in the range of -100 kPa to -1500 kPa, and PEG solutions were chosen as a suitable osmoticum to induce these moderate water stresses. The 0 kPa water potential indicates a fully saturated leaf, while a high negative water potential indicates a relatively dry leaf. The transmittance shows the expected high to low trend as the water potential ranges from -1500 kPa to 0 kPa.

Both curves have the same form showing an increased sensitivity of the transmission to changes in water potential at the lower end of the negative water potential range. This correlates well with plots of RWC against water potential, the transmission being more directly related to RWC.

Analysis of the results indicates an instrument precision of 1% in transmittance. The increased error bars for the leaf measurements are the result of thickness variations in the leaves.

4. Conclusions and suggestions for further work

We have demonstrated the potential of using millimetre-wave transmissometry for measuring leaf water content. A primary advantage of the technique is that leaf surface irregularites are small compared to the wavelength, and the measured transmission is averaged over the relatively large beam cross-section. The technique is also non-destructive and non-invasive. A novel quasi-optical null-transmissometer used to carry out the measurements has been described.

There remains, however, a relatively large uncertainty in the measured leaf water content due to thickness variations between leaves and over the surface of large leaves. While the water content and thickness are interrelated, the transmission amplitude is primarily dependent on the water content and the transmission phase shift is primarily dependent on the thickness. We intend to explore the possibility of compensating for leaf thickness variations by phase measurement in future experiments.

The sensitivity to water content could be maximised by operating at a frequency at which water shows a stronger absorption. Working on the edge of one of the water vapour absorption lines at 558 GHz or 754 GHz should allow the detectivity to be optimised. In the future we intend to carry out wide-band transmission measurements using a Martin-Puplett interferometric spectrometer to determine the optimum operating frequency for a leaf water content measurement system. We also intend to improve the precision of the transmissometer by using a feedback system to achieve null balancing.

Thermocouple psychrometers and pressure chambers will be used to further evaluate the instrument performance. A greater range of water potential could be explored by replacing the PEG solutions with NaCl solutions which are more appropriate for inducing responses in xerophytes (-3 to -4 MPa) and halophytes (-10 MPa).

The ultimate aim is to develop an intelligent integrated hand-held water content sensor based on these techniques. We are collaborating with the Universities of Bath, Nottingham and Leeds in the development of suitable integrated technology as part of the EPSRC-funded Terahertz Integrated Technology Initiative (TINTIN).

5. Acknowledgements

The authors would like to acknowledge Professor D.H. Martin for originally suggesting the concept of a quasi-optical null-transmissometer. Furthermore, the technical help from Mr. C. King is greatly appreciated.

6. References

[1]. Slatyer R.O., 1958. The measurement of diffusion pressure deficit in plants by a method of vapour equilibrium, *Australian Journal of Biological Sciences,* **11,** 349-365.

[2]. Scholander P.F., Hammel H.T., Bradstreet E.D, and Hemmingsen E.A., 1965. Sap pressure in vascular plants , *Science,* **148,** 339-346.

[3]. Fensom D.S. and Donald R.G., 1982. Thickness Fluctuations in Veins of Corn and Sunflower detected by a Linear Transducer. *Journal of Experimental Botany*, **33,** 1176-1184.

[4]. McBurney T., 1992. The Relationship Between Leaf Thickness and Plant Water Potential, *Journal of Experimental Botany* , **43,** 327-335

[5]. Meidner H., 1952. An Instrument for the Continuous Determination of Leaf Thickness Changes in the Field, *Journal of Experimental Botany*, **3,** 319-325

[6]. Syversten J. and Levy Y., 1982. Diurnal Changes in Citrus Leaf Thickness Leaf Water Potential and Leaf to Air Temperature Difference, *Journal of Experimental Botany*, **33,** 783-789.

[7]. Burquez A., 1986. Leaf thickness and water deficit in plants, *Journal of Experimental Botany*, **38,** 109-114.

[8]. Hadjiloucas S., Karatzas L.S., Keating D.A. and Usher M.J., 1994. Plastic Optical Fibres for Measuring Leaf Water Potential, *POF'94, Third International Conference on Plastic Optical Fibres and Applications*, Yokohama, Japan.

[9]. Hadjiloucas S., Karatzas L.S., Keating D.A. and Usher M.J., 1995. Optical sensors for measuring water uptake in plants, *IEEE Journal of Lightwave Technology,* Special Issue (in press).

[10]. Martin, D.H. and Bowen, J.W., 1993. Long-Wave Optics, *IEEE Transactions Microwave Theory Technology,* Special Issue on Quasi-Optics, **MTT-41,** 1676-1691.

[11]. Lesurf, J., 1990. *Millimetre-wave Optics, Devices and Systems*. (Bristol: Adam Hilger).

[12]. Thomas Keating Ltd., Station Mills, Billingshurst, West Sussex, RH14 9SH.

A dual-channel fibre optic system for monitoring rotating shafts.

M.L.Everington and A.T.Augousti.

School of Applied Physics, Kingston University, Kingston-upon-Thames, Surrey, KT1 2EE.

Abstract

A non-contact fibre-optic sensor for the measurement of angular velocity, angular acceleration and torsional strain of rotating shafts is described. A device was developed which could monitor variations in the intensity of reflected light produced by the surface contours of the shaft as it rotates. A second identical channel has been added so that data can be collected at a different point along the axis of the shaft. The two sets of data are cross-correlated and changes in the relative phase of the two points may be observed. This information may then be used to determine the angular acceleration of the shaft.

1. Introduction

A non-contact method for measuring the characteristics of rotating shafts is desirable because the normal functioning and properties of the shaft are not affected in any way. The utilisation of fibre-optic cable means that the set up is flexible and especially suited to environments which are inaccessible, hazardous or prone to electro-magnetic interference. The system directly utilizes variations in the level of reflected light caused by the surface contours of the shaft without the need for more complicated techniques such as speckle or interferometric analysis which require extra optical components and more data processing.

2. Development of the single-channel device.

The overall system configuration is shown in Figure 1. Two separate optical fibres are jacketed together and polished down simultaneously to form a head which is aimed at the shaft being monitored. The two free ends of the optical fibres are connected to a control unit which houses the electronics. The output from an IR LED is coupled to one of the fibres thereby illuminating the shaft. Part of the reflected light is captured by the other fibre and transmitted back to the photodiode. The voltage signal produced by the photodiode is amplified and fed through a low pass filter. The electronic signal from the control unit is connected to an ADC card residing in a PC. It is in the PC that the processing to determine the fundamental frequency of the signal is carried out. The analogue signal is converted to a sequence of digital data by the ADC at a sampling rate (F_s) which can be set by the controlling software.

The algorithms for windowing and Fast Fourier Transform (FFT) analysis, which are described in more detail later, may be applied as required, being implemented by software written in C^{++} language. The core of the program which performs the FFT analysis is quick enough to evaluate the result three times a second on sample lengths (N) of 2048 points with the program endlessly capturing series of data and repeating the processing. The frequency of rotation is determined by scanning the frequency spectrum produced by the FFT analysis to find which value has the greatest power for each data series and the result is printed on the screen. This is updated after each series of data is processed and the computer displays graphically a trace of the waveform and the frequency spectrum produced by the FFT algorithm as required.

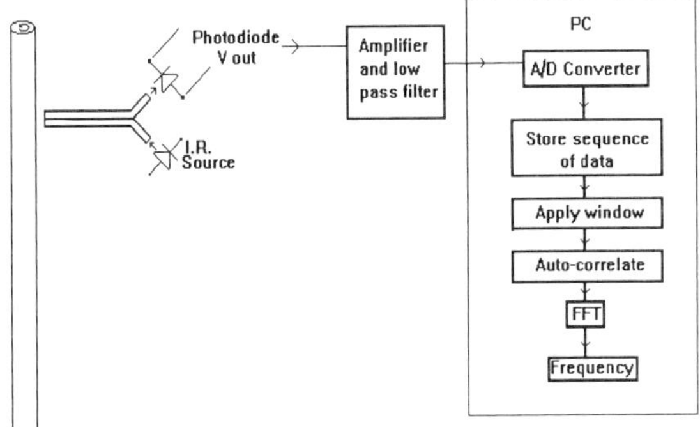

Figure 1. Opto-mechanical arrangement of system.

2.1. Frequency Determination.

A good comparison of the various techniques used in the measurement of frequency is given by Cain and Yardim [1]. Fourier transforms are a useful method of representing time domain signals in the frequency domain and involve integrating a time domain signal to work out the energy density spectrum in the frequency domain. The integration is performed using a finite mesh. However, if a sequence is periodic it is known that it can be represented by a series of discrete summations known as the Discrete Fourier Transform (DFT) such that

$$X(k)=\sum_{n=0}^{N-1} x(n)e^{-j(2\pi/N)nk}, \qquad k=0,1,2,....,N-1 \quad (1)$$

The theoretical justification of this is covered in detail by Strum and Kirk [2] and by Rabiner and Schafer [3].

The calculation of the DFT requires a number of operations which increases with the square of the number of samples in the sequence being processed. This would prove time consuming even on a PC if a reasonable sample length is to be used. However Fast Fourier transforms reduce the number of calculations required considerably by recognising that many of the calculations are repeated and redundant.

2.2. Techniques for Improving Signal Clarity.

When a sequence of data points has been obtained through the sampling of a continuous signal, effectively a rectangular window of uniform strength has been placed over a portion of that signal, i.e the sampled data represents the product of the original signal (over all time) and a rectangular function in time of unit height. The frequency spectrum produced by the FFT analysis of the sequence is modified according to the response of the Fourier transform of that window, causing spectral leakage of frequency components within the passband of the transform of the window. Also the process of sampling with the rectangular window introduces discontinuities at the start and finish in the frequency components whose periods do not exactly fit into the window. When

Fourier Transforms are taken the consequence of this is also the spectral leakage of the power associated with one frequency into other frequency values. By altering the weighting of the coefficients of the sampling window the discontinuities can be nullified by tapering the values of the end points towards zero and its Fourier transform can be adapted to gain the most beneficial and accurate results when the FFT analysis is performed [4]. The Hanning window which incorporates a raised cosine was applied to the data series during the processing. The bandwidth of the Hanning window is twice that of the rectangular window and it causes much greater attenuation outside the passband. A side effect of the broader bandwidth is greater dispersion [5] of frequency components which causes a broadening or smudging of the individual frequency peaks.

2.3 Choice of Sampling Rate

The highest speed that the ADC is capable of operating at is in the order of 75 kHz. If the shaft being investigated is rotating at 50 Hz, i.e. 3000 rpm, with the sampling rate at 75 kHz then 1500 data points would be captured per revolution meaning that only one and a third revolutions would be captured in each data series of 2048 points. At least two complete periods of the lowest frequency component present are required to provide a reliable result, so the effects of reducing the sampling and altering the sample lengths were investigated so as to obtain the optimum values. F_s can be altered as required by changing the delay loop in the computer program.

A benefit of reducing the sampling rate is improved resolution in the frequency spectrum but this is gained at the expense of the detail of the surface features. The frequency resolution (Δf) of the FFT analysis, and therefore the overall system, equals F_s/N. The arrangement that was originally tested was with F_s = 10 kHz and N = 1024 giving Δf = 9.75 Hz, which is a poor resolution since the normal rates of shaft rotation encountered are between 10 Hz and 100 Hz. The sampling rate was reduced by up to a factor of 10 but it was found that the sample lengths had to be reduced as well to ensure that the rate of update of results was not lowered. Much lower sampling rates of 1 or 2 kHz were tried to assess the effects on the accuracy of the system. One benefit of shorter sample lengths is that there is less chance of a variation in the signal frequency causing uncertainty in the result of the FFT analysis.

One danger of reducing the sampling rate is aliasing [6] whereby signals whose frequencies are more than half that of F_s are mistaken for frequencies of less than half F_s. This is avoided by using the low pass anti-aliasing filter to remove the unwanted higher frequencies. The cut-off point of the filter is variable as required to suit the ranges of angular velocity of the shaft being monitored.

2.4 Results and discussion.

Figure 2 shows the reconstruction of a typical series of 2048 data points with F_s = 10kHz taken from a shaft revolving at 45 Hz. It demonstrates a clear periodicity but higher order harmonics can be seen imposed on the signal. Figure 3 displays the frequency spectrum obtained by the application of FFT analysis to that data, and has picked out the components present clearly. The scale of the y-axis is arbitrary showing the relative amounts of power for each frequency component present in the signal; the x-axis shows the frequency in steps of k; the value of k is equal to F_s divided by the sample length. This also defines the resolution of the spectrum. Figure 4 shows the spectrum when the same signal is sampled using the Hanning window. The higher harmonics are slightly weaker and the peaks are broader than in Figure 3. Figure 5 shows the frequency spectrum obtained from the same signal using a Hanning window with F_s = 1 kHz and 256 points. This gives a resolution of 4 Hz per frequency component and shows that the results have not degraded whilst a better resolution has been achieved with no loss of the rate of result update.

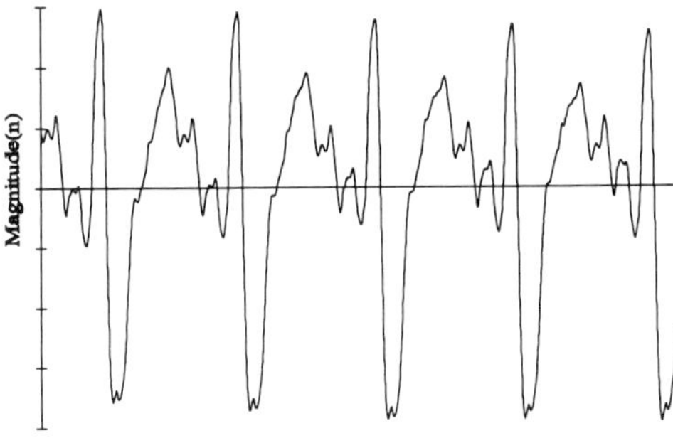

Figure 2. Reconstruction of typical data sequence.

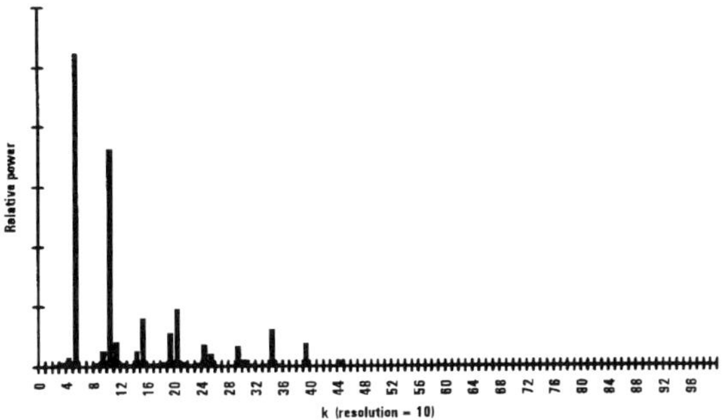

Figure 3. FFT spectrum of waveform shown in Figure 2.

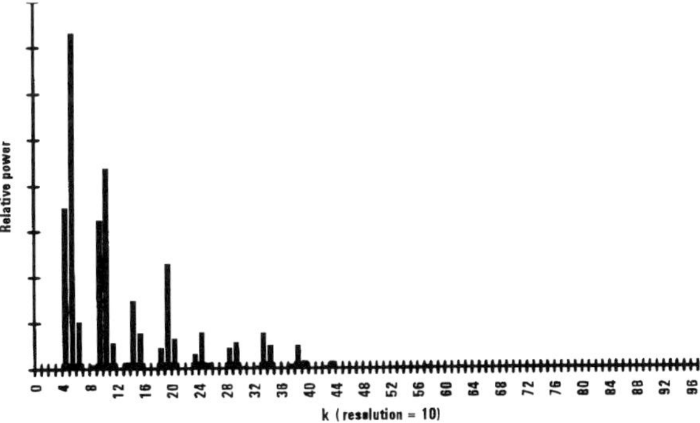

Figure 4. FFT spectrum of data sampled using Hanning window.

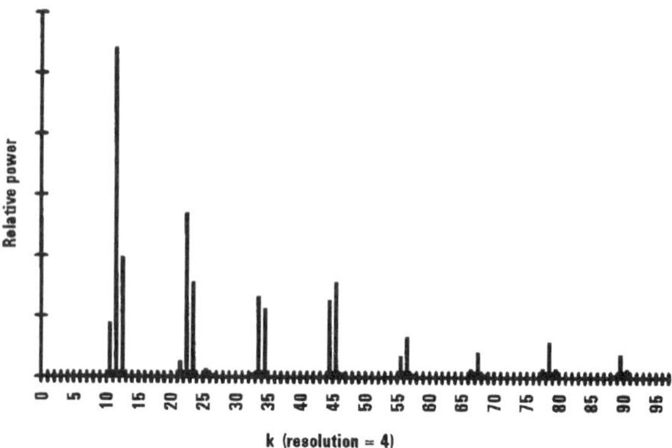

Figure 5. FFT spectrum of signal with F_s = 1 kHz from 256 points.

3. Dual channel system.

The dual channel system operates using two channels identical to that described previously. This enables the signals from different points along the axis of the shaft to be cross-correlated or FFT analysis may be performed on one channel to determine the frequency of rotation.

3.1. Cross-Correlation

The cross-correlation function is defined as follows

$$R_{x_1x_2}(p) = \sum_{m=-\infty}^{m=+\infty} x_1(m)x_2(p+m) \quad \cdots\cdots\cdots\cdots \quad (2)$$

where x_1 and x_2 are the two signal data sequences being cross-correlated. When the signals obtained from the axially separated points on the shaft are cross-correlated the shift of the central peak of the cross correlation function from the zero position shows the phase difference between the two. If one end of the shaft is subjected to an external torque causing angular acceleration or deceleration, then strain will be induced which will produce twist in the shaft. This will cause one of the signals to lag behind the other and this can be detected by changes in the position of the peaks of the cross-correlation function.

3.2. Results

Figure 6 shows two traces; the broken line shows the cross-correlation function obtained when the shaft is rotating normally and the bold line shows the function when a friction brake of the order of a few Newton.m is applied to the end of the shaft. The second function shows that the period of rotation has increased and that there is a phase difference between the two functions.

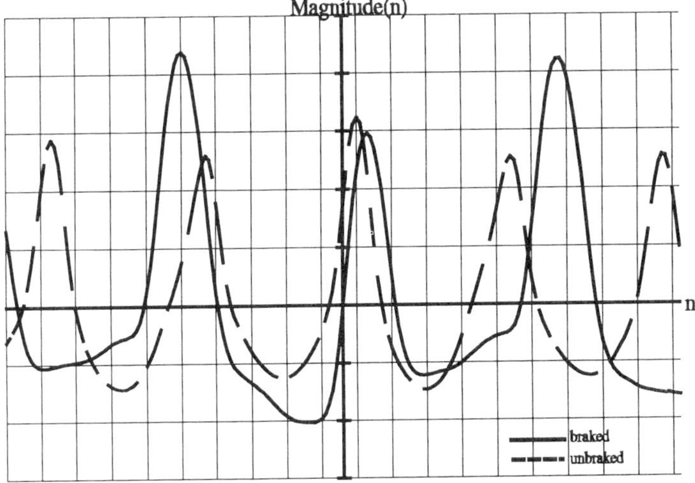

Figure 6. Graph demonstrating change in phase of signal.

4. Conclusion.

A robust fibre optic shaft monitor has been constructed, interfaced to a PC and its characteristics evaluated. The PC displays a graphical output in real time either in the time domain or the frequency domain. The response of the device depends on the level of resolution required and response times of under a second with a resolution of 1 Hz using FFT analysis are achievable for sample sizes up to 1024 points. With a second channel added it has been shown that changes in the phase of the cross-correlation function at different points along the axis of the shaft can be detected. It is planned that given greater computing power a system will be developed to measure angular acceleration and hence torque in untreated rotating shafts.

5. References.

1. **Cain G.D and Yardim.A,** "*Digital Estimation of Doppler Frequency in Waveband Signalling Applications*" Fifth International Conference on Digital Processing of Signals in Communications, Loughborough 1988.
2. **Strum R.D. and Kirk D.E.** *First Principles of Discrete Systems and Signal Processing*, Massachusetts: Addison-Wesley 1988
3. **Rabiner L.R. and Schafer R.W.** *Digital Processing of Speech Signals*, New Jersey: Prentice-Hall, 1978
4. **Harris F.J.**, *On the use of Windows for Harmonic Analysis with the Discrete Fourier Transform* Proc IEEE,66,No.1,pp.51-83,Jan 1978.
5. **Durrani R.S.,** "*Data Windows for Digital Spectral Analysis,*" Proc. IEE.,119, No 3,March 1970, pp.343-352.
6. **Ifeachor E.C. and Jervis B.W.,** "*Digital Signal Processing: A Practical Approach*", Addison-Wesley Publishing, 1993.

Humidity measurements using a plastic optical fibre sensor.

S. Hadjiloucas, L.S. Karatzas, D.A. Keating, and M.J. Usher.

Department of Cybernetics, University of Reading, P.O. Box 225, Whiteknights, Reading. Berks. RG6 2AY, U.K.

Abstract. A plastic optical fibre reflectance sensor that makes full use of the critical angle of the fibres is implemented to monitor dew formation on a Peltier cooled mirror surface. The optical configuration permits better light coupling between the mirror and the detecting fibre, giving a better signal of the onset of dew formation on the mirror. The application of feedback further reduces the possibility of contamination of the sensor head.

1. Introduction

Typical dew point sensors comprise a mirror cooled by a small Peltier unit. The current fed to the cooler is adjusted until dew begins to form on the mirror; this point is detected by a capacitive [1] or optical [2] sensor and the temperature of the mirror surface then measured by a suitable temperature transducer.

The major limitations of such systems are the tendency of the mirror to become contaminated by aerosol particles which shift the condensation point to higher temperatures and the relatively slow time response of the system. Some systems include detection and correction circuits that eliminate the effect of optical contamination using an optical balance (Model Dew 10, General Eastern Co., USA). The problem of slow time response can be reduced by using a cycling cooling system [3] consisting of a first stage of rapid cooling of the mirror to approximately 1.5 °C above the last known dew-point, a second stage of slow cooling rate until dew formation is detected, a third reheat stage and a final wait stage.

Alternative techniques for humidity measurements include: Absorption methods that depend on the sorption equilibrium of hydroscopic materials, but in which the diffusion of water molecules is a relatively slow process with hysteresis and poor reproducibility. Gas hygrometry which, despite being a direct method, requires the use of a reference gas and finally accurate vapour removal or saturation processes which are difficult and preclude sensor miniaturisation.

The system described here employs a novel optical fibre method for detecting the formation of dew and a feedback system to control the temperature of the mirror in a narrow range about the dew point. This produces an improved sensitivity in the detection of the onset of dew, reducing the possibility of contamination. The response time of the system is significantly reduced and the overall accuracy similarly improved.

In many industrial applications the sensor must be capable of measuring dew-points approaching 100 °C in environments where the air might also be static. A standard method of isolating the opto-electronic components from the sensing head has been the use of a bifurcated fibre-optic cable with a random distribution of emitting and receiving glass fibres (Protimeter Inc., UK). In order to improve the efficiency of the transmitting and receiving fibres and to cover a larger area of the detector, many thin optical fibres must be used. An alternative technique is therefore proposed for sensing dew formation using plastic PMMA 1 mm fibres.

2. Sensor description

The dew detection sensor comprises a novel dual optical fibre reflective sensor. Because the reflection coefficient of dew on the mirror is a function of incident angle, it is beneficial to emit and receive light at an angle with respect to the vertical of the mirror. A simple configuration that achieves this requirement involves cutting the emitter (T_X) and receiver (R_X) fibres at an angle (α), to make full use of the critical angle of the fibre, (Figure 1). This ensures improved coupling and therefore better signal to noise ratio. The condition for α is the one for which the ray emerging from the fibre at the most acute angle is parallel to its sloping face, and using Snell's law to cut the fibres gives $n_{air}\sin 90^{o}=n_{core}\sin(18.24^{o}+\alpha)$ so $\alpha=23.84^{o}$. However, a significant improvement in performance is obtained by cutting the fibres twice, at angles of 24° and 66° as shown in Figure 2 [4]. The advantage of the double-cut configuration is the further decrease in the stand-off distance of the sensor's head and the improved reflectivity of the mirror, until the size of dew becomes similar to the wavelength of light. Using a thin opaque film (paint) between the two fibres minimises light directly coupled from the emitter to the receiver in the double-cut configuration.

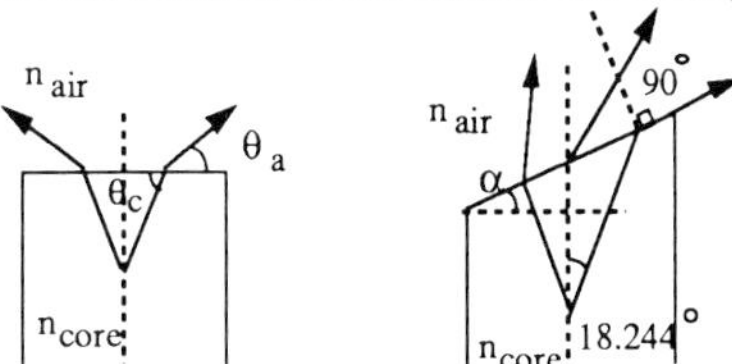

Figure 1. Plastic optical fibre reflectance sensor making full use of the critical angle of the fibres.

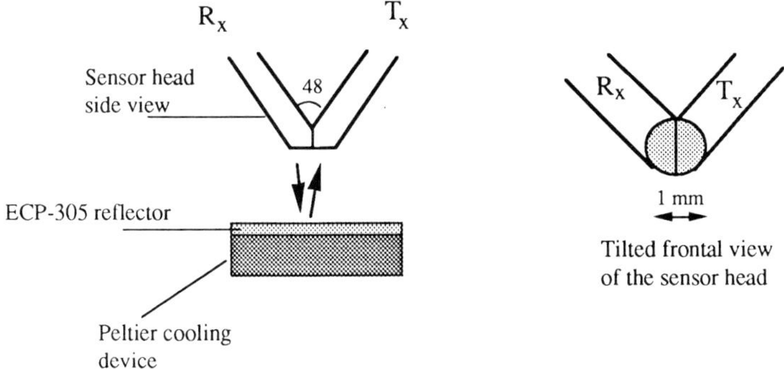

Figure 2. Dew-point reflectance sensor utilising double cut plastic optical fibres.

Since the high porosity of the ceramic material of the Peltier precludes direct polishing and direct gold deposition, a reflector must be used. Solid gold-plated thin mirrors prepared by gold plasma vacuum deposition were initially used. Their performance was found to be inferior to self-adhesive PMMA coated, silvered, reflecting film ECP-305 (3M Co., USA). The film's performance when measured spectrophotometrically, was found to be better than 97% of pure aluminium in the range 300-1000 nm. The good thermal conductivity of the film minimises the time lag for the evaporation of dew making it particularly suitable for feedback configurations. The hydrophobic properties of the PMMA protective coating reduces the Kelvin effect and the possibility of surface contamination. Furthermore, accelerated tests of ECP-305 samples at 80 °C and 75% relative humidity [5] have shown excellent long term stability.

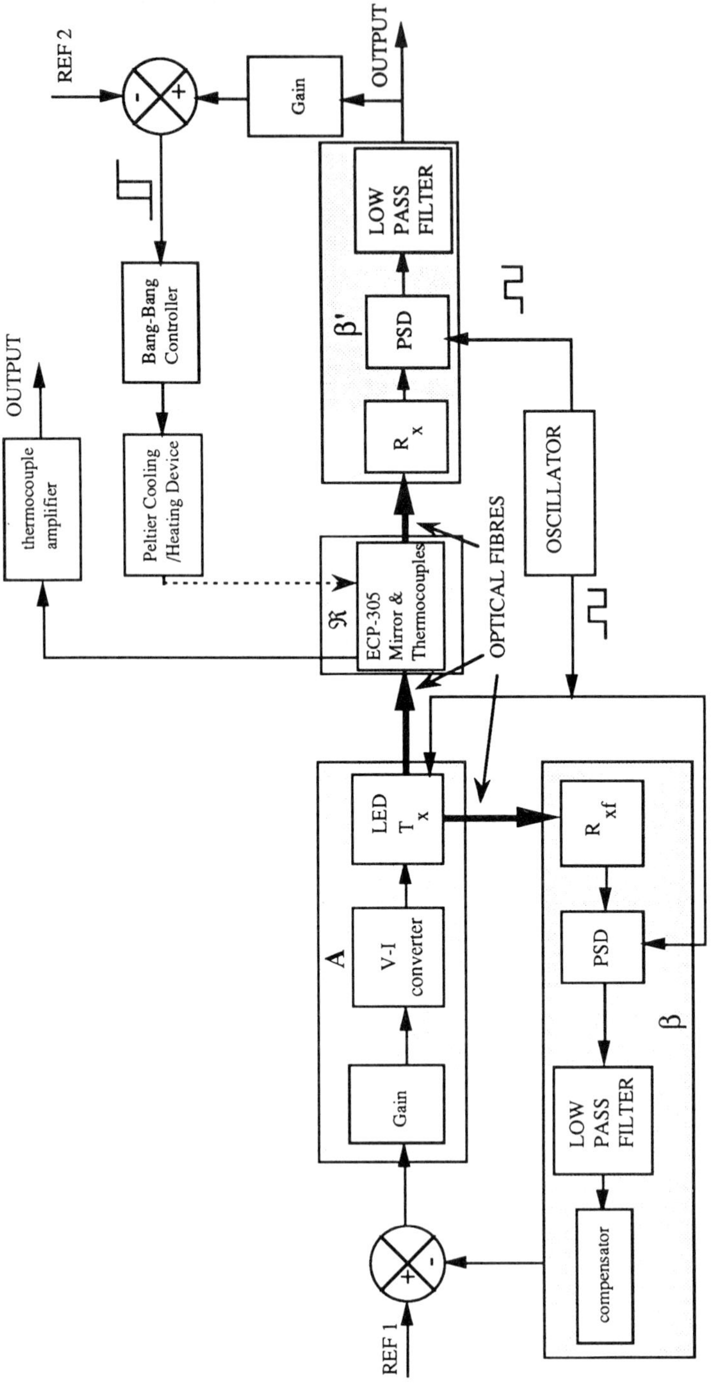

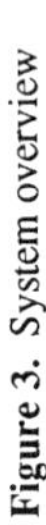

Figure 3. System overview

The need for an optical balance was further eliminated with the use of a cyclic heating/cooling feedback technique. An overview of the complete system is shown in Figure 3. The signal from the optical fibres is amplified and compared with a reference using a comparator (20 mV hysteresis). The output from the comparator is used to reverse the current flow in the Peltier using four solid state power MOSFETs, producing a bang-bang controller operating around the point of dew formation on the reflector. The temperature of the mirror is sensed using thin, rapid response time thermocouples. A feedback circuit that thermally stabilises the LED light source is also incorporated in the design and the technique of amplitude modulation, intensity referencing and phase sensitive detection is used to minimise environmental noise. The forward path of the circuit comprises a differential amplifier, a high gain amplification stage and a voltage to current converter (V-I). The feedback path contains a photodiode (R_{xf}), a current-to-voltage converter (I-V), a phase sensitive detector (PSD) with low pass filtering, and a compensator. When applying negative feedback, for high loop gains, the system response is governed by the characteristics of the feedback path β. Therefore, for a very linear feedback characteristic, the non-linearities in the forward path A will be eliminated, and using transfer function notation for Figure 3:

$$G=\frac{\text{OUTPUT}}{\text{REF 1}}=\frac{A}{1+A\cdot\beta}\cdot\Re\cdot\beta' \quad \text{and as} \quad A\cdot\beta>>1,\ G\rightarrow\Re\cdot\frac{\beta'}{\beta} \tag{1}$$

Using the closed loop configuration, the response of the voltage-to-current converter is not critical for good thermal stability of the emitter. The two receivers with the corresponding PSD and filter, β and β' however, must be matched as closely as possible if a unity gain of the overall system is to be achieved. Such arrangement ensures that the output of the system is only a function of reflectivity and the reference voltage:

$$\text{OUTPUT}=\Re\cdot\text{REF 1} \tag{2}$$

3. Results

Complete cycles during the operation of the sensor (double-cut configuration) are shown in Figure 4. The lower trace represents the change in temperature of the PMMA film on the Peltier, whereas the upper trace shows the change in reflectance due to dew formation. Under closed-loop operation the surface of the reflective film is controlled to 50 mV (lower trace) which corresponds to 0.1 °C.

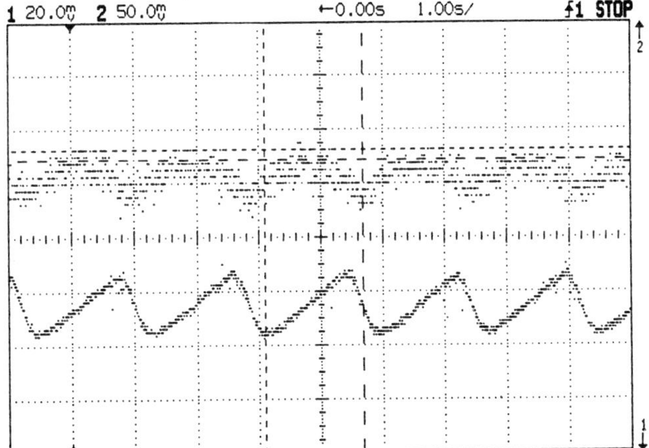

Figure 4. Closed loop control of the reflector's temperature by reversing the currents in the Peltier. The upper trace shows the change in reflectance and the lower trace shows the change in temperature.

The merits of the double-cut configuration for this application are shown in Table 1, where the corresponding theoretical rms variation in reflectance (equivalent to the self-noise of the optical sensor) and the optimal stand-off distances for different fibre-optic configurations are shown. The dominant noise source in the system is the shot noise of the output photodiode. This is given by $i_n=(2eBi)^{1/2}$ where e is the electronic charge, i is the diode current, B is the bandwidth (5.3Hz) of the filter after the PSD and i_n is the rms noise current. In the present noise analysis, the thermal noise is also taken into consideration, using $i_t=(4KTB/R_f)^{1/2}$ where K is Boltzmann's constant, T is the absolute temperature, and R_f (100K) is the feedback resistor in the current to voltage converter. After the equivalent noise voltage of the photodiode was calculated from i_n and R_f, this was converted to absolute changes in reflectance (%) providing an equivalent rms variation in reflectance due to the shot noise. The frequency of oscillation (duty cycle) and the temperature variation on the surface of the reflector when the Peltier is operated in feedback mode are also shown. The 20 mV hysteresis in the comparator corresponds to a change in reflectance signal of 8.4% for the uncut configuration. This can be compared to changes in reflectance signal of only 1.9% and 1.0% for the single cut and double cut configurations respectively.

Table 1. Collective results of theoretical rms variation in reflectance, fibre's stand-off distance, duty cycle frequency and reflector surface temperature variation, for the three sensor configurations. Results shown are for the ECP-305 reflector.

sensor type	rms variation in reflectance (%)	stand-off distance (m)	duty cycle (Hz)	reflector's temperature variation $\Delta\theta$ (°C)
uncut (90°) fibres	8.300×10^{-6}	1800×10^{-6}	0.26	0.5
single-cut (24°) fibres	3.964×10^{-6}	1200×10^{-6}	0.45	0.1
double-cut (24°), (66°) fibres	2.841×10^{-6}	400×10^{-6}	0.58	0.1

Results from calibration experiments using a humidity generator have shown that the dew point measured, was particularly sensitive to the reference reflectance setting (REF2) in the comparator. The problem can be alleviated by monitoring the rate of change in reflectance instead of the absolute reflectance. This is achieved by introducing a lead-lag circuit after the PSD. Such a circuit differentiates the signal and provides a low pass filtering for the higher frequencies amplified by the differentiator. The point of inflexion is therefore used to set the reference signal in the comparator.

4. Discussion

Preliminary results have shown that a very good control of temperature on the reflector can be achieved using the signal from the highly responsive double-cut fibres. Operating around the point of inflexion of the slope of the reflectance function (on-set of dew) eliminates associated problems of system sensitivity to reflectance setting. Although many dew or frost point hygrometers have been described in the literature since the 1960's [6], such good temperature control of the reflector surface was only recently possible using capacitive sensors and feedback.

A major limitation of the plastic optical fibres used, is their restricted operation to maximum temperatures of +85 °C. When a sample of plastic fibre is left exposed to high temperature, a transmission loss occurs from the oxidised degradation of the core polymer. Carbonyl groups are formed, and cross-linking and double-bonding of the polymer chains occur due to molecular dissociation or elimination reactions. These, in turn, induce an increase in the electron transition absorption at the 660 nm and an adaptive shift towards a longer-wavelength band. It is possible to use heat resistant plastic optical fibres with a thermoplastic resin for the core and silicone elastomer for the cladding (Furukawa Electric Co., Japan). These fibres have a wavelength window ranging from 730 nm to 820 nm, and are hot-humidity resistant showing an increase in transmission loss of only 0.2 dB/m after 500 hours of exposure at 80 °C and 95 % RH environment [7].

Furthermore, new polymer optical fibres fabricated by Fujitsu Co., (Japan) using ARTON™ (Japan Synthetic Rubber Co., Ltd.) with high thermal stability up to 150 °C, good transparency at the visible wavelength, high flexibility and high tensile strength properties may be implemented in the new humidity sensor. The minimum optical transmission loss spectrum is 0.8dB/m at 680 nm and 1.2 dB/m at 780 nm [8]. The thermal expansion of ARTON™ is much lower than the one of conventional or low water absorption PMMA and can be considered negligible at high humidity.

The new dew point sensor is suitable for meteorological observations, measurements of moisture content in crops and food products (as long as the absorption isotherm is known), or water activity in agricultural products. One particular advantage is that it can function at high humidities, a characteristic of environments such as tropical forests where other sensors are often less reliable. The relatively small size of the sensor makes possible its use for localised CO_2 and H_2O exchange measurements in plant canopies and for photosynthesis research.

5. Acknowledgements

The authors would like to express their appreciation to Les Comley and Chris King for technical help. Many thanks also to Roger Hunnemann (Cybernetics, Infrared Research Group) for partially metalising glass surfaces and providing a humidity generator for calibration purposes.

6. References

[1] P.D. Harris and M.K. Andrews, 1993. A miniature dew-point hygrometer based on capacitance measurement. *"Sensors VI, Technology, Systems and Applications,"* KTV Grattan, AT Augousti (Eds.), IOP Publishing, pp.435-440

[2] P.P.L. Regtien, 1981/1982. "Solid state humidity sensors", *Sensors and Actuators,* **2,** 85-95.

[3] F. Dadachandji, 1992. Humidity measurements at elevated temperatures, *Measurement + Control,* **25,** March.

[4] S. Hadjiloucas, L. S. Karatzas, D. A. Keating, and M. J. Usher, 1994. A new Plastic Optical Fibre Displacement Transducer. Proceedings of *"Trends in Optical Fibre Metrology and Standards"* Viana do Castelo, Portugal, O.D.D.Soares, Editor, NATO ASI Series, (*In press*).

[5] P. Schissel, G Jorgensen, C. Kennedy, R. Goggin, 1994. Silvered PMMA reflectors, *Solar Energy Materials and Solar Cells,* **33,** 183-197.

[6] A.W. Brewer, 1965. " The Dew- or Frost-point Hygrometer", Principles and Methods of Measuring Humidity in Gases, Vol.1, Ed. A Wexler, Reinhold, NY, pp. 135-143.

[7] S. Irie and M. Nishiguchi, 1994. Development of the heat resistant plastic optical fiber. *Third International Conference on Plastic Optical Fibres and Applications. POF '94*. The European Institute of Communications and Networks, (EICN) Yokohama, Japan, pp. 88-91.

[8] T. Sukegawa , M. Hirano, M. Tomatsu, T. Otsuki, H. Shinohara, Y. Hara and A. Tanaka, 1994. New Polymer Optical Fiber for High Temperature Use. *Third International Conference on Plastic Optical Fibres and Applications. POF '94*. The European Institute of Communications and Networks, (EICN) Yokohama, Japan, pp. 92-94.

DEVELOPMENT OF AN OPTICAL FIBRE BASED COLORIMETRIC AMMONIA SENSOR.

M. Fneer, W. J. O. Boyle and K.T.V. Grattan.

Measurement and Instrumentation Centre

Department of Electrical, Electronic and Information Engineering,

City University, Northampton Square, London EC1V 0HB, UK.

Abstract

The construction and performance characteristics of a novel optical fibre-coupled optrode for ammonia concentration measurement are described. The operation of the optrode is analogous to the common electrochemical ammonia sensor, where changes in the concentration of ammonia that arise from its diffusion through a gas permeable membrane are monitored by their effect on the pH of an ammonium sulphide buffered solution. In the optrode, changes in pH are measured optically rather than electrochemically. These alterations in the optical absorption characteristics of the pH indicator dye (phenol red) which result from the changes in ammonia concentration are observed using a dual optical wavelength scheme to provide a reference channel. The buffer solution of the optrode is isolated from the solution under test by a Teflon membrane, and results obtained from the optically addressed colorimetric sensor for ammonia are described and discussed.

Key words: Fibre optic ammonia sensor, phenol red, membrane, pH, colorimetric, optical probe.

Introduction

The measurement of ammonia is of a particular importance, not only in biomedical science but also in environmental, forensic, food sciences as well as in the study of enzymatically degradable nitrogenous compounds. Numerous methods for this purpose have been introduced, and some of them are commercially available[1] . However, so far none of these methods satisfies all the common needs for a simple, reliable, cheap and stable ammonia sensor. During the same development period similar needs for a glucose sensor have been fulfilled by the production of a device based on the amperometric method. This kind of sensor is now commercially available as a disposable sensor strip, equipped with a pen-sized potentiostat[2]. This shows the adaptability of an amperiometric method for miniaturisation and mass production. The extension of this technology for the development of multi-functional sensors is easily envisioned. Bearing these facts in mind, it is useful to explore the use of colorimetric techniques for the development of optical fibre-based sensors for the measurement of the concentration of different chemical species, of which ammonia is among the most important.

Ammonia sensors

The detection of ammonia, either as a vapour or in aqueous solution, may be carried out in a similar manner to the measurement of pH, i.e. based on the use of pH indicators. Shahriari et al.[3] and Zhou et al.[4] have constructed sensors as shown in figure 1, based upon optical absorption by the ammonia indicator bromocresol purple, suitable for ammonia detection.

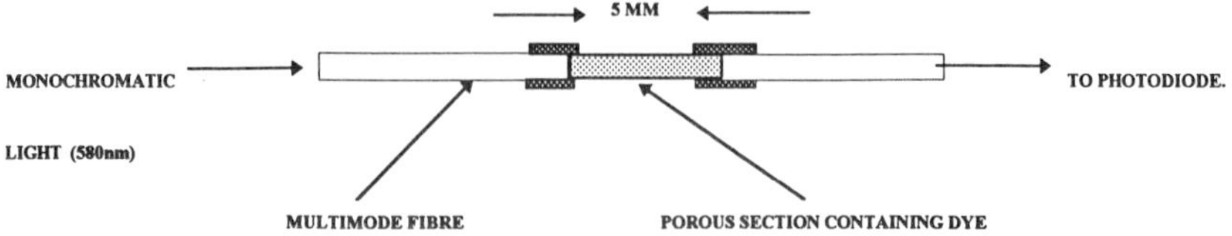

Fig.1.: Ammonia sensors of Shahriari et al. and Zhou et al.

The dye is adsorbed into a section of porous fibre, into which the **NH_3** enters and interacts with it. For this reaction to occur, adsorbed water must be present to create NH_4OH, and so the devices are sensitive to humidity as well as ammonia. Shahriari's device uses porous glass made by heat treating and etching the borosilicate glass, giving a pore size of 80 - 150 nm. It can detect an ammonia level down to 0.7ppm, and shows a linear relationship between adsorption coefficient and concentration up to 3ppm. Zhou's sensor is less sensitive, working in the range 10 - 90ppm, utilising a porous plastic (a treated polymethylmethacralyte) with a pore size of ≈10nm. Besides the sensitivity to humidity, both devices have the major disadvantage of very long response time, typically 20 minutes.

In recent years, some investigations on coated optical waveguide spectrophotometric and fluorescence sensors for ammonia vapour have been described(5,6). Each of these investigations has followed a different technique in measuring the ammonia vapour concentration. Further development to an earlier work on pH optical fibre sensors,(7) in a new scheme to measure ammonia concentration in water using a fibre optic probe, is described here.

The work presented lies within the context of continuing development of an ammonia optical fibre sensor, based on the pH change induced by dissolved ammonia. In this sensor, ammonia changes the equilibrium balance of a buffer solution (NH_4SO_2) causing H^+ ions to diffuse through a gas permeable membrane, hence modifying the H^+ concentration and the colour of the phenol red indicator. This work reports on the development of the optical and data acquisition system, and the evaluation of this system in measuring pH. Thus the sensor is analogous to the electrochemical sensor of the Severinghaus electrode(9).

The separation between the water sample and the phenol red dye is achieved by the use of a Teflon membrane(8). The response of the probe is dependent on the variation in the concentration of the ammonia in water.

Instrument Aspects

A schematic diagram of the experimental instrumentation employed is shown on figure 2. The absorption change of the dye solution used, phenol red, is monitored using the change in its absorption spectrum at 565nm, using light from an ultra-bright green LED (40 nm FWHM) of output intensity 120 mcd. The diode is housed in a standard SMA fibre optic housing connector. The reference wavelength is provided by light from an infrared (IR) LED at 810nm (40 nm FWHM) as, at this wavelength, there is very small absorption by the indicator dye, and tests carried out showed that it is not dependent on pH value at this wavelength. The LEDs are operated at room temperature with a fixed current to allow equalisation of their temperature. More active stabilisation of temperature and thus the spectral characteristics of LEDs is an aspect of the requirement of the system.

Light from the green LEDs was coupled into 6 x 600μm diameter plastic clad silica fibre (PCS) of which three fibres were coupled into each green diode. The infrared LED was coupled to a single fibre of 600μm diameter. This bundle of fibres led to the sensor head (or probe), which was approximately 1.5m from the instrument control box. The probe was manufactured from a solid plastic core of internal length 2cm, the ends of which were made from stainless steel and closed and demounteable for easy cleaning and re-polishing, if necessary. One end was polished to provide a reflective surface, and the other end was drilled with 7 small holes at an angle of 7.1 degrees to the Centre. The fibres were inserted and glued to the holes and the ends could also be polished. The materials used are such that they can easily be polished if necessary and re-calibration with respect to standard buffer solutions can be easily performed. At the fibre input end, the light which had travelled across the solution and had been reflected was received by one fibre of 1000μm (1mm) diameter which was surrounded by the other transmit fibres. This fibre was connected to a silicon light detector PIN diode where the intensities of the signals on each wavelength band were determined. For all the fibre couplings, expanded, but otherwise standard, SMA - compatible stainless steel connections were used. Fibre bundles were made from individual fibres stripped of the plastic coating, then fixed with epoxy resin and polished. This combination is inert and resistant to strongly acid or alkaline solutions.

The device was tested by placing the probe in a solution containing ammonium sulphide (NH_4SO_2) and the indicator dye is separated by a membrane in the probe chamber and its absorption, integrated over the emission spectrum of the green LED, is measured as being indicative of the pH.

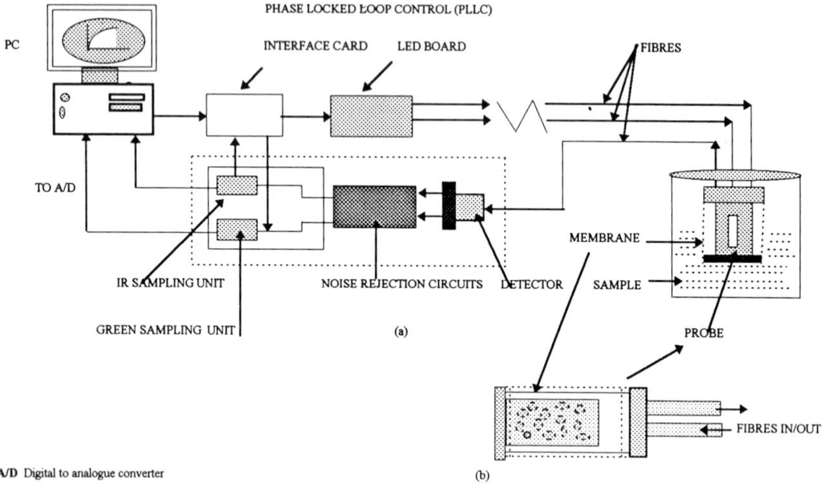

Figure 2. :(a) Schematic of the ammonia sensor showing the electronic system and the fibre optic Probe construction. **(b)** Expanded view of probe.

Electronic Aspects.

As shown schematically in fig.2, the hardware system is controlled by a PC. The interface consists of two units, one to control the LEDs and the other to control the 8 bit analogue-to-digital (A/D) converter which is connected to the PIN diode detector. The interface to the LEDs is controlled via the parallel printer port, whilst the A/D converter is located in the PC on a bus interface card.

This hardware is controlled by software written in Turbo Pascal. This software provides a train of pulses that alternately turn on-and-off the green and the infrared LEDs, and it also reads the digital values generated by the A/D converter from the signal produced by the optical detector. With the present software configuration, details of the pulse timing can be controlled and the time-varying signal from the detector can be displayed graphically and recorded to file, for later analysis by spreadsheet software. With the system as it is, the dynamic response of the sensor to changes in ambient chemistry can be expected to be recorded. With further development of the sensor systems, the computer hardware and software may be simplified to be suitable for a single chip microprocessor controller solution.

The detector circuit used is based on a silicon PIN diode with integral amplifier (RS type 308067). Such a device is low noise and of adequate bandwidth for this application (linear to a few kHz). Additionally a three-stage, low noise amplifier is used further to boost the small signal level. Three simple type 071 operational amplifiers were used with the gain selected by appropriate resistors. The phase selective detector circuit, through the use of sample and hold elements (LF398H), synchronised with the LED signals rejects the d.c. level and enables the intensities of the infrared and green signals to be measured. These outputs (green and infrared outputs) were then measured and displayed and the ratio of the infrared to green signal intensity obtained. This parameter is directly related to the pH of the solution via a pre-determinated calibration, which is performed prior to be used of the system.

Indicator and reagent.

Phenolsulfonphthalein (phenol red) was the indicator dye employed in this work as it can be used to measure pH and thus ammonia concentration over a wide range. It exists in two tautometric forms, each having a significantly different absorption spectrum.

As the characteristics of the solution vary, the relative size of the optical absorption of each tautometer varies in response to the changing relative concentrations of the acid and base forms of the indicator. This can conveniently be monitored at the peak of the absorption at 565nm, corresponding to the green LED emission while the small infrared absorption is independent of that change. Fig 3 show the absorption spectra of the indicator dye for sample solutions of varying pH, obtained in a conventional spectrophotometer in 1 cm sample cell.

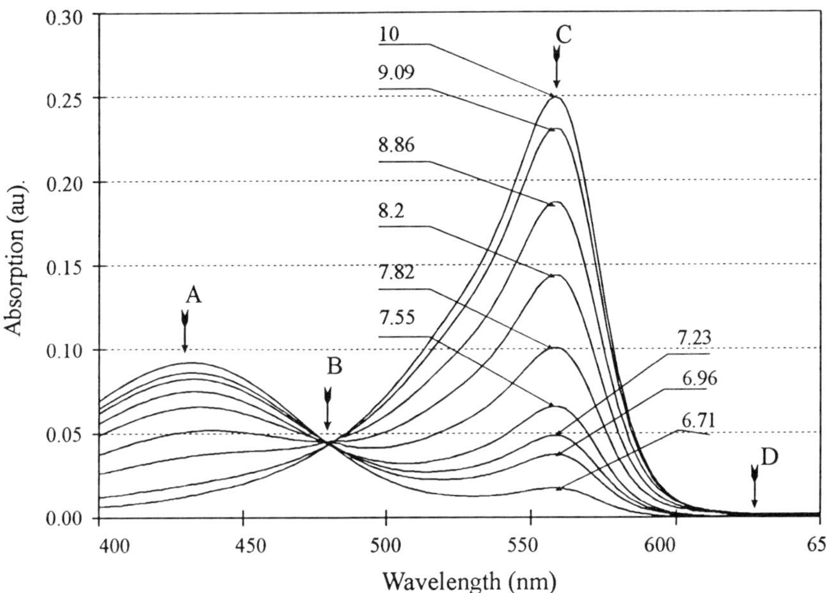

Figure 3: Absorption Spectra of phenol red as a function of wavelength for different pH values, related to ammonia concentration in water.

Calibration.

The response of the device, i.e. the ratio of the signals from the two LEDs, is plotted as a function of the output from an electrochemical pH meter which itself was referenced to standard buffer solutions, carefully prepared in the normal way. A typical set of calibration curves is shown on fig.4 over the pH range from 6 to 10 for solutions which were prepared containing ≈0.5 ml indicator and stirred continuously.

The accuracy of the sensor as used is 0.05 pH, this arising from the error in the measurement of the peak at 565nm, due to the electronic background noise in the system, which corresponds to an error in the reading of approximately ± 1%. The stability of the reading is very satisfactory showing a negligible change over 1h for a solution of constant pH. The temperature stability of the device is very good, corresponding to $0.04pHK^{-1}$ over the range 22 °C to 50 °C, in the pH region around 7.8 units.

The response of such a device is very fast and diffusion limited only, due to the rapid responses from the modulated LEDs and the electronic processing. This compares with a conventional electronic meter which may require several seconds to re-stabilise after the pH of the solution is changed.

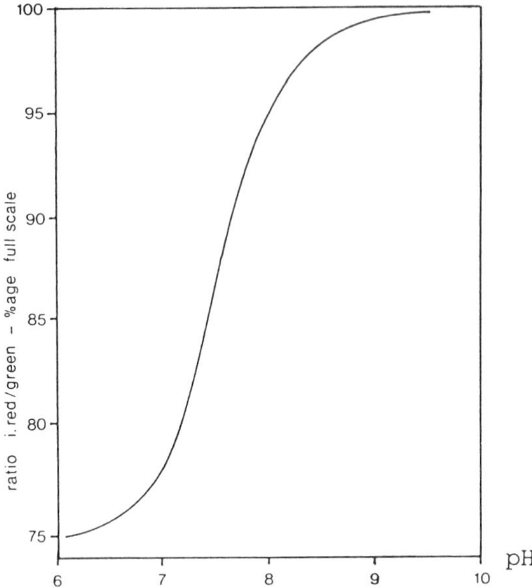

Fig 4 : Calibration curve of the device, over the pH range 6 to 10.

Discussion.

The preliminary stage in the development of fibre optic ammonia sensor, for use in aqueous solutions, has been described. The first stage of construction of the sensor has been evaluated by measuring the performance in measuring pH. Further work on the development of the sensor in which it is evaluated with various gas permeable and ammonia selective membranes will be presented. The objective of demonstrating the use of optical and data acquisition was achieved through the use of two LED emitting at 565 nm and 810 nm, for the signal and reference light channels and PIN diode as a detector. Further, a software program was used to control the device with a PC.

References.

1. Manufactured by Medicines inc., Abingdon, England and Cambridge, Mass., USA. Cooper A.J.

2. Plum F. "Physiol Rev., 61, 440 519. (1987).

3. Shahriari et. al., "Porous optical fibres for high sensitivity ammonia vapour sensors", Op.Lett. 13, 407 (1988)

4. Zhou et. al. "Porous plastic optical fibre sensors for ammonia measurement", Appl. Op.28, (2022 (1989).

5. Giuliani et. al., "Reversible optical waveguid sensor for ammonia vapours", Op. Lett. 8, 54 (1983).

6. Day, R. A., and A. L. Underwood: "Quantitative Analysis," 6th ed., Prentice-Hall, Englewood Cliffs, NJ, (1991).

7. Grattan, K.T.V., Mouaziz, Z. and Palmer, A.W. " Dual wavelength optical fibre sensor for pH Measurement" Biosensor 3, 17-25, (1987).

8. Day, R. A., and A. L. Underwood: "Quantitative Analysis," 6th ed., Prentice-Hall, Englewood Cliffs, NJ, (1991).

9. Janata, J. "Principal of Chemical Sensors" chapter 4, Plenum Press 1990.

A hybrid optoelectronic temperature sensor - simplified configuration

J. Mason and A.T. Augousti

School of Applied Physics, Kingston University, Kingston, Surrey KT1 2EE, UK

Abstract

A simplified configuration of an earlier hybrid optoelectronic temperature sensor has been developed. The new configuration requires no signal processing to generate a self-oscillatory system, being in effect an optically-driven flip-flop constructed at the probe head. The oscillation time depends on the alignment response time of the liquid crystals in an LCD element, which is temperature dependent. The output is the frequency domain, and is essentially intensity-independent.

Introduction

The authors have previously explored the use of optically-energised and addressed LCD elements as temperature transducing elements in a number of alternative configurations [1-5]. Most hybrid optoelectronic sensors, which are based by definition based on the conversion of optical power to electrical power at the probe head, usually operate in a regime which places great demands on the power that must be transmitted optically [6]. This can cause problems, since recent work on the safety limitations on the use of optical sensors in hazardous environments has suggested that no more than 1mW of optical power be permitted to be transmitted in the fibre [7]. The use of liquid crystals in a display is thus attractive due to the very low power consumption at the probe head, allowing the use of optical energy conveyed by an optical fibre as the primary power source.

Most of these sensors of this type developed by the authors previously are configured so that light from an optical fibre which is reflected from the LCD segment is used as a signal in a closed-loop configuration, rendering the system unstable either in the activated or inactivated state, and thereby causing it to oscillate between the states with a frequency which depends on the alignment response of the liquid crystals. The authors have previously shown that this is primarily through the viscosity of the liquid crystals, which is strongly dependent on the temperature [3]. As a consequence, a signal is retrieved in the frequency domain, and which is essentially intensity-independent.

In order to generate the appropriate control signal to the LEDs which are used to energise the photovoltaic elements which activate the LCD segments, it has been necessary to apply signal processing techniques of varying sophistication to the signal obtained from the reflected light intensity (ie the signal which monitors the present status of the LCD). Whilst this is perfectly acceptable, it is perhaps more elegant to configure the probe head in such a way that it oscillates without external intervention, other than continuous-wave illumination for both monitoring and energising. This work reports the development of such a system, and describes the characterisation which has been performed.

Experimental arrangement

Figure 1 shows a block diagram of the system. The two LCD elements located in the probe head are in fact two different segments of a single display that has been modified for this application. The commercial 3½ digit display (RS 589-250) was originally designed to be transflective, but the partially silvered mirror and polariser were removed. To ensure one segment would operate in the 'normally dark' mode and the other in the 'normally light' mode, polarisers were fixed to the back of the display. Their transmission axes were parallel to, and perpendicular to, the transmission axis of the front polariser respectively.

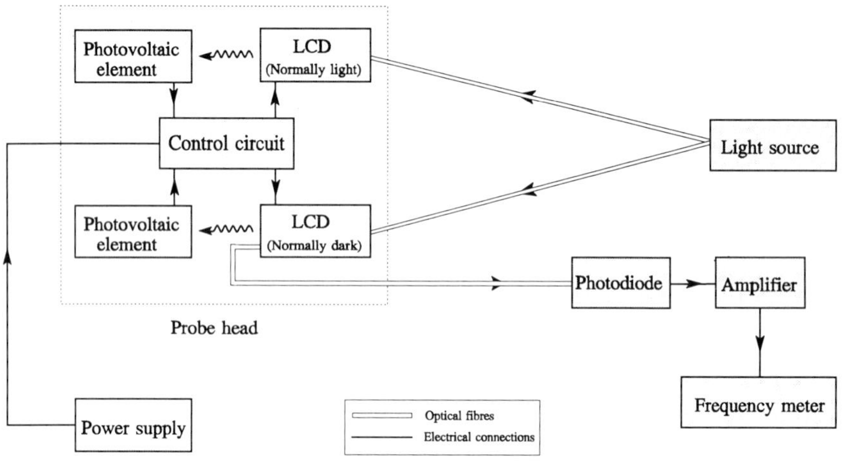

Figure 1 Block diagram showing the optoelectronic arrangement of the system

The photovoltaic elements situated behind each of the LCD segments were used to monitor their light transmission characteristics and provide an input voltage to the control circuit. The authors have pioneered the use of light emitting diodes (LEDs) in this application, and their photovoltaic properties have been previously documented by the authors [5]. The light sources used to illuminate the photovoltaic elements were also LEDs, identical to the photovoltaic elements, thereby spectrally matching source and detector. Plastic 1mm diameter fibres were used to transmit light from each source to the probe head, and permit the transmitted light from one of the elements to be monitored using a photodiode and associated circuitry.

One may observe from the diagram that in this instance the control circuit located within the probe head is powered by a remote power supply unit. It is envisaged that this will be replaced by a photovoltaic stack of elements illuminated by a single optical fibre. The authors have previously demonstrated that a photovoltaic stack can be constructed capable of delivering the drive currents and voltages required by the probe head circuitry when illuminated by low light levels [5].

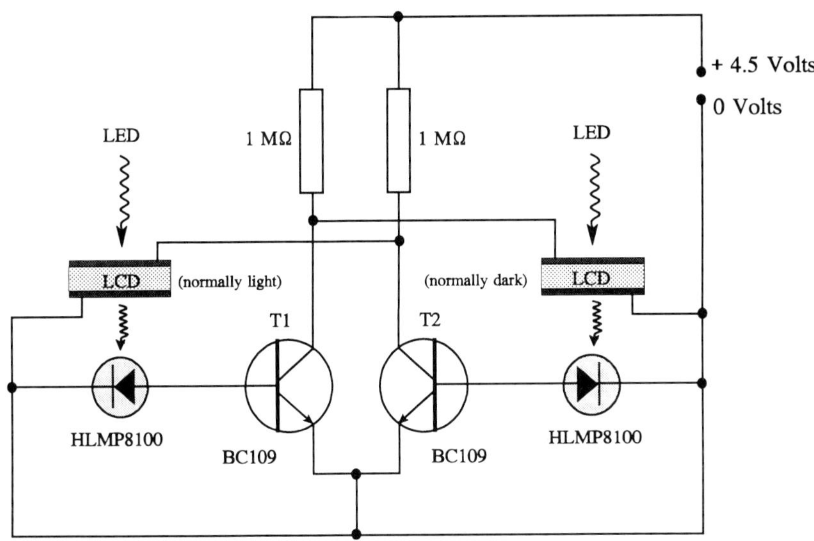

Figure 2 A diagram showing the electrical connections at the probe head

Mode of Operation

Figure 2 shows the electrical components and their connection inside the probe head. Consider the situation when the sensing system is first energised. Light from the source is incident on both the normally light and normally dark display segments. The rise time of the photovoltaic elements is long compared to the propagation delay of the transistors and so the collectors of T1 and T2 both float up to the supply voltage (+4.5 volts). The normally light segment begins to darken and the normally dark segment begins to lighten. The photovoltaic element behind the normally light segment sees a reduction in intensity. The photovoltaic element behind the normally dark element however receives more and more light until the voltage generated approaches the base-emitter conduction voltage (of around 0.6 volts) and transistor T2 is turned on, effectively grounding the collector of T2. When this occurs the field across the normally light segment is removed and consequently it begins to lighten allowing more transmitted light to fall on the photovoltaic element behind it. When it reaches a state where sufficient light can pass through to generate a voltage to turn on T1 the voltage across the normally dark segment is removed since the collector of T1 is now grounded. The normally dark segment then attenuates the light and the whole process is repeated. The rate at which the cycle is repeated depends on the response time of the liquid crystal both to the application and removal of an electric field.

The current generated by each of the photovoltaic elements is very small (around 1μA) and so the 1MΩ resistor ensures that the collector voltage is grounded when a small base current flows.

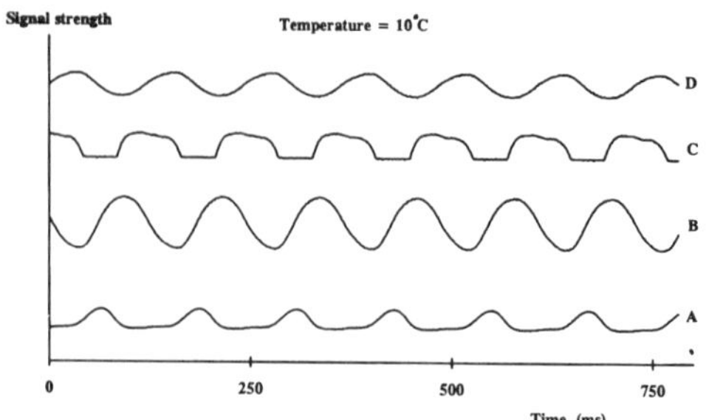

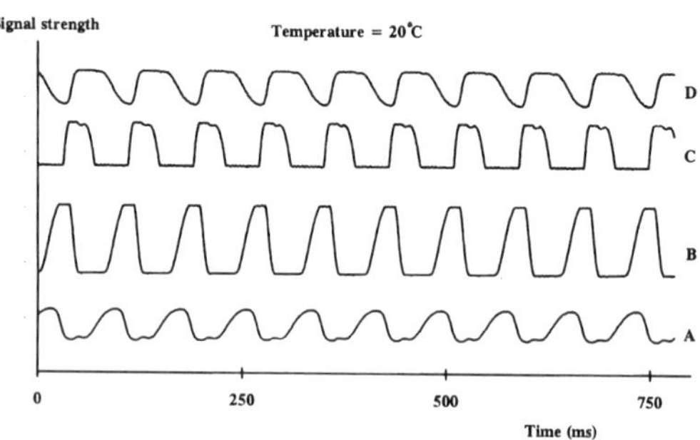

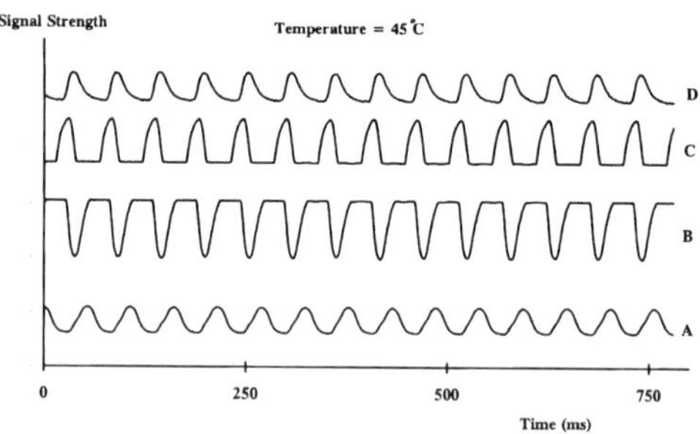

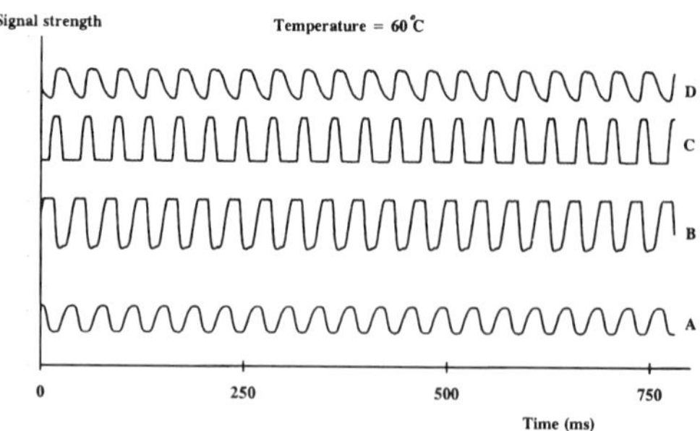

Figure 3 A series of graphs showing the variation with temperature of the signal amplitude at various locations in the probe head. The characteristic labelled A is the value of the collector voltage applied to the normally light segment, B is the same for the normally dark segment, C represents the intensity of the light transmitted through the normally dark segment, and D is the same for the normally light segment.

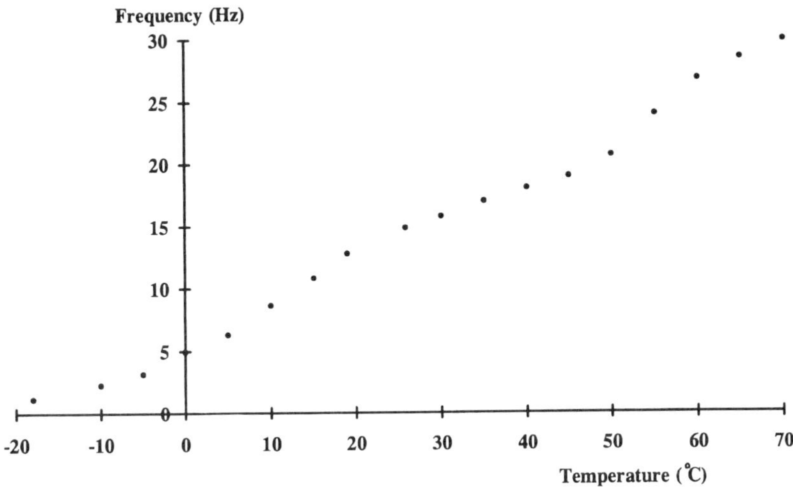

Figure 4 Graph of the oscillation frequency versus temperature

It is also worth noting that a threshold voltage is required to turn on the transistor which means that when the field is either removed or applied across the LCD segments the molecules can pass their equilibrium position before switching occurs. This ensures that the base current is sufficiently large to remove the potential difference applied across the display and hence any possible charge build up. This is desirable since it is well known that dc components in the LCD drive signal will eventually cause chemical degradation of the display. This represents a significant improvement over the original unipolar configuration [1] (ie voltage applied to LCD swinging between positive voltage rail and ground), which has so far been circumvented by the bipolar configurations [2-5].

Results

Figure 3 shows the results when the system is subjected to different temperatures. One may note that the system is remarkably stable, with little noise detectable in the signals, and the frequency of oscillation of the system increases monotonically from about 1.2Hz to 30Hz over a temperature range from -20°C to 70°C (Figure 4). Little attempt has been made to optimise the signal, and the wide dynamic range of the system leads to acceptable temperature resolution. One may correlate the voltages produced by the photovoltaic sources (and hence the associated base currents), as given by curves C and D in the family of graphs in Figure 3, with the resultant collector currents given by curves A and B. It is interesting to note that the signals are not essentially symmetrical (which in this case would mean that the two curves C and D simply differ in phase by 180°), which conforms with what one might expect based physical reasoning, since the two LED segment states (normally light and normally dark) are not exact complements of each other in terms of the amount of light transmitted for any particular applied voltage.

Conclusion

A simplified hybrid optoelectronic temperature sensor has been constructed. The system is self-oscillatory, giving an output in the frequency domain, and is almost intensity-independent. This configuration demonstrates once more the potential of liquid crystals as transducing elements in hybrid optoelectronic sensors. Additionally, it illustrates the advantages of LEDs as alternative photovoltaic sources compared to more orthodox methods employed to energise hybrid optoelectronic sensors [8].

References

1 An optical resonant system for measurement of liquid crystal display response times **J.Mason and A.T.Augousti** *Optical Engineering* **31**(8) 1992 p1663-6

2 A novel hybrid optoelectronic sensor for temperature measurement **A.T.Augousti and J. Mason and M.A. Koenders** Proceedings of the *International Conference on Electronic Measurement and Instruments,* 20-22 October 1992, Tianjin, China pp337-42

3 Mathematical modelling of a resonant optical device based on liquid crystals **A.T.Augousti and J. Mason** SPIE **2025** p322-32 Proceedings of *Nonlinear Optical Properties of Organic Materials VI*, 11-16 July 1993, San Diego, California, USA

4 Temperature measurement using a resonant hybrid optoelectronic system **A.T.Augousti and J. Mason** p317-322 *Sensors VI: Technology, Systems and Applications* **K.T.V. Grattan and A.T. Augousti (Eds)** IOP Publishing 1993

5 Temperature sensing using a hybrid optoelectronic multivibrating system under full computer control **A.T.Augousti and J.Mason** *Meas. Sci. Technol.* **8** 1994 p736-40

6 Fibre optic sensor system, preferably for measuring physical parameters **G.H. Sichling** *Europ. Patent* 0 053 790

7 Optical equipment in flammable gases and vapours **P. McGeehin** *OSCA Doc No 94/56* 1994

8 Optical power for sensor interfaces **J.N. Ross** *Meas. Sci. Technol.* **3** 1992 p651-5

Intensity based optical fibre sensors for damage detection in advanced fibre reinforced composites

R A Badcock, G F Fernando*, S A Hayes, A R Martin and C Nightingale

* To whom correspondence should be addressed

Brunel University, Sensors for Materials Group, Kingston Lane, Uxbridge, Middlesex UB8 3PH, UK

Abstract.

Advanced Fibre Reinforced Composite (AFRC) materials are being extensively used in load-bearing structures because of their superior specific properties when compared to other engineering materials. However detection of internal damage is vital to the safe use of these materials, and can be difficult to detect using conventional non-destructive techniques (NDT).

Embedded optical fibre sensors (OFS) have been identified as being suitable for damage detection in AFRC, as they have minimal interference with the composite bulk properties. Intensity based OFS have been designed and evaluated for damage detection in composite laminates and filament wound tubes.

These sensors have the ability to monitor strain or the propagation of damage in the material. In this study the sensors have been embedded within cross-ply carbon/epoxy and E-glass/epoxy laminates, and a 3 cover +/-75 E-glass/epoxy filament wound tube. The response of the sensors have been monitored within these composites.

1. Introduction

Advanced Fibre Reinforced Composite (AFRC) materials are being extensively used by the aerospace industries because of their excellent specific mechanical properties and ease of manufacture. However, damage detection from events such as matrix cracking, fibre breakage and debonding are difficult to detect using conventional non-destructive testing (NDT) techniques. Furthermore conventional techniques such as surface mounted strain gauges, thermography[1], penetrant enhanced X-rays[2], acoustic emission[3,4] and ultrasonic C-scanning[2,5] are not suitable for on-line health monitoring in AFRC.

Optical fibre sensors (OFS) offer a number of advantages for in-situ health monitoring in AFRC. For example, (a) the OFS can be incorporated into the composite at specified locations during the manufacturing stage. (b) The OFS can be designed to fulfil a number of roles such as, cure monitoring during processing and health monitoring during service. (c) The interfacial properties of the OFS can be optimised to suit the chemistry of the matrix material. (d) The OFS is immune to the effects of EMI and with the correct selection of material can be used at elevated temperatures or in aggressive environments. This paper describes the use of two novel intensity based OFS for fatigue and impact damage detection in AFRC.

2. Experimental

2.1. Displacement strain sensor

The details of the displacement strain sensor theory and construction have been reported elsewhere[6]. Figure 1 shows a schematic illustration of the displacement strain sensor.

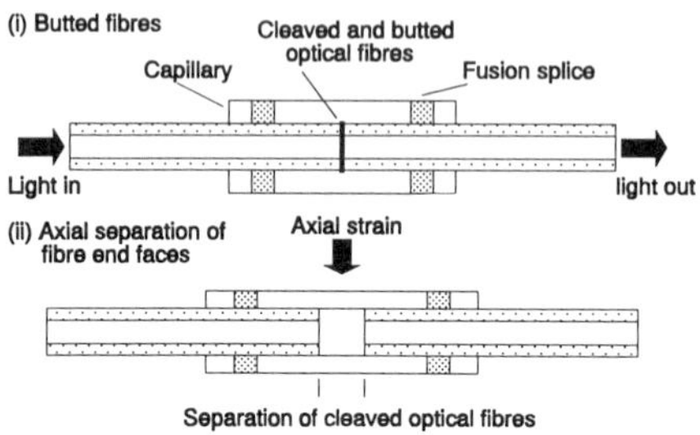

Figure 1 Schematic illustration of the mode of operation of the displacement strain sensor.

With reference to figure 1, since the gauge length of the sensor is specified, it is possible to relate the end-face separation to attenuation and to the applied strain by Equation (1);

$$\textit{Attenuation}\ \alpha \left(\frac{1}{\varepsilon}\right) + F(\varepsilon) \quad (1)$$

Where F(ε) is a correction factor that can be applied to account for reflective coupling of light from the capillary tube if the separation of the fibre ends becomes very large. The sensors used in this paper were designed so that reflective coupling did not take place. It has to be pointed out that the displacement strain sensor can be tuned for sensitivity and strain range as desired.

2.2. The profile strain sensor

This sensor design is currently under development by the authors and consists of a sinusoidal waveform being created on the optical fibre. This profile can be created when the optical fibre is drawn or subsequent to manufacture. The light transmission characteristics of the sensor will be influenced by the applied strain.

2.3. Optical fibre crack detector

Optical fibres can be used as a crude crack detector[7-9]. However, in order for these crack detectors to be reliable and effective a detailed knowledge of the failure mechanisms of the AFRC, the interfacial bond strength and chemical and physical compatibility between reinforcing and crack detection fibres is required. In this study, 50 μm optical fibres were used to investigate the feasibility of using a crack detection system to study impact damage in filament wound tubes.

2.4. Manufacture of AFRC

16 ply carbon fibre reinforced epoxy resin pre-pregs (Ciba-Geigy T300/920)[10] were fabricated in a single specimen mould. A $(0,90_2,0_2,90,0,90)_s$ stacking sequence was used in this study and the laminated pre-pregs were cured using the cure schedule recommended by the manufacturer. Test specimens for tensile and fatigue tests were end-tabbed soon after production using previous grit-blasted aluminium of dimensions 20 mm x 50 mm x 1.6 mm. The test specimens were stored in a desiccator until required.

Filament wound tubes were manufactured on a Josef Baer No. 3:300927 using E-glass fibres and Ciba-Geigy MY750/HY917 epoxy resin. A total of 3 'covers' were laid down to produce a 325 mm long and 100 mm diameter tube. A cover is defined as laying down enough circuits to completely cover the surface area of the mandrel with wetted fibre tows. The optical fibres were wound simultaneously with the reinforcing fibre. For the purpose of this study the profiled OFS were manually laid down onto the mandrel after a specified number of covers had been laid.

2.5. Fatigue and impact damage

The tensile test specimens were subjected to fatigue loads at stress ratios of 0.25, 0.1, 0.0, -0.1, -0.25, and -0.33 on an Instron 8500 servo-hydraulic dynamic testing machine. The fatigue tests were carried out using a sinusoidal loading waveform with a constant rate of stress application.

A Rosand instrumented impact testing machine was used to induce impact damage on the filament wound tubes. The tubes were then subjected to hydrostatic testing.

2.6. Surface properties

A Cahn (model no. DCA-322) dynamic contact angle analyzer was used to characterise the wetting of Ciba Geigy MY-750 epoxy resin, HY-917 hardener and DY-063 accelerator on 50/125 μm silica optical fibres with and without buffer coatings. The equipment uses a modified Wilhelmy plate method consisting of a micro-balance to measure force changes as an optical fibre is immersed and withdrawn from the resin. This immersing/withdrawing cycle produces a hysteresis force loop from which the advancing and receding contact angles are calculated. The equation for the contact angle is given by[11]:

$$\cos q = \frac{F}{p * (st)} \qquad (2)$$

where (q) is the contact angle, (F) is the force in Pa, (p) is the perimeter of the optical fibre and (st) is the surface tension of the resin. The surface tension is measured by immersing a clean glass plate into the resin and correcting for buoyancy effects. The calculated contact angles can then be used to predict wetting between the resin and the optical fibre.

3. Results and discussion

3.1. The displacement strain sensor

Fatigue damage development in the cross-ply AFRC investigated in this study typically started as matrix damage until matrix crack saturation density was reached. This is followed by crack growth and linkage. The actual damage modes under fatigue loading will depend on the stress regime imposed on the specimen[12]. A typical example of a fatigue damaged specimen is shown in Figure 2.

Damage within the AFRC leads to stiffness changes of the specimen[13-14]. The stiffness decay of a fatigue test specimen can be related to progressive damage development in the composite[13-17]. Therefore an embedded OFS for monitoring strain would be an ideal candidate for monitoring fatigue damage progression in AFRC. Figure 3 illustrates a typical stiffness decay curve for a T300/920 composite fatigue tested at a maximum stress of 0.238 GPa, at a loading rate of 250 KN/s and at a stress ratio (R) of -1. A surface mounted extensometer was used to measure the stiffness decay. The compressive stiffness was found to decrease very rapidly after a few cycles to a relative threshold value much lower than that observed for the tensile stiffness.

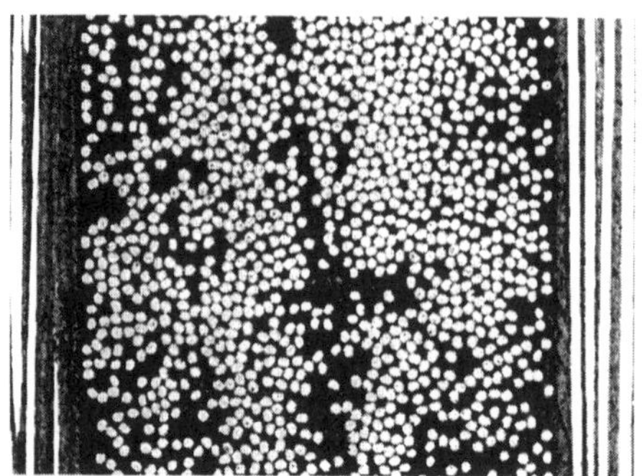

Figure 2 Micrograph of a fatigue damaged composite.

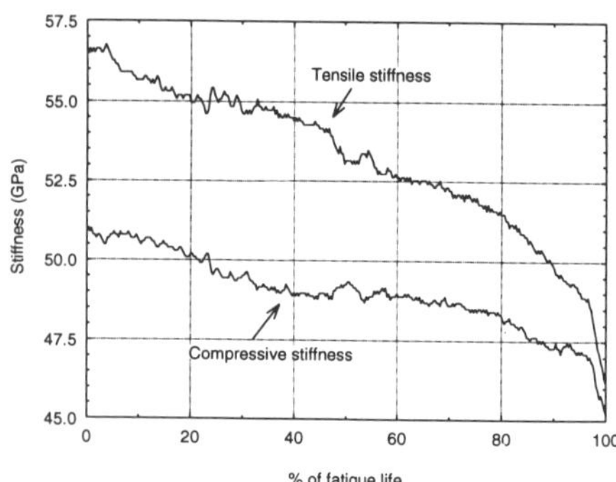

Figure 3 Stiffness decay of a T300/920 composite undergoing tension/compression fatigue at R = -1.0, peak stress = 0.238 GPa.

The stiffness from both the tensile and compressive phases were found to decrease at an approximately linear rate up to 75% of fatigue life. Following this, a very rapid decay in stiffness was observed culminating in final failure.

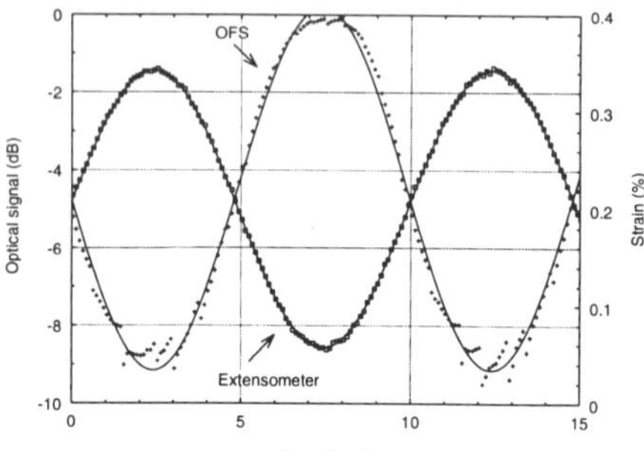

Figure 4 Response of the OFS compared to extensometer over a period of 15 seconds, tested at a frequency of 0.1 Hz and R = 0.1.

Figure 4 illustrates the excellent correlation between the surface mounted extensometer and the embedded displacement strain sensor at a frequency of 0.1 Hz and a stress ratio of 0.1.

The dynamic response of the embedded OFS tested at a frequency of 1.0 Hz and a stress ratio of 0.25 and -0.1 are shown in Figures 5 and 6 respectively. Once again it can be seen that the response of the OFS follows very closely to that observed for the surface mounted extensometer. It has also been found that the sensor still performs satisfactorily after more than 570,000 fatigue cycles.

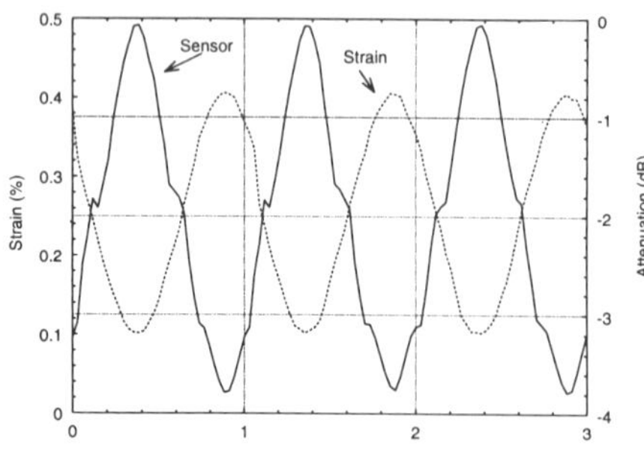

Figure 5 Dynamic response of OFS embedded in cross-ply composite, tested at a frequency of 1.0 Hz and R = 0.25.

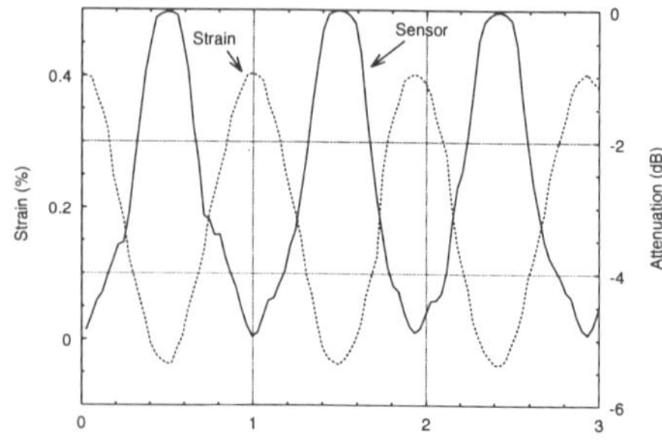

Figure 6 Dynamic response of OFS embedded in cross-ply composite, tested at a frequency of 1.0 Hz and R = -0.1.

3.2. *The profile sensor*

The profiled optical fibre strain sensor which has been designed and evaluated has a linear response to applied strain. This sensor can also be fabricated so that the attenuation of light through it either increases or decreases as it is strained. Figures 7 and 8 show the static response of the sensor when they were embedded within a 16 ply T300/920 composite coupon. Figure 7 shows the sensor fabricated to demonstrate decreasing attenuation with strain whereas Figure 8 shows the sensor fabricated to demonstrate increasing attenuation with strain. The results clearly show that the latter, i.e. the sensor fabricated to increase attenuation with strain, demonstrates a greater sensitivity to straining with a very good linearity which is useful for strain monitoring.

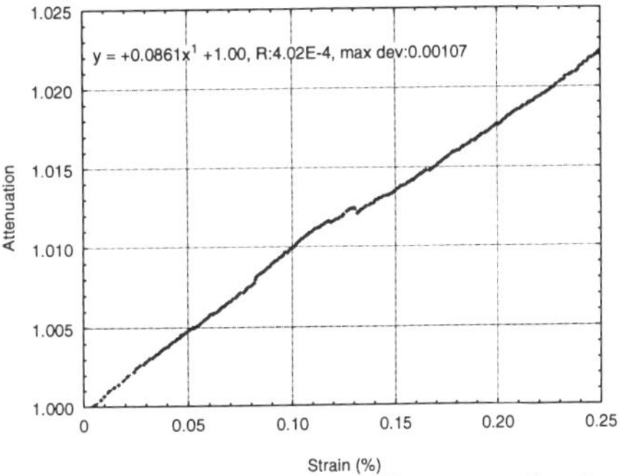

Figure 7 Response of the profile sensor showing decreasing attenuation with applied strain.

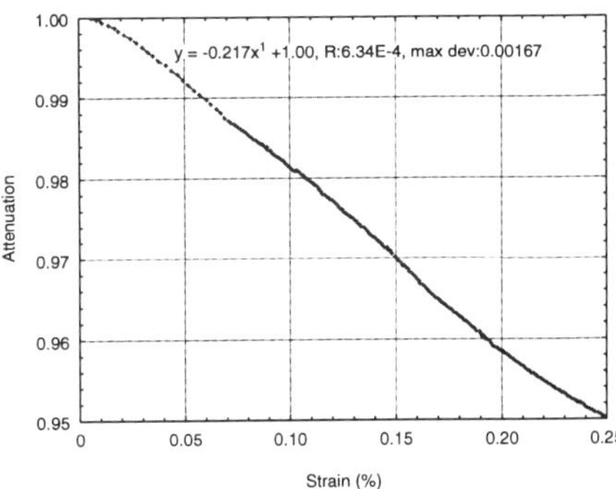

Figure 8 Response of the profile sensor showing increasing attenuation with applied strain.

Impact damage in filament wound tubes can reduce the burst strength below safe working levels[18]. The amount of damage sustained by the tube is influenced by a number of factors which include; i) impact projectile geometry and energy, ii) matrix and fibre type, iii) wind angle of fibres, iv) Relative volume fraction of fibres, v) dimensions of tube, vi) environment and vii) angle of impact[19-20].

The majority of damage in filament wound tubes occurs directly beneath the impact point, however in E-glass composite tubes, cracks can appear remote from the impact point. At certain impact energies, termed 'critical' the residual burst strength of the filament wound tube falls below a safe working level. The level of damage termed 'critical' in AFRC is often difficult to see without detailed examination and hence the need for NDT methods.

Impact damage also causes compliance changes in the filament wound tubes which will affect the strain response of both the tube and any embedded optical fibre sensors[21]. By internally pressurising the tube (hydrostatically) the strain response of a damaged and undamaged tube can be characterised using embedded optical fibre strain sensors.

The profile sensor has also been used to detect compliance changes in low-velocity impact damaged filament wound tubes. Initial tests were carried out to compare the sensitivity of this sensor to that of an optical fibre (no sensing element). These were embedded within a 16 ply E-glass/913 composite coupon. Figure 9 clearly shows that the optical fibre (no sensing element- trace a) has very poor sensitivity compared to the sensor (trace b).

Figure 10 shows the results of these sensors surface mounted on a filament wound tube. The sensor was able to detect the compliance change caused by impacting a tube with a 50 mm hemispherical tup. The level of damage sustained by the tube constituted 'critical' levels and was barely visible to the naked eye. In non-transparent composites this type of damage would be very difficult to observe visually. The optical fibre sensor was however capable of detecting the effect of the impact damage.

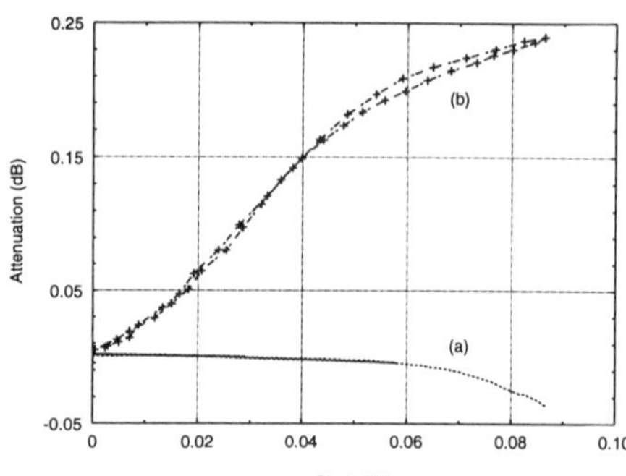

Figure 9 The signal response of a polyimide coated optical fibre (a) and a profiled optical fibre strain sensor (b).

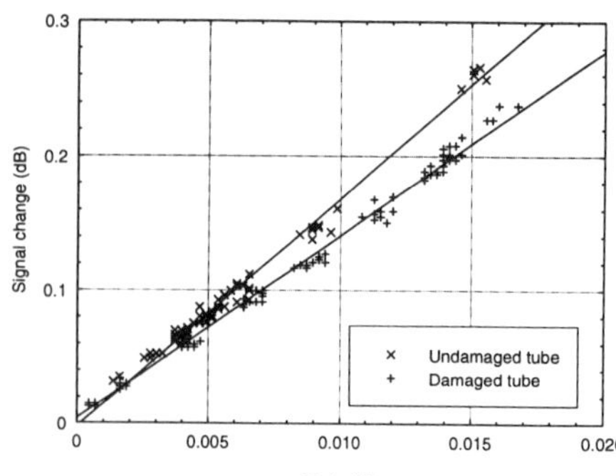

Figure 10 Damage characterisation using a surface mounted profiled optical fibre strain sensor.

3.3. Crack detection

The effect of impact damage on the composite and surrounding optical fibre sensors form the basis of the crack detection technique. In the case where an crack intercepts and fractures an optical fibre, a large increase in attenuation can be easily observed visually or detected using simple instrumentation. Careful selection of the optical fibre, its orientation and surface treatments are required to ensure reliable crack detection

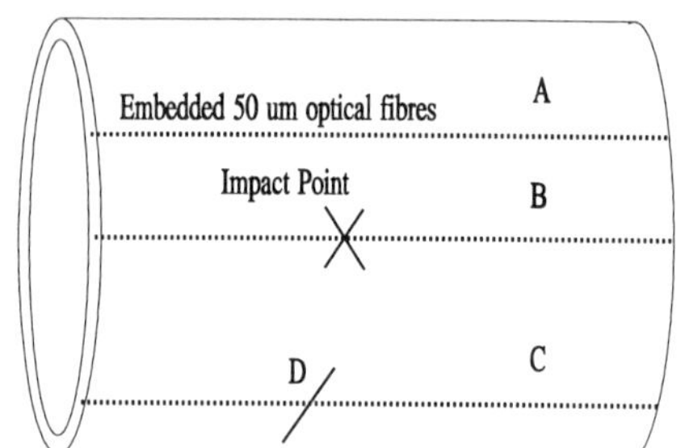

Figure 11 A schematic illustration of the crack detection system.

Figure 11 shows the layout of 50 µm optical fibre bundles (5-6 optical fibres) embedded into E-glass/MY750 +/- 75° filament wound tubes. These optical fibres were used instead of standard 50/125 telecommunication grade optical fibres to reduce the volume fraction of the optical fibres within the composite. The filament wound tubes were impacted at point X at a range of impact energies from 2-12.5 J using a 50 mm hemispherical tup.

After each impact a white light source was shone down each length of optical fibre bundle to observe bleeding light coming from fractured optical fibres.

Table 1 Summary of impact test results for the crack detection system.

Impact energy J	Notes
2,4,5	No bleeding light.
6	Small amount of bleeding light at point (X).
7	Complete fracture of all optical fibres at point (X).
8,9	No bleeding light remote from impact point i.e. along (A) and (C).
10	Extensive bleeding light at point (D) where a critical crack has intercepted the optical fibres.
11,12.5	Still no bleeding light along (A) and still some light transmitting from the optical fibres embedded along (C).

The results of these tests are summarised in Table 1 and show that the optical fibres were able to detect both direct impacts and also the propagation of cracks remote from the impact point at impact energies close to 'critical' levels.

3.4. Surface properties

For any embedded sensor the interface between the sensor and the host material will play a vital role for good stress transfer. To this end the effects of surface treatments on the wetting of silica based optical fibres by epoxy resins was investigated using dynamic contact angle measurements. The method used was based upon a Wilhelmy plate method.

From dynamic contact angle measurements, the advancing and receding contact angles can be predicted. For reference, the surface tension of the resin system was measured at 50.14 dynes/cm which agrees with published data. The results of the test are shown in Figure 12.

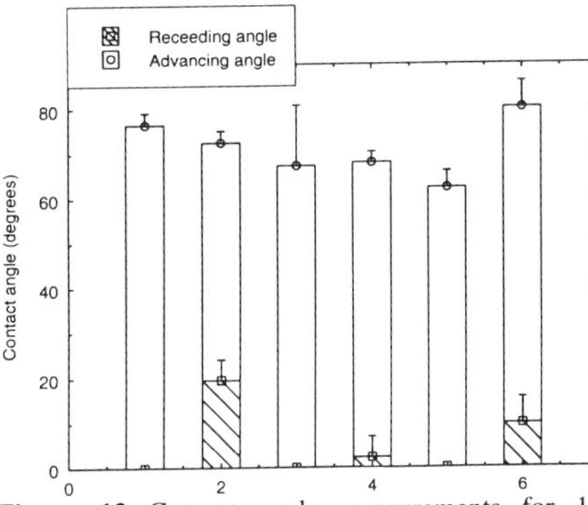

Figure 12 Contact angle measurements for 1) acrylate coated, 2) polyimide coated 3) mechanically stripped, 4) flame stripped 5) silane treated and 6) etched optical fibres.

All of the results suggest some form of wetting has occurred for all the optical fibres (with and without surface treating them). The effect of stripping off the buffer coating generally increased the degree of wetting although silane treating the stripped optical fibres had little effect. However the influence of the various surface treatments on the interfacial properties will be presented in a subsequent publication.

There was significant scatter in the results obtained for the mechanically stripped optical fibres. This scatter is consistent with the acrylate buffer not being completely removed from the silica optical fibre. The etched optical fibres also showed significant scatter in the results, although this may be due to variations in the diameter along the length of the optical fibre (caused by etching) which were not accounted for in calculating the contact angles. The main conclusions which can be drawn from this is that for all the optical fibres tested (with and without surface treatments) some degree of wetting occurred.

4. Conclusions

Three forms of intensity based OFS for damage detection in AFRC have been demonstrated. However these sensing techniques can be applied for use in other engineering materials such as polymers, metal matrix composites and ceramics. These sensors can be embedded or surface mounted for health monitoring purposes.

This study has shown excellent correlation between conventional strain sensing techniques and the profile and displacement strain sensors. The use of these embedded intensity based OFS has been shown to be viable for dynamic loading conditions in AFRC structures. Furthermore it was possible to detect barely visible impact damage via the use of these sensors.

5. Acknowledgements

This work has been supported by EPSRC, The Royal Society, DRA Fort Halstead and DRA Farnborough. The authors wish to acknowledge the support given by Professors P. Curtis and B. Ralph, Drs M. Kemp and K. Hale, A. Howard, J. Curtis, J. Felgate and P. Dodia.

6. References

1. T. Jones and H. Berger. "Thermographic detection of impact damage in graphite-epoxy composites," *Materials Evaluation*, pp. 1446-1453, December 1992.

2. J. J. Nevadunsky and J. J. Lucas. "Early fatigue damage detection in composite materials," *Journal of Composite Materials*, Vol. 9, pp. 394-408, 1975.
3. H. Bolduc and C. Roy. "Evaluation of impact damage in composite materials using acoustic emission," *ASTM Special Technical Publication, Fatigue and Fracture*, No. 1156, pp. 127-138, 1993.
4. M. R. Bhat and C. R. L. Murthy. "Fatigue damage stages in unidirectional glass-fibre-epoxy composites: identification through acoustic emission technique," *International Journal of Fatigue*, Vol. 15, pp. 401-405, 1993.
5. O. U. Anders. "Ultrasonic C-scan as an affordable technique to quantitatively assess damage in laminate composite materials," *International SAMPE Technical Conference*, Vol. 23, pp. 1083-1096, 1991.
6. R. A. Badcock and G. F. Fernando. "Fatigue damage detection in carbon fibre reinforced composites using an intensity based optical fibre sensor," 1995 North American Conf. on Smart Structures and Materials, W. B. Spillman, Editor, SPIE Proc. Vol. 2444, San Diego, 1995.
7. R. Measures. "Fibre optic sensing for composite smart structures," Composite Engineering, Vol. 3, No. 7-8, pp. 715-750, 1993.
8. A. R. Martin, S. A. Hayes, G. F. Fernando and K. F. Hale. "Impact damage detection in filament wound tubes utilizing embedded optical fibres," 1995 North American Conf. on Smart Structures and Materials, W. B. Spillman, Editor, SPIE Proc. Vol. 2444, San Diego, 1995.
9. K. F. Hale. "An optical-fibre fatigue crack-detection and monitoring system," *Smart Material Structure*, Vol. 1, pp. 156-161, 1992.
10. Information Sheet No. FTC 118e, Ciba-Geigy Composites, UK, October 1991.
11. Cahn Instruments Inc. Instruction manual DCA-322, 1992.
12. G. Fernando, R. F. Dickson, T. Adam, H. Reiter and B. Harris. "Fatigue behaviour of hybrid composites: part 1 carbon/kevlar hybrids," *Journal of Materials Science*, Vol. 23, pp. 3732-3743, 1988.
13. R. Talreja. "Stiffness properties of composite laminates with matrix cracking and interior delamination," *Engineering Fracture Mechanics*, Vol. 25, pp. 751-762, 1986.
14. Z. Hashin. "Analysis of stiffness reduction of cracked cross-ply laminates," *Engineering Fracture Mechanics*, Vol. 25, Nos. 5/6, pp. 771-778, 1986.
15. T. K. O'Brien and K. L. Reifsnider. "Fatigue damage evaluation through stiffness measurements in boron-epoxy laminates," *Journal of Composite Materials*, Vol. 15, pp. 55-70, January 1981.
16. S. M. Spearing, P. W. R. Beaumont and P. A. Smith. "Fatigue damage mechanics of composite materials part IV: prediction of post-fatigue stiffness," *Composites Science and Technology*, Vol. 44, pp. 309-317, 1992.
17. Y. W. Kwon. "Calculation of effective moduli of fibrous composites with micro-mechanical damage," *Composite Structures*, Vol. 25, pp. 187-192, 1993.
18. P. Meyer, "Low velocity hard-object impact of filament wound Kevlar/epoxy composites," *Composite Science and Technology*, Vol. 33, pp. 279-293, 1988.
19. P. Soden, R. Kitching, P. Tse and Y. Tsavalas, "Influence of winding angle on the strength and deformation of filament wound tubes," *Composite Science and Technology*, Vol. 46, No. 4, pp. 363-378, 1993.
20. M. Bader and P. Manders, "Simulated impact damage in filament wound glass-fibre reinforced tubes," Experimental Programme, 4th Report, University of Surrey, November 1980.
21. V. L. Chen, H.-Y.T. Wu and H.-Y. Yeh, "A parametric study of residual strength and stiffness for impact damaged composites, "*Composite Structures*, Vol. 25, Iss: 1-4, pp. 267-75, 1993.
Marcel Dekker Inc., 1990.
22. P. W. Atkins, Physical Chemistry, Edn. 5, Oxford University Press, 1994.

Cure monitoring of epoxy resins using optical fibre sensors

P. A. Crosby, G. R. Powell, G. F. Fernando[#], C. M. France[*], D. N. Waters[§], R. C. Spooncer[*]

Brunel University, Department of Materials Technology, Kingston Lane, Uxbridge, Middlesex, UB8 3PH, UK.
[*]Department of Manufacturing and Engineering Systems, [§]Department of Chemistry.
[#]To whom correspondence should be addressed.

Abstract. Two methods utilizing optical fibre sensors to follow refractive index changes and chemical changes during epoxy resin cure are described. A laser diode source, whose wavelength was remote from any interfering absorption bands of the resin was used to observe changes in the light guiding characteristics through a stripped optical fibre embedded in a curing resin. Results were compared to data obtained from transmission FTIR spectroscopy measurements on the same sample. Good correlation was obtained between the two techniques.

The use of an optical fibre evanescent wave sensor to follow changes in near-infrared absorption bands during epoxy cure was demonstrated. The depletion of a N-H absorption band at 1542nm was followed for the room temperature cure of a resin cured with an amine hardener. Spectra showed effects due to the increase in material refractive index as well as absorption band depletion.

1. Introduction

Epoxy resin and amine hardener systems are utilized extensively in advanced fibre reinforced composites (AFRC). The mechanical properties of the AFRC in general are dictated by the reinforcing fibres. However the matrix dominated properties such as absorption of moisture[1, 2], compressive properties[3], interlaminar shear strength[4, 5], failure strain[6] etc. can be influenced by the resin's crosslink density. The crosslink density in turn can be affected by the chemical state of the resin prior to cure, eg. presence of moisture in resin[7] and cure schedule used to process the material[8]. Hence it is important to have tools to monitor the cure kinetics of epoxy-amine resins.

AFRC are being increasingly utilised by the aerospace industries for larger, more complex and thicker sections. However in order to ensure homogeneous curing of these structures, on-line process monitoring and process optimisation techniques are necessary. Conventional techniques such as differential scanning calorimetry (DSC), viscosity measurements and conventional infrared spectroscopy are not suitable for on-line cure monitoring. Optical fibre based techniques which have been reported in the literature as suitable for in-situ process monitoring include optical fibre fluorescence[9, 10], Raman[11, 12] and near-infrared spectroscopy[13, 14]. Non-optical fibre based techniques include electrical resistance and dielectric analysis[15], ultrasonic wave propagation[16, 17], heat flux measurements[18] and viscosity measurements[19]. This paper reports on the in-situ cure monitoring of epoxy resins using two different optical fibre based techniques, namely, refractive index monitoring and evanescent spectroscopy. These results have been compared with conventional FTIR spectroscopy.

2. Experimental

2.1. Simultaneous optical fibre refractive index sensor and FTIR spectroscopy

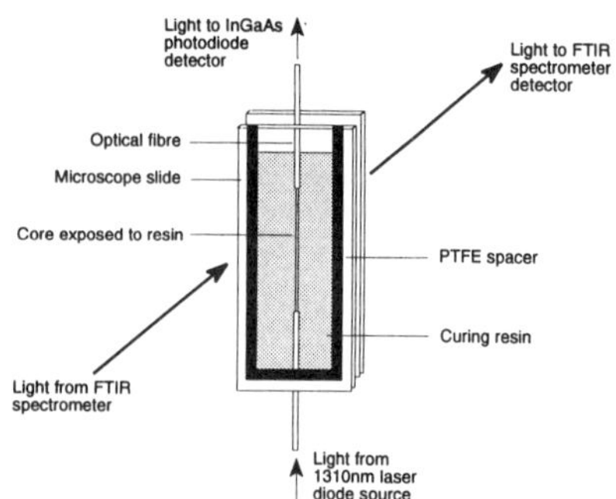

Figure 1 Experimental set-up inside the FTIR spectrometer sample compartment for simultaneous optical fibre cure monitoring and FTIR spectroscopy.

The multimode optical fibre used for refractive index measurements consisted of a high refractive index glass core (1.7) surrounded by a silicone resin cladding. To prepare the sensor, a 30cm length of fibre was used. A small section of fibre cladding (25mm) was removed by soaking the fibre in sulphuric acid at 60°C for 5 minutes. Following removal of the cladding, the fibre was washed thoroughly with distilled water then isopropyl alcohol and allowed to dry. An infrared transmission cell was constructed from two microscope slides separated by a PTFE spacer, giving a pathlength of approximately 1mm. The near-infrared transmission characteristics of the microscope slides showed low absorption in the range 1500-2500nm. The optical fibre sensor was placed between the two slides so that the entire sensor region was within the contained resin area (Figure 1). A Tefzel sleeve was placed over the fibre away from the sensor region to protect the fibre from damage. Rubber sealing tape was used to ensure that no resin leakage occurred. The empty transmission cell was placed in the sample holder of a Nicolet 710 FTIR spectrometer and the trailing ends of the optical fibre were connected via silica fibre patchcords to a 1310nm laser diode source and InGaAs photodiode detector. Spectra were obtained using 32 scans and 2cm^{-1} resolution over the range 7000-4000cm^{-1} (1429-2500nm). A background spectrum was collected with the empty transmission cell located in the sample holder. The cell was filled with a stoichiometric mixture of Shell Chemicals Epikote 828 + hexane 1,6-diamine, 98% (Aldrich Chemicals) and the acquisition of fibre sensor data was started. Infrared spectra were obtained at 10 minute intervals over a 6 hour cure schedule at room temperature.

2.2. Optical fibre evanescent wave cure sensor

A 50cm length of high refractive index optical fibre was used to obtain evanescent wave spectra during cure. The 17cm stripped sensor region was prepared using the same method as for refractive index cure monitoring. The fibre was inserted in a resin container which allowed all of the sensor length to be in contact with the resin. Monochromatic light was launched into the fibre using a stabilised, chopped tungsten quartz halogen source and a Bentham Instruments grating monochromator with optical fibre attachment. The free end of the fibre was connected to an InGaAs photodiode detector and lock-in amplifier. Wavelength selection and collection of the amplifier signal were undertaken using a PC and DT-Vee acquisition software. A background spectrum was obtained with the stripped fibre exposed to air over the range 1390-1650nm, using 1nm resolution. A stoichiometric mixture of Shell Chemicals Epikote 828 + triethylenetetramine, 60% (Aldrich Chemicals) was poured over the sensor region and spectra were acquired at 10 minute intervals during cure at room temperature. Schematic diagrams of the experimental set-up and the sensor arrangement within the resin container are shown in Figure 2 and Figure 3 respectively.

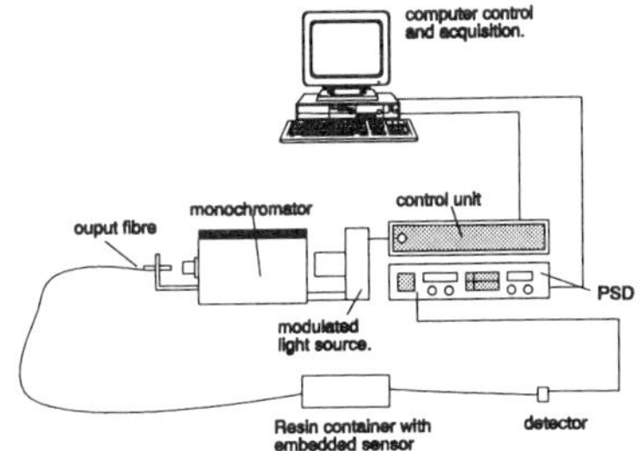

Figure 2 Schematic diagram of fibre spectrometer and evanescent wave sensor experimental set-up.

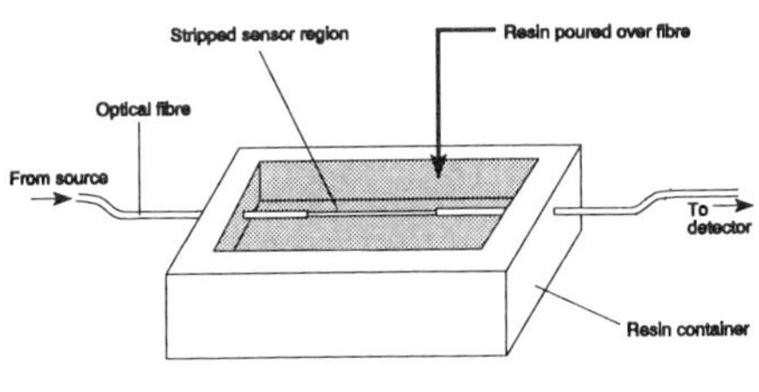

Figure 3 Schematic diagram showing sensor arrangement within the resin container.

3. Results and discussion

3.1. Simultaneous optical fibre refractive index sensor and FTIR spectroscopy

Infrared spectra of the two peaks used for data analysis are shown in Figure 4. Peak area measurements were made for the epoxy peak centred at 2208nm and the aromatic C-H peak centred at 2136nm from each spectrum acquired. The epoxy peak area was ratioed against the aromatic C-H peak area as an internal standard.

Since epoxy area/aromatic C-H area is directly proportional to epoxy group concentration, the relationship of the refractive index sensor data to epoxy concentration can be investigated. Figure 5 shows data obtained for optical fibre measurements and epoxy area/aromatic C-H area from FTIR spectroscopy. Peak area ratios have been normalised so that first and last data points correspond with optical fibre sensor data. It can be seen that both sets of results compare well over the cure schedule, indicating a relationship between the refractive index of the resin and the concentration of epoxy groups. Further work is being undertaken to investigate the relationship in more detail.

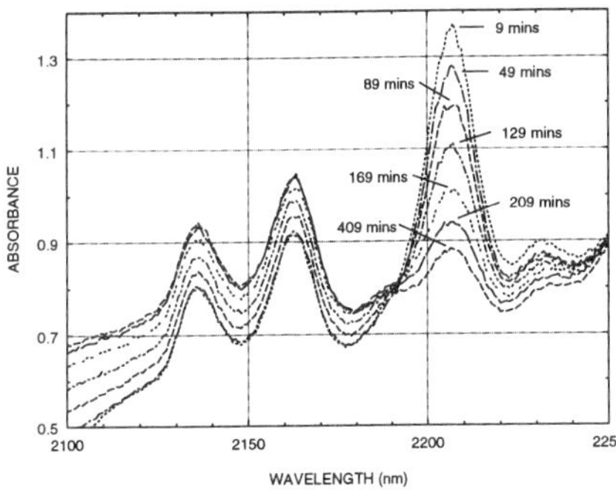

Figure 4 Overlaid near-infrared spectra obtained during cure of Epikote 828 + hexane 1,6-diamine cure at 25°C.

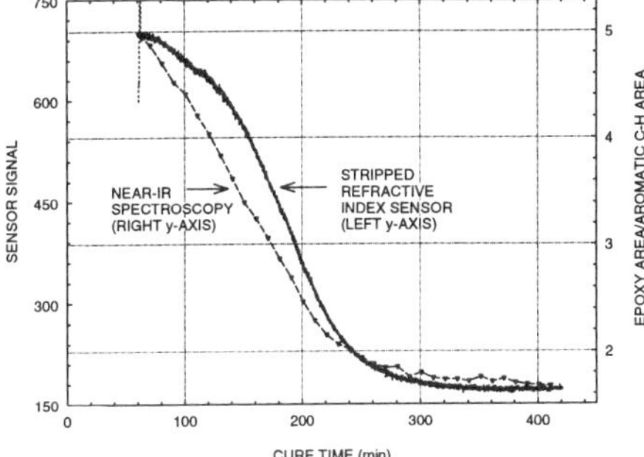

Figure 5 Results obtained from simultaneous FTIR spectroscopy and optical fibre sensor cure monitoring of Epikote 828 + hexane 1,6-diamine at 25°C.

3.2. Optical fibre evanescent wave cure sensor

Raw data obtained from optical fibre evanescent wave cure monitoring experiments were converted to absorbance by ratioing to a baseline scan (a scan from the sensor in air) using the equation:

$$A = (\log_{10}\frac{I_0}{I}) \tag{1}$$

where, I_0 is the intensity of light with the sensor exposed to air and I is the intensity of light with the sensor exposed to the resin. Figure 6 shows overlaid spectra for the N-H absorption peak at 1542nm obtained from the optical fibre evanescent sensor during cure. It can clearly be seen that the magnitude of the peak at 1542nm decreases with time as the amine hardener is consumed. It is also apparent from Figure 6 that the overall absorption level has increased. This effect is due to an increase in the resin refractive index during cure. Peak area measurements were undertaken for each cure spectrum obtained by constructing a baseline to the peak. Results obtained for N-H peak area as a function of cure time for the middle part of the cure period are presented in Figure 7.

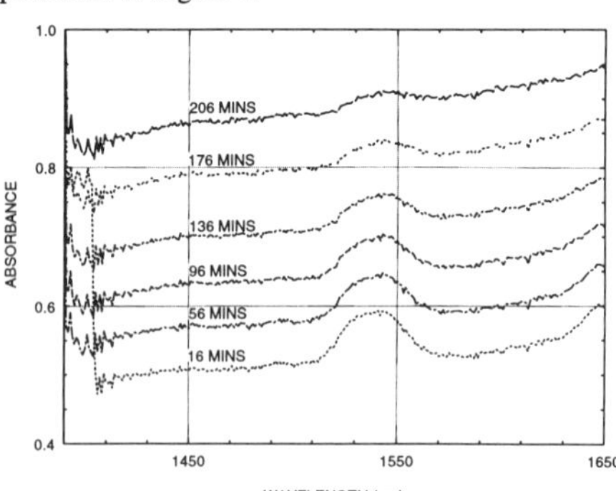

Figure 6 Evanescent absorption peak at intervals during resin cure at 25°C (Epikote 828 + triethylenetetramine).

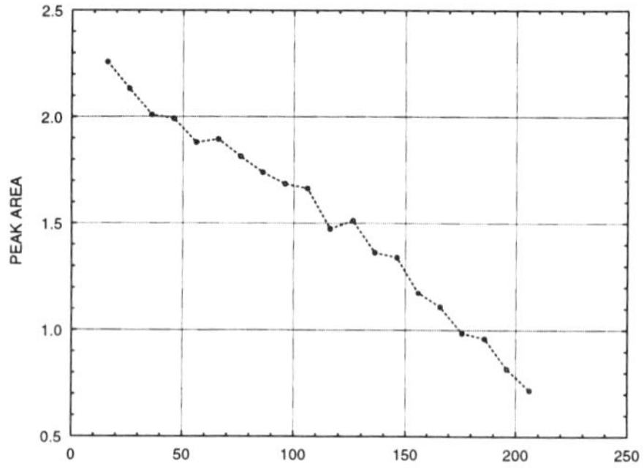

Figure 7 N-H absorption peak area during room temperature cure of Epikote 828 + triethylenetetramine.

4. Conclusions

An optical fibre sensor was demonstrated which can follow the cure of an epoxy-amine resin system by two methods; refractive index tracking and infrared absorption band depletion. The results were compared to measurements obtained using FTIR spectroscopy. In order to evaluate further the optical fibre sensor results, a technique was developed which enabled simultaneous optical fibre refractive index cure monitoring and FTIR spectroscopy to be undertaken. Sensor data obtained by this method were compared with epoxy absorption peak results during cure and showed good agreement.

The depletion of the amine hardener absorption peak during cure was observed using the sensor as an evanescent, or ATR, cell. The data obtained showed agreement with conventional spectroscopic data. These methods are being incorporated into a system for on-line cure monitoring and process optimisation.

5. Acknowledgements

The authors would like to thank Dr. S. Hitchen and Dr. J. Barton at DRA Farnborough for their support and encouragement for this programme. This work was supported by the Defence Research Agency, Farnborough and the Department of Trade and Industry. G. Fernando acknowledges the receipt of a Royal Society Grant for optical fibre sensor research. The authors are grateful to Professor B. Ralph for his advice given during the course of this work and to J. Felgate and T. Banerjee for their assistance.

6. References

1. J. M. Barton, D. C. L. Greenfield, "The use of dynamic mechanical methods to study the effect of absorbed water on temperature-dependant properties of an epoxy resin-carbon fibre composite." *British Polymer Journal*, Vol. 18, No. 1, pp. 51-56, 1986.
2. A. Apicella, L. Nicolais, "Effect of water on the properties of epoxy matrix composite." *Advances in Polymer Science*, Vol. 72, pp. 69-77, 1985.
3. C. Jordan, J. Galy, J. Pascault, "Measurement of the extent of reaction of an epoxy-cycloaliphatic amine system and influence of the extent of reaction on its dynamic and static mechanical properties." *Journal of Applied Polymer Science*, Vol. 46, pp. 859-871, 1992.
4. K. J. Bowles, S. Frimpong, "Void effects on the interlaminar shear strength of unidirectional graphite-fibre-reinforced composites." *Journal of Composite Materials*, Vol. 26, No. 10, pp. 1487-1509, 1992.
5. H. K. Soni, R. G. Patel, V. S. Patel, "Structural, physical and mechanical properties of carbon fibre-reinforced composites of diglycidyl ether of bisphenol-A and bisphenol-C." *Die Angewandte Makromoleculare Chemie*, Vol. 211, pp. 1-8, 1993.
6. R. J. Morgan, "Structure-property relations of epoxies used as composite matrices." *Advances in Polymer Science*, Vol. 72, pp. 1-43, 1985.
7. R. F. Brady, Jr., J. M. Charlesworth, "Influence of water on the reaction between amidoamine and epoxy resins." Polymeric Materials and Engineering, Proceedings of the ACS Division of Polymeric Materials Science and Engineering, Vol. 67, pp. 274-275, 1992.
8. B. A. Rozenberg, "Kinetics, thermodynamics and mechanism of reactions of epoxy oligomers with amines." *Advances in Polymer Science*, Vol. 75, pp. 113-165, 1985.
9. S. D. Schwab, R. L. Levy, "In-service characterization of composite matrices with an embedded fluorescence optrode sensor." Fiber Optic Smart Structures and Skins II, SPIE Vol. 1170, Ch. 54, pp. 230-238, 1990.
10. W. Deng, N. Sung, "In-situ cure monitoring of diamine cured epoxy by fiberoptic fluorimetry using extrinsic reactive fluorophore." *Polymer Engineering and Science*, Vol. 34, No. 9, pp. 707-715, 1994.
11. S. K. Sharma, C. L. Schoen, T. F. Cooney, "Fiberoptic remote raman probe design for use in monitoring processes in a high-temperature oven." *Applied Spectroscopy*, Vol. 47, No. 3, pp. 377-379, 1993.
12. K. C. Hong, T. M. Vess, R. E. Lyon, K. E. Chike, J. F. Aust, M. L. Myrick, "Remote cure monitoring of polymeric resins by laser Raman spectroscopy." 38th International SAMPE Symposium, May, pp. 427-435, 1993.
13. G. A. George, P. Cole-Clarke, N. St. John, G. Friend, "Real-time monitoring of the cure reaction of a TGDDM/DDS epoxy resin using fibre optic FTIR." *Journal of Applied Polymer Science*, Vol. 42, pp. 643-657, 1991.
14. G. R. Powell, P. A. Crosby, G. F. Fernando, C. M. France, R. C. Spooncer, D. N. Waters, "In-situ cure monitoring of advanced fibre reinforced composites." 1995 North American Conf. on Smart Structures and Materials, W. B. Spillman, Editor, SPIE Vol. 2444, San Diego, 1995.
15. F. I. Mopsik, S. S. Chang, D. L. Hunston, "Dielectric measurements for cure monitoring." *Materials Evaluation,* Vol. 47, No.4, p. 448, 1989.
16. A. Davis , M. M. Ohn, K. X. Liu , R. M. Measures, "A study of an opto-ultrasonic technique for cure monitoring." Fiber Optic Smart Structures and Skins IV, SPIE Vol. 1588, Ch.35, pp. 264-274, 1991.
17. A. Davis , M. M. Ohn, K. Liu, R. M. Measures, "Composite cure monitoring with embedded optical fiber sensors." Structures Sensing and Control, SPIE Vol. 1489, Ch.25, pp. 33-43, 1991.
18. M. J. Perry, L. J. Lee, C. W. Lee, "Online cure monitoring of epoxy graphite composites using a scaling analysis and a dual heat-flux sensor." *Journal of Composite Materials*, Vol. 26, No.2, pp. 274-292, 1992.
19. J. S. Kim, D. G. Lee, "Online cure monitoring and viscosity measurement of carbon-fiber epoxy composite-materials." *Journal of Materials Processing Technology*, Vol. 37, pp. 405-416, 1993.

The instrumental engineering of a fibre drop analyser for both quantitative and qualitative analysis with special reference to fingerprinting products for industrial alcohol and sugar manufacture.

N.D.McMillan, M.Baker, S.Smith, D.Lane, R.Corden, J.Hanrahan, K.Thompson, K.Boylan
Drop Tech Ltd., Unit 4, Cookstown Enterprise Park, Belgard Road, Tallaght, Dublin 24

M.Bree, Max -Planck -Institut für Kolloid- und Grenzflächenforschung, Berlin-Aldershof, Germany

ABSTRACT

The fibre drop analyser (fda) detailed here has been designed for general laboratory and industrial application. Details are presented of general design features necessary for the production of a commercial instrument. The measurement capabilities of the instrument are indicated and the results of some development measurements obtained with stepper pump and gravity feed delivery using a single wavelength (850nm) system are outlined together with some colorimetric results obtained from the multi-wavelength operation. Details of results on what is referred to as fda fingerprinting are discussed. This approach could be used for a wide range of qualitative sample determination. The software for this fda is described briefly and the potential industrial applications of a multiwavelength fda presently under construction are indicated with special reference to the wine, beer, sugar and food industries.

1 FORM OF THE FDA DATA

The fda is an instrument based on the fiber drop head shown in Figure 1 for the simplest case of a two fiber system. Light from a tungsten or led source is injected into the drop through the source fibre and the signal picked up by the collector fiber is delivered to a ccd or phototransistor/photodiode detector at the end of the second return fibre.

The three principal drop head designs used in the present instrument are referred to here as, the flat silica fibre, flat polymer fibre and concave polymer fibre drop heads, detailed design drawings of which are shown in Figure 1. The size and design of the drop head is very critical in the present work of multi-measurand characterisation. Ideally the head should be designed such that it wets (i.e. the suspended liquid covers the entire lower surface of the drop head) when liquid is delivered to this head without any mechanical stimulation or human intervention. The second required characteristic of a drop head for multi-measurand analysis is that the signal known as the fibre drop trace (fdt) should exhibit both rainbow and fda peaks for all liquids, and not just the single fda peaks, as is the usual situation for most heads.

Figure 2 shows a typical fdt taken a water that shows the desired features in the trace in the first part of the signal in which the stepper pump was delivering this reference liquid at a constant rate. The labelled features of the fdt are the separation peak, separation vibration, rainbow peak, fda peaks and drop period. This figure also shows a series of vibration drop trace (vdt) with the stepper pump stopped half way up the leading edge of an fda peak that will be discussed below.

The fdt is a temporal optoelectronic trace that arises from the modulation of the light as it is coupled between source and detector fibre inside the drop during various drop growth and drop vibration processes. It is produced by the changes in the reflected coupling light paths as the drop grows as liquid is added to the drop head by the stepper pump. The liquid drop on the fibre head develops gradually in size until it falls off to be replaced by a second growing drop and so forth. For measurement reasons the drop growth during the recording of an fdt should be slow such that the drop remains throughout as much of the drop cycle in a quasi-equilibrium state. The drop is not in this state during the small time interval involved in the separation process. In practice this means drop periods should exceed 40 seconds.

The fdt, which is a reproducible signal, may be considered as a fingerprint for the liquid because it has a unique structure that qualitatively identifies the sample. To date, no two liquid samples from the hundreds tested, have yet given exactly the same fdt and it is presumed that this is indeed a unique signature of the sample. The differences in an fdt arise from changes between samples in the drop period, peak positions, peak

heights and peak widths, and usually differences in all four characteristics. A multi-wavelength array of fdt can be obtained with the multi-wavelength fda. Such a data array obviously contain additional information as there are differences in the fdt from wavelength to wavelength
A second fda trace, the vdt, seen in Figure 2 for water in the second part of the trace can be recorded by stopping the drop growth at a selected drop volume or at some designated fda signal position and then to produce a modulation of the coupled optical signal by vibrations of the drop base excited by a simple mechanical shock. The signal that arises is complex and its analysis is discussed below, but as can be seen from Figure 3, which shows the vibration fdt of vdts recorded on toluene and 40% w/w sucrose solution obtained for the concave polymer fibre drop head. The vdt is also a characteristic of the solution and may therefore be used to fingerprint a sample.
The science of fiber drop analysis can perhaps thus be most simply stated as the extraction of physical, chemical and product information from the instrumental fdts and vdts. This analysis can be done largely by the software that will be described briefly below.

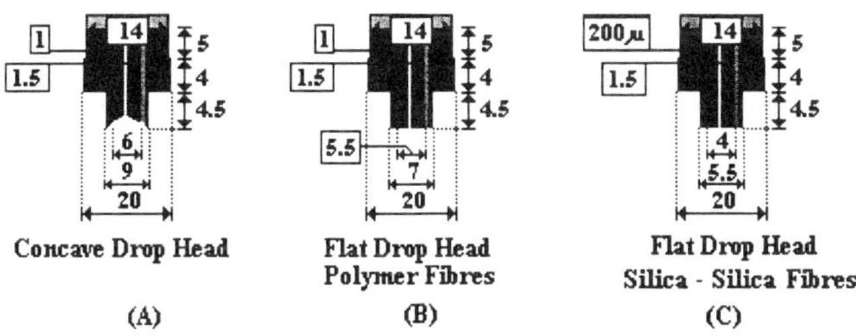

Fig 1. Drop Head Optrode Designs.

2. INSTRUMENT DESIGN

Figure 4 shows the schematic diagram of a fully functional fda that can be deployed for both multi-wavelength and led operation. The drop head is fitted onto the end of a steel tube shown in this figure. This drop head is pushed from above into a temperature block which has been designed to provide also an environmental chamber for the drops as they are delivered at the drop head. The drop head tube shown here in this figure is such that the actual drop head at the end of the tube just pushes from above into the glass environmental tube that is fitted into the heater block from below. The tube can be charged with the specific liquid under test if the bleed valve on the glass is shut so as to produce a saturated vapour around the drop and thereby to reduce evaporation for volatile liquids. The glass fitting is vented to the atmosphere to stop the build up of pressure in the chamber as liquid is delivered to the drop head. Any increase in pressure change effects the drop shape and consequently the form of the fdt.
The liquid can be delivered to the drop head via a Paar Scientific DPRT density meter. The optional density meter is also housed in an aluminium block that is bolted onto the main temperature block and this is thereby maintained at the same temperature as the drop head. The liquid can alternatively be delivered from the constant head apparatus by selecting the appropriate position of valve V3. A short piece of teflon HPLC 2.8 mm teflon tubing is fitted above the drop head which ensures that the liquid flow from the constant head is such that even for the maximum head the liquid drops are slowly delivered and drop times are in excess of 40 seconds as required for the quasi-equilibrium conditions of drop growth.
The multi-wavelength set up is the same as for that shown in Figure 4 with the exception that the source and detector box is replaced by the Ocean Optics Sources and detectors. The light that is injected into the drop for multi-wavelength operation is from the Ocean Optics LS-1 tungsten halogen lamp, the Ocean Optics PX-1 xenon flash lamp, or from a number of LEDs housed in a specially constructed source/detector box that are discussed more fully below. The light coupled back from the detector fiber in the drop head is connected either to the Ocean Optics SD1000 ccd fibre spectrometer, or to one of the several specially constructed

photodiode/phototransistor detectors which are housed in the source/detector box. In all cases the amplified signal from the detector is passed to a Metrabyte DASH 16 A/D card which digitises the signal for analysis by the computer. The signal acquisition is triggered when a drop falls by the infra-red beam of the optical eyes being intercepted. The optical eyes are fitted to the glass unit in the temperature block which also records the drop period by timing drop fall events.

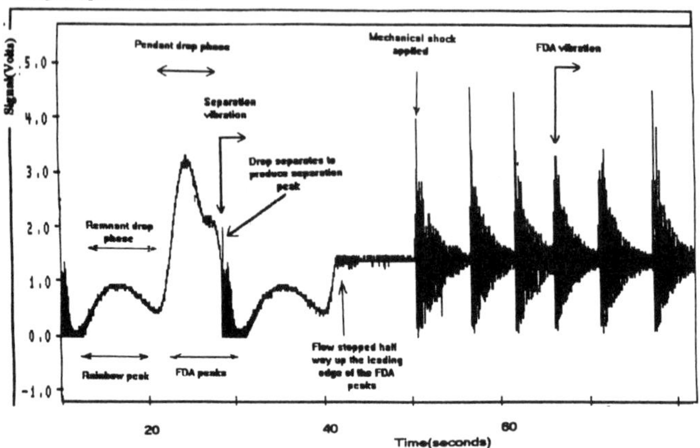

Fig.2 FDT of water at 20 C showing the characteristic features of the trace. The abscissa is time and the ordinate the optoelectronic signal. The pump delivery was stopped halfway up the FDA peak of the second FDT and six vdt

The various components are all mounted in a specially designed instrument housing and to the front left is mounted the Hamilton Microlab M OEM stepper motor pump that is used to deliver the liquid sample The sample to be measured is placed in a buzz bath situated in a recess in the housing just in front of the OEM pump and is used for degassing the sample before this is injected into the system. The sample to be analysed, is placed in the bath just below the pump. The instrument panel has controls to switch on the instrument, to allows the selection of the heater block temperature and to operate of the buzz bath via a push button.
The heating block/environmental chamber which is bolted to the main floor of the housing. This unit is stood on Farrat CR11 spring mounted anti-vibration feet that are each loaded with more than 4.0 kg so as to load these and provide limited vibration isolation removing frequencies above 3.4 Hz effect. These mounting only give the drop partial vibration protection since drops typically have resonance vibrations of 10 Hz, and it is impossible to totally isolate the drop in a portable apparatus given that drop resonance frequencies are so low. The housing is mounted on screw feet that allow the heater block to be levelled via a spirit level and this may be viewed through the small hinged lid on the top of the instrument housing.
The drop head can be mechanically excited by a drop hammer attached to the drop head situated on the top of the stainless steel tube which pushes onto the chamber. This design for the drop head mounting is shown in Figure 4 and the stainless tube is filled with a thermal medium to ensure that the liquid delivered to the drop head is heated to the measurement temperature by the heater block before it is delivered onto the drop head.

3. SOFTWARE

The program is written in C++ and Visual Basic and links dynamically with DADiSP, a commercial package written by the DSP Corporation, USA.
A User Interface is of the standard windows type and provides header menu options. The full range of menu options available on this fda package are (1) Data Capture (2) Cleaning and Pump Priming (3) Drop Period Measurements (4) The Paar Density Meter (5) FDT (6) VDT (7) Hardware Set-up (8) Fingerprinting.
The special feature of this software is the fingerprinting option which allows the analysis of both the fdt or vdt of a sample. A library archive has been devised that can be built up by the user to store the information on samples in a logical way that can sensibly separate sample types. There is cross referencing between the fdt and vdt archives to allow a double check for the identification if the archive contains data on both the fdt and vdt for any sample under investigation. The user is encouraged to build up their own archive in a way that is design for their own specific application. Once the archive is in place, this computer record can be deployed for identification of unknown samples. Once a file is entered for evaluation then this may be analysed using a general search of the archive without any restrictions, or alternatively a search can be restricted to a specific

library section or subset of library sections. The software finds the best fits to the unknown sample and presents the closest matching data sets which are visually presented to the user on a tiled screen. The user can then overlay data to check the fit and can obtain quantitative values on the goodness of fit for any offered option.

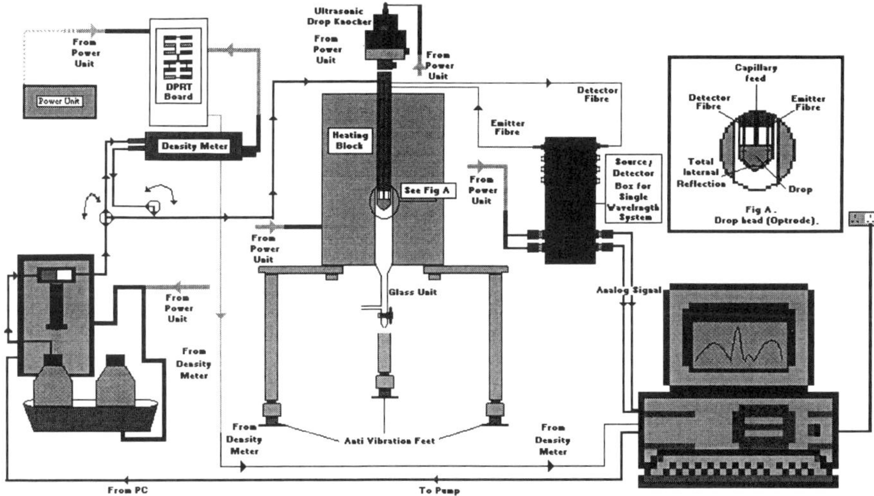

Fig -3. Schematic diagram of the Fibre Drop Analyser

4. EXAMPLES OF RESULTS

The measurement time for a liquid sample is typically about 60s per drop and it might be felt useful in applications testing to perform a number of tests using in total several drops of liquid samples requiring a sample volume of perhaps a few millilitres. In an actual application, very small volumes perhaps of no more than 250 microlitres will be required. A full test programme including flushing out a system and putting in a new sample which may take typically take 8 minutes. For quality control of liquids of the same type it is not of course necessary to flush a system and a simple prime cycle can be used which takes typically no more than 2 minutes.

(1) FINGERPRINTING

The early testing of the fingerprinting capability of the fda has shown that it is able to identify different alcohols and sugars. A test on pharmaceutical samples has shown that the instrument is capable of differentiating aspirin from paracetamol. At this early stage the evidence is that these trivial fingerprinting trials give no real indication as to the enormous potential of the instrument for qualitative sample analysis and a large applications test programme is presently underway. It would be expected that this technique could identify one specific aspirin from another if there was any difference in manufacture.

(2) COLOUR

The only test done to date on the fda colorimetric capability is some preliminary work done on dyes and test results on rhodamine-b shows all the usual features of this UV-VIS spectrum of this dye. A sensitivity test on fda colour measurement has shown that the fda is capable of performing at least to the sensitivity standard of a traditional UV-VIS laboratory measurements.

(3) REFRACTIVE INDEX

The work to date on refractive index measurement capability of the fda has shown that for a range of pure liquids and aqueous solutions that refractive index is related to the size of the rainbow peak of a sample. This peak is measured relative to the corresponding size of the rainbow peak of water selected as a reference for this measurement. The instrument test programme has shown that the fda can perform at least to the standard of an Abbe refractometer for this measurand for a wide range of liquids with refractive index above 1.30.

(4) VISCOSITY/MOLECULAR WEIGHT

A viscosity measurement based on the delivery of the sample from a constant head has been shown to depend in a somewhat complex way on both surface tension and viscosity. The technique, based on drop period measurements, has given viscosity measurement for a very wide range of liquids and to an accuracy better than 1% and repeatability of 0.1%.

(5) SINGLE DROP VIBRATION STUDIES

The vdt obtained from a series of aqueous alcohol and sugar samples were analysed on the basis of the theoretical propositions put forward by McMillan et al, Instrumentation Science & Technology,22(4), 375-395(1994). Recent work has suggested that this trace can be used to obtain quantitative information on the viscosity of a sample that can be forced into vibration which in practice means for liquids with viscosities below about 40 cP. No performance figures are yet available for this method which should also give information on the surface tension and density of the sample.

(6) TURBIDITY

The fda has been employed to obtain a measure of turbidity obtained from the relative size of the fda peak for a water blank and a series of solutions of milk. These measurements are obtained using a turbidity function for the fda derived from the log ratio of the ratios of the fda peaks of the sample divided the peak for the blank reference of water. No performance figures have yet been properly established for this measurement which is not independent of absorbance measurements.

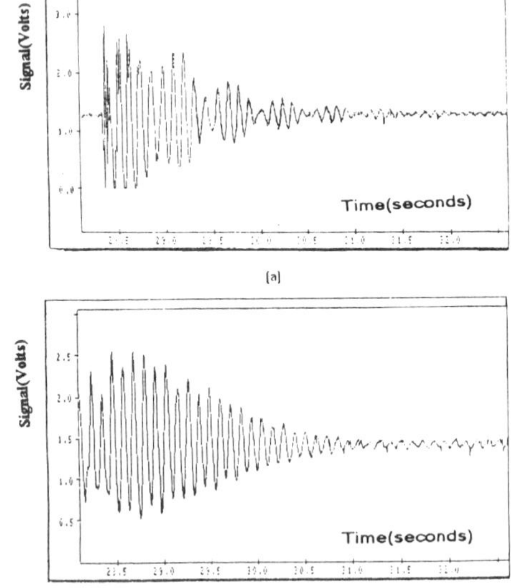

Fig.4. (a) A vdt for toluene recorded on the concave drop head.
(b) A vdt for 40%w/w sucrose solution recorded on the concave drop head.

5. TEST PROGRAMME AND COMMERCIALISATION ROUTE

The fda has been successfully tested for a range of measurements but only to date on a limited set of liquids. The only application for which the fda has been successfully tested so far has been the measurement of the properties of labelled synovial fluid for the diagnosis of disease in this body fluid and the successful results of this study have been published in a refereed journal. Some applications measurements on sugar have been done for Tate and Lyle samples but these were only preliminary results on a single wavelength fda which must be extended now that an advanced fully engineered multiwavelength system is available.

6. FEATURES, BENEFITS AND ADVANTAGES OF THE FDA

FEATURES

1) Small volume analysis, less than 200 ml is feasible.
2) The fda is a multi-measurand instrument giving simultaneous information on a range of physical and chemical properties of the sample including colour, refractive index, surface tension, viscosity and probably other measurands.
3) The fda software can be designed for a specific product such as wine to give product analysis rather than a measurement of sample properties.
4) Refractive index, surface tension and viscosity are all temperature dependant and the apparatus can give further information on these measurands in the temperature domain.
5) The fda is based on HPLC technology that is highly advanced but which is also rapidly advancing in technical performance and these developments will ultimately feed through to further improve the performance of the fda.
6) The fda measures in drop aliquots and this feature could be a useful basis for certain monitoring and measurement applications which require either new samples to be analysed, or changes between aliquots to be dynamically study.
7) The viscosity measurement sensitivity is good for the constant head method for this measurand and consequently provides an excellent approach to molecular weight measurement of polymer and macromolecules based on the measurement of the so called Limiting Viscosity Ratio.

BENEFITS

1) The fda has easy sample handling which requires little of no cleaning.
2) The information obtained can be tailored to the specific needs of the user.
3) No great expertise or skill is required of the operator.
4) There is a short analysis time when compared to qualitative GC and HPLC identification.
5) The fda research capabilities would seem to be very significant and the capability of the fda has yet to be explored in the vast percentage of applications.

ADVANTAGES

1) The fda provides in one instrument the measurement capability of several instruments and can potentially offer the complete analysis of the sample as it has now been combined with a very accurate density meter.
2) The analytical capability of the fda for refractometry is very significant and can provides a respectable measurement. It also enables refractive index (dispersion) measurements to be made on a small sample in the ranges 200 - 1100 nm for wavelength and ambient to 50C in temperature.
3) The fda colour measurement is one that correlates well with a standard UV-VIS, but which in trials has been shown to be somewhat more sensitive than this laboratory instrument.
4) The fda provides a good standard stalagmometric surface tension measurement, but when the instrument is used with a calibration graph of drop volume versus surface tension it is capable of providing a very much more accurate measurement.
5) The fda provides information from the vdt on the liquids viscosity, surface tension and density which may be capable of providing quantitative determinations of some of these measurands.
6) The instrument has an integrated control software with the whole experimental measurement capability of the fda. It would appear to be difficult to find a set of instruments that would rival the measurement capability of the fda and such an array of instruments would inevitably use large amounts of liquid sample.
7) The small volume measurement and research capability of the fda may be unique.
8) The fda fingerprinting capability appears to be absolutely unrivalled in that the present instrument provides both fdt and vdt fingerprinting capability. In addition the fingerprint library has the additional dimensions of wavelength and temperature for the fdt and temperature for the vdts.

NEW MAGNETOOPTIC DC CURRENT SENSOR FOR POWER GENERATION NETWORKS OF MUNICIPAL ELECTROTRANSPORT

S.V. Lebedev

Laboratory of Applied Magnetooptics, Moscow Engineering Physics Institute,
Kashirskoe shosse 31, 115409 Moscow, Russia

Abstract. A simple magnetooptic dc current sensor to be used on power substations of municipal electrotransport is presented. The compensation scheme permits current measurements from 0 up to 5000 A with 0,2% accuracy in wide temperature range. It is demonstrated that while using in high-unified systems our sensor is a real alternative to known galvanomagnetic or fiber optic ones meaning the set of their operating and commercial characteristics. Sensors mounting geometry is optimized providing its maximal immunity to external interferences from several possible sources.

1. Introduction

Review of scientific publications of last years shows that the development of new contactless dc current sensors remains up to now a problem of high priority. Two approaches to decision of problem of contactless measurements are known [1]. In first ones galvanomagnetic phenomena in semiconductors and metals are used as physical base for dc current sensors. Their alternative are magnetooptic effects in ferromagnetic and paramagnetic media, when the light, passing through the sample placed into magnetic field, undergoes the rotation of polarization.

On method of measurements all sensors are to be classified as devices of direct transformation, where the measured value is really amplified electrical signal of sensitive element /SE/ or of photodetector (for optical sensors), and as that of compensatory type, where the field of current to be measured is compensated by magnetic field of external coil, and SE is in fact "null-detector", included in negative feedback circuit.The sensors of compensatory type have as a rule smaller temperature drift and are more precise in measurements. The sensitive elements used in them should have only odd transfer function, but its linearity is not required.

On constructive features the sensors also can be conditionally separated on devises with closed and open transformation contour. In first case the signal of SE is proportional to integral of magnetic field on closed contour, containing inside the bus with current. Thus, according to Maxwells equation , the sensors output signal is absolutely insensitive both to conductor displacement inside the contour or to changing of its geometry, and to influence of external currents (these factors give rise really to nonzero error of measurements). The formation of integrating contour is possible, in particular, by application of ferromagnetic core with small gap, where SE is to be located [2]. The key disadvantage of such construction is, as a rule, large weight and dimensions of magnetic core, which cross-section is chosen in order to ensure no saturation of magnetic circuit at a maximum possible value of measured current. The sensors with open transformation contour require both recalibration by changing of current geometry, and account of influence of adjacent buses.

The specific class of sensors form fiber optic devices. They successfully combine the high level of useful signal because of significant SE length, linearity of transformation in wide range of magnetic fields (currents), high immunity to interferences, electrical passivity of optical head mounted on the bus. Recently the fiber optic sensors are considered to be the most universal and perspective ones. They are the object of investigation in suppressing quantity of publications devoted to contactless measurements of electrical currents. However there are serious problems alongside with described advantages, such as strong sensitivity to mechanical perturbations, negative influence of linear birefringence [3,4]. It happens because the fiber optic sensors belong to the class of direct measurements devices with all inherent them defects.

Except for application of optical fiber of significant length the using of epitaxial garnet films as SE leads to strong increase of slope of own transfer function of primary optoelectronic converter. However, in offered technical decision of current sensor [5] no measures were undertaken for compensation of magnetooptic crystals own coercivity, that considerably limits the accuracy and range of measured fields (currents). Furthermore all known us sensors use garnet films with strong uniaxial anisotropy, what doesn't allow to obtain extremely high sensitivity.

2. Principles of operation

The objective here were the development and testing of simple magnetooptic dc current sensor with in-plane anisotropic garnet film to be used on power substation of municipal electrotransport. The investigated peculiarities of magnetic reversal of such films in dc and ac magnetic fields [6] permitted to hope, on the one hand, on slope increasing of transfer function as compared with uniaxial anisotropic films, and, on the other hand, on realization of simple and cheap device, having high operational characteristics, including immunity to external interferences both owing to unique anisotropy of chosen SE and because of specific geometry of standard mutual location of current buses of power generation networks.

Perfectly understanding unpopularity of idea of sensors that are to be calibrated, we think nevertheless, that in most cases our sensor presents the real alternative to mentioned above. It relates to power generation networks with buses of standard cross-section and their standard mutual location, as it takes place, for example, on all power substations of municipal and railway electrotransport. In this case the sensor is to be calibrated when produced and is to be mounted by consumer in strong accordance with developers recommendations.

The 5-mkm thick epitaxial garnet film of $(BiLuCa)_3 (FeGe)_5 O_{12}$ grown on (111)-oriented $Gd_3 Ga_5 O_{12}$ single-crystal substrate was chosen for using as SE of sensor. It was shown [6], that except for the standard hysteresis curve obtained at longitudinal magnetizing along easy axis of magnetization in absence of other magnetic fields (Fig. 1a), there is such regime of high-frequency longitudinal and constant transversal magnetizing, when the central part of hysteresis loop is closing, but its slope remains rather high (Fig. 1b). As it was stated in [7], such regime of external magnetizing seems to be optimum for designing of portable magnetic field sensor, that was used and now by developing of dc current sensor.

Crystals anisotropy sets unambiguously the mutual location of current bus and SE: their base planes

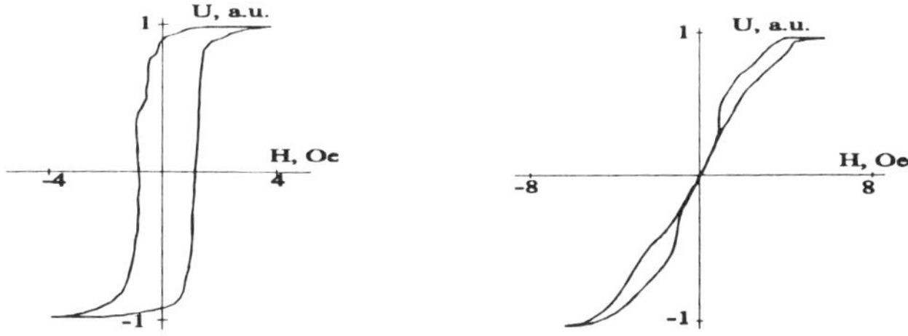

Fig.1. Longitudinal hysteresis of garnet film with anisotropy "easy axis in plane":
a) no external magnetizing;
b) dc transversal (2 Oe) and high-frequency longitudinal (1 Oe) magnetizing.

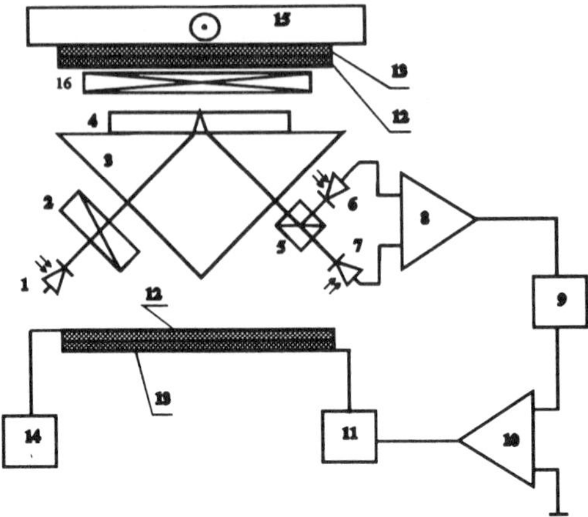

Fig.2. Block diagram of a magnetooptic current sensor.

should be parallel and easy axis of ferrite film should be directed along width coordinate of bus and parallel to magnetic field induced by current. The constructive impossibility to use the experimental geometry of unitary passing, when the source and light detector are placed on different sides from plane of sensitive element, has defined the choosing of prism geometry of magnetooptic converter (Fig. 2). The garnet crystal (4) is bonded with transparent glue to hypotenuse side of triangular glass prism (3). The linearly polarized by input polaroid film (2) light from LED (1) enters the prism through one side, undergoes the total internal reflection on the interface "film - air" and leaves the prism through other side. The geometry of oblique light incidence provides its interaction with magnetization of garnet, which orientation and distribution is determined by external magnetic field, induced by current in bus (15), it results in rotation of plane of light polarization. Analyzer (5) transforms the light modulation from asimutal to amplitude one, that is subsequently registered by photodetector. Taking into account the small value of Faraday signal as compared with direct transmission of pair "analyzer-polarizer" the balance scheme with two photodiodes (6,7) was applied, where two orthogonal polarizations of output beams are oriented at angle of 45^0 to input one.

While developing the sensor we used the compensatory scheme of measurements, when the external effect on sensitive element is compensated by magnetic field of reference coil (13) and the measured value is really the current in this coil. As it was mentioned above, the using of compensatory technique excludes completely the negative influence of light source drift on operational characteristics of sensor. It is possible to neglect the temperature dependence of Faraday rotation of sensitive element, that is serious problem in developing direct measurement devices. To maintain the best uniformity of compensating field and maximal efficiency of current amplifier (11) the coil is reeled-up on hollow rectangular former with metal walls for best heat removal, that results in sensors dynamic range increasing to higher magnetic fields (currents). On the same former is the special coil (12) for high-frequency magnetizing reeled-up, powered by appropriate generator (14) and being applied to decrease dynamic coercivity of magnetooptic crystal. For elimination from feedback circuit the high-frequency signal the low-pass filter (9) was applied, coupled to the output of differential amplifier (8). Thus the frequency of ac magnetizing as well as roll-off of LPF are chosen so that - firstly, to supply the given time response of the sensor, and, secondly, to suppress the output level of high-frequency signal to the value of allowable error. The standard negative feedback scheme is further applied: the output of LPF is applied to the first input of comparator (10) with grounded second input, and the output of comparator is applied to a coil driver amplifier (11), connected with compensatory coil (13). The adduced scheme uses the generator of high-frequency magnetizing with dc bias voltage of output to compensate the small value of external (laboratory) magnetic field being present at mounting point of the sensor. Electromagnetic coil (16) provided the necessary value of dc perpendicular magnetizing.

3. Results of industrial testing

According to adduced block scheme the brassboard model was produced, intended for operation on standard 100x10 mm^2 buses, that are exclusively in use on power substations of municipal electrotransport in Moscow. It consists of the optical measuring head, mounted at once on the bus, and measuring block, containing control circuit and source of power supply.The industrial tests:on real substation have demonstrated its perfect operating characteristics:

- the measuring range 0-5000 A;
- the overall accuracy of measurements 0,5% (in laboratory - 0,2%);
- own time response on current jump of maximal amplitude not more than 50 ms (can be considerably reduced);
- dielectric strength test - 5 kV/50 Hz/1 min;
- high overload capability and immunity to external interferences.

The slope of transfer function of our magnetooptic measuring head is on input of first amplifier cascade around 20-50 V/T, and can reach 1000 V/T, that much more exceeds the parameters of galvanomagnetic devices. It permits to lower the requirements to sensors input amplifier. Our investigations have shown, that dc perpendicular magnetizing effectively suppresses the main noise of SE - magnetic Barkhausen noise to level below of several millivolts in wide frequency band, that puts forward magnetooptic measuring heads in leaders among other sensors of magnetic fields and currents according to signal-to-noise ratio. The conversion coefficient of actually received contactless current transformer is 2500:1.

4. Discussion

The necessity of calibration is closely connected with the problem of sensors sensitivity to its mounting accuracy on the bus. Our calculation shows, that for the sensor being spaced from the central part of bus surface perpendicularly by 1 cm the change of field (and, accordingly, of sensors indications) on 1% is corresponded to the displacement of measuring point on 10 mm along surface or on 1 mm perpendicularly to it. Obviously, one can provide more accurate sensors mounting producing them serially with fastening unit under standard bus.

While discussing the problem of influence of adjacent buses on current measurements accuracy in given bus one should pay attention, that the sensor is practically insensitive to fields of orthogonal to the plane of film orientation and is comparatively immune to magnetic fields, oriented in plane of film along hard axis of magnetization. In both cases only the change in the slope of SE's transfer function occur, however the specific anisotropy of used garnet films is such, that the appreciable increase of own noise and decreasing of sensors sensitivity take place in magnetic fields of some hundreds and some tens Oe, respectively. The occurrence of fields of such strength is completely excluded in conditions of mutual location of current buses on real substation.

Thus, the unique source of interferences are magnetic fields of the same orientation, as the measured is. The results of simple theoretical investigation of immunity to interferences of sensor under consideration for three various geometries of interference sources are presented below:

a) single bus with current;

b) the system of two equal and opposite directed currents, flowing in parallel buses, spaced by a distance d from each other - the kind of current dipole with axis oriented along easy axis of garnet film; c) current dipole with axis, oriented perpendicularly to hard axis of garnet film.

The isolines of constant interference of definite relative value were calculated with SE as origin of coordinate system XOY using approximation of point currents. X-axis is directed along easy axis of garnet film and Y-axis - perpendicularly to its plane. The current of interference source was assumed

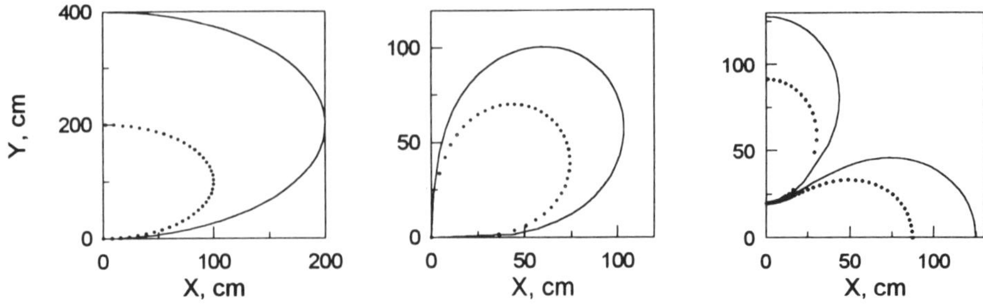

Fig.3. Isolines of constant interference from various sources:
a) single bus;
b) current dipole oriented along easy axis of garnet film;
c) current dipole oriented perpendicularly to films plane.

to be of maximum allowable value (5 kA). The results of calculation of isolines of one and two-percentage error for d=40 cm are shown on fig. 3 with solid and dotted lines, respectively. In accordance with used approach they are applicable for evaluation of influence of adjacent buses, spaced from SE by a distance, considerably exceeding the characteristic size of bus (10 cm), that always takes place. It is to emphasize, that the mutual location of current buses on power substation is highly unified and, as a rule, corresponds the geometry of interference source (b) (see above), with polar angle of 0^0 or 90^0. In this case, as it results from presented data, the adjacent buses don't influence practically on accuracy of sensors measurements, that was observed while carried out industrial tests. More better results may be obtained obviously by using two sensors, placed on the opposite sides of the bus.

5. Conclusion

Thus, present work describes simple, high-sensitive and broadband magnetooptic sensor of compensatory type, intended for measurements of dc current. Being real alternative for use in unified power generation networks, it exceeds on two order in sensitivity galvanomagnetic devices and is considerably cheaper than the fiber optic sensors are. The conditions of mutual arrangement of sensor and sources of external interference are defined, ensuring given accuracy of measurements. The sensor can be used both for current measurements, and for further developing on its base systems of operating control of high-voltage power generation networks, and in the first place - on municipal and railway electrotransport.

The research described in this publication was made possible in part by Grant No.M6Q000 from International Science Foundation.

6. References

[1]. J.E. Lenz. - Proc. of the IEEE, 1990, v.78, No.6, p.973.
[2]. Patent of Switzerland No. 679527, G01R15/02, 1992.
[3]. Day G.W., Deeter M.N., Rose A.H. - Advances in Optical Fiber Sensors. Bellingham, Washington: SPIE, 1992, p.11-26.
[4]. Bucholtz F., Koo K.P., Kersey A.D, Dandridge A. - SPIE, 1986, v.718, p.56-65.
[5]. Patent of USA No. 4.947.107, G01R33/032, 1990.
[6]. Lebedev S.V. - J.Tech.Fyz. v.63, No.11, p.72-82 (in Russian).
[7]. Lebedev S.V. - Digests of International Magnetics Conference (INTERMAG'93). Stockholm, Sweden, 1993, BP-12

Section E

Flow Measurement and Process Tomography

An Experimental Comparison Of Three Electrode Configurations For Electrical Impedance Tomography

M J Booth, I Basarab-Horwath

School Of Engineering Information Technology, Sheffield Hallam University, City Campus, Pound Street, Sheffield S1 1WB

Abstract. The goal of electrical impedance tomography(EIT) is to obtain the electrical impedance distribution within an area of interest via measurements made between pairs of electrodes placed at the surface. There are several applications of E.I.T where the electrodes need not be placed on the periphery of the area to be imaged. For this study, three electrode configurations have been assessed using three measures of performance and by examining the image quality produced by a basic sensitivity algorithm for each electrode configuration.

1. Introduction

Electrical impedance tomography(E.I.T) is used to determine the spatial distribution of electrical impedance within either a two or three dimensional object. This spatial distribution of electrical impedance is traditionally calculated from measurements made at the surface of the object[1]. Voltage measurements are recorded between different pairs of electrodes placed at the surface of the object and various excitation and measurement techniques have been developed[2]. The technique of using a constant current supply and voltage measurements is used in the work presented in this paper. A constant current is injected between one pair of electrodes and the voltages developed between the remaining pairs measured, the current is then injected between the next pair of electrodes and the voltages at the remaining pairs again measured. This is repeated for all electrode pairs connected to the surface of the object. A diagram of the EIT system used in the investigation is shown in Figure 1.

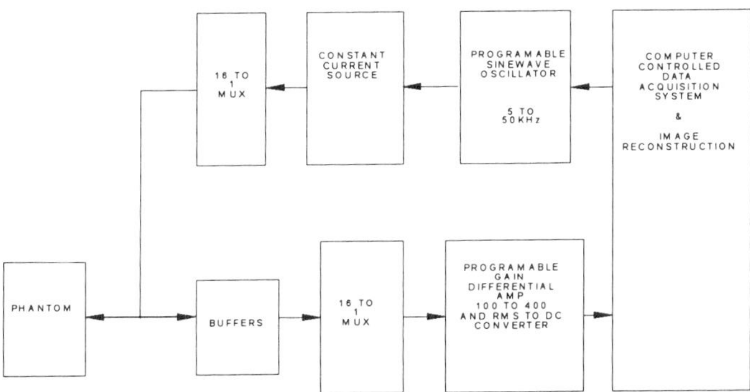

Figure 1 Block diagram of EIT system

The reconstructed images of the electrical impedance distribution are used to infer information about other parameters within the area being interrogated, i.e. concentration profiles within stirred mixing vessels[3],in vivo images of the human body[1], pressure distribution across a piezo-resistive material [4].

There are several applications of EIT where the electrodes need not be placed at the boundary of a circular object, which is the usual electrode configuration. Applications exist in areas such as the measurement of distributed pressure, where a piezo-resistive planar sensor element can have electrodes placed behind it[4] and in process mixing, where the electrodes can be mounted on the centrally placed stirrer(s).

It is known that the peripheral mounted electrode systems have limited resolution towards the centre of the phantom[5] (phantom is the term used to describe the area being interrogated and is a term adopted from medical imaging), and it has been shown that placing electrodes near to the centre of the phantom improves the resolution towards the centre[3]. It is known that the classical EIT problem is ill-conditioned and that placing electrodes near to the centre reduces this ill-conditioning[6]. Therefore by evenly distributing the electrodes across the full phantom area the resolution should be improved.

2. Experimental Procedure

The three electrode configurations examined were the 'classical' circular arrangement, with the electrodes placed on the periphery of a circular phantom, a square arrangement with the electrodes placed on the periphery of a square phantom and a distributed arrangement where the electrodes are evenly distributed over a square phantom. The three electrode configurations are shown in Figure 2.

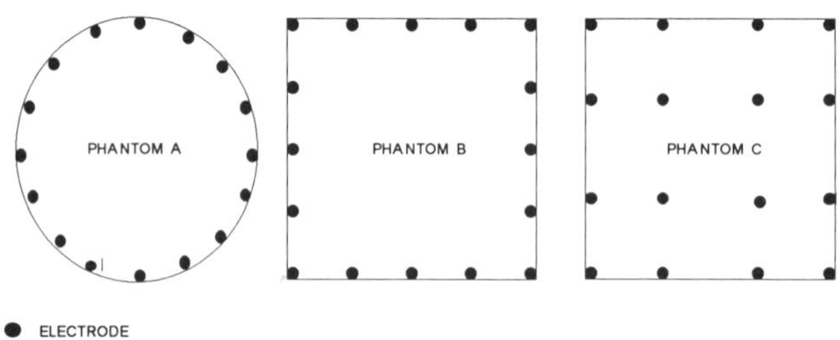

Figure 2 Electrode Configurations

Phantom A classical peripheral configuration
Phantom B peripheral configuration square phantom
Phantom C distributed configuration square phantom

The three phantoms are perpex tanks, filled to a depth of 3cm with saline solution. The electrodes continue to the full height of the tanks so that only two dimensional effects need be considered. In order to compare the three electrode configurations the size and conductivity of a square object placed at the centre of the tank is varied. The conductivity of the object or anomaly is varied between 1 mS cm^{-1} and 2S cm^{-1} and the area of the anomaly varied between 0.5% and 5% of the total phantom area, defined as normalised area(A).

3. Measure of Performance

Before defining the measures of performance used to compare the performance of the two electrode arrangements the terms used are described:

(i) α, conductivity contrast, is the ratio of the anomaly conductivity (σ_a) to the background conductivity (σ_b), that is

$$\alpha = \sigma_a / \sigma_b \ ,$$

(ii) V is the measured boundary voltage with an anomaly present in the phantom

and (iii) V_u is the measured boundary voltage with no anomaly present; that is, a uniform conductivity distribution

The measures of performance are defined as:

(a) $\delta V/V$, the fractional change in boundary voltage is

$$\delta V / V = (V - V_u) / V_u$$

(b) Q, the visibility of an anomaly is [5]

$$Q = \frac{(V - V_u)}{(V + V_u)}$$

For both $\delta V/V$ and Q the average value is calculated, over all boundary voltage measurements for each of the electrode configurations.

(c) V_{rms}, the rms value of the difference between the boundary voltages with an anomaly present and with no anomaly present is defined as

$$V_{rms} = \sqrt{\frac{1}{R}\sum_{i=1}^{R}(V_{(i)} - V_{u(i)})^2}$$

where R is the total number of boundary voltage measurements.

4. Image Reconstruction:

The algorithm for image reconstruction is based on filtered backprojection using sensitivity coefficients [7][8]. The area to be imaged is divided into 225 squares, each square representing one pixel in the reconstructed image. The sensitivity coefficients are calculated using an experimental procedure. A conducting rod is inserted into each pixel location in turn and a set of boundary voltages is recorded for that pixel location. The sensitivity coefficients are then defined from the boundary voltages as

$$S_{(p,m,n)} = V_{(m,n)} - V_{u(m,n)}$$

where p is the pixel number, m is the projection number (there is one projection for each injecting current electrode pair, giving a total of M projections) and n refers to the nth element of the measured voltage vector for a particular projection (there are a total of N voltage readings for each projection). Note that N is different for each projection for the distributed electrode configuration

This procedure generates a large amount of data for any particular electrode configuration; the total amount of data is calculated by p x M x N. For the peripheral electrode configuration, this gives a total of 46,800 readings and for the distributed electrode configuration, a total of 100,800 readings. Each pixel's sensitivity matrix ($S_{(p,m,n)}$) describes the systems response to an input at that pixel location. The reconstruction algorithm is defined as :

$$P_{(p)} = \sum_{m=1}^{M}\sum_{n=1}^{N} S_{(p,m,n)} \times (V_{(m,n)} - V_{u(m,n)})$$

Where P is the pixel value calculated for the pth pixel.

5. Results and Conclusions

Figures 3 to 5 show the average visibility, fractional change in voltage and the rms difference plotted against conductivity contrast for an anomaly of 5% normalised area. The results presented show that the classical arrangement performs less well that the square peripheral layout which in turn was 'out-performed' by the evenly distributed electrodes. There is quite a significant improvement in all three measures for the latter arrangement compared to either of the two other layouts. Also from figures 3 to 5 it can be seen that with a conductivity contrast greater than one a much larger change in magnitude for all measures used is recorded than for conductivity contrasts less than one. Reconstructed images are shown in Figure 6 for the three electrode configurations, with five anomalise place in the phantom. Four of the anomalies are placed towards the edge with the fifth at the centre. The superiority of the evenly distributed configuration can be clearly seen in it's ability to distinguish or detect small centrally placed anomalies and multiple anomalies.

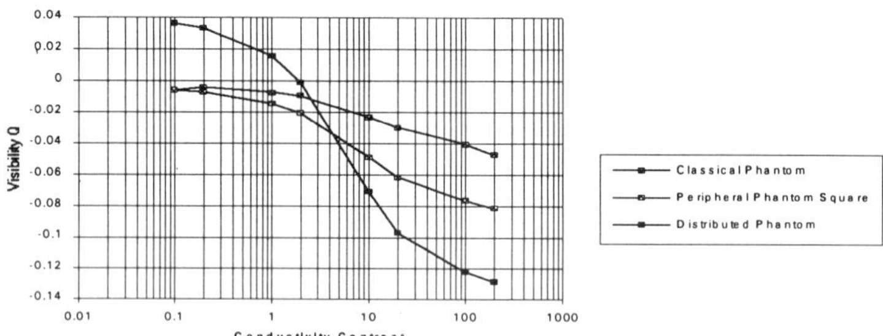

Figure 3 visibility Q against conductivity contrast for anomaly with 5% normalised area

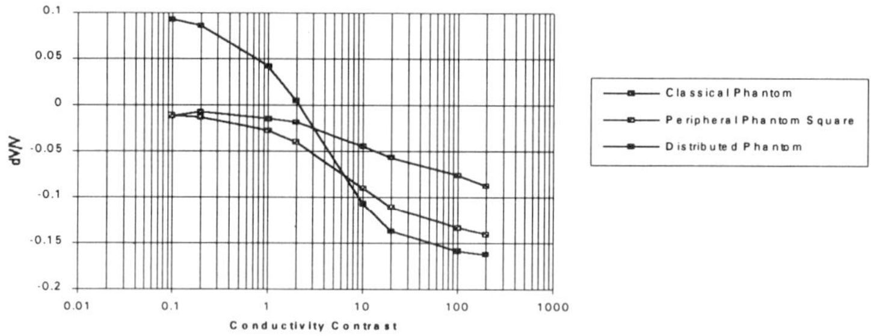

Figure 4 $\delta V/V$ against conductivity contrast for anomaly with 5% normalised area

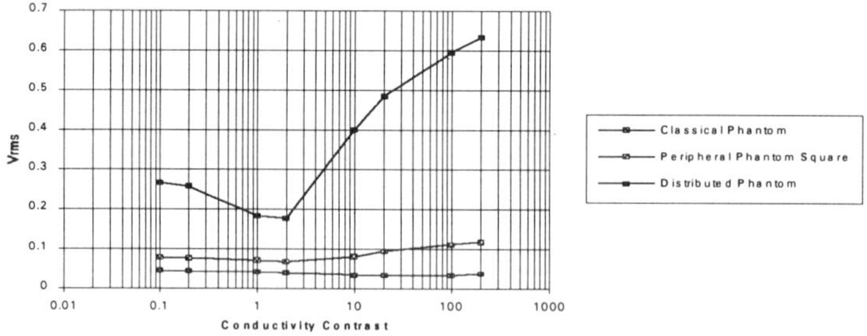

Figure 5 Vrms against conductivity contrast for anomaly with 5% normalised area

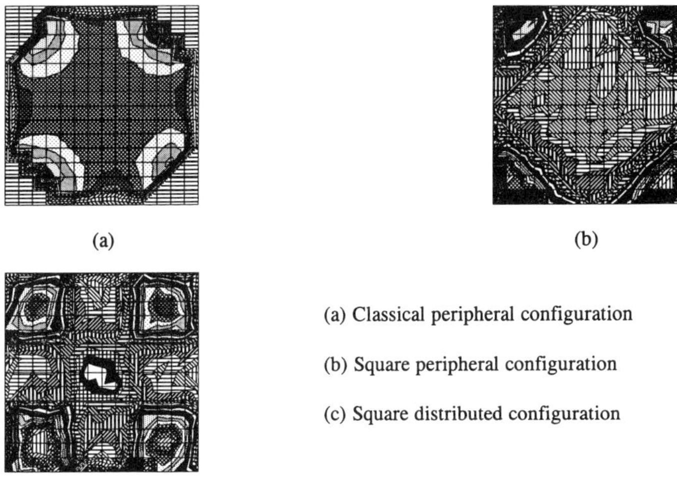

(a) (b)

(a) Classical peripheral configuration

(b) Square peripheral configuration

(c) Square distributed configuration

(c)

Figure 6 Reconstructed Images

6. References

[1] Barber D C and Brown D H, Applied Potential Tomography, J.Phys.E: Sci. Instrum., 1984, 17, pp723-33.

[2] Electrical Impedance Tomography, Ed. J.G Webster , Pub. Adam Hilger,1990. ISBN 0-85274-304-1.

[3] Lyon G M and Oakley J P, A Simulation Study Of Sensitivity In Stirred Vessel Electrical Impedance Tomography, Proc E.C.A.P.T Conf.Manchester, 1992, pp 104 -113.

[4] Lacey P and Basarab-Horwath I, 1993, An EIT-based Distributed Pressure Sensor - Further Results, Process Tomography - A Strategy For Industrial Exploitation, ISBN 0-9523165-01, pp 202-5.

[5] Segar A D, Barber D C and Brown B H, Theoretical Limits To Sensitivity and Resolution In Impedance Imaging, Clin. Phys. Physiol. Meas., 1987, 8, Suppl. A, pp 13 - 31.

[6] Fulton W.S and Lipczynski R.T, Body-Support Pressure Measurement Using electrical Impedance Tomography, Proc 15th International conf. of the IEEE Engineering In Medicine and Biology Society, 1993, 15, pp98-99

[7] Kotre C J, 1989 A Sensitivity Coefficient Method For The Reconstruction Of Electrical Impedance Tomograms, Clin. Phys. Physiol. Meas., 10, N0. 3, pp 275 - 281.

[8] Yu Z Z, Peyton A T, Beck M S, Conway W F and Xu L A, Imaging SystemBased On Electromagnetic Tomography (EMT), Electronic Letters, 1989,29, No. 7, pp 625-626.

Towards optimising the sensitivity distribution of electrical tomography sensors

Z Z Yu, G M Lyon, S Al-Zeiabak, H P Tan, A J Peyton, M S Beck

Department of Electrical Engineering and Electronics, UMIST, PO Box 88, Manchester M60 1QD.

Abstract: The sensitivity and resolution of electrical tomography systems including electrical resistance tomography, electrical capacitance tomography and electromagnetic tomography, are generally lowest in the centre of the object space. One method of improving the sensitivity at the centre is by the design of an appropriate excitation field distribution. This paper suggests that the optimum excitation field distribution for a non invasive sensor array is uniform and parallel. The practical means of achieving this distribution are also described.

1. INTRODUCTION

Electrical tomographic techniques such as Electrical Impedance Tomography (EIT), Electrical Capacitance Tomography (ECT) and Electromagnetic Inductance Tomography (EMT), have received increasing attention over the previous decade for a wide range of industrial, scientific and medical applications [1-3]. The major attractions of these electrical techniques is that they are relatively inexpensive, non-invasive, non-hazardous and fast (frame rates of over 100/s are possible). However, compared to traditional *hard field* tomographic methods (such as X-ray for example where the interrogating field follows straight lines), electrical techniques are *soft field* systems and consequently have relatively low resolution.

An overview of a typical electrical tomographic system is shown in figure 1. A particular problem associated with these systems is that the spatial and contrast resolution at the centre of the object space is generally significantly lower than that near the edges. The reasons for this poor central sensitivity may be understood by treating the field within the object space as two components, namely the excitation field and the object field. The excitation field is applied to the space in order to interrogate the object material present. When objects are present, an object field is created which is superimposed onto the excitation field. It is effectively measurements on the object field strength at the periphery which are used to reconstruct the image. As a result there are two main reasons for the reduced central sensitivity. First the excitation field strength is low at the centre because overall this is furthest away from the individual excitation elements, and second all the sensing elements are then also furthest away from the source of the object field.

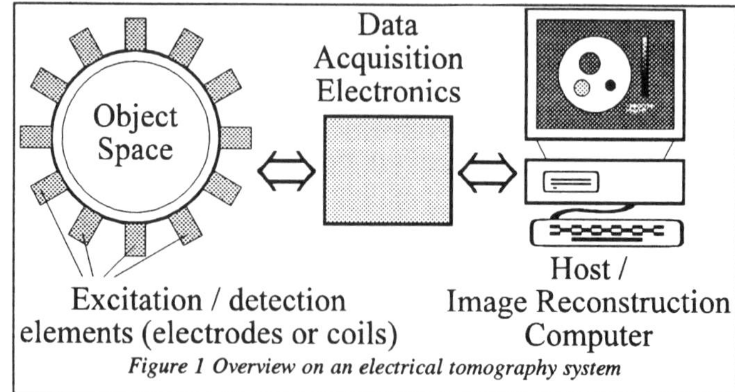

Figure 1 Overview on an electrical tomography system

This paper is concerned with the design of the excitation field. As the sensitivity is proportional to the excitation field strength it would be desirable to energise the object space with a field which is stronger in the centre than at the edges. Unfortunately for a non-invasive sensor this is impossible, as shown in the following pages. The optimum field distribution for maximising the sensitively in the centre with respect to the edges is shown to be a uniform parallel field. The practical methods for achieving such a field are described.

2. ANALYSIS

The three electrical tomographic techniques (EMT, ECT and ERT) are considered separately In the following three sections to establish their generic excitation field relationships, which are then considered to determine the optimum case.

2.1 Magnetic field equations for EMT

EMT systems [4-6] are excited by magnetic fields which are described by Maxwell's equations as:

$$\nabla \cdot \mathbf{B} = 0 \tag{1}$$

$$\nabla \times \mathbf{H} = \mathbf{J} + \frac{\partial \mathbf{D}}{\partial t} \tag{2}$$

where **B** is the magnetic flux density vector, **H** is the magnetic field strength vector, **J** is the current density vector, and $\frac{\partial \mathbf{D}}{\partial t}$ is the displacement current density vector. In general EMT sensors operate in the near field region and so the frequency is not high enough for the $\frac{\partial \mathbf{D}}{\partial t}$ term to come into effect. Thus,

$$\frac{\partial \mathbf{D}}{\partial t} = 0 \tag{3}$$

As the sensors are non-invasive, the excitation current must be arranged outside the object space. In addition by considering only the excitation field we can assume that the space is empty. Thus within the space,

$$\mathbf{J} = 0 \tag{4}$$

and

$$\mathbf{B} = \mu_0 \mathbf{H} \tag{5}$$

Therefore the description of excitation magnetic field inside the object space can be rewritten as:

$$\nabla \cdot \mathbf{H} = 0 \tag{6}$$

$$\nabla \times \mathbf{H} = 0 \tag{7}$$

2.1 Electrical field equations for ECT

ECT uses electrical field excitation [7] which according to Maxwell's Equations is described by:

$$\nabla \cdot \mathbf{D} = \rho \tag{8}$$

$$\nabla \times \mathbf{E} = -\frac{\partial \mathbf{B}}{\partial t} \tag{9}$$

where **D** is the electrical field flux density vector, **E** is the electrical field strength vector, ρ is the charge density, and $\frac{\partial \mathbf{B}}{\partial t}$ is the induced electric field vector. ECT systems are generally excited at a sufficiently low frequency such that the induced electric field term, $\frac{\partial \mathbf{B}}{\partial t}$ makes little contribution to the electrical excitation field. For this analysis this part can be ignored. Thus,

$$\frac{\partial \mathbf{B}}{\partial t} = 0 \tag{10}$$

An empty object space in ECT (air based) also has a constant permittivity, e.g. $\varepsilon_r = 1$. Thus the field flux density and field strength follow a simple form.

$$\mathbf{D} = \varepsilon_o \mathbf{E} \tag{11}$$

As the sensor is non-invasive, the excitation current and voltage must be arranged outside the object space. In addition the space is empty, consequently there is no charge build up inside the region of interest. Thus

$$\rho = 0 \tag{12}$$

Therefore the description of the ECT excitation field over the empty object space can be rewritten as:

$$\nabla \cdot \mathbf{E} = 0 \tag{13}$$

$$\nabla \times \mathbf{E} = 0 \tag{14}$$

2.2 Electrical field equations for ERT

Another technique that uses an electrical excitation field is ERT [8]. Here the excitation field is current density, and obeys Poisson's equation:

$$\nabla \cdot \mathbf{J} = 0 \tag{15}$$

As the media has a constant conductivity σ, according to the continuity laws, the electrical field and current density have a simple relation:

$$\mathbf{J} = \sigma \cdot \mathbf{E} \tag{16}$$

Equation (14) is also satisfied, as the electrical field is mainly set up by the voltages on the electrodes. For a similar reason as ECT, at low frequency, $\frac{\partial \mathbf{B}}{\partial t}$ makes little contribution to the electrical field. Thus:

$$\nabla \times \mathbf{J} = 0 \tag{17}$$

2.3 Derivation of the optimum field distribution

Equation pairs (6 & 7), (13 & 14) and (15 & 17) are general descriptions of the magnetic, electric and current density excitation fields used in three kinds of electrical and magnetic non-invasive tomographic sensors. No specific source or excitation arrangement is assumed. The equations do not depend on any spatial dimensions, voltage or current levels, or shield configurations. So the equations are quite general. The main limitation is that the object space is entirely empty or homogenous, so that only the excitation field is under the consideration.

Comparing equations (6) (7) and (13) (14) and (15) (17), it is obvious that the magnetic, electric and current density excitation fields follow the same form of equations. Therefore mathematically their optimum excitation fields follow the same rules. Therefore only one case, say the magnetic case for EMT, needs to be considered.

As we are interested in the cross section of the object space, the proof can be made over 2 dimensional space, thus the field can be expressed as:

$$\mathbf{H}(x,y) = H_x\mathbf{i} + H_y\mathbf{j} \tag{18}$$

where $\mathbf{H}(x, y)$ is the vector of field strength at point (x, y), **i** and **j** are unit vectors in the X and Y directions. H_x and H_y are scalars of field strength at point (x, y) in x and y directions respectively.

The amplitude of the field strength is then:

$$|\mathbf{H}| = \sqrt{H_x^2 + H_y^2} \tag{19}$$

We can prove that excitation field strength $|\mathbf{H}|$, which satisfies (6), (7), (18) and (19), has no maximum point in the object space. The detail of the proof is quite complicated and listed in [5], and due to space limitations, the main results are presented only:

The proof has been made on the points that satisfy the necessary condition for the existence of extreme points of the field. At these points, we test the sufficient condition for the existence of maximum points of field.

Extreme points occur at (x_i, y_i) where

$$\left.\frac{\partial |\mathbf{H}|}{\partial x}\right| = 0, \qquad \left.\frac{\partial |\mathbf{H}|}{\partial y}\right| = 0 \qquad (20) \quad (21)$$

If at one of the extreme points

$$\left(\left.\frac{\partial^2 |\mathbf{H}|}{\partial x \partial y}\right|_{(x_i,y_i)}\right)^2 - \left.\frac{\partial^2 |\mathbf{H}|}{\partial x^2}\right|_{(x_i,y_i)} \cdot \left.\frac{\partial^2 |\mathbf{H}|}{\partial y^2}\right|_{(x_i,y_i)} < 0 \tag{22}$$

is satisfied, the point is a maximum.

If at all of the points

$$\left(\left. \frac{\partial^2 |\mathbf{H}|}{\partial x \partial y} \right|_{(x_i, y_i)} \right)^2 - \left. \frac{\partial^2 |\mathbf{H}|}{\partial x^2} \right|_{(x_i, y_i)} \cdot \left. \frac{\partial^2 |\mathbf{H}|}{\partial y^2} \right|_{(x_i, y_i)} \geq 0 \tag{23}$$

there are no maximum extreme points.

The final derivation shows that for any points (x_i, y_i) which satisfy (6), (7), (18), (19), (20) and (21):

$$\left(\frac{\partial^2 |\mathbf{H}|}{\partial x \partial y}\right)^2 - \frac{\partial^2 |\mathbf{H}|}{\partial x^2} \cdot \frac{\partial^2 |\mathbf{H}|}{\partial y^2} = \frac{H_x^{\,2}[(\frac{\partial^2 H_x}{\partial x \partial y})^2 + (\frac{\partial^2 H_y}{\partial x \partial y})^2] + H_y^{\,2}[(\frac{\partial^2 H_y}{\partial x \partial y})^2 + (\frac{\partial^2 H_x}{\partial x \partial y})^2]}{H_x^{\,2} + H_y^{\,2}} \geq 0 \tag{24}$$

This equation shows that a maximum point does not exist. This proves that there is no point inside source-free region at which the electrical or magnetic field strength due to the source(s) outside this region is stronger than the field near the source. So for the non-invasive electric and magnetic tomography sensors considered here, the excitation field in the central region cannot be stronger than that near the walls. It therefore follows that the optimum condition is the maximally flat case. This occurs when the excitation field is both uniform and parallel. The next question is how such a distribution can be created in practice.

3. THE SOURCE ARRANGEMENTS FOR A PARALLEL UNIFORM EXCITATION FIELD IN A CIRCULAR OBJECT SPACE

It can be shown that in order to create a uniform parallel excitation field over a two dimensional circular object space, the source arrangement must have a sinusoidal relation with angular position. Figure 2 shows the case for an EMT sensor. Here the current density J perpendicular to the page at angle α is proportional to $\sin\alpha$. A proof that this current density produces a parallel uniform field from the Biot-Savart Law is contained in [5].

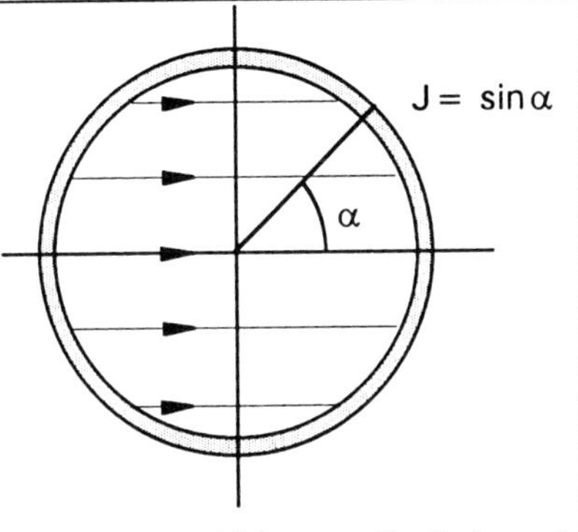

Figure 2. Sinusoidal current distribution and its magnetic field inside the space it enclosed

For ERT and ECT a similar arrangement can be used [8] and a simulation for an ECT sensor is shown in the next section. For ERT, Gisser has shown that a sinusoidal current injection can provide a parallel uniform electrical current field. The table below summarises the optimum excitation requirements for the three tomographic sensors:

Sensor type	Source type	arrangement
ERT	Current injection	sinusoidal
ECT	Voltage source	sinusoidal
EMT	Current density	sinusoidal

Table 1 Excitation source arrangement for three kinds of sensors

Note that in theory, only when the condition along the third dimension (in this case the dimension along the axis of the object space) is constant, can a problem be considered as simply two dimensional. So in practice the length of the excitation field in the Z direction should be significantly longer than the diameter of the object space, in order to ensure a nearly parallel uniform field.

4. SIMULATION RESULTS

An ERT system can use the so called adaptive method [8] which produces the "best currents" to distinguish objects of different conductivities. This is an example of the optimum uniform excitation field. It has been shown that with sinusoidal current injection a nearly uniform electrical current field can be created over a circular 2D cross-section. Figure 3 and Figure 4 show the simulation results of a sinusoidal voltage arrangement for ECT and a sinusoidal current arrangement for EMT. The simulation was done by using the SLIM finite element package (courtesy GEC Alsthom).

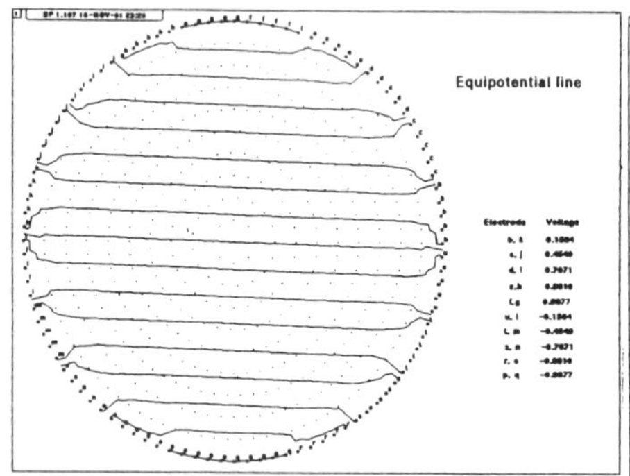

Figure 3. Two dimensional simulation of parallel uniform electrical field for ECT created by sinusoidal voltage arrangement outside a circular object space

Figure 4. Two dimensional simulation of parallel uniform magnetic field for EMT created by sinusoidal current arrangement outside circular object space

In the ECT simulation shown in Figure 3, 20 source segments were simulated with each segment given a sinusoidal voltage assignment. The field in the centre is parallel and uniform. However, as the source is a discrete sinusoidal distribution, the field near the walls is not as uniform as in the centre. The EMT simulation shown in Figure 4 assigned a sinusoidal current distribution to each element in a circular source region. The field again is parallel and uniform. Both simulations show that a sinusoidal sources distribution around a circular space can provide a parallel uniform field.

As an example of sensor excitation construction, the assembly structure of UMIST EMT system is shown in Figure 5. The object space (140 mm in diameter) is surrounded by two sets of field coils (X field coils and Y field coils) which have a sinusoidal turns distribution in order to provide the necessary sinusoidal current density distribution described in Figure 2. Surrounding the coils is the magnetic confinement shield shown at the top of Figure 5. The magnetic confinement is made of 30 ferrite rings (inside diameter 155 mm) slightly larger than the excitation coil assembly.

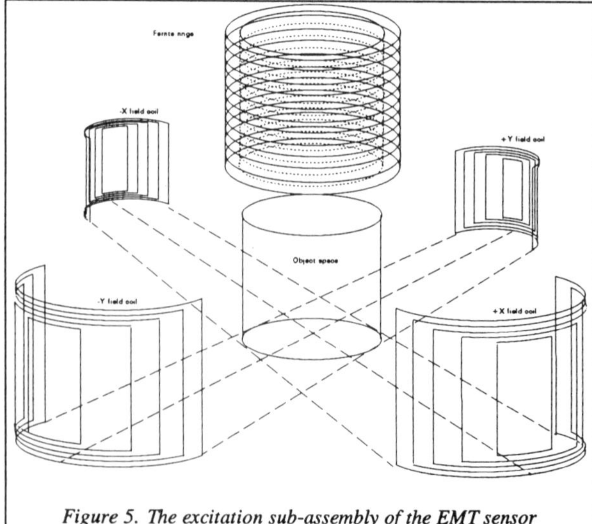

Figure 5. The excitation sub-assembly of the EMT sensor

The two sets of excitation field coils are arranged in orthogonal positions. Both can generate a parallel field. By energising relatively different currents in the X and Y coils, the excitation field can be set to any desired angle. The excitation field strength and direction are computer controlled via twin 12-bit DACs. Theoretically it is possible to apply 16777216 (i.e. 4096^2) projections at different strengths and angles. In practice, only the angle is changed while maintaining a constant field strength. Also, as the sensitivity of the detectors is limited, the incremental angle of each projection need to be reasonably large, such as 1°. Therefore the number of effective projections applied is less than 360. This is adequate for process applications. The number of independent measurements is the product of the number of projections and the number of detection channels. By applying more projections more measurements can be obtained. This is important for improving image resolution in the future development of the system.

5. CONCLUSIONS

This paper introduces another approach to the centre space sensitivity problem in electrical and magnetic tomographic sensors. The problem is divided into two aspects, the excitation field and the object field. This paper is directly concerned with the excitation field and presents an optimum excitation field theory for non-invasive electrical and magnetic tomographic sensors.

6. REFERENCES

[1] Beck MS, Campogrande E, Morris M, Williams RA and Waterfall RC (eds) "European concerted action on process tomography", conference proceeding, CEC Brite-Euram, Manchester UK, March 1992.

[2] Beck MS, Campogrande E, Morris M, Williams RA and Waterfall RC (eds) "European concerted action on process tomography", conference proceeding, CEC Brite-Euram, Karlsruhe Germany, March 1993.

[3] Beck MS, Campogrande E, Hammer E. A., Morris M, Williams RA and Waterfall RC (editors) "European concerted action on process tomography", conference proceeding, CEC Brite-Euram, Proto, Portugal, March 1994.

[4] Xie C G, "Review of process tomography image reconstruction methods", Sensors VI: Technology, Systems and Applications, IOP conference, Manchester 1993 (Ed. Grattan K.T.V. and Augousti A.T.) pp341-346.

[5] Yu Z Z, "Non-intrusive inductance based electromagnetic tomography (EMT) techniques for process imaging" Ph.D. thesis, UMIST, 1994.

[6] Peyton AJ, Yu ZZ, Al-Zeibak S, Saunders NH and Borges, “Electromagnetic imaging using mutual inductance tomography: potential for process applications”, Part. Syst. Charact., Vol 12 (1995) in press.

[7] Xie CG, Huang SM, Hoyle BS, Thorn R, Lenn C, Snowden D and Beck MS. "Electrical capacitance tomography for flow imaging: system model for development of image reconstruction algorithms and design of primary sensors". IEE Proceeding-G, Vol.139, No.1, Feb.1992,pp.89-98.

[8] Gisser D G, Isaacson D and Newell J C 1987, "Current topics in impedance imaging" Clin. Phys. Physiol. Meas. 8 Suppl. A 39-46.

[9] Yu ZZ, Peyton AJ, Beck MS, Conway WF and Xu LA, "Imaging system based on electromagnetic tomography (EMT)", Electronics Letters, 1st April 1993 Vol.29 No.7, pp625-6.

[10] Yu ZZ, Peyton AJ, Beck MS and Xu LA, "Electromagnetic tomography (EMT): A new process imaging system", Prod. of Sensor and Applications VI 12-15 ,Sept. 1993, Manchester, UK, pp347-52.

A high frequency two coil electromagnetic inductance tomography system

S Al-Zeibak[+], D Goss[+], A J Peyton[+] and N H Saunders*

[+] Department of Electrical Engineering and Electronics, UMIST, P O Box 88, Manchester M60 1QD, UK
* Department of Physics, University of Wales Swansea, Singleton Park, Swansea SA2 8PP, UK

Abstract. An experimental high frequency (1 - 10 MHz) electromagnetic inductance tomography (EMT) system using a two coil arrangement is described. The potential of the system for differentiating between conductive and ferromagnetic materials, as well as between materials of different conductivities has been studied and preliminary experiments, using simple saline and metallic objects, are discussed. An investigation has been carried out to evaluate the exact nature of the coupling between the coils, and to determine the optimum screening arrangement that can reduce the capacitive coupling between the coils to a minimum. Images generated using back projection and filtered back projection methods of simple conductive objects are presented.

1. Introduction

The application of computed tomography has progressed from medicine to many other fields, (for example, astronomy, molecular biology, geophysics, industry, etc.). The techniques employed to image objects include x-ray, gamma ray, ultrasound, microwave, nuclear magnetic resonance (hailed as the major innovation in medical imaging of the 1980's), and more recently electrical methods (Bates et al 1985, Webb 1988, Webster 1990). The latter have received growing attention because of their relatively low cost and high speed of operation. Electrical tomographic methods include; electrical impedance tomography (EIT), which is based on resistance/reactance measurements to generate conductivity (σ) /permittivity (ε) images, electrical capacitance tomography (ECT) which is based on capacitance measurements to generate ε, and possibly σ, images and electromagnetic inductance tomography (EMT) which is based on extracting data on permeability (μ) and σ distributions. As a result, electrical techniques can be considered as a complementary group of methods which each measure a sets of either one or two of the passive electrical quantities (σ, ε or μ) in order to generate an image related to these values (Peyton et al 1994).

An EMT system intended for process applications was introduced in the proceedings of the previous Sensors VI conference (Yu et al.). The aim of this paper is to report the further developments in one area of EMT. The paper describes the work carried out on a high frequency two coil type of EMT system, both, at the University of Wales, Swansea (Al-Zeibak and Saunders 1993) and at UMIST, Manchester (Al-Zeibak, Goss, and Peyton). The potential of the system for biomedical applications, in particular to differentiate between fat and fat-free tissue, was investigated at Swansea, while the potential industrial applications, for example to distinguish between ferrite and conductive, materials are being investigated at UMIST. An important feature of the system is its high excitation frequency (which enables relatively low conductivity materials such as ionised water and biological tissue to be detected.

2. System operation

The system operates by interrogating the object space with a high frequency (1-10 MHz) alternating magnetic field, generated by a small excitation coil. The variations in the field due to the presence of an object are then measured with a small solenoidal pick-up coil. The two coils are coaxial and their dimensions are small compared with their separation, which in turn is very small compared with the electromagnetic wavelengths used. Therefore, the EMT system operates entirely within the electromagnetic near zone.

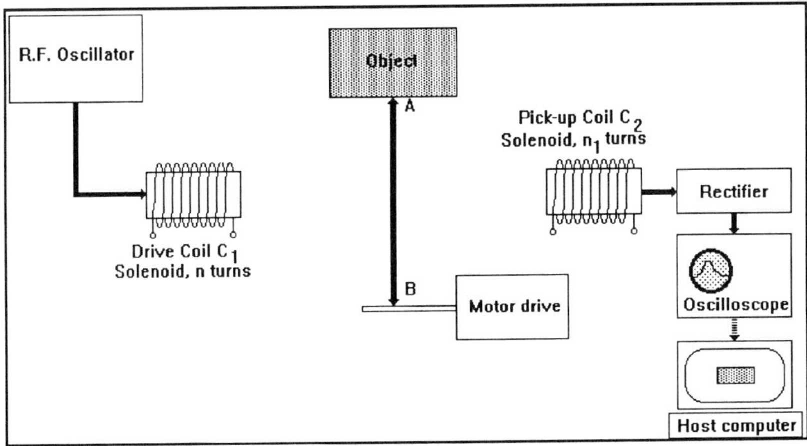

Figure 1. Schematic diagram of the two coil EMT system.

Figure 1 shows a schematic diagram of a two coil EMT system. The object to be imaged was placed on a turntable which could be drawn at uniform speed in a horizontal plane along the line of symmetry, AB, between the two coaxial solenoidal coils. Coils C_1 and C_2 were used for excitation and detection purposes respectively. The coils were typically a few centimetres long, 1 to 2 cm in diameter, separated by 30 cm, and had up to 200 turns (depending on the operating frequency). Both coils had earthed internal and external electrostatic screens, as described later. The coil C_1 was driven from a low power sinusoidal supply (output voltage 20 V) at the frequency at which coil C_2 and its associated cable were resonant (typically 1 to 10 MHz). The number of turns on C_2 was used to determine the resonant frequency. The object was scanned between the coils at typically few cm/s to produce data for one profile and the turntable was then rotated and the same procedure was followed several times to obtain a set of the desired number of profiles. The data set was fed into an image reconstruction computer and images were generated using back projection and filtered back projection algorithms.

3. Preliminary assessment

A series of experiments were carried out with different objects. These experiments can be divided into those which indicate the sensitivity of the system to different kind of conductive materials, such as human tissue, and those which show how the measured signal depends on the geometry of the intervening object. First, the ability of the system to distinguish between materials of different conductivities was assessed using various phantom objects containing different concentrations of saline solution (the saline solution consisted of NaCl in de-ionised water at room temperature). The output signal from C_2 was measured for a range of concentrations correspond to that of normal human tissues (figure 2). For example, 0.01 mole/l and 0.1 mole/l correspond to the saline concentration in fat and fat-free tissues respectively (Pethig 1987, Duck 1990). The results in figure 2 illustrate the relation between the output signal and salinity (conductivity), which is reasonably sufficient to discriminate between, for example, fat and fat-free tissues (each datum point is the mean value of ten measurements). Second, the ability of the system to distinguish objects of different geometry was assessed by placing a vertical cylindrical containers symmetrically between the two coils. The containers were filled to varying levels with 0.1 mole/l saline solution.

The experimental results in figure 3 demonstrate how the system senses the geometrical shapes of four cylindrical objects, with diameters D1 to D4. A linear relationship between signal and the length of the saline column can clearly be seen. The diameters of the cylinder, D, also has a significant effect on the output signal.

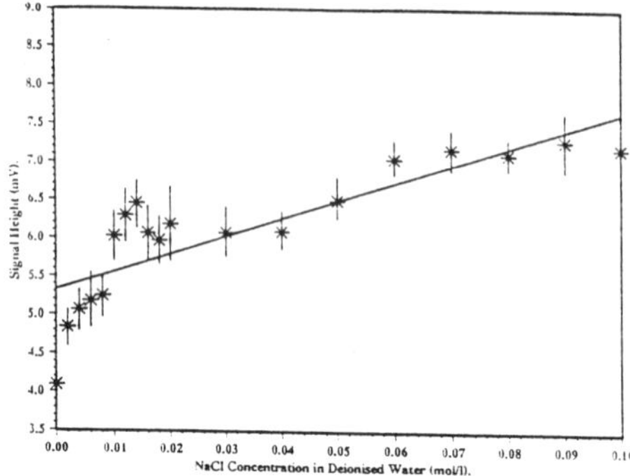

Figure 2. System sensitivity to changes in conductivity.

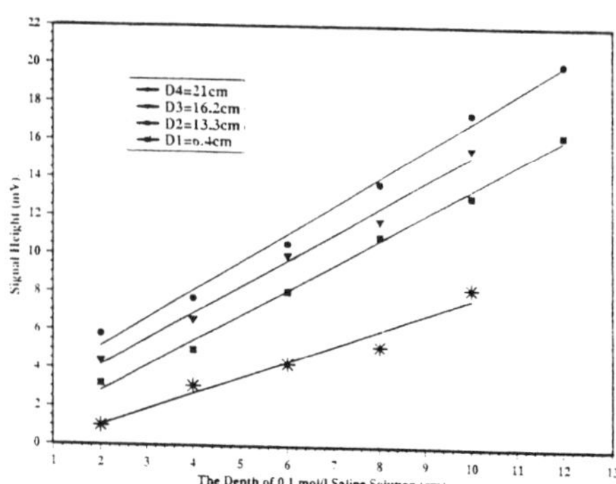

Figure 3. Signal in C_2 versus the depth of saline columns for several cylindrical containers.

4. The effect of screening on the system sensors

In an EMT system it is desired to reduce the electric field components to negligible levels, to eliminate the unwanted effects of capacitive coupling. The effect of capacitive coupling becomes a problem in practical systems particularly as the frequency increases above approximately 1 MHz. In order to minimise the effect of capacitive coupling, various shield arrangements were investigated including shields made from copper laminate, aluminium mesh and combination of both. Electric and magnetic fields probes were used together with the detection coil C_2 in order to map both field components. The measurements showed that a combination of external and internal copper screening foil was essential, otherwise the effects of capacitive coupling could dominate the performance of the system. In extreme cases, without the screening, the system would not be able to discriminate between conducting and ferrite object material, which should normally produce signal changes in opposite directions.

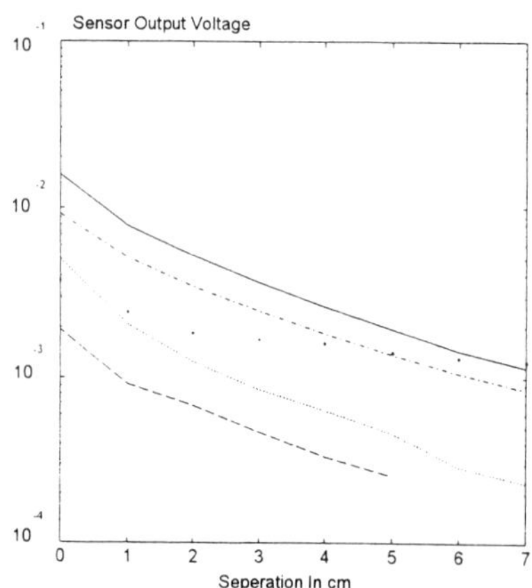

Figure 4. Electric field changes with longitudinal distance from C1.

Straight line (no screening). Dotted line (external screening). Dash-dot line (internal screening). Dashed line (external and internal screening). Circles (external and internal screening outside screened box).

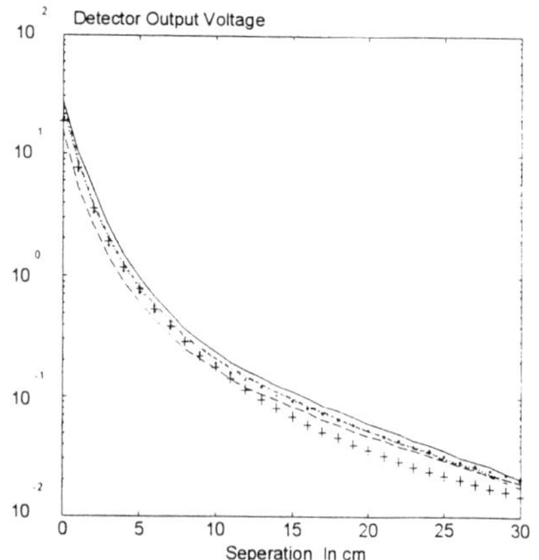

Figure 5. Magnetic field changes with longitudinal distance from C1.

Straight line (C1/C2 no screening). Dotted line (C1 internal screening/C2 none). Dash-dot line (C1 external and internal screening/C2 none). Dashed line (C1 external and internal screening/C2 internal). Circles (C1/C2 external and internal screening). Crosses (C1/C2 external and internal screening outside screened box).

To illustrate this point, figures 4 and 5 show the effects of using different screening arrangements. From figure 4, it can be seen that the combination of both internal and external screening has reduced the capacitive coupling to an almost negligible level for all practical separations over 5 cm. Figure 5 shows that for any combinations of screening there was no substantial affect on the inductive coupling.

5. Image reconstruction methods

This section describes briefly the reconstruction procedures used to generate images from the measurements. For more information about image reconstruction methods see, for example, Brooks and Di Chiro 1976, Rosenfeld and Kak 1982, Wang 1986 and Hinz 1988.

5.1. Back projection

Scanning an object between the coils at a uniform speed and fixed angle along the line AB (figure 1) produces one profile (or projection). Each profile presents a plot of output signal in the pick-up coil against position along AB. By changing the object orientation and following the same procedure, a number of desired profiles can be produced. The back projection technique is performed by back projecting each profile across the plane, i.e. each image pixel is allocated a value of the appropriate output signals sum from each profile (Brooks an Di Chiro 1976). This image reconstruction method is easily implemented without any sophisticated algorithms. However, it can generate images with unwanted blurred artefacts.

5.2 Filtered back projection

The filtered back projection method relies on filtering, or modifying, each profile before being back projected. The filtration, which is usually done using a 1D fast Fourier transform procedure (FFT), produces a filtered profiles containing negative and positive components which when back projected help to cancel out blurred artefacts and, as a result, improve the reconstruction of the image (Shepp and Logan 1974, Brooks and Di Chiro 1976, Chellappa and Sawchuk 1985).

6. Images of different simple objects

Several simple object shapes were scanned and the image reconstruction methods described in section 5 were evaluated, together with the effect of increasing the number of profiles on the image quality. The objects were a rectangular aluminium box of dimensions 11.25 cm × 8.5 cm, a circular steel box of diameter 9.5 cm, a rectangular saline-filled (0.1 mole/l) plastic box of dimensions 14 cm × 10.5 cm with a 8.3 cm diameter aluminium cylinder positioned in the middle, and a square saline-filled (0.1 mole/l) plastic box of dimensions 22.5 cm × 22.5 cm with two aluminium cylinders of diameters 4.5 cm and 8.3 cm positioned inside it. The conductivity contrast between the metallic cylinders and saline solution is greater than that between different human tissues. Nevertheless, the last two objects were used to assess, as a first step, the ability of the system to sense internal structure.

Figures 6-9 show images of the above objects. The shaded regions present ten equal steps in pixel density grey level. Figures 6 and 7 present images of the rectangular and circular metallic boxes generated from only two profiles at 0° and 90° using the back projection method. The ability of the system to sense internal structure and the change in the image as a result of increasing the number of profiles, four at 0°, 45°, 90° and 135°, is demonstrated in figure 8. Similar conclusion can be drawn from figure 9 which shows images from twelve profiles, at intervals of 15°, of the above square object with the two cylinders inside it using the improved image reconstruction method, the filtered back projection technique.

In comparing the images produced, the simple object structures are clearly recognisable. An improvement in the images of figures 8 and 9 over those of figures 6 and 7 can be seen as a result of increasing the number of profiles, as well as between figures 8 and 9 by using the filtered back projection procedure. Furthermore, the potential of the system to produce quantitative images was examined. It is been observed that the actual relative geometries of the objects are in reasonable agreement with the relative dimensions and positions of objects estimated from the images.

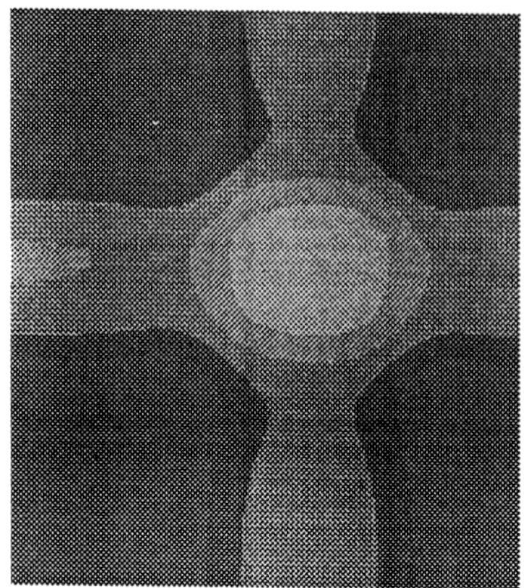

Figure 6. Image of rectangular metallic box by back projection from two profiles

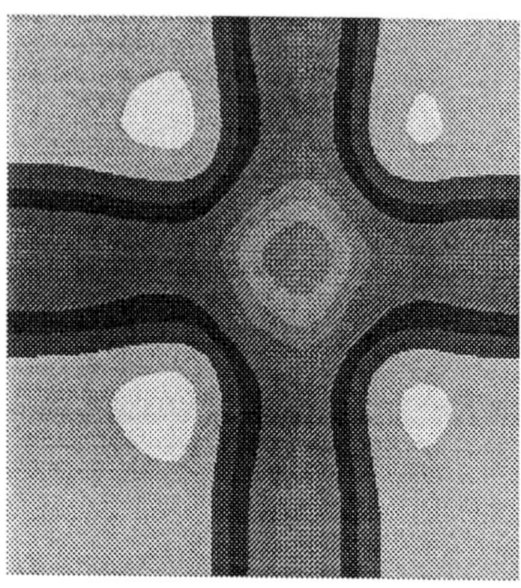

Figure 7. Image of circular metallic box by back projection from two profiles.

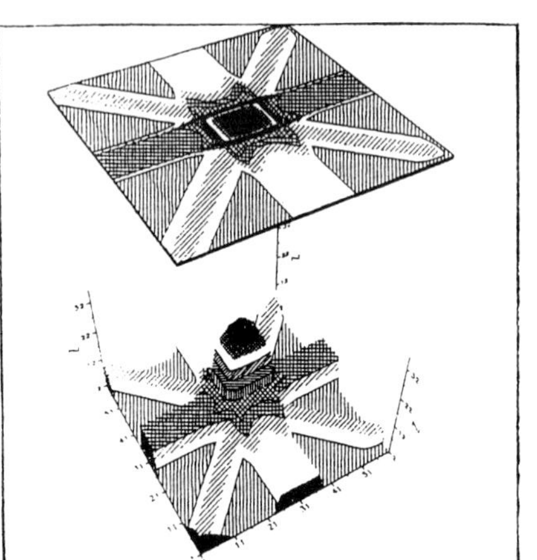

Figure 8. Image of saline-filled rectangular box containing centred metallic cylinder by back projection from four profiles

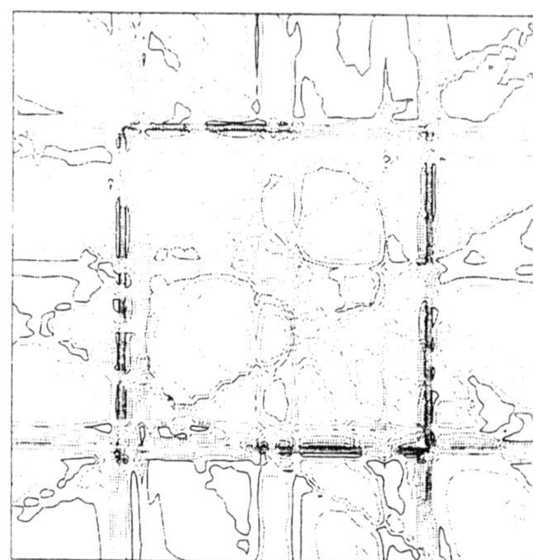

Figure 9. Image of saline-filled square box containing two different diameters metallic cylinders by filtered back projection from twelve profiles.

7. Conclusions

The two coil EMT system can discriminate between ferrite and conductive materials, as well as materials of different conductivities such as human tissues. The external and internal copper screening of the sensors enhance the sensitivity of the system by reducing capacitive coupling to a small level, and permitting a distinction between ferrite and conductive materials. Images produced using simple shapes and structured objects are encouraging, although they were generated from relatively small profiles numbers. The system is simple, inexpensive, and non-intrusive and has the ability to generate reasonable images of ferrite and conductive objects which, in turn, demonstrate its potential for medical and industrial applications.

8. Acknowledgements

The authors would like to thank the EPRSC for financing the UMIST part of this work.

9. References

Al-Zeibak S. and Saunders N. H., 1993, A Feasibility Study of *In Vivo* Electromagnetic Imaging, *Phys. Med. Biol.* **38,** 151-160.

Bates R H S, Garden K L and Peters T M., 1985, Overview of Computerised Tomography with Emphasis on Future Developments, *Digital Image Processing and Analysis* eds R Chellappa and A A Sawchuk (Silver Spring, MD: IEEE Computer Society).

Brooks R A and Di Chiro G., 1976, Principles of Computer Assisted Tomography (CAT) in Radiographic and Radioisotopic Imaging, *Phys. Med. Biol.* **21,** 689-732.

Chellappa R and Sawchuk A A (eds)., 1985, *Digital Image Processing and Analysis:, Vol.* 1: *Digital Image Processing* (Silver Spring, MD: IEEE Computer Society).

Duck F A., *1990, Physical Properties of Tissues: A Comprehensive Reference Book* (London: Academic).

Hinz T., 1988, Utilisation of reconstruction Algorithms in Transmission and Emission Computed Tomography, *Imaging Techniques in Biology and Medicine* eds C E Swenberg et al (London: Academic).

Pethig R., 1987, Dielectric Properties of Body Tissues, *Clin. Phys. Physiol. Meas.* **8** Suppl A 5-12.

Peyton A J, Yu Z Z, Al-Zeibak S, Saunders N H and Borges A R., 1994, Electromagnetic Imaging Using Matual Inductance Tomography: Potential for Process Applications *Process Tomography - A Strategy for Industrial Exploitation*, eds M S Beck et al (Manchester: UMIST).

Rosenfeld A and Kak A C., 1982, *Digital Picture Processing Vol 1* (New York: Academic).

Shepp L A and Logan B F., 1974, The Fourier Reconstruction of a Head Section, *IEEE Trans. Nucl. Sci.* **NS-21** 21-43.

Wang T T., 1986, Image Reconstruction from Incomplete Projections and some Applications for Industrial CT Scanners, *PhD Thesis* , University of Pittsburgh.

Webb S (ed)., 1988, *The Physics of Medical Imaging* (London: Hilger).

Webster J G (ed)., 1990, *Electrical Impedance Tomography* (Bristol: Hilger).

Yu Z Z, Peyton A J, Beck M S and Xu L A., 1993, Electromagnetic tomography (EMT): a new process imaging system, *Sensors VI conference proceedings,* pp 347-52, IOP Press.

Preliminary result in tomographic imaging system using optical fibres for pneumatic conveyors

R. Abdul Rahim[1 3], M.J. Nordin[1,4], N. Horbury[1], Dr. F. J. Dickin[2], Prof. R. G. Green[1], B.D. Naylor[1],

1 School of Engineering Information Technology, Sheffield Hallam University, Pond Street, Sheffield S1 1WB. FAX 0742 533306
2 Process Tomography Unit, UMIST
3 Control Dept, Faculty of Electrical Engineering, Universiti Teknologi Malaysia
4. Computer Science Department , Universiti Kebangsaan Malaysia.

It is a major problem to make measurements in two components flow such as gas/solid and gas/liquid mixtures because most sensors are flow regime dependent. These measurements are required to improve production and to increase safety during manufacture in process industries, e.g. North Sea oil production. The new technique of process tomography is being evaluated for many industrial applications such as pneumatically conveyed solids in the chemical industry where it is providing important measurements which have never been available in the past.

This paper describes the initial work which has been completed on a tomographic imaging system using optical fibre sensors for application to pneumatically conveyed solids.

The overall aim of the project is to produce tomographic images relating to particle size distribution and concentration distribution over the cross section of the measurement section. The system under development uses an array of 32 transducer pairs, aligned in a square matrix.

Practical tests have been carried out, on a prototype system built into a metal pipe of 100 mm of diameter. The present excitation illumination system uses a single quartz halogen light source to provide a large beam area providing illumination for all of the optical transmitter fibres which are arranged in a bundle. The light arriving at the receiver fibres is modulated by the flowing solids and converted to an electrical signal by PIN diodes. The electronic transducer is surrounded by an earthed screen to minimise the effects of electrical and optical noise. The output is amplified and signal conditioned for the data capture system.

A series of experiments have been carried out which investigate the effect of particle size and particle concentration on the received signals. The range of the particle sizes are from 100 μm to 2 mm.

Elastic beam deflection flowmeter for environmental applications.

R. Chandy, R. Morgan, P. J. Scully
School of Electrical Engineering, Electronics and Physics,
Liverpool John Moores University
Byrom Street, Liverpool. L3 3AF.

Abstract. This paper describes a simple, cheap and rugged multi-directional flowmeter that measures airflow velocity by sensing the deformation of an elastic beam with attached strain gauges, when subjected to stresses induced by drag forces in the presence of the airflow. Using strain gauges, the deflection of a square section beam can be measured in two orthogonal directions, enabling vectorial addition and hence deduction of the magnitude and direction of the deflection and hence the airflow. Experiments performed in a wind tunnel showed that the sensor was unaffected by its orientation to the airflow, and so is ideal for measurement of environmental fluid flow, with sensitivity of 0.0062 $\pm$ 0.0001 units of strain s/m and resolution of 0.64 m/s.

1. Introduction

Flow measurement is a widespread and fundamental parameter required for many industrial processes and environmental applications.There is an ever increasing need for accurate and reliable flow measurement at realistic cost. Over the years, a multitude of flow sensors for both air and water have been developed. The best known is the Venturi tube which measures fluid velocity as a function of a pressure differential induced in the fluid by a constriction. Other examples include orifice plate and pitot tube flowmeters, the vortex shedding flowmeter, and flow sensors using optical and ultrasonic interactions with turbulence to deduce velocity via the Doppler effect [1-4].

An important application of flowmeters involves environmental monitoring in areas such as meteorology, to measure turbulence close to the surface; also hydrology and for maritime studies. Here, the fluid flows are sporadic and multidirectional, compared with the controlled flows confined in pipes in industrial situations.

Existing flowmeters have a number of disadvantages associated with their manufacture and application. Magnetic flowmeters require individual calibration at the manufacturing stage due to variations in the properties of magnetic materials. Doppler systems are simple but depend on effective optical or ultrasonic transmission through the fluid. Techniques based on pressure differentials, restrict and obstruct fluid flow and require a pipe structure.

2. Unusual features of the sensor

This paper describes the development of a simple, cheap and rugged flow meter for environmental applications that measures wind speed by sensing the deflection of an elastic beam using strain gauges. The beam is clamped in place at one end and the other end is free to move in the airstream.

The flowmeter is compact, and requires no infrastructure, other than a rigid support. The advantages of this device over rotating flowmeters are lower inertia enabling measurements of gusts and rapid changes in windspeed, absence of mechanical parts, and robustness. It is easy to maintain, and can cope with fluids containing solid matter, such as sludges or slurries. Compared with vortex shedding and Doppler devices based on ultrasonics, the proposed instrument is cheaper, allowing multiple-headed sensing. Hot wire anemometers are affected by raindrops, but this instrument is rugged and less susceptable to moisture. The compactness of the device, and the fact that both speed and direction are measured by the same transducer at the same place and time, offers advantages for the measurement of flow which varies over small distances, such as winds close to the ground.

3. Principle of operation:

From elasticity theory and incompressible fluid dynamics, it can be shown that the extension or compression (ε) at the surface of the beam at its longitudinal centre is given by

$$\varepsilon = (3\rho v^2 S C_D L) / 2 a^2 b \qquad (1)$$

where ρ is the fluid density, v is the fluid speed, S is the area of the drag element, C_D is the drag coefficient, L is the longitudinal length of the beam, ε is the Young's modulus, a is the beam thickness, b is the beam width. Putting in typical figures, for a beam of dimensions 100 mm x $(10\text{mm})^2$ with modulus of 5×10^6 N m $^{-2}$, and a drag element of area 10 cm^2 with a drag coefficient of 1.0, an air flow of 10 m s^{-1} would give a strain of 0.04 or 4%, which is about the largest strain that can be measured by polymide/nichrome strain gauges. Equation 1 indicates that strain is a square-law function of fluid speed.

Since wind speed is proportional to the square root of strain, it is necessary to relate wind speed to strain measured in two orthogonal directions. The wind speed, U has orthogonal components u_1 and u_2 which are proportional to $strain_x$ and $strain_y$ respectively. Since $u_1 \propto \sqrt{strain_x}$ and $u_2 \propto \sqrt{strain_y}$, then

$$U = (u_1^2+u_2^2)^{1/2}$$

thus $$U \propto (strain_x + strain_y)^{1/2} \qquad (2)$$

Therefore, wind velocity, U, is proportional to the square root of the orthogonal strain components, $strain_x$ and $strain_y$.

4. Design of the sensor

A drag element, such as a spherical ball, is attached to the free end of an elastic beam, which is positioned in the airflow. The moving fluid exerts a force on the drag element, which deflects the beam. The resultant compression and extension of the beam is recorded by strain gauges (RS 308-102) attached to opposite sides of the beam surface[5,6]. Using a square-section beam, the deflection can be measured in orthogonal directions, enabling magnitude and direction of the force and thus the velocity, to be recorded.

The beam material is chosen to match the fluid; for wind speed measurement, a rubber beam is selected. A Wheatstone bridge arrangement (Figure 1) is used to measure the resistance changes produced, when the orthonormal gauges are exposed to strain. The change in resistance is converted into a voltage signal in the order of millivolts[7]. Strain gauges G_1 and G_3, mounted on opposite sides of the rubber beam and undergoing extension and compression respectively, generate voltage signal v_X. G_2 and G_4 are mounted orthonormally, on the other two sides of the square section beam. and generate voltage signal v_y. R_1, R_2, R_3, and R_4 are 1 KΩ 0.1% wire wound resistors. Voltage signals v_X and v_y are input to a low noise, strain gauge amplifier (RS 308-815), where they are amplified by identical factors to give voltage signals V_x and V_y.

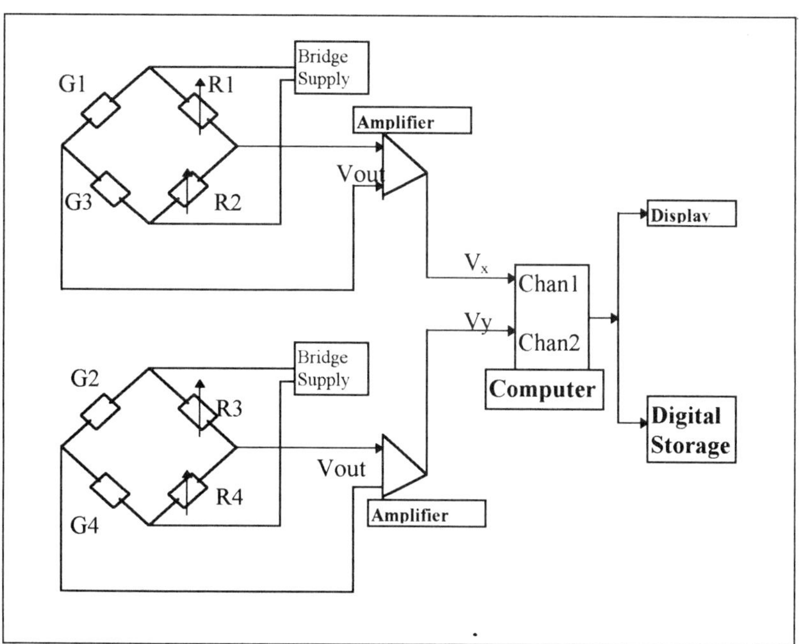

Figure 1: Diagram of signal conditioning and analysis circuitry

The acquisition and analysis of the voltage signals V_X and V_y are performed using a PC based virtual instruments software package, Labview (National Instruments Corporation). The signals are inputted to a 486 DX2 PC via a data acquisition board Lab PC+. Labview was programmed to sample and store the voltage signals, and to convert them to units of strain, using equation 3.

$$\text{Strain} = -2V_r/G\,(1+R_l/R_g) \tag{3}$$

where $V_r = (V_{sg} - V_0)/V_{ex}$

V_{ex} is the excitation voltage used which is 12V. V_0 is the unstrained voltage of the strain gauge after it is mounted in its bridge configuration. V_{sg} is the voltage read from the strain gauge. G is the gauge factor of the strain gauge, which is set to 2. R_l is the strain gauge lead resistance and R_g is the strain gauge nominal resistance value in ohms. Strain_x and strain_y are summed and square rooted as indicated by Equation 2 to give a resultant square rooted unit of strain. The Labview block diagram is shown in Figure 2 and a block diagram of the signal processing is shown in Figure 3.

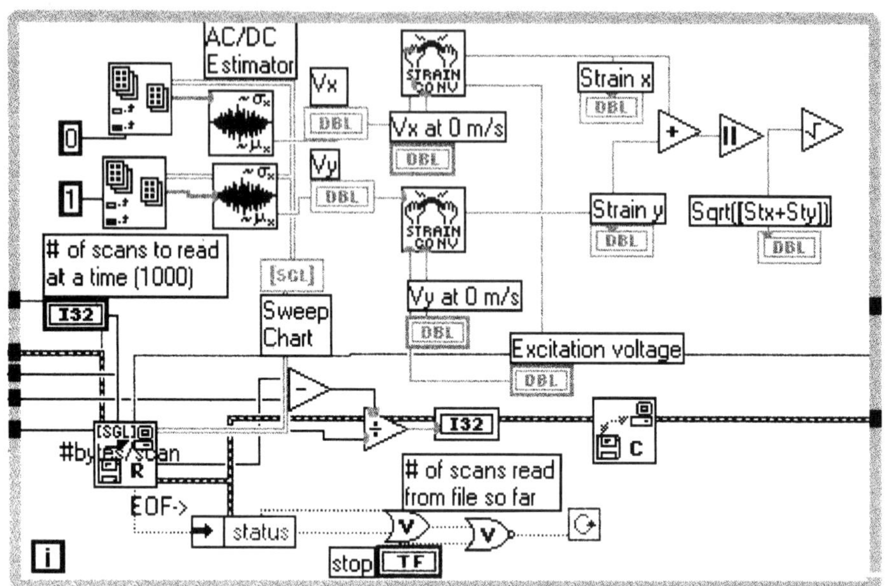

Figure 2: Block diagram of Labview showing the acquisition and analysis of the signals - V_x,V_y from the two channels.

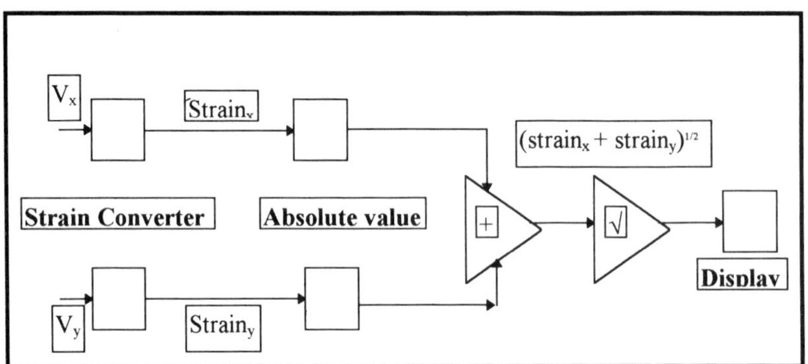

Figure 3: Block diagram showing processing of voltage signals Vx, Vy using Labview.

5. Experiments in the Wind tunnel

The sensor was mounted in a 460 x 460 mm closed circuit wind tunnel with a maximum fan speed of 1500 revolutions/minute. Using an anemometer and a stop watch, the wind speeds for fan speeds from 0 rpm to 1500 rpm were recorded and tabulated. The wind tunnel speed was varied from 0 m/s to 23 m/s and voltage signals V_x and V_y were sampled and processed using Labview.

The device was clamped on a turn table and this was rotated about its longitudinal axis from 0^0 to 90^0 at 15^0 intervals. At each angle V_x and V_y were recorded as a function of wind velocity, over a range from 0 to 23 m/s. Figure 4 shows a plot of $(strain_x+strain_y)^{1/2}$ versus wind speed when the elastic beam was oriented at 45^0.

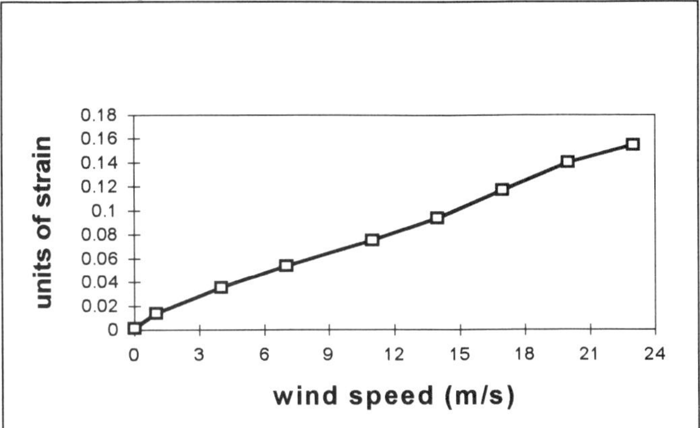

Figure 4: Units of strain plotted against wind speed when the flowmeter is oriented at 45 degrees to wind flow.

The graph shown in Figure 4 has a linear gradient, indicating that wind speed is indeed proportional to square root of strain, thus verifying Equation 2. A linear regression analysis indicates a sensitivity of 0.0065 ± 0.0001 units of strain s/m, and a resolution of 0.43 m/s. In this first prototype of the sensor, the range was limited by noise signals due to beam oscillations at speeds greater than 23m/s, causing the strain gauge wires to touch each other, and causing spurious signals. In principle, much greater speeds upto 50 m/s can be monitored by this sensor, if the strain gauges are mounted effectively.

Consideration of Equation 1 indicates that the gradient of the graph in Figure 4, is dependent on fluid density, the area of the drag element, and the dimensions and material properties of the beam. Therefore, by matching the beam material and dimensions to the fluid, the sensor can be optimised for particular applications.

Figure 5 shows plots of $(strain_x + strain_y)^{1/2}$ versus wind speed, when the elastic beam was oriented at different angles to the air flow. Consideration of the 95% confidence limits for each plot indicate that the plots are coincident, and that the sensor output is independent of angle of orientation of the elastic beam to the direction of wind flow. Therefore, all the data presented in Figure 5 can be used to obtain a calibration curve for the sensor, which will be valid for multi-directional airflow. When a linear regression was performed on all the data, the sensitivity was found to be 0.0062 ± 0.0001 units of strain s/m and the resolution to be 0.64 m/s.

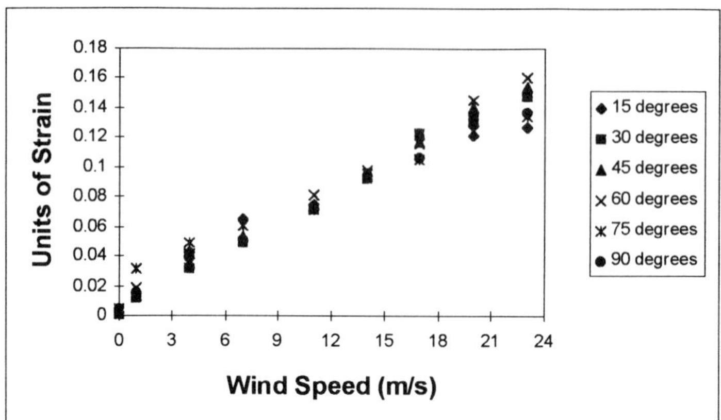

Figure 5: Units of strain plotted against wind speed with the flowmeter oriented at various angles to wind flow.

6. Conclusion

A simple, cheap and rugged flowmeter has been presented, which has been configured to measure airflow over a range of 0 to 23 m/s, with a sensitivity of 0.0062 ± 0.0001 units of strain s/m, a resolution of 0.64 m/s and a repeatability of ±1 %. Experiments have shown that the output of the sensor is independent of the direction of airflow, so it is ideal for environmental applications where wind, and water flow can change speed and direction. Depending on the deflection of an elastic beam for its operation, the dimensions and material of the sensor beam can be tailored to suit the fluid. The signal processing performed by Labview will be replaced by a compact, portable, battery driven purpose built micro-controller based processing circuitry, enabling the sensor to be used in the field. Additional work includes field trials, obtaining directional information about the wind velocity from the strain gauges signals, and replacing the conventional strain gauges with optical fibre strain gauges, to ensure electrical noise immunity and intrinsic safety for use in hazardous environments [8].

7. References

[1] Brain T J S and Scott R W W(1982); 'Survey of pipeline flowmeters'- Journal of Physics E: Instrument Science and Technology, Vol 15.
[2] Holzbock W G; Instrumentation for measurement and control.
[3] Measurement and Control, 'Flow measurement Special issue'. June 1986.
[4] Jones, B E; Instrumentation, Measurement and feedback (1978).Tata McGraw-Hill.
[5] Window L, Holister G; Strain Gauge Technology. Applied Science Publishers.
[6] Perry,C C and Lissner, H R; The Strain Gauge Primer. Mc Graw-Hill.
[7] Open University - Instrumentation Units 1 and 2 pp, 10-46.
[8] Schmitt N.F., Morgan R., Scully P.J., Lewis E., and Chandy R. (1995). Flow sensor using optical fibre strain gauges. Chemical, Biological and Environmental Fibre Sensors VII, Munich, June 19-23.

Analysis of Tomographic Data Using Transform Techniques - A Preliminary Study

I. Basarab-Horwath, M. J. Booth

School of EIT, Sheffield Hallam University,
Pond Street, Sheffield, S1 1WB, United Kingdom

Abstract Data collected from electrical impedance tomography measurement systems is conventionally used as the input to a computationally intensive image reconstruction algorithm. It is this algorithm that forms the system bottleneck in terms of speed of response and is also a major factor in limiting the accuracy and spatial resolution of these systems. Many processes do not necessarily require a reconstructed image. An alternative method of analysing the data set is to treat it as an image and to use zonal filtering or masking in the Fourier domain in order to segment the image of the data set. These segmented images show consistent measurable changes with changes in the input to the tomographic system. This method of processing the data can be used as a pre-processing stage for other processing techniques.

1. Introduction

In a tomographic system a particular modality or emanation (e.g. x-rays, light, acoustic energy, electrical current) is impressed onto a two or three dimensional body. These emanations from outside the body penetrate into and through the body and are affected by the distribution of a parameter or parameters within that body. The resultant effects of the emanations are measured at the boundary of the body. The derivation of the distribution of the parameter(s) within the body from the boundary measurements is one example of the inverse problem. Tomography is therefore concerned with the solution of the inverse problem and it is the solution of this problem that is one of the main factors in reducing the accuracy and resolution of the tomographic systems. Various parameters can been measured using tomographic techniques; for example, density, electrical conductivity and permeability (1).

The impressed modality - in the case of electrical impedance tomography or EIT, electrical current - results in a set of measured boundary values - in the case of EIT, differential voltages. These measured values are then sampled and recorded before being used as the input to a reconstruction algorithm. This algorithm uses a great deal of a-priori information as well as the input data set to construct an image of the distributed parameter within the interrogated region. Typically this image is displayed on a screen for visual analysis. Standard image analysis techniques can also be applied to the image to produce a set of useful process parameters.

However, not all processes need an image (or an analysed image). For example, it could well be that only changes in, say, the distribution of one phase of a multi-phase flow needs to be monitored. Or, in the case of an applied distributed pressure, the changes in the amplitude and position of applied load need to be measured.

These changes must be present in the matrix of boundary values since the a-priori information is the same for different data sets. The process of reconstruction (with a-priori information) is a deconvolution process with filtering. It is very computationally intensive process and further analysis is needed if the required information is to be extracted from the (analysed) image. An alternative possibility is to look at the original data set - the matrix of boundary values - and to treat this as an image in its own right. In fact, this has been done previously for data from a prototype distributed pressure sensor (2). The rows and columns of the matrix were summed and the vector formed from these sums can be seen to change with changes in the magnitude and position of the applied load. For a constant length vector this is equivalent to taking the average or Fourier dc term of the matrix rows and columns.

In a sixteen electrode EIT system current is usually injected through two adjacent electrodes and the resulting differential boundary voltages are measured. Each adjacent pair of electrodes are then used in turn to supply current to the object; there are therefore sixteen pairs of independent current drive electrodes. Three differential voltage readings are not taken out of the possible sixteen because of the effect of the interface resistance between the current carrying electrodes and the boundary. The boundary voltages are arranged to form a 16 by 13 matrix which can be considered to form an image. Two methods for analysing an image are (3) :

(i) standard image analysis methods, such as forming histograms of the amplitude distribution within the image, or

(ii) transform techniques. Transforms of the image give the frequency domain characteristics of that image; image features map into the frequency domain and these features can be extracted using zonal filtering in the selected transform space. Any transform can be used, for example, Fourier, Haar, Hadamard or the more fashionable wavelet transform. In this paper results using the Fourier and Daubechies wavelet transforms.

The zonal filter is also called the feature mask; it is simply a slit or another similar aperture; for example, angular slits are used to detect the orientation of features in images. This allows the image to be segmented; each component or segmented part of the image can then be analysed separately.

2. Results and Discussion

For brevity, only the results for one data set are shown in this paper. The data set is shown in Figure 1 and represents the boundary voltage values for an extremely good conductor placed in a saline solution in the centre of a circular tank. The electrodes are distributed evenly around the circumference of the tank. The conductor has a normalised radius of 0.25. Note that the data set forms a U shaped valley; this is because the first and last readings of each

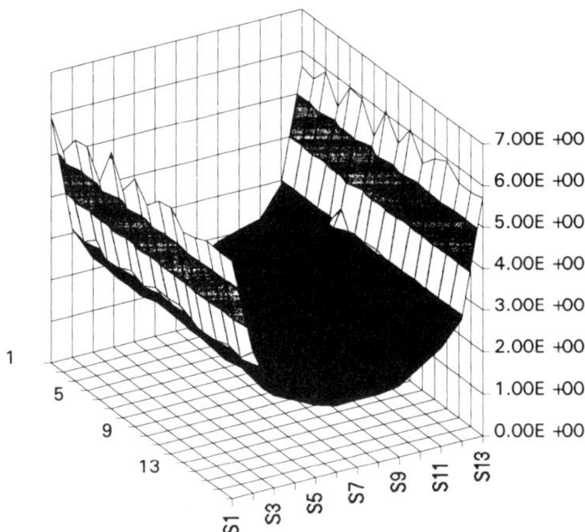

Figure 1. Data set for a conductor of normalised radius 0.25.

row of the matrix are each adjacent to one of the current injection (drive) electrodes and are therefore of a higher value than the remaining readings. This data set is mapped to the Fourier domain and a zonal filter applied; the result of this in the data space or image domain is shown in Figure 2. This shows the same broad U-shaped valley as in Figure 1, but this time it is smooth. Applying a different mask allows us to see (Figure 3) the U-shaped valley and a particular set of perturbation to the smooth U-shaped valley. The difference between this last segment of the image and the original image produced without using a mask (Figure 1) is shown in Figure 4.

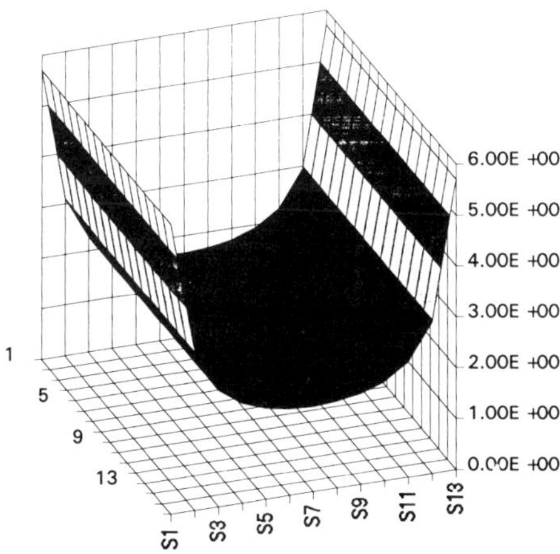

Figure 2 Data set with a zonal filter applied so as to leave only the main feature.

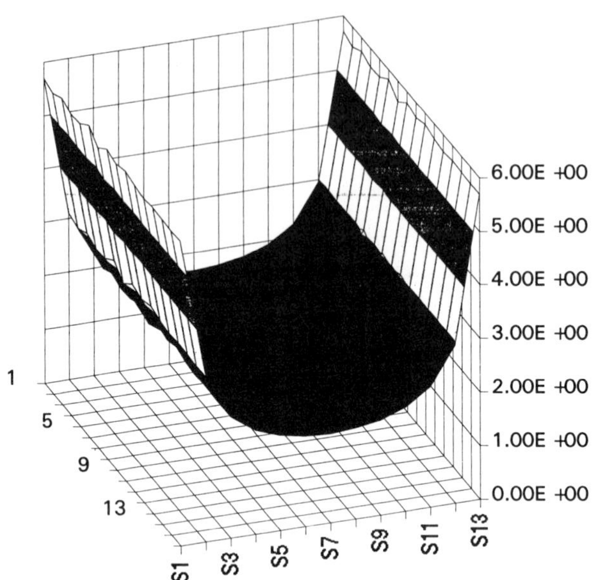

Figure 3. An alternative zonal filter applied to the data set to show its two main features

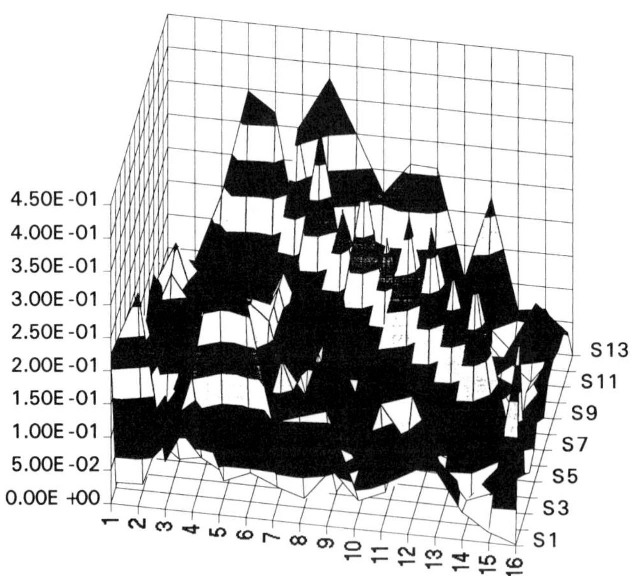

Figure 4. The small, third noise-like component of the data set produced using a third mask.

This figure has been produced using a third mask. It can be seen quite clearly that the use of zonal filtering is a powerful technique; it enables the extraction of relatively small amplitude variations from an image with a much larger amplitude.

Each segment can be analysed independently; for example, there is a clear correlation between a calculated shape index for the valley and the position and material composition (conductivity) of the anomaly.

The use of the Fourier domain can be seen to be extremely useful. However, the Fourier transform suffers from at least one well known major disadvantage in that it can be considered to be a convolution with a continuous sinusoid and it therefore matches the entire signal with the sinusoid. This does not allow any localisation in space (or time) of the particular frequency component of interest - it is this localisation in space or time that gives wavelets their major advantage over Fourier techniques (4). Some preliminary results have been obtained using the wavelet transform. Figure 5 shows the two-dimensional Daubechies 4th order transform (DAUB4) for the data set. This transform changes for different data sets; these changes can be related to changes in the data.

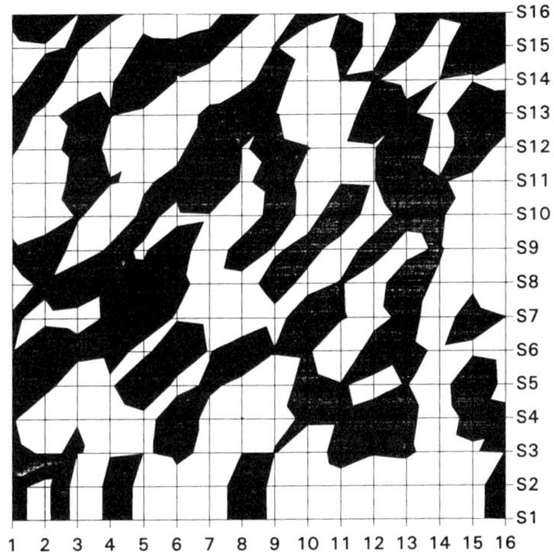

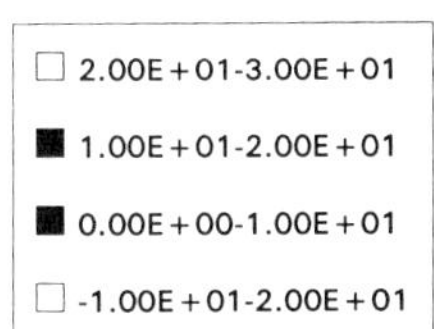

Figure 5. A two-dimensional wavelet transform of the data set.

3. Conclusions

Preliminary results have been presented for the use of transform techniques in analysing the data set from electrical impedance tomography. It has been shown that zonal filtering allows the image to be segmented; the various components can then be analysed independently. It is of course possible to use zonal masking for the data set as a pre-processing stage for other processing methods; for example, neural nets.

4. References

(1) C. G. Xie, Review of Image Reconstruction Methods for Process Tomography, Proc. Process Tomography Workshop, Karlsruhe, 1993, Eds. M. S. Beck, E. Campagrande, M. Morris, R. A. Williams, R. C. Waterfall, UMIST, pp 115 - 119.

(2) I. Basarab-Horwath, P. Lacey, A distributed pressure sensor which utilises electrical impedance tomography. Sensors VI, Technology, systems and applications, pp 381-386, Ed. K. T. V. Grattan and A. T. Augousti, IoP, Bristol, 1993.

(3) A. K. Jain, Fundamentals of digital image processing, Prentice-Hall, 1989.

(4) Meyer, M. Wavelets, Algorithms and Applications, SIAM, Philadelphia, 1993.

Section F

Biomedical Sensors

Optical imaging and localised spectroscopy of tissue using near infrared light.

D. T. Delpy

Department of Medical Physics & Bioengineering
University College London
Shropshire House, Capper Street
London WC1E 6JA

Near infrared spectroscopy (NIRS) of intact organs using light in the wavelength range 700-1000 nm was first described by Frans Jobsis[1] in 1977. In this wavelength range, oxygenation dependent changes in attenuation can be observed which arise from variations in the concentration of oxy and deoxyhaemoglobin (HbO_2, Hb) in the blood, and of oxidised cytochrome oxidase in the cell mitochondria. Since this initial report, considerable instrument development has taken place, and it is now possible to measure attenuation changes across many centimetres of tissue[2]. These instruments have largely been used to examine oxygenation and haemodynamic changes in muscle[3] or brain[4].

The relative transparency of tissue to NIR light which makes spectroscopic studies possible, has lead to considerable interest in the possibility of imaging through tissue, the major applications being imaging of the breast to detect tumours, and imaging of the brain to detect regions of oxygen deprivation. Unfortunately, the considerable scattering of light that takes place in tissue[5] makes achieving a clinically acceptable image resolution (<5mm in the breast, <1cm in the brain) extremely difficult. Early attempts to image through the breast using simple light sources and cameras[6] gave resolutions of 1-2 cm, which did not improve significantly with the use of simple collimation and filtering techniques aimed at rejecting some of the scattered light. However, since the late 1980's, significant progress has been made through a combination of technical and theoretical developments. The approaches taken towards imaging fall broadly into three different categories:

a) The first of these techniques is not a true "imaging" methodology, but relies upon careful positioning of the NIRS optodes to restrict the detected signal to light that has travelled through a limited volume of tissue. This technique is more suitably termed "localised spectroscopy". Because of the high degree of light scatter in tissue, it had initially been thought that such localisation would not be possible. However, improved modelling of light transport in tissue[7] showed that the detected light is in general confined to a "banana" shaped region between the optodes. These predictions have been further supported by NIRS studies of the response of the cerebral visual and motor cortex to applied stimuli, where the NIRS data can be compared with images acquired simultaneously using magnetic resonance imaging[8]. These indicate that with an optode spacing of 3-4 cm, the sampled tissue volume is confined to a region between the optodes and extending approximately 1 cm either side of the line between them.

b) The use of simple back projection image reconstruction schemes using data from a collimated light source and detector which are scanned co-linearly across the tissues. Several techniques have been used in conjunction with this method to try and reduce the blurring due to the detection of multiply scattered light. These have included the use of ultrashort pulsed (ps) light sources and fast time gated detectors to enable only that light which has travelled the shortest (and hence straightest) path to be detected[9], or the use of frequency modulated light and the detection of mean phase shift (equivalent to the mean optical pathlength)[10].

c) The use of mathematical models of light transport to predict the signal detected at all points on the tissue surface, and the use of this "forward model" in an iterative image reconstruction algorithm where its results are compared with the experimentally measured data. The mathematical modelling involved has varied from the relatively simple techniques, often based on an assumption of approximate tissue homogeneity[11], to very complex schemes employing Finite Element Modelling of tissue heterogeneity[12]. With these schemes, any data may be used as the input data set including measures of total surface intensity, time gated intensity, mean phase shift etc.

At UCL, all three approaches to imaging are being studied. Localised spectroscopy has been used in volunteers to quantify the changes in HbO_2 and Hb concentration in the visual cortex in response to a visual stimulus[13]. The use of time gated detection is being studied for imaging of the breast, where the geometry of the breast in compression during a mammographic study lends itself to a rectilinear scanning and linear projection image reconstruction methods. By fitting a model of light transport to the time resolved data, it has been possible to predict the light intensity at very early photon arrival times where in practice no photons are detectable. Using this predicted data, a resolution of approximately 5mm has been achieved[14]. The iterative imaging scheme is being pursued for the imaging of changes in brain oxygenation in premature babies. Theoretical studies have shown that a fast and accurate forward model is needed for successful reconstruction, and that the effects of the surface boundary must in particular be accurately taken into account[15]. The studies have also shown that the mean optical pathlength data is less susceptible to errors from this source, and that images reconstructed from this data contain more detail of structures deep within the tissues[16].

References.

1) Jobsis, F.F. 1977, Science, 198, 1264-1267.
2) Cope, M. et. al. 1988, Med. & Biol. Eng. & Comp., 26, 289-294.
3) De Blasi, R.R et. al. 1993, Eur. J. Appl. Physiol., 67, 20-25.
4) Elwell, C.E. et. al. 1994, J. Appl. Physiol., 77, 2753-2760.
5) Cheong, W.F. et. al. 1990, IEEE J. Quant. Elec., 26, 2166-2185.
6) Monsees, B. et. al. 1987, Radiology, 163, 467-470.
7) Wilson, B.C. et. al. 1992, Proc. IEEE, 80, 918-930.
8) Obrig, H. et. al. 1994, Proc 2nd Int Meeting SMR (San Francisco, Aug 6-12), 67.
9) Hebden, J.C. et. al. 1993, 32, 372-380
10) Gratton, E. et. al. 1993, Bioimaging. 1, 40-46.
11) Benaron, D.A. et. al. 1994, Adv. Exp. Med. & Biol., 361, 215-222.
12) Arridge, S.R. et. al. 1993, Med. Phys., 20, 299-309.
13) Meek, M. et al. 1995, Proc. Roy. Soc. Biol. Sci. (in press)
14) Hebden, J.C. et. al. 1994, Opt. Letters, 19, 311-313.
15) Schweiger, M. et. al. Med. Phys., (in press).
16) Arridge, S.R. et al. 1995, Proc SPIE, 2389 (in press).

A fluorescence sensor for investigating vascular damage

P D Goodyer, J C Fothergill, A H Gerschlick, N B Jones and D P de Bono

Department of Engineering, University of Leicester, Leicester LE1 7RH, U.K.

Abstract. An optical fibre probe is being developed to detect radiation from fluorescein attached to endothelial cells using biological macro molecules. Endothelial cells grown in vitro, when damaged, have been successfully labelled with macro molecules carrying fluorescein. Two different methods of exciting the fluorescein are reported, external illumination via a 40W halogen light and in vivo illumination from an Argon Ion laser. Measurements have been made in simulated arteries, in animal models, and in human umbilical artery. The experiments indicate that an in vivo probe for monitoring cell damage is possible.

1. Introduction

1.1. Endothelial cell damage in arteries

The cardiac artery is a complex structure consisting of several types of cell. The innermost layer is made up of endothelial cells. These arteries can become obstructed by atheromatous stenoses which can often be treated by Balloon Angioplasty. This operation damages the endothelial cells and the return of vascular obstruction after operation can occur with drastic consequences. It is important to monitor the healing process of these cells both for research purposes and for prognostic reasons (Preisack and Karsh 1993).

This project aims to find means to monitor the damage caused by this operation and the subsequent healing process. It is proposed that antibodies specific to endothelial cells be introduced into the bloodstream and that these antibodies are labelled in such a way that the extent of damage can be measured.

1.2. Measurement of damage

If labels can be attached to the damaged cells using monoclonal antibodies or other macro molecules it would then be necessary to show that a scheme can be devised to detect the markers. The substance chosen for the markers was sodium fluorescein since this is well tolerated by the body and has a peak spectral response in a low absorption window of blood.

The solution proposed is to excite the fluorescein by injection of light energy and to observe the fluorescent response with a fibre optic probe. The displacement of the probe in the artery and the intensity of the response can thus be used to estimate the degree and range of cell damage.

1.3. Demonstration of principle

In order to demonstrate the feasibility of the proposed scheme it is necessary to establish whether the markers can be successfully attached to the damaged cells and also that the fluorescein can be detected in the concentrations expected. It must be noted that the measurements need to be made in the presence of oxygenated blood and so account has to be taken of attenuation due to that.

Two sets of experiments are reported; one which shows that fluorescein can be attached to damaged endothelial cells in preference to undamaged endothelial cells and the fluorescence detected, the other which shows that fluorescein can be detected in suitable concentrations in simulated and real arteries.

2. Experiments

2.1. The detection of damaged cells in vitro

2.1.1. Preparation of the cells

Experiments by Cuello et al (1982) have shown that it is possible to internally label monoclonal antibodies without impairing their binding characteristics. To examine this finding in the present context endothelial cell specific antibodies were prepared. Endothelial cells were then grown in culture. The cell layer was scratched on the culture plate and the plate was washed with a suspension containing the monoclonal antibodies labelled with fluorescein.

2.1.2. Measurement of fluorescent response

Fluorescein absorbs energy at 490 nm and emits energy at 550 nm. A scheme was devised in which a filtered light source was used to illuminate the plate and an optical fibre was drawn across the plate to detect the fluorescent response. Appropriate filters and detectors were designed to measure the detected light at 490 nm and 550 nm and the ratio taken. Figure 1 shows the apparatus for controlling the position of the fibre and for measuring the light energy. Figure 2 shows the ratio of the intensity at the two wavelengths as a function of distance. The increased response at the location of the damage is clearly seen. Further details are available from Goodyer et al (1992).

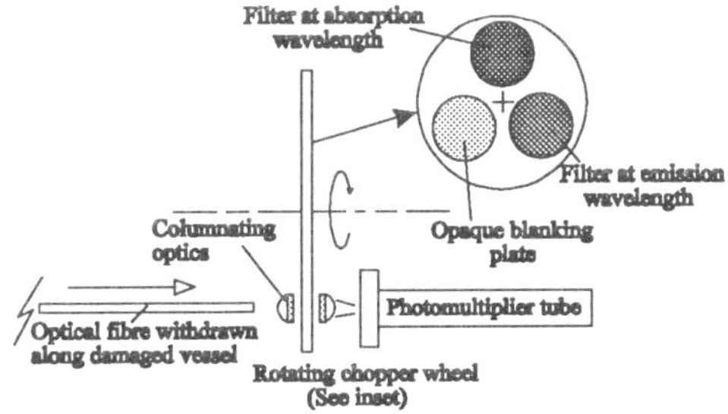

Figure 1. The apparatus for controlling the position of the optical fibre and acquiring the light energy.

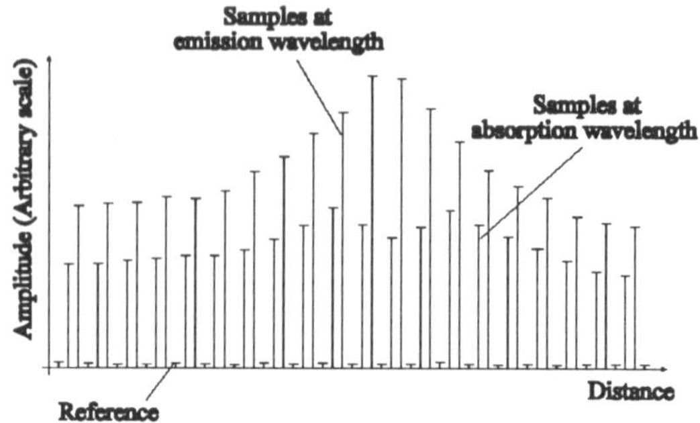

Figure 2. The intensity of the two wavelengths as a function of distance.

2.2. Measurements of fluorescence in simulated cardiac arteries

2.2.1. External illumination

In these experiments a false vessel was constructed from a length of translucent tubing comparable in size to that of a typical human cardiac artery (≈2mmØ ID) and a helix of narrower gauge tubing (0.25mmØ ID) arranged around its circumference in several turns. The helix could be filled with a spectrum fluorescein solution of variable concentration and the simulated vessel perfused with red blood cells as a first approximation to a functioning real artery. A probe consisting of a single polymethylmethacrylate optical fibre was then placed into the vessel, illuminated and withdrawn at a constant velocity whilst sampling at the appropriate wavelength (490 nm and 550 nm). Using the artificial artery the system was able to detect a minimum of 45 picomoles of fluorescein.

Figure 3 shows a specimen fluorescein response as a function of displacement. Further details are available from Goodyer et al (1994).

Following confirmation that a fluorescent material can be detected with red blood cells present, experimentation was extended to involve real mammalian tissue.

Experiments were conducted with a length of human umbilical cord devoid of blood and a vessel in a recently sacrificed rat in order to investigate the problems associated with using an external light source to excite the fluorescein. In each case a subject area of tissue in close proximity to the vessel wall was injected with 0.1 ml of 622μMole/l solution, the neighbouring tissues were then replaced prior to being illuminated from the halogen light source. 0.04μMoles of solution could be detected, the poor efficiency being attributed to the high attenuation of the tissues.

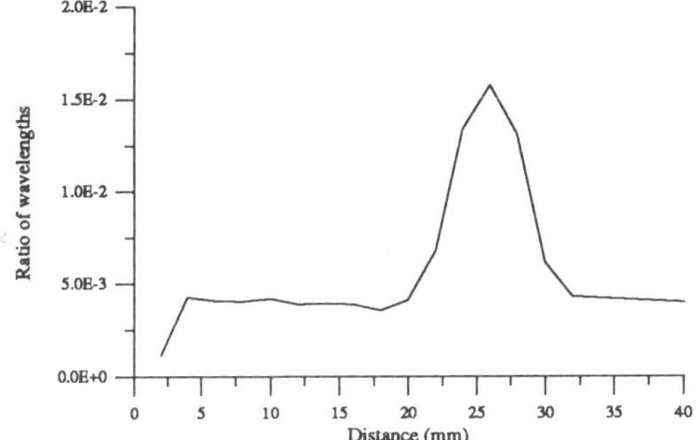

Figure 3. The ratio of wavelengths corresponding to fluorescence (5μMole/l dilution).

2.2.2. Internal illumination

The in vivo light source chosen was an Argon Ion laser having a principle wavelength of 488nm, just below that of the optimum absorption wavelength of the fluorescein at 493nm. The laser light is introduced via a fibre optic. The laser has advantages over a broad band light source in that complications from localised heating damaging the end of the optical fibre do not easily occur and the high degree of parallelism permits the beam to be readily focused onto the face of the optical fibre. The system sensitivity can be improved by increasing the power of the laser, practical limitations being only that of the risk of further vascular damage as a result of the laser.

Fluorescein was linked to heparin in varying dilutions and introduced into the aorta of a live rabbit by a perforated angioplasty balloon. It was hoped that the heparin would adhere to the vessel wall and be detectable by the fluorescent probe once the vessel had been harvested from the animal. Using this technique the peaks corresponding to 60IU/ml and 400IU/ml concentrations of fluorescein were detected[1]. Figure 4 shows that a good response has again been observed in this case.

The fluorescein plus heparin solution was also investigated using the aforementioned simulated artery. The dilutions of 500IU/ml, 310IU/ml and 80IU/ml were easily detected using a saline filled simulated artery.

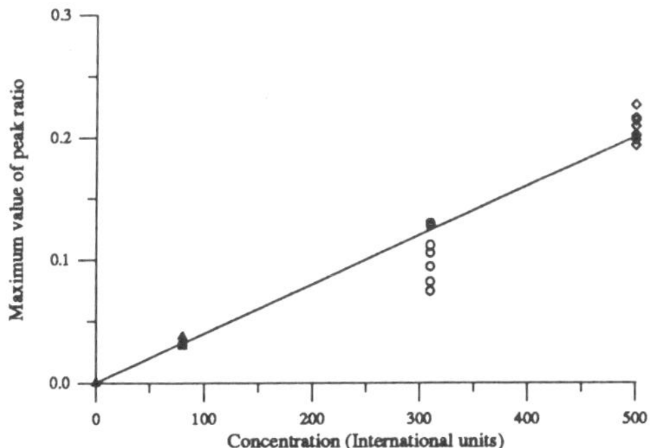

Figure 4. The peak of the ratio of the emission to absorption wavelengths using varying dilutions of fluorescein with heparin.

---

[1]The technique for linking fluorescein and heparin is yet to be refined. The authors suspect that only 1/28th of the concentration stated was actually present.

3. Conclusions

These pilot experiments have shown that it is possible to get biological macro molecules labelled with fluorescein to bond with damaged endothelial cells. Further it has been shown that it is possible to detect the fluorescein in appropriate dilutions in situations and conditions very similar to those expected in living human cardiac artery. The use of heparin as a carrier of the fluorescein is possible. Exciting the fluorescein by external illumination is possible but may not be practicable due to the high attenuation of the tissue. In vivo excitation by Argon Ion laser looks like a more promising alternative approach.

4. References

A.C. Cuello *et al* (1982), Immunocytochemistry with internally labelled monoclonal antibodies, Proc. Nat. Acad. Sc. (USA), **29**(1), pp.665-669

P.D. Goodyer *et al* (1994), Fibre optic radiation probe and the study of vascular damage, EPSRC Research Grant Report: GR/F80777, Dept. of Engineering, University of Leicester

M.B. Preisack and K.R. Karsch (1993), The paradigm of restenosis following percutaneous transluminal coronary angioplasty, European Heart Journal, **14**(1), pp.187-192.

APPLICATION OF FLUORIMETRIC BIOSENSORS FOR THE RAPID DETECTION OF FOOD PATHOGENS BY DNA- AND IMMUNO- ASSAYS.

NJC Strachan, DI Gray, IG Millar, JC Neidziela and PG John

CSL, Food Science Laboratory Torry, PO Box 31, 135 Abbey Rd., Scotland, UK.

Abstract

The reported outbreaks of sporadic food poisoning are continuing to increase and this is of considerable public health concern.

Listeria spp. may be found at low levels in a food and consequently, a DNA assay was selected because by using the polymerase chain reaction (PCR) it is possible to amplify the target DNA from low numbers of *Listeria* previously isolated from foods. The product which was fluorescently labelled during the PCR process was modified to make it single stranded. This was then detected by a DNA hybridisation assay using a fibre-optic biosensor.

Staphylococcus enterotoxins were also detected using the fluorimetric based biosensor. Antibodies are commercially available against these toxins and these were fluorescently labelled. In this method, the optical fibre was replaced with disposable 100 μm particles upon which a sandwich immunoassay was performed. This took place within a capillary which is situated inside the final lens of the fluorimeter. The immunoassay takes about 8 minutes to perform and the sensitivity is 10 ng/ml in milk.

1. Introduction

Food poisoning caused by bacterial action can arise through two main mechanisms. The first involves ingestion of the microorganism which causes poisoning (e.g. *Listeria* and *Salmonella*) and the second involves a microorganism producing a toxin in the food, which is then consumed (e.g. *Staphylococcus* and *Clostridium botulinum*).

DNA amplification techniques incorporating the polymerase chain reaction (PCR) have been developed which can amplify target DNA from as few as 1-10 microorganisms (Bej and Mahbubani, 1992). Currently PCR products are detected by flat bed agarose gel electrophoresis (Sambrook *et. al.*,1989) which can be time consuming and gives only an indication of the presence and size of the products. Precise identification of the product can be achieved by subsequent blotting and hybridisation methods using probes (Sambrook *et. al.*,1989) but these increase the time for the whole procedure to 1-2d.

Recently, fibre-optic fluorimetric sensors which utilise the evanescent wave principle have been developed to rapidly detect DNA hybridisation (Graham *et al.*, 1992) and antigen/antibody binding sandwich immunoassays for the detection of botulinum toxin (Ogert *et al.*, 1992 and Kumar *et al.*, 1994). The sensor works on the principle that when light is reflected within a fibre-optic an electromagnetic wave is generated in the medium outside of the fibre. This is the evanescent wave and because it penetrates beyond the optical interface it can be used to stimulate fluorescent labels which can either be attached or in close proximity to the surface of the glass fibre. This fluorescence is propagated back up the fibre and detected by a solid state sensor.

Currently, immunoassays for the detection of staphylococcus toxins are mainly performed using ELISA (Enzyme Linked Immunosorbent Assay) methodology. This involves carrying out the immunoassay on a microtiter plate and takes between 3-7 hours. The food industry requires a more rapid method be developed, in order to assay more suspect food samples.

A problem with performing immunoassays on a fibre-optic sensor is the regeneration of the fibre. This is usually performed using either acid or alkaline solutions and is only partially successful. As a result of this a particle based system has been developed by Sapidyne Inc. (Idaho, USA). In this system the fibre-optic is replaced by disposable particles and hence no regeneration is required. For DNA hybridisation assays regeneration is easily performed on the fibre-optic sensor by simply raising the temperature of the fibre-optic to above the melting temperature of the DNA.

Here, biosensors are described for the detection of *Listeria* spp., by DNA hybridisation using the fibre-optic sensor and staphylococcus enterotoxin B, by immunoassay using the particle based system.

2. Biosensor Instrumentation

A simple fluorimeter (Sapidyne Inc., Idaho, U.S.A) was used to perform the assays. The instrument is configured as in Glass *et al.*,(1987). The output voltage from the sensor was digitised and stored on a personal computer. For the DNA hybridisation assay, the glass fibre-optic

(7mm x 1mm diameter) was encompassed by the sample cell. This consisted of a glass-jacketed stainless steel flow cell, with screw fit plastic end caps (Omnifit, Cambridge, UK). The temperature was controlled by pumping water from a heated reservoir through the glass jacket and the temperature was monitored by means of a thermocouple. Hybridisation was performed at 60°C and was reversed by raising the temperature to 80°C.

The particle based system comprised a capillary contained within the final lens of the fluorimeter (Fig. 1.) in which 100 μm particles were massed against a filter. Putting the flow cell inside the lens ensured collection of approximately 50% of the light emitted. A PC was used to control a syringe pump which flowed the particles, samples and other reagents into the sample cell.

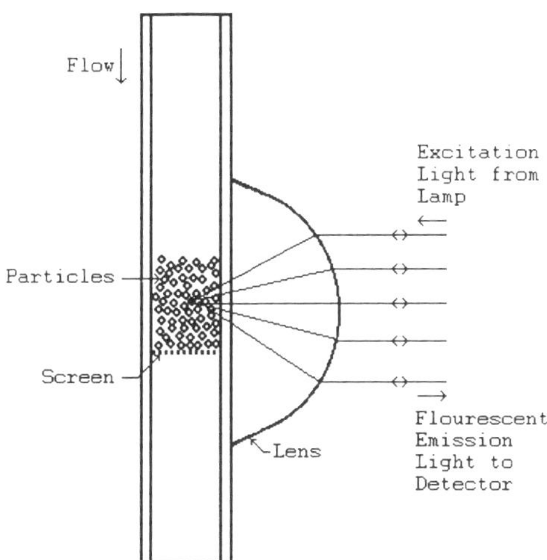

Figure 1: A schematic representation of the particle based system.

3. DNA hybridisation assay to detect *Listeria* spp.

Genomic DNA was isolated from *Listeria monocytogenes* using Instagene (Biorad, U.K.) . This was then used to PCR amplify a 200 base pair region of the genus specific fla A gene (Gray and Kroll, 1995). The PCR was used to incorporate fluorescein and biotin labelled primers into the final product. The strand which contained the fluorescent label was then isolated using the method described by Strachan and Gray, (1995). This was then added to pre-hybridisation buffer (Strachan and Gray, 1995) and passed into the flow cell.

Two experiments were performed (Fig. 2). This involved covalently bonding the amino terminated oligomers (mer) M2980 and M2979 (Table 1) onto glass fibre optics as described by Graham *et al.* (1992). In the first experiment, a central portion of the single stranded fluorescently labelled product was allowed to hybridise onto the distal 20 complementary bases of the 40-mer M2980. The steady increase in fluorescence detected by the sensor (Fig. 3a and 3b.) showed that hybridisation was taking place. The second experiment was the negative control. It involved attempting to hybridise the single stranded fluorescein labelled product to the non-complementary 20-mer M2979. The sensor showed no significant increase in fluorescence and therefore no hybridisation occured.

Table 1: Oligonucleotides used during this investigation

Number	**Sequence**	**Comment (Numbers in brackets correspond with position in <u>fla</u>A gene sequence)** (Dons *et al.*,1992)
M2980	5'-amino-[GTTCTCTTGATGACGCTGCT]-* CAATCTTGCAACGTATGCGT-3'	40-mer with distal 20 bases complementary to the central portion of the single stranded PCR product (112-131)+(253-272)
M2979	5'-amino-[GTTCTCTTGATGACGCTGCT]*-3'	Negative control with amino label (112-131)

* spacer sequence consisting of non-complementary sequence to the <u>fla</u>A gene region under study.

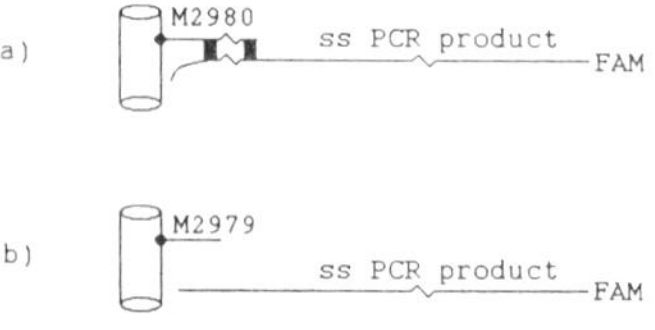

Figure 2: Experiments performed on the fibre optic biosensor.
a)Hybridisation of central region of ss PCR product to distal 20 bases of 40-mer (M2980).
b)Negative control oligo bound onto fibre which should not hybridise to ss PCR product

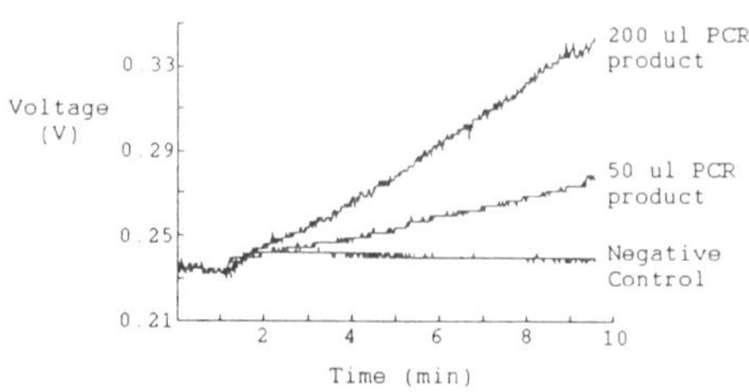

Figure 3: Biosensor output showing hybridisation of central region of ss PCR product

4. Immunoassay to detect Staphylococcus Enterotoxin B.

Figure 4 shows the format of the sandwich immunoassay. The 100 μm diameter polymethylmethacralate beads (200 mg) (Bangs, U.S.A.) were coated with 1 ml of antibodies against staphylococcus enterotoxin B (1 mg/ml) (Sigma, U.K.) for 1 hour at 37°C on a rocker. They were then added to 29 ml phosphate buffer saline (Sigma, U.K.) and stored at 4°C for up to 4 weeks prior to use. Solutions of staphylococcus enterotoxin B was made up in milk and left overnight at 4°C. Antibodies against staphylococcus enterotoxin B were conjugated with fluorescein using the method of Harlow and Lane (1988).

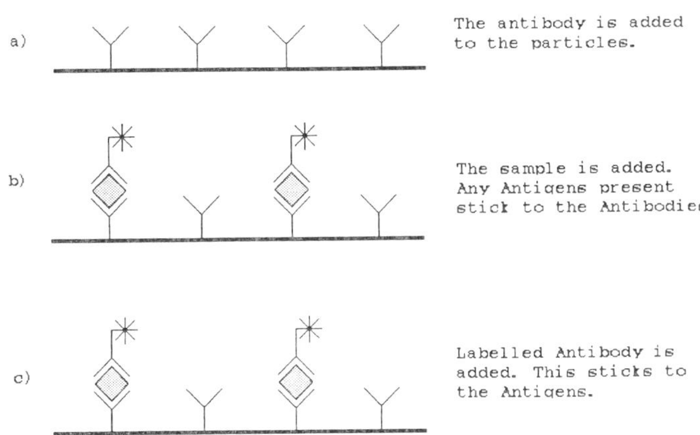

Figure 4: Schematic representation of a Sandwich Immunoassay

Approximately 100 antibody coated particles were pumped into the flow cell and trapped on the filter. This was followed by the sample (4 ml) and fluorescently labelled antibody (3 ml) to yield the sandwich. Then PBS (4.5 ml) was flowed passed the particles to wash away excess label.

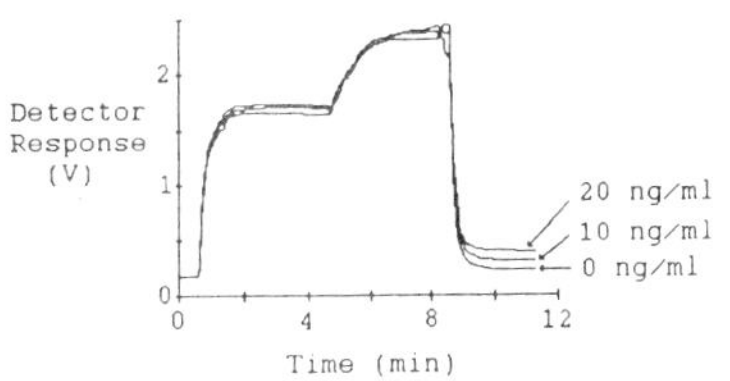

Figure 5: The detection of Staphylococcus Enterotoxin B in milk.

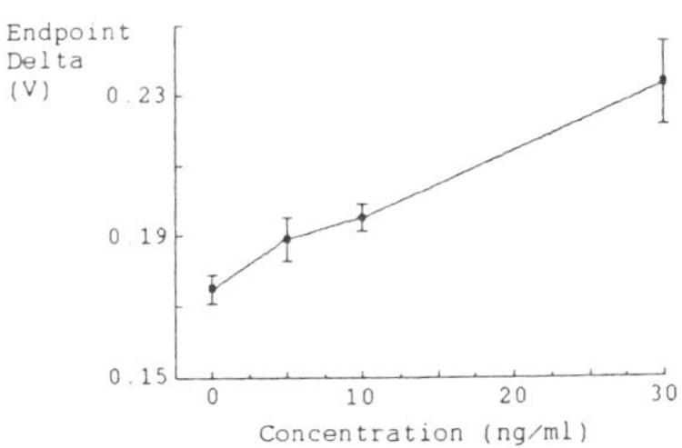

Figure 6: Sandwich immunoassay calibration plot for Staphylococcus Enterotoxin B in milk.

Figure 5. shows that the output from the biosensor increases as the amount of toxin in the sample increases. The amount of binding to the sensor can be calculated by measuring the difference in detector response (Volts)

between the start and end of the assay (endpoint delta). Three replicates of the assays in Figure 5 were performed and a calibration curve was generated (Fig. 6). The calibration curve shows that 10 ng/ml of the toxin could be detected within 8 minutes.

5. Discussion and Conclusions

The protocol described here demonstrates how rapid and specific biosensor-based methods are for the identification of unique gene sequences, exemplified here for the flaA gene from *Listeria*. This method is of general application and could be used for the specific identification of any PCR product and replace the requirement for electrophoresis and DNA blotting and probing.

The immunoassay to detect staphylococcus toxin in milk is approximately 20 times faster than ELISA methods and this makes it a suitable technique for the rapid assaying of suspect food samples. The sensitivity of the method (10 ng/ml) is less than that obtained by the authors from a commercial ELISA kit (Ridascreen,) which achieved between 1-5 ng/ml. This may be due to the reactivity of the particular antibodies being used in this work. Further research using another source of antibodies is currently being performed.

6. References

Bej, A. and Mahbubani, M. (1992) Meth. Appl. **1**(3), 151-159.

Dons, L., Rasmussen, O.F. and Olsen, J.E. (1992).Molecular Microbiology, **6**, 2919-2929.

Glass, T.R., Lackie, S. and Hirschfeld, T. (1987).Applied Optics, **26**, 2181-7.

Graham, C.R., Leslie, D. and Squirrell, D.J. (1992) Biosensors and Bioelectronics, **7**, 487-493.

Gray, D.I. and Kroll, R.G. (1995) Letters in Applied Microbiology **20**, 65-68.

Harlow, E. and Lane, D. (1988). Antibodies- A laboratory manual. Cold Spring Harbour Laboratory Press, USA.

Kumar, P., Colston, J.T., and Chambers, J.P. (1994). Biosensors and Bioelectronics, **9**, 57-63.

Ogert, R.A., Brown, J.E., Singh, B.R., Shriver-Lake, L.C. and Ligler, F.S. (1992). Analytical Biochemistry, **205**, 306-312.

Sambrook, J., Fritsch, E.F. and Maniatis, T. (1989) Molecular cloning a laboratory manual (2nd Ed.). Cold Spring Harbor Laboratory Press, USA.

Strachan, N.J.C. and Gray D.I. (1995) Letters in Applied Microbiology. In press.

A Rationale for Using a Fibre Optic Toxicity Sensor in the Automatic Control of a Sewage Treatment Plant - Some Measurements of the Inhibition Effects of Metal Contaminants

J Grabowski[1], P R Baker[1], M Latosińska[1], P J Scully[2] and R Edwards[2]

[1]Environmental Physics Research Group, Institute of Physics, Poznań Technical University, ul. Piotrowo 3, 60-965 Poznań, Poland.
[2]Department of Chemical and Physical Sciences, Division of Engineering and Science, Liverpool John Moores University, Byrom Street, Liverpool, L3 3AF, UK.

Abstract

It has recently been suggested that sewage treatment in the presence of metal contaminants could be optimized if the toxicity and inhibition mechanism of the contaminants are known. In this paper we demonstrate that it is possible to obtain this information by measuring the rate of fluorescein diacetate hydrolysis. In experimental studies using activated sludge and single metal contaminants we obtained toxicity data which correlates well with other published results and also found examples of competitive, non-competitive and uncompetitive inhibition. Recent attempts at developing a fibre optic toxicity sensor may eventually allow such an optimization of the sewage treatment process to be automated.

1. Introduction

The metal concentrations in the waste waters entering 239 treatment plants in the U.S.A. have been studied by Minear et al (1981). For over 200 of the plants the maximum concentrations of metals such as Pb, Zn, Cu, Cr and Ni were typically in the range of 100-200 mg/L. The presence of heavy metals during the biodegradation of organic contaminants is of concern because it inhibits the growth of the activated sludge bacteria thus lowering the efficiency of the sewage treatment process.

Vavilin V A and Vasiliev V B (1979) proposed a sewage treatment system in which the influent waste water is distributed between two aeration tanks which are in a row (Fig. 1). Grabowski et al (1992) showed that it would be possible to optimize the value of the flux distribution η between the two tanks if the toxicity and the inhibition mechanism of the contaminants was known.

Both the toxicity and inhibition mechanism can be monitored by measuring the rate of fluorescein diacetate (FDA) hydrolysis. FDA does not fluoresce after being dissolved in water. However when living cells are present in the water the FDA molecules, being non-polar, diffuse into the cytoplasm. In the cytoplasm the FDA becomes hydrolyzed by the enzymatic action of an esterase. The hydrolyzed fluorescein molecules are polar and remain in the cytoplasm. The metabolic activity of the cells is affected by the toxicity level which affects the rate of enzymatic conversion and hence the rate of increase of the fluorescence emission from the fluorescein. FDA hydrolysis has previously been effectively used to determine the microbial activity in soil and litter (Schnürer and Rosswall 1982) and to determine the effect of pollutants on the biocenosis in rivers (Obst et al 1988).

A theoretical analysis of the inhibition mechanisms is very complex (Dixon and Weeb 1964). Despite this simple empirical relationships are often observed in practice. Three main mechanisms of inhibition can be recognized by so called Lineweaver-Burk plots (Goueli and Steer 1981). These mechanisms have been called competitive, non-competitive and uncompetitive inhibition. Lineweaver-Burk plots can be generated by measuring the rates of increase of fluorescence emission for different contaminant levels and different FDA concentrations.

In this paper we will demonstrate the use of FDA hydrolysis to measure the toxicity level and to indicate the mechanism of inhibition of a number of metal contaminants which are typically present in a sewage treatment plant.

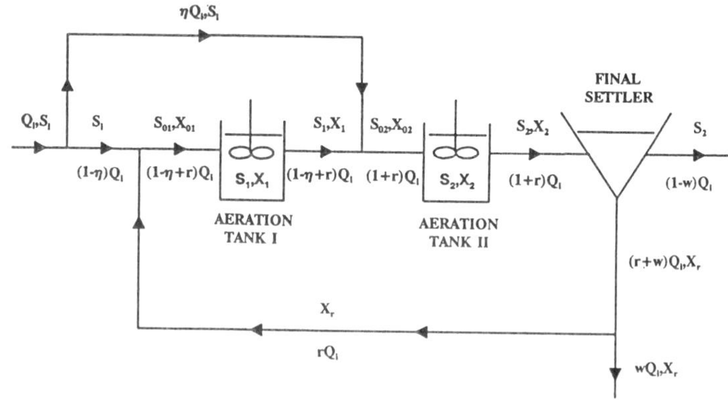

Fig. 1 A scheme for a sewage treatment system using two aeration tanks in a row.
Q - sewage flux (m^3/h);
r - returned sludge coefficient;
w - rejected sludge coefficient;
μ - distribution coefficient between the two reactors;
S - concentration of biodegradable organic substances (kg BOD/m^3);
X - MLSS (mixed liquor suspended solids) concentration (kg/m^3);

2. Materials and Methods

2.1 Activated Sludge Sample Preparation
The activated sludge microorganisms were grown at room temperature as a continuous culture. The nutrient for the microorganisms was a solution of 4 g/L of meat peptone (P.P.P.P. Bacutil) and 10 ml/L of mineral salts. The composition of the mineral salts was 10 g/L $NaHCO_3$ in H_2O, 7 g/L K_2HPO_4 in H_2O, 20 g/L NH_4NO_3 in H_2O, 0.7 g/L $MgSO_4.H_2O$ in H_2O and 0.3 g/L NaCl in H_2O. The pH in the growth chamber was 6.8. The samples were prepared by diluting the concentration of the microorganisms such that the extinction measured at 530 nm on a spectrofluorimeter was 0.1. Before the measurement each sample was further diluted by a factor of two using a phosphate buffer (0.1M; pH 6.0).

2.2 Metal Toxicity
The examined metal contaminant was then added at the examined concentration and after 10 minutes 1 μl of a 1 mg/ml FDA in acetone stock solution was added to each 20 ml sample. The sample was transferred to a 2 ml cuvette in a spectrofluorimeter where the measurement of the rate of increase of the fluorescence emission intensity was started immediately. The fluorescence excitation wavelength was 480 nm and the observation wavelength was 525 nm. The metal contaminants studied were Hg^{2+}, Cu^{2+}, Pb^{2+}, Cd^{2+}, Zn^{2+} and Ni^{2+}. For each metal measurements were made at three different metal concentrations. A measurement for a sample without metal contamination was also taken. The rate of increase of the emission intensity was determined by linear regression.

2.3 Inhibition Mechanisms
The above experiment was carried out for each metal using FDA stock solution additions of 1 μl, 2.5 μl and 10 μl. For each metal a Lineweaver-Burk plot was prepared. The y-axis of the plot is the ratio of the rate of increase of the fluorescence without the metal to the rate of increase of the fluorescence with the metal. The x-axis is the reciprocal of the FDA concentration. The plot contains one line for each metal concentration. The slopes and intercepts of the lines were determined by linear regression. The form of this graph was used to predict the inhibition mechanism. For competitive inhibition the lines converge to meet at a point on the y-axis, for non-competitive inhibition the lines are converging but meet after crossing the x axis and for uncompetitive inhibition the lines are parallel (Goueli and Steer 1981).

3. Results

3.1 Metal Toxicity
For all the cases studied the presence of the metals inhibited the intracellular conversion of FDA into fluorescein. This is demonstrated by the decrease in the rate of the measured fluorescence emission intensity at 525 nm (Fig. 2a). From the data shown in Fig. 2a a second type of plot (Fig. 2b) can be constructed which directly relates the inhibition effect $\Delta F/F_0$ to the metal concentration. From such plots the following sequence of EC50 metal toxicity was obtained: $Hg^{2+} > Cu^{2+} > Pb^{2+} > Cd^{2+} > Zn^{2+} > Ni^{2+}$. This sequence agrees with the findings of Laland et al (1977 and 1978).

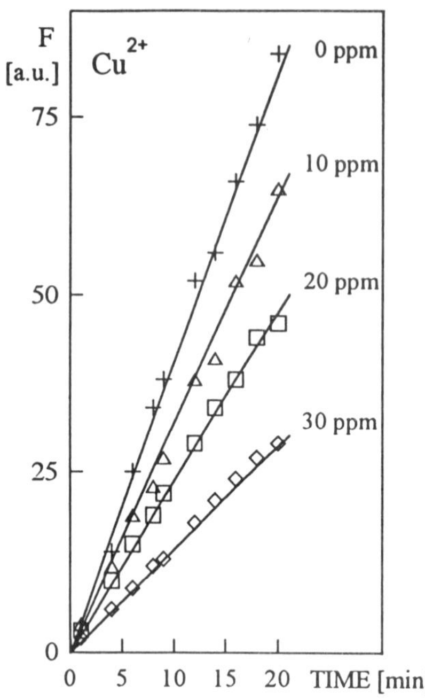

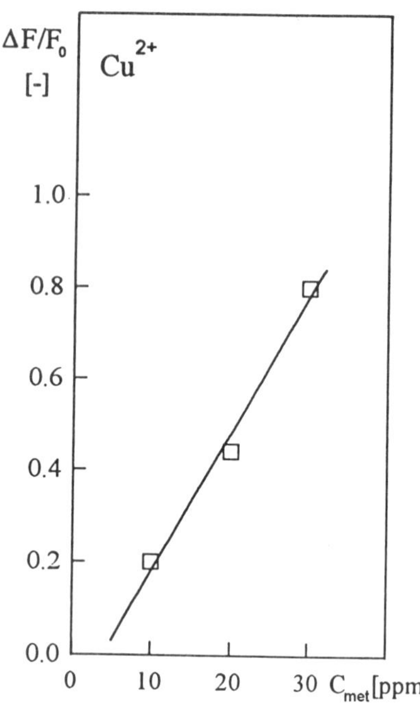

Fig. 2 (a) The effect of different levels of Cu^{2+} contamination on the time dependence of the fluorescence emission intensity at 525 nm from fluorescein formed from intracellularly hydrolysed FDA.

Fig. 2 (b) The data from Fig. 2 (a) plotted as the $\Delta F/F_0$ dependence on the Cu^{2+} concentration. $\Delta F = F_0 - F$ where F_0 is the fluorescence without the metal contamination and F is the fluorescence with the metal contamination.

3.2 Inhibition Mechanisms

Fig. 4 shows Lineweaver-Burk plots for Hg^{2+}, Pb^{2+} and Cd^{2+} which result in competitive, non-competitive and uncompetitive inhibition respectively. A non-competitive inhibition mechanism was also indicated by such plots obtained for Co^{2+} and Zn^{2+}. Note that to determine the rate of increase of the fluorescence about ten independent measurements were taken.

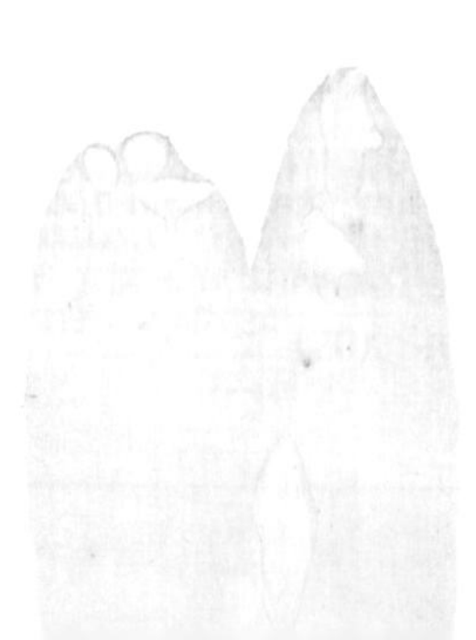

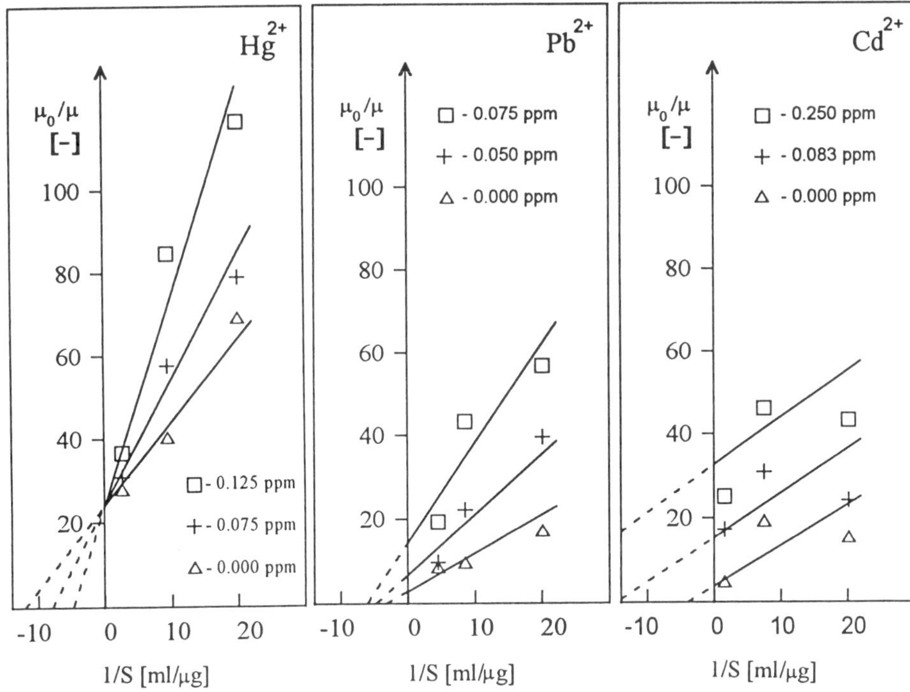

Fig. 3 Lineweaver-Burk plots for Hg^{2+}, Pb^{2+} and Cd^{2+}. μ_0/μ is the ratio of the rate of increase of the fluorescence intensity without the metal contaminant to that with the contaminant. S is the FDA concentration.

4. Conclusions

Real sewage is of course contaminated by a mixture of different metals and synergistic and antagonistic effects may occur. It is probably not theoretically possible to predict the toxic effects of mixtures of metals from the effects of single metals. More research is therefore required into the nature of the inhibition mechanisms occurring in a practical situation.

Currently such toxicity measurements need to be performed in an analytical laboratory. However recent attempts at developing a fibre optic toxicity sensor (Edwards et al 1995) may eventually allow such an optimization of the sewage treatment process to be automated.

Further work will also be required to find the optimum fluorogenic substrate. Fluorescein diacetate is only one of a number of compounds which may be appropriate (Holzapfel-Pschorn et al 1987).

5. Acknowledgement

This work was supported by Poznań Technical University grant number DS 62-112/02.

6. References

Dixon M and Weeb E C 1964 *Enzymes* (London:Longmans Green) 54-359

Edwards R, El-Saadawy S, Gander M, Scully P, Young A, Baker P and Grabowski J 1995 *The Development of a Biological Toxicity Based Test for Water Quality Using a Fibre Optic Sensor* Private Communication

Goueli S A and Steer R C 1981 *Some Representative Mathematical Plots in Biological Research* in *Advanced Cell Biology* ed L M Schwartz and M M Azar (New York:Van Nostrand Reinhold) 112-125

Grabowski J, Hansen P D and Vasconcelos G 1992 *Toxicity is an Important Process in the Biodegradation of the Organic Contaminants of Waste Waters by the Activated Sludge Method* in *Aplicacoes Technologicas no Saneamento Basico (Residuos Solidos - Efluentes - Reciclagem)* ed A Gamma-Xavier and C E Vargas-Lopez (Covilha, Portugal:University of the Biera Interior) 113-122

Holzapfel-Pschorn A, Obst U and Haberer K 1987 *Fresenius Z. Anal Chem.* **327** 521-523

Laland H V, Lulona S N and Wilkes D J 1977 *J. Wat. Pollut. Control Fed.* **49** 1340-1369

Laland H V, Lulona S N, Elder J E and Wilkie D J 1978 *J. Wat Pollut. Control Fed.* **50** 1469-1514

Minear R A, Ball R O and Church R L 1981 *United States Environmental Protection Agency Research and Development Report* **EPA-600/S2-81-220**

Obst U, Holzapfel-Pschorn A and Wiegand-Rosinus M *Toxicity Assessment* **3** 81-91

Schnürer J and Rosswall T 1982 *Applied and Environmental Microbiology* **43** 1256-1261

Vavilin V A and Vasiliev V B 1979 *Mathematical Modelling of the Biodegradation of Organic Contaminants of Sewage by the Activated Sludge Method* (Moscow:Science) p 69

Optical measurement of respiration rates

A. Raza and A.T. Augousti

School of Applied Physics, Kingston University, Kingston, Surrey KT1 2EE, U.K.

Abstract

We report the development of an optical analogue of the RIP; the Fibre Optic Respiratory Plethysmograph (FORP). A dual-channel device has been constructed and preliminary testing has been carried out. Each channel is capable of monitoring relative changes in the circumference of the ribcage and abdomen respectively. The device has been interfaced to a dedicated PC and has analytical software that is operated by a 'push button' menu system. The device is able to show clear distinctions between deep and shallow breathing and also detects the movement of air solely between the ribcage and abdomen when the nasal and oral passages are occluded (isovolume breathing).

Keywords: Fibre optic, Plethysmograph, Respiration, RIP, FORP.

Introduction

Earlier work has demonstrated that the measurement of respiratory volumes using invasive methods invariably give rise to an abnormal type of respiration [1]. Another significant drawback of these devices is that they are often uncomfortable and cumbersome. Thus non-invasive methods which measure ribcage or abdominal movements and hence infer respiration volumes are desirable. In 1967 it was proposed breathing may be modelled using two degrees of freedom of motion only; that of the abdomen and that of the rib cage [2]. Thus

$$V_{TOTAL} = k_{RC}V_{RC} + k_{AB}V_{AB}$$

Where V_{TOTAL} = total volume change
V_{RC} and V_{AB} = volume changes of the ribcage and abdomen respectively
k_{RC} and k_{AB} = calibration constants.

When the mouth and nose are occluded, air can only be shifted from the abdomen to the ribcage and vice-versa. Thus a given volume change in the abdomen will result in an equal and opposite volume change in the rib cage. Consequently non-invasive methods such as the use of magnetometers [3] and more notably the respiratory inductive plethysmograph (RIP) [4,5] have been developed. Both of these monitors, however, use electromagnetic transduction principles and are therefore inapplicable for operation in harsh electromagnetic environments, such as the interior of a Magnetic Resonance Imaging (MRI) scanner.

It is therefore a logical step to develop a monitor based on an all-optical transducing principle, and we report here the continued development of a dual channel optical analogue of the RIP, the Fibre Optic Respiratory Plethysmograph (FORP). This device is capable of operation in a wider range of environments than the RIP. Preliminary comparison of the device to a standard spirometer has been undertaken and the results are presented below. The construction of the device and typical data recorded from a test subject under laboratory conditions are presented.

Optomechanical Design

Each channel within the sensor consists of a light emitter and associated detector, as well as an electronics interface and a sensor band. Two channels were used to monitor the movement of the ribcage and abdomen separately. A PC was used to display a range of information in real time: the respective signals from each of the two bands, a weighted average of the two signals (termed the FORP signal), and a separate signal from the spirometer. Figure 1 shows the experimental arrangement with a detailed enlargement of the sensor band illustrated in figure 2.

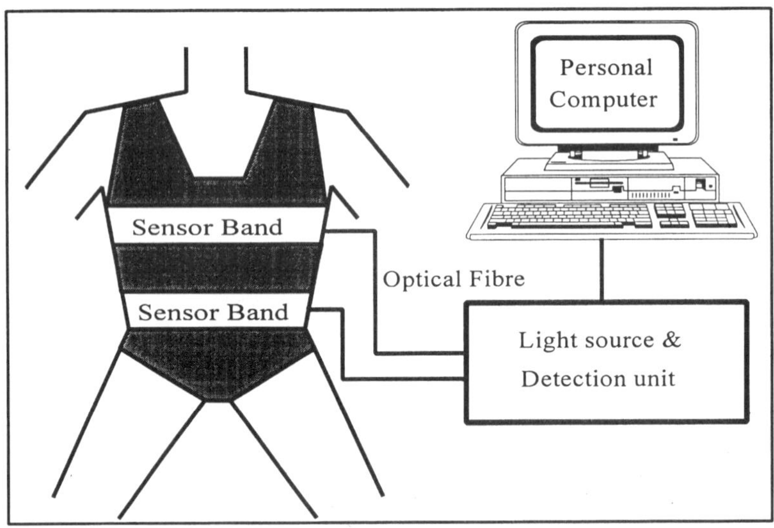

Figure 1 Experimental arrangement. The length of optical fibre used in each channel is approximately 3m.

Light from an infrared LED centred on a wavelength of 880nm is guided from the LED to the sensor band and then on to a photodiode via a single 100μm core diameter glass (PCS) optical fibre. In an early configuration of this device reported previously [6], the fibre had been sandwiched in a series of loops between two strips of elasticated bandage (Tubigrip) which made up the complete sensor band. However, even though this gave adequate results the device was prone to breaking and lacked reliable reproducibility. The revised band uses only one section of elastic bandage onto which are placed a series of vertical collars, (see Figure 2). These collars were made from Teflon tubing and are attached by first covering each one with heat-shrink tubing allowing the entire collar to be glued onto the bandage. The fibre is threaded through each of the collars in the manner shown, forming loops. The size of the loops is carefully chosen to optimise the sensitivity of the band, whist maintaining an adequate signal strength: too large, and the band is insensitive to variations in the loop size;

too small and the signal attenuation is too great to provide a viable signal. Masking tape is placed on the two ends of the fibre that protrude out of the band acting as a stop, preventing the fibre from retracting when the band expands.

Each sensor band is wrapped around either the chest or the abdomen and secured with a Velcro fastening. As the test subject inhales, the torso diameter increases, pulling on the two ends of the fibre and stretching the band. This in turn causes the loops to diminish in size, reducing the light propagating along the fibre and hence causing a decrease in signal strength. Conversely, exhalation causes the loops to increase in size and results in an increase in signal strength. Signals are captured by the detector unit and then, via an 8-bit internal A/D converter, are processed by an interface package written in Visual Basic for DOS running on a PC.

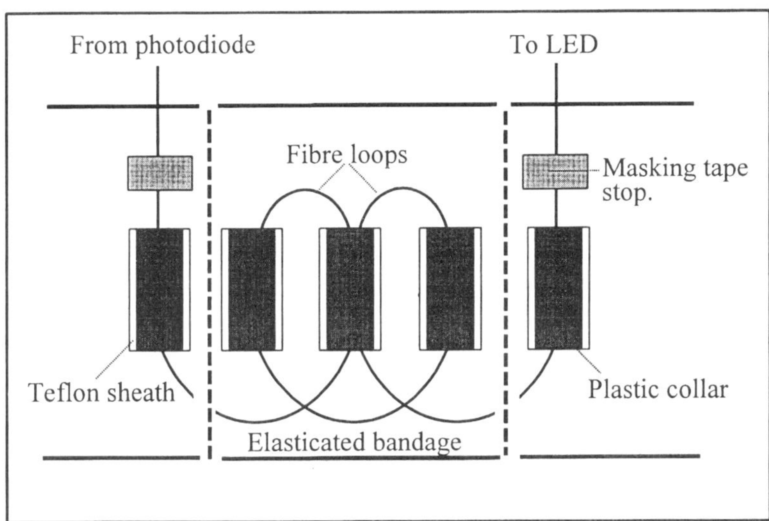

Figure 2 A schematic showing the arrangement of the loops within the Teflon collars. This was optimised in order to minimise sticking and non-uniform strain in each of the loops.

The interface allows easy alteration of a number of parameters which determine the form of the display. These include the number of readings which are averaged to produce a single graphical data point, and the values of the signal gradient which are used to distinguish between deep and shallow inhalations or exhalations, and when the breath is being held. Most importantly, one is able to mix the signals from the abdominal channel and the ribcage channel with a weighting that is selected using a simple scroll bar, thus enabling calibration to be performed under a range of conditions. During use, the software displays a series of graphs in real time. A maximum of four graphs can be chosen and displayed simultaneously. Two of these show the individual movements of the ribcage and abdomen as detected by their respective sensor bands. A weighted sum of these two graphs gives the FORP signal. The final graph represents the movements of a standard spirometer used in the calibration of the FORP. A simple algorithm based on the slope of the weighted average of the two signals is used to estimate the depth of the current breathing status. A colouring scheme is also used to further represent and record this status. All the graphs display relative volume changes in the appropriate areas of the torso in real time and can be recorded and played back later.

With the test subject standing, a series of slightly exaggerated breaths was recorded. A healthy test subject was monitored in a standing position, with a sensor band placed around both ribcage and abdomen, encompassing the areas where visible movement during breathing was most apparent. For display purposes, a series of four exaggerated 'normal' breaths, three exaggerated 'isovolume' movements and then three more 'normal' breaths were recorded (Figure 3). Isovolume movements [7] occurred when the mouth and nose were occluded and internal air only was forced to travel between the ribcage and abdomen 'compartments'. The abdomen and ribcage traces then appeared as complements although on different scales due to the different overall sensitivity of each band. This variation in sensitivity was compensated by weighting the two signals, and the process of isovolume breathing was chosen as the mechanism for selecting the weighting factor. The weighting factor was selected that gave the smallest variations in the combined signal in isovolume breathing. In this case it was found that a weighting of 56% to 44% in favour of the ribcage gave the optimum output.

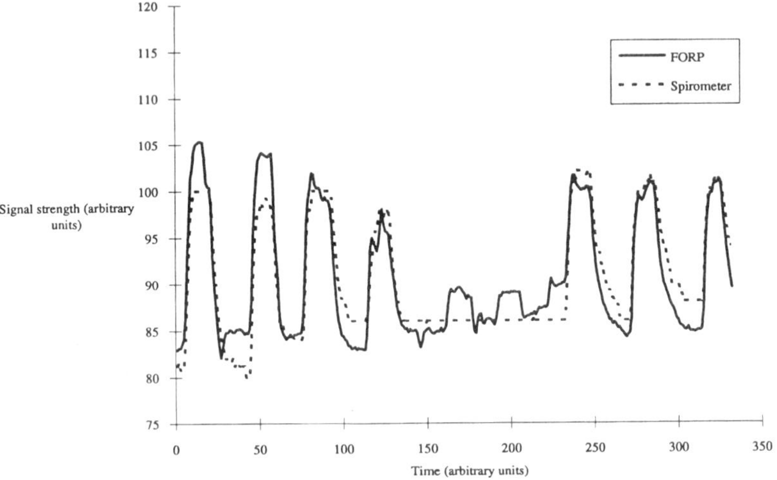

Figure 3 A graph showing a comparison between the FORP and spirometer signals. The graph shows 4 normal breaths, 3 isovolume movements, then three more breaths. The two signals have been scaled for comparative purposes, so that they appear similar in size. In practice, the FORP signal is approximately twice as large as the spirometer signal.

Results

Figure 3 shows a graph of the FORP and the spirometer signals superimposed. In the case of the spirometer, the isovolume trace is a horizontal line, as one would expect, since there is no volume change during this motion. The FORP monitors movement in both the ribcage and abdomen which move out of phase during isovolume movement. This should also give rise to a horizontal line, although in practice the individual response by each band to variations in diamter is somewhat non-linear. However, even with a perfectly linear response, one would not expect total compensation, since the compartments are of different sizes, hence the same volume of air would cause a different proportional change and thus only local circumference changes are registered. Notwithstanding this, the rejection of spurious trunk movements is greatly improved by the inclusion of a second channel, since the variations indicated by each channel alone are almost as large as a genuine breath.

Whilst the compensation would probably be improved by adding a third band, the device starts to become unwieldy and it is probably not necessary given the intended levels of accuracy. More important to the accuracy is the size of the envelope which contains the individual breathing trace, an example of which is shown in Figure 4. This shows an isolated breath taken from Figure 3 in order to determine the correlation between the spirometer and the FORP. This the error obtained in the FORP output during isovolume movements can be used to estimate the uncertainties in inspired volumes measured during a range of different breathing regimes. Given the test data obtained with this relatively unsophisticated implementation of a microbending sensing principle, the accuracy is estimated to be about 15% of overall lung capacity.

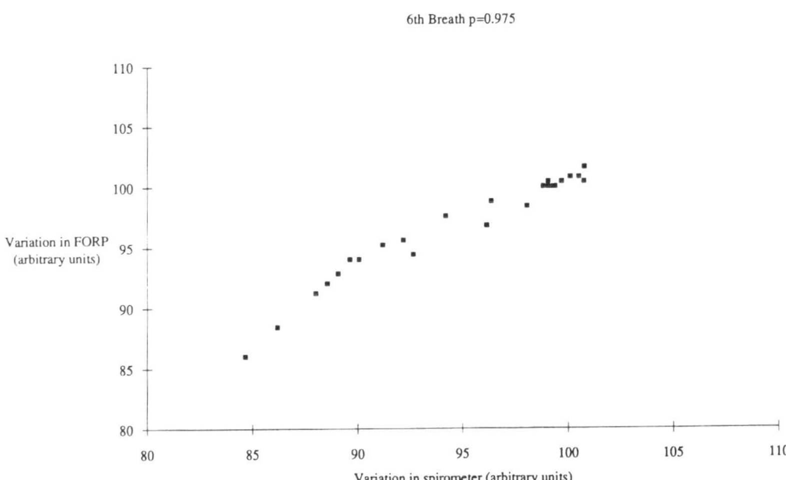

Figure 4 A graph of FORP output versus spirometer output for calibration purposes. The data is taken from the sixth breath shown in Figure 3, and the plot has a correlation coefficient of 0.975.

Discussion

The early version of the FORP [6] has been extended to monitor two channels and the system software and signal processing has been further refined. A full scale working prototype has been constructed capable of displaying changes in both ribcage and abdomen in real time. A more advanced menu system allows the operator to alter the measuring regime of the device by selecting the criteria for shallow and deep breathing, by choosing the limiting values of the signal gradient which determine the category of breathing.

The interface package can be used to optimise the FORP using isovolume movements and then it closely mirrors the output of the spirometer, but has the extra advantage of being able to detect movement in the torso when no air has been inspired (gagging). It is interesting to note that the FORP was able to affirm an effect documented prevously using the RIP [3], namely that during inhalation with the test subject upright and coupled to the spirometer movement in the ribcage precedes movement in the abdomen, although that data is not presented here. Without the nose occluded, this phenomenon is not so pronounced but is still apparent.

At the moment, the data analysis is limited to displaying the relative depths and patterns of breathing, but it is possible to envisage the device determining more precisely the actions of the subject (frequency and number of breaths, apnoeas and sudden bending or twisting of the trunk, for example), through the development of more intelligent software. Future work will aim to refine the device - in particular to extend the dynamic range and modulation depth, which are far from optimised. Although acceptable, the accuracy of the device is not very high, and improvements in the signal-to-noise ratio through better coupling of the LEDs to the fibres will allow better calibration to volume changes, thereby giving diagnostic information about lung capacities and capabilities in a manner similar to the RIP. It is also intended to apply the system to provide a gating signal during MRI recordings.

References

1. **Dolfin T, Duffy P, Wilkes D, England S, Bryan H.** Effects of a face mask and Pneumotachograph on breathing in Sleeping Infants. *Amer. Rev. of Resp. Dis.* **128** 977-79. 1983.
2. **Kono K, Mead J** Measurements of the separate volume changes of rib cage and abdomen during breathing. *J. Appl. Physiol.* **22** 407-422 1967.
3. **Grimby G, Bunn J, and Mead J** Relative contribution of rib cage and abdomen to ventilation during exercise. *J. Appl. Physiol.* **24** 159-66 1968.
4. **Cohn MA, Watson H, Weisshaut R, Stott F and Sackner MA** A transducer for non-invasive monitoring of respiration. *ISAM Proceedings of the 2nd International Symposium on Ambulatory monitoring,* London, Academic Press 119-28 1975 .
5. **Sackner JD, Nixon AJ, Davis B, Atkins N and Sackner MA** Non-invasive measurement of ventilation during exercise using a respiratory inductive plethysmograph. *Amer. Rev. of Resp. Dis.* **122** 867-71 1980 .
6. **Augousti AT and Raza A.** Development of a Fibre Optic Respiratory Plethysmograph (FORP), Sensors VI Technology, Systems and Applications: ISBN 0-7503-0316-6 401-06 1993.
7. **Cohn MA, Rao ASV, Broudy M, Birch S, Watson H, Atkins N, Davis B, Stott FD, Sackner Ma.** The Respiratory Inductive Plethysmograph: A new non-invasive monitor of respiration. *Bull. Europ. Physiopath. Resp.* **18** 643-658 1982.

A NOVEL OPTICAL-FIBRE LASER-INDUCED-FLUORESCENCE DETECTION SYSTEM FOR CAPILLARY ARRAY ELECTROPHORESIS

S. Bourin[1] , D. Mcstay[1], P. Kong Thoo Lin[1], W.J. Martin[2]

1.The Robert Gordon University , School of Applied Sciences, St Andrews Street, Aberdeen, AB11HG, Scotland.

2.The Rowett Research Institute, Greenburn Road, Bucksburn, Aberdeen, AB2 9SB Scotland.

Abstract:

A novel laser excited fluorescence detection system based on optic fibre technology for use with capillary electrophoresis has been developed. The system is easy to automate and extendable to a large number of capillaries. Initial results from the characterisation of the system are presented.

1.INTRODUCTION:

In the last decade, standard microbiology identification tests based on biochemical reactions have been phased out because they have a limited sensitivity, are labour intensive and are time consuming . They are gradually being replaced by DNA assays[1,2]. Although molecular biology and medicine have made immense progress in recent years, manipulation of large numbers of DNA samples, for example in hybridisation assays, PCR amplification, sequencing etc is slow and very labour intensive and additionally the results are often unreliable. The above factors and the initiation of the Human Genome Project for the mapping and sequencing of the whole human genome have stimulated research into the development of powerful, rapid and automated instrumentation for DNA analysis. One of the major challenges of the Human Genome project is to increase the sequencing output by a factor 100 compared to the present gel based sequencing techniques. A good candidate for achieving this increased output is Capillary Electrophoresis which has proven to have essential qualities for that purpose by overcoming most of the problems encountered with PAGE (PolyAcrylamide Gel Electrophoresis)[3,4,5].

In this paper, we describe a novel laser induced fluorescence detection system for capillary array electrophoresis based on optic fibre technology. The system has the advantage of being simple, robust and easily extendable to a very large number of capillaries.

2. Capillary electrophoresis :

In capillary electrophoresis the sample is introduced into a capillary that has been filled with a buffer for free solution capillary electrophoresis or with a sieving matrix (e.g. agarose, polyacrylamide) for gel filled capillary electrophoresis, which is typically used for the separation of DNA . A schematic representation of a basic CE system is shown in figure 1.

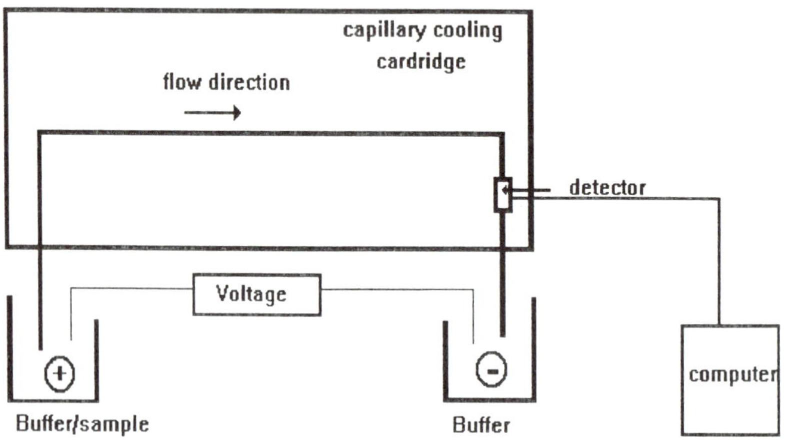

Figure 1: Schematic representation for capillary electrophoresis.

In a typical CE system , both ends of the capillary and electrodes are placed into buffer solutions and up to 30kV is applied to the system. As electrophoretic separation occurs, molecules pass by a detector where information is collected and stored by an appropriate data acquisition system. The main advantage of CE over slab gel electrophoresis is the use of a small inside diameter capillary, to which high electric fields can be applied without a significant increase in temperature. Higher electric fields allow a shorter separation time (minutes instead of hours with PAGE).

3. DNA analysis by CE.

DNA is a negatively charged molecule due to the phosphate groups it contains. Their charge density is linear, if for all DNA molecules the mass to charge ratio is constant. As a result, a sieving matrix is used (such as polyacrylamide and agarose), where the DNA mobility depends on the molecular weight. One problem has been the fragility of gel filled capillaries, however, this has been partially overcome by the development of replaceable gels[6,7] , which improve the capillary life time, reducing cost and labour.

The detection of the DNA molecules in a CE is performed using a variety of techniques, however, the most commonly employed are absorption and fluorescence[10].

In recent years, fluorescence based schemes have found favour in Capillary Electrophoresis due to their high sensitivity and ease of use. In the simplest fluorescence based systems a single fluorescent reporter molecule e.g. Fluorescein is attached to the DNA chain. More recently improved fluorescence detection schemes based on intercalating dye molecules have found favour. An intercalating dye is a flat molecule that undergoes sandwiching in between the stacked bases of a double stranded nucleic acid chain. By doing so its fluorescence quantum yield is increased by a significant factor.
The main advantage of using such molecules is the large number of fluorescent molecules being attached per DNA chain as compared to classical tagging strategies, that usually have only one fluorescent molecule per chain. Their usefulness for detecting dsDNA during CE separation has been proven in a number of publications[7,8,9].

4. Fluorescence spectra of intercalating dyes.

In the present work the intercalating dyes to be used are YOYO and TOTO (Molecular Probes Inc). For these dyes an increase of 1,100 and 3,200 respectively for the fluorescence yield has been observed experimentally on intercalation with ds DNA.

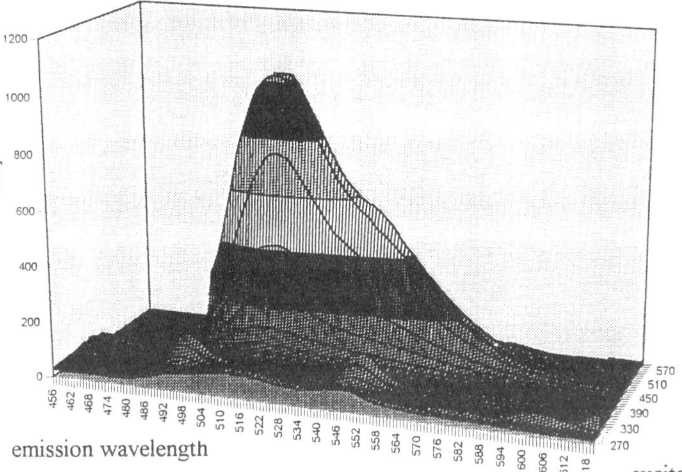

Figure 2: fluorescence spectrum of YOYO intercalated to DNA in Beckman CE buffer.

The measured excitation and emission spectra for YOYO intercalated to DNA in some Beckman capillary electrophoresis buffer is shown in figure 2. From the figure it can be seen that the intercalated YOYO can be readily excited using an argon ion laser.

5. capillary array detection system:

The principle drawback with CE is the limited number of samples which can be processed at the same time as all the systems commercially available have only one capillary. The use of capillary arrays is one route by which the processing power of CE systems may be increased. The development of detection systems and array designs for such systems is actively being pursued by a number of researchers[11-13].

In general the designs fall into two categories: 1). the excitation and detection optics are mechanically scanned across the capillary array[11] .2). a laser beam is directed from the side through a linear array of capillaries and detection is performed using an imaging detector[13]. Instruments based on 1) are generally slow while those based on 2) are limited in the number of capillaries which can be reliably and easily incorporated as in for example the system developed by Kambara and Takahashi[13] which supports 20 capillaries. Additionally in such a design it is difficult to achieve a uniform excitation of all the samples, potentially resulting in bleaching for some of the capillaries and insufficient excitation for some others.

To overcome the problems associated with the systems discussed above an optical fibre based excitation/detection system for use with CE arrays has been developed. In this system light from an argon laser is amplitude modulated and launched equally into a number of optical fibres. This is achieved by mounting the fibres in a purpose-built holder. The far end of each optical fibre is positioned close to a section of a capillary from which the plastic coating had been removed. In this way the optical power reaching each capillary was the same. As can be seen on figure 3 each capillary is mounted adjacent to a channel of an eight channel photomultiplier array (Hamamatsu). An interference filter is mounted between the capillary and the detector to ensure that only the laser induced fluorescence reaches the detector. The output of the photomultiplier is connected to a lock-in amplifier board, the output of which is logged by a computer.

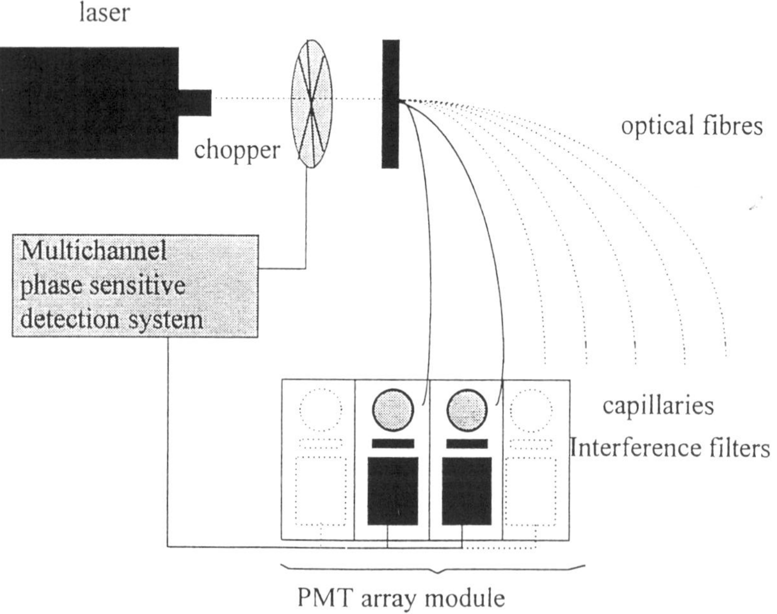

Figure 3: 8 capillary array detection system.

In order to test the optical fibre system a Beckman PA/CE 2200 CE system was modified to support a version of the new detection system. Varying concentrations of the dye Rhodamine B were then pumped through a capillary (75 µm ID 375 µm OD Composite Metal Services Ltd) and the resultant fluorescent signal recorded. From figure 4, which shows a typical result it can be seen that the system is capable of measuring concentrations of rhodamine B down to 10^{-6}M.

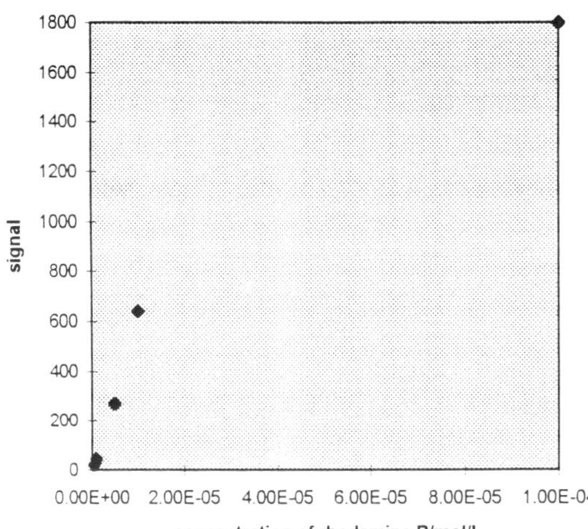

Figure 4: Fluorescence signal for different concentrations of Rhodamine B.

Although this sensitivity is not as good as many existing single channel systems, it demonstrates the practicality of the design. The main reason for the current limited resolution was scattered laser light, which produces a large background. Several techniques including polishing the capillary and the use of index matching liquids are currently being investigated to overcome this problem. The system is currently being tested for the detection of DNA sequences.

REFERENCES:

1. J.Walker and G.Dougan (1989): DNA probes: a new role in diagnostic microbiology, *Journal of applied bacteriology*, **67**, 229-238(1989).

2.Heller M.J. and Morisson L.E (1985): In Rapid detection and identification of infectious agents ,pp245-256, Academic press, New York.

3.T.J.Kasper *et al* (1988): Separation and detection of DNA by capillary electrophoresis, *J. of Chromat.* , **458**, 303-312(1988).

4.C.A.Monnig and R.T.Kennedy (1994): Capillary electrophoresis (review), *Anal.Chem.*, **66**, 280R-314R (1994).

5.J.P.Landers *et al* (1993): Capillary electrophoresis : A powerful microanalytical technique for biologically active molecules, *Biotechniques*, **14**, 98-111 (1993).

6.M.Ruiz Martinez *et al*(1993): DNA sequencing by CE with replaceable linear polyacrylamide and laser induced fluorescence detection, *Anal. chem.*, **65**, 2851-2858(1993).
7.D.Figeys *et al* (1994): Use of the fluorescent intercalating dyes POPO-3,YOYO-3 and YOYO-1 for ultrasensitive detection of double stranded DNA separated by CE with hydroxypropylmethyl cellulose and non-cross-linked polyacrylamide, *J.of Chromat. A*, **669**, 205-216(1994).
8.A.Glazer and H.Rye (1992): Stable DNA-dye intercalation complexes as reagents for high sensitivity fluorescence detection, *Nature*, **359**, 859-861 (1992).
9.K.Srinivasan *et al* (1993): Enhanced detection of PCR products through use of TOTO and YOYO intercalating dyes with laser induced fluorescence, *Appl.Ther. Electroph.*, **3**, 235-239(1993).
10.J.Z.Zhang *et al* (1991): High sensitivity laser induced fluorescence detection for capillary electrophoresis, *Clin.Chem*, **37/9**, 1492-1496(1991).
11.X.C.Huang *et al* (1992): Capillary array electrophoresis using laser excited confocal fluorescence detection, *Anal.Chem*, **64**, 967-972 (1992).
12.J.A.Taylor and E.A.Yeung (1993): Multiplexed fluorescence detector for capillary electrophoresis using axial optical fibre illumination , *Anal.Chem*.**65** , 956-960(1993).
13.H.Kambara and S.Takahashi (1993): Multiple sheathflow capillary array DNA analyser,*Nature* , **361**, 565-566 (1993).

Acknowledgements:

This project is part of an EC project funded by the CEC Biomedical and Health Research work program (area 3 in "human genome analysis).

Phosphorometric Monitoring of Different Enzymatic Conformeric States

L. Yang[1], D. McStay[1], P.J. Quinn[2], A. Rubtsov[3], A.A. Boldyrev[3]

1 School of Applied Sciences, The Robert Gordon University, Aberdeen, U.K.
2 Biochemistry Section, Division of Life Sciences, King's College London, UK
3 Department of Biochemistry, Moscow State University, Moscow, Russia

ABSTRACT

Successfully application of phosphorometric method to probe the different conformeric states of Caesium pump (Ca^{2+}-ATPase) in sarcoplasmic reticulum membrane is reported. The enzymatic conformation transition plays a very important role during the enzyme catalytic function, and can previously be monitored by fluorescence quenching measurement. The phosphorometry provides a new approach for monitoring / study of the enzymatic conformation states, and may provide more detailed internal information about the nature of the protein conformation transitions, which is difficult to obtain using fluorescence measurement alone.

1. INTRODUCTION

Many biological processes are mediated by the interaction between molecules, often in a highly specific manner. The formation of such complexes is known to underlie a range of responses observed at the cellular level and is fundamental to molecular recognition processes. The aggregation of cell surface receptors or redistributions of receptors on the surface of a cell are well characterised events. Rotational relaxation mechanisms are altered as a consequence of interactions between molecules and measurements of rotational correlation times provides a method for investigating the formation of complexes or molecular interactions.

Cellular components that can be examined conveniently using rotational characteristics are intrinsic membrane proteins. These molecules experience segmental motion and lateral diffusion in the plane of the membrane but rotational motion is restricted to an axis perpendicular to the plane of the membrane. Measurements of rotational motion of such proteins can provide information about their function. Rotational dynamics may be measured by the time-averaged technique of saturation transfer electron paramagnetic resonance, or flash photolysis.[1-2] Of the optical photolysis methods, transient absorption dichroism, phase-modulation fluorescence, fluorescence photobleaching recovery, and particularly the more sensitive light emission techniques (phosphorescence and delayed fluorescence), have been exploited extensively in the study of protein rotational dynamics. Such studies have included the interaction between myosin heads in muscle cross-bridges, human serum lipoproteins, the band 3, anion transport system of human erythrocyte membranes, and Ca^{2+},Mg^{2+}-activated adenosine triphosphatase (ATPase) and other proteins.[1-4]

Structure of Ca^{2+}, M^{2+}-ATPase

Ca^{2+}-activated adenosine triphosphatase (ATPase) is one of the most important cation transport proteins in the plasma membranes of all eukaryotic cells. The protein has been studied intensively since its first identification thirty years ago.[5] Various methods have been employed to investigate the relationship of the function and structure of the enzyme and much progress has been made since.[6] Earlier studies on the enzyme structure indicated that the Ca^{2+}-ATPase has an α subunit in its molecular constituent, which has been identified as the minimum functional unit of the enzymatic function of the protein (monomer). There is some evidence suggesting that the enzyme may form dimmers or higher oligmeric complexes in native membrane for functioning purpose[7] while others interpret the functioning unit in Ca^{2+}-ATPase as α monomeric protein.[8] Figure 1 shows a schematic diagram of the membrane-bound protein monomer. Protein-protein interactions have been proposed to play an important role in the functioning of the enzyme. During the catalytic function, the protein has been identified to experience a two conformation (E_1 and E_2 conformations) transition cycle, and the enzymatic conformation has been believed to coupling with its molecular structure changes, which has been proved by using fluorescence measurement. The fluorescence was found to be quenched / enhanced when the enzymatic conformation changed. In this paper we discuss the extended application of the phosphorometric method to the study of conformeric states of the cation transport protein Ca^{2+}, M^{2+}-activated adenosine triphosphatase (ATPase) in sarcoplasmic reticulum membranes.

2. MATERIALS AND METHODS

Preparation of the enzyme: The light fraction of sarcoplasmic reticulum (SR) membrane consisting of Ca^{2+}-ATPase was isolated from rabbit hind leg white muscles by differential centrifugation, as described by Rubtsov previously.[9] In the final preparations of the sarcoplasmic reticulum membranes, the purity of the Ca^{2+}-ATPase was determined using a method described by Liaemmli in 1970,[10] and was found that the Ca^{2+}-ATPase accounted for about 75 % of the total protein constituents present in the membranes. The SR preparations suspended in a solution containing 0.3 M sucrose, 30mM imidazole (pH 7.0) were kept in frozen in liquid nitrogen, at -70 °C, until prior to the labelling with eosin.

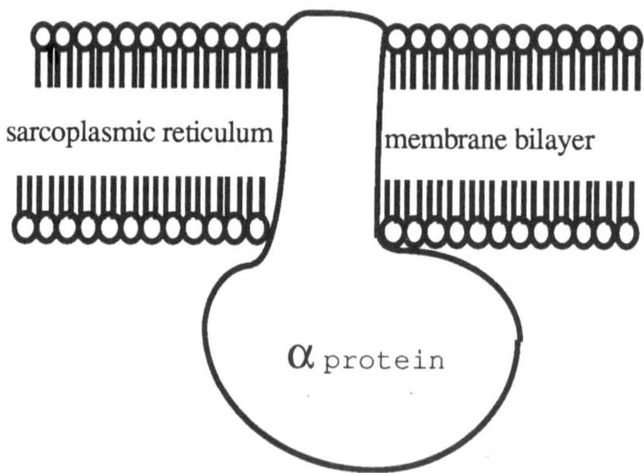

Figure 1 Simplified model for Ca^{2+}, Mg^{2+}-ATPase: a schematically presented α monomer.of the protein All binding sites are omitted from the figure.

The triplet probe eosin 5'-isothiocyanate was purchased prom Molecular Probes Inc (Eugene, OR, USA), and the oxygen reduced argon gas was obtained from the British Oxygen Company Ltd (Surrey, UK). Other chemicals were the purest grade available from Sigma or Aldwich Chemical Company Ltd.

Labelling the protein with eosin: Specifically labelling the phosphorescent probe eosin 5'-isothiocyanate to the putative ATP binding site of Ca^{2+}-ATPase of sarcoplasmic reticulum preparations was performed according to a method described previously[11].

Briefly, a suspension of sarcoplasmic reticulum vesicles in buffer, containing 30 mM Tris-HCl (pH 8.7), 5 mM MgCl and 100 mM KCl, (protein concentration 4 mg/ml), was reacted with 2.5 ~ 100 μM eosin-5'-isothiocyanate (to achieve a different stoichiometry of probe:protein). The reaction mixture was incubated in the dark for 45 min at 20 °C. After addition of 5 volumes of cold (0°-4°C) buffer containing 30 mM MOPS' (pH 7.0), 0.3 M sucrose and 1mg/ml BSA, the membrane suspension was incubated on ice for 20 min and then centrifuged on a Beckman centrifuge at 60,000 xg for 40 min. The supernatant containing the unreacted dye was discarded and the pellet washed once with cold 30 mM MOPS' (pH 7.0) and 0.3 M sucrose and finally resuspended in the same buffer. The stoichiometry of the protein : probe was determined by measurement of the protein concentration using the method of Lowry et al[12] and the dye label concentration by estimating the maximum absorption of the preparations at 530 nm. Preparations with approximately 1:1 stoichiometry of protein to probe were used to perform the experiments.

De-oxygenation of sample: To avoid possible interference from time-dependent denaturation and other protein-protein interactions of the Ca^{2+}-ATPase the eosin labelled SR samples were thawed immediately prior to the experiments. The enzyme was then shifted into its E_1 or E_2 conformeric state by addition of appropriate cation compositions into the SR preparation. Afterwards, the oxygen was removed from the eosin-labelled protein samples enzymatically using a method described by Eads et al[13] by addition of 100 μg/ml glucose oxidase, 15 μg/ml catalase, and 5 mg/ml glucose into sample cuvette containing 0.67 mg/ml sarcoplasmic reticulum protein, 30 mM MOPS', 100 mM KCl, 0.5 mM EGTA, and 0.3 M sucrose (pH 7.0). After the addition of the enzymes the air inside the cuvette was replaced by gently blowing Ar gas into the cuvette, and the cuvette sealed. The sample mixture was incubated for 20 min in the dark to deoxygenate at (0°-4 °C), prior to the phosphorescence measurement.

Measurement of the time-resolved phosphorescence: After de-oxygenation the eosin labelled enzyme samples in different conformeric states were excited by a narrow laser pulse from a frequency doubled Nd:YAG laser (λ = 532 nm, FWHM = 5 ns), and the resulting orthogonally polarised phosphorescence emission components measured using a T-format dual-channel phosphorimeter, as described elsewhere.[14]

The time-resolved phosphorescence anisotropy was calculated from the above data using the definition of $r = (I_{//} - I_{\perp}) / (I_{//} - 2I_{\perp})$, and the rotational dynamics studied by analysis of the time-resolved phosphorescence depolarisation procedure using a Marquardt non-linear curve fitting procedure.

3. RESULTS AND DISCUSSION

Emission intensities: The total intensities of the phosphorescence emissions from the Ca^{2+}-ATPase samples in E_1 and E_2 conformations were measured 10 μs after the laser flash, and recorded over the temperature range 2°-42 °C (fig 2). The phosphorescence intensities were found to be influenced by the protein conformation. The Ca^{2+}-ATPase in E_1 conformation had a relatively high phosphorescence emission intensity, while protein in E_2 conformation had a relatively low intensity of phosphorescence emission. The observation coincides well with the previous fluorescence and internal phosphorescence studies of the enzymatic conformations of the Ca^{2+}-ATPase.[15] The quenching of the phosphorescence intensity by conformational transitions in the protein was found to be strong at low temperatures (26% ~ 36%, at 2 °C), and became progressively weaker as the temperature increased (15% ~ 17%, at 40 °C). The trend was analysed by linear regression analyses and found to be -6.6 mV/°C and -9.6 mV/°C for the E_2 form and E_1 form ATPases respectively, with the same regression coefficient of 0.975. A plot of the linear regressions is presented in Fig 2. Previous studies have indicated that the Ca^{2+}-ATPase undergoes conformational changes following the temperature changes,[3] tending to form E_2 form ATPase at high temperature.[3] The decrease in the difference of the quenching of the total phosphorescence intensities of the eosin labelled Ca^{2+}-ATPase by the protein conformations following the increase of temperature, observed during this experiment agrees with this argument of the temperature-dependent conformation transition of the enzyme.

Initial emission anisotropy: The initial phosphorescence anisotropy obtained 10 μs after laser flash was recorded over the temperature range 2°-42 °C (fig 3). Both anisotropy data

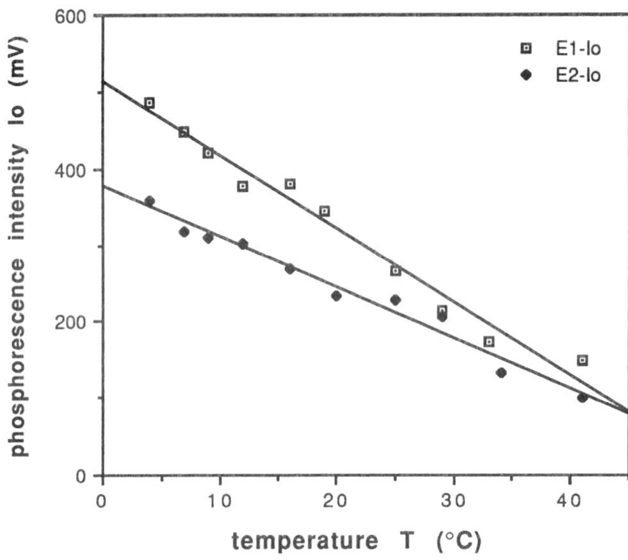

Figure 2 Total phosphorescence emission intensities of the E_1 and E_2 form Ca^{2+}-ATPases obtained over temperature span of 2 ° - 42 °C. Solid lines are the results of linear regression of the experimental data, indicating the general trends of the temperature-dependence of the data.

exhibit similar trends of temperature-dependence with an increase in value following the increase of temperature when the temperature is lower than 15 °C and a decrease when the temperature is greater than 15 °C. However, the changes of the anisotropy signals from different conformeric proteins are obviously different with the anisotropy values of the E_2 conformer higher (about 0.22 peak value) than that of the E_1 conformer (about 0.18 peak value) as shown in the graph. It was also found that increasing temperature resulted in a convergence in the initial anisotropy values of the phosphorescence of the enzyme samples in two different conformations, which could also be due to the conformation changes in either of the two sets samples, possibly the E_1 type ATPase. This conformation convergence would naturally result in the convergence in the characteristics of the phosphorescence emissions. Considering the fact that the emission intensities of the phosphorescence from the same samples also showed the same trend of temperature-dependent convergence, this may suggest that the Ca^{2+}-ATPase in SR membrane does undergo conformational transitions following the increase of temperature.

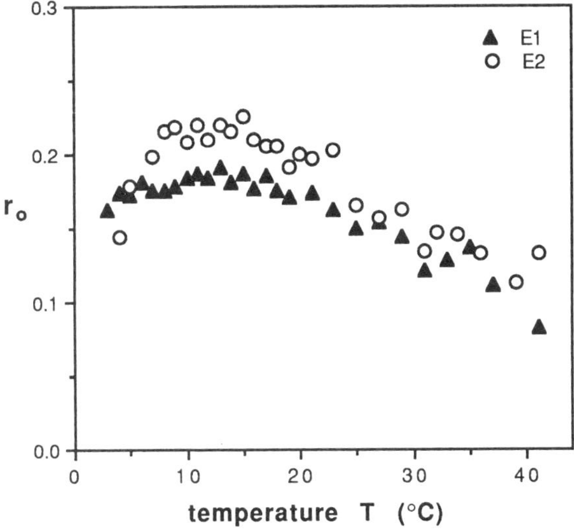

Figure 3. Phosphorescence anisotropy of Ca^{2+}-ATPase in E_1 and E_2 conformeric states. The data were collected 10 μs after the laser flash

As indicated above both the intensities and the anisotropies of the phosphorescence obtained from two sets of enzyme samples in different conformations confirmed that the characteristic features of the phosphorescence of Ca^{2+}-ATPase in different conformations are different. The temperature-dependent convergence of such characteristics of the phosphorescence of the two sets of samples, observed during the experiments, may suggest that there was a convergence of the sample conformations, which could be attributed to the conformation changes in either of the samples, or maybe, in both samples.

parameters / conformarion	rotational correlation times		effective radii		orientation Dm=1/nΦ
	Φ s (μs)	ΦL (μs)	as (Å)	aL (Å)	
E1-Ca-ATPase	10.6 (±3)	86.1 (±15)	17.1 (±2.6)	48.7 (±4.5)	n = 1
E2-Ca-ATPase	10.9 (±2.5)	116.0 (±25)	17.3 (±2.1)	56.5 (±6.5)	

Table 1. Rotational correlation times of phosphorescence and the associated effective radii of the rotating species of different conformeric Ca^{2+}-ATPase in sarcoplasmic reticulum. The calculation are performed using the data obtained at 20 °C, and the microviscosity and thickness of the membrane are 2.6P and 45 Å respectively, as reported previously.

The rotational dynamics analysis has found that in either the E_1 or E_2 conformation the Ca^{2+}-ATPase sample consists of, at least, two different single-exponential decay processes with their rotational correlation times taking values of tens and 100 of μs. According to the above argument, this may be explained as there were, at least, two different rotating species consisting of either of the two different conformeric protein samples. The volumes of these rotating species estimated using a cylindrical rotator model is presented in Table 1. From the table it can be found that both species were very close to those of the protein monomer and dimmer. Due to the real orientation of the probe being unknown, the above calculations were performed by assuming that the probe orientation satisfied the condition $D_m = 1/\phi$ (n = 1, and Dm is rorational diffusion coefficient). However, the probe orientation may also satisfy $D_m = 1/4\phi$ (n = 4) in practice, therefore, the real sizes of the rotating species could be even bigger (double the size of the sizes in table 1, if n = 4) depending on the real orientation of the eosin probe. It was also found that the fast rotating component of a sample in E_1 conformation contributes more to the total signal than that of a sample in the E_2 conformation does. However, this difference mainly occurred at low temperature (<20 °C), and steadily reduced to zero at high temperature, i.e. the different contributions by the E_1 and E_2 type ATPase samples converged to the same contributions following the temperature increases.

5. CONCLUSIONS

The results presented demonstrate the utility of the phosphorometric monitoring to the study of enzymatic conformeric states. In general, the results suggest that both E_1 and E_2 type Ca^{2+}-ATPase enzymes are a heterogeneous equilibrium of different protein aggregates, which perform a temperature-dependent association or disassociation, in other words, protein-protein interactions. The enzyme conformation also underwent temperature-dependent transitions, coupling with the protein-protein interactions.

5. REFERENCES

1. D.D. Thomas, T.M. Eads, V.V. Barnett, K.M. Lindahl, D.A. Momot, & T.C. Squier, Spectroscopy and Dynamics of Molecular Biological System (ed. Bayley, P.M., & Dale, R.E.), pp.79-95, London: Academic Press, (1985).
2. R.J. Cherry, Spectroscopy and Dynamics of Molecular Biological System, (ed. Bayley, P.M., & Dale, R.E.), 79-95, London: Academic Press, (1985).
3. T.M.Jovin and W.L.C.Vaz, *Methods in Enzymology*, **172**, 471-513, 1989.
4. E.D.Matayoshi and T.M.Jovin, *Biochemistry*, **30**, 3527-3528, (1991).
5. E.T.Dunham, I.M.Glynn, *J. Physiol.*, **156**, 274-293, 1961.
6. E.Carafoli, *Annu. Rev. Physiol.*, **53**, 531-547, 1991.
7. J.P.Andersen, P.Fellmann, J.V.Møller, and P.F.Devaux, *Biochemistry*, **20**, 4928-4936, (1981).
8. J.P.Andersen, U.Gerdes, and J.V.Møller, In Structure and Function of Membrane Proteins, (eds. E.Quagliariello & F.Palmieri), Elsevier/North-Holland, Amsterdam, 57-61, 1983.
9. A.M.Rubtsove, *Biokhimiya* (USSR), **46**(6), 1045-1054, 1982.
10. U.K.Liaemmli, *Nature* (London), **227**, 680-685, 1970.
11. W.Birmachu and D.D.Thomas, *Biochemistry*, **29**, 3904-3914, 1990.
12. O.H.Lowry, N.J.Rosebrough, A.L.Farr, and R.J.Randall, *J. Biol. Chem.*, **193**, 256-275, 1951.
13. T.M.Eads, D.D.Thomas, and R.H.Austin, *J. Mol. Biol.*, **179**, 58- 81, 1984
14. L. Yang, D.McStay, A.Sharma, P.J.Quinn, & A.J.Rogers, *Proc. SPIE*, **1640**, 513-519, 1992.
22. U. Pick, and S.J.D.Karlish, *Biochim. Biophys. Acta*, **626**, 255-261, 1980.
15. J.M.Vanderkooi, S.Papp, S.Pikula, and A.Martonosi, *Biochim. Biophys. Acta*, **957**, 230-236, 1988.

Biological Surface Probing Using Nonlinear Optical techniques

D. McStay[1], L. Yang[1], P. J. Quinn[2],

1 School of Applied Sciences, The Robert Gordon University, Aberdeen, U.K.
2 Biochemistry Section, Division of Life Sciences, King's College London, UK

ABSTRACT

The successful application of a surface second harmonic generation (SSHG) technique for biological surface sensing and analysis is reported. The SSHG technique has been used to study surface samples with biological molecular coatings, i.e. the protein bovine serum albumin (BSA), the antibody (rat Ig G anti-BSA antibody), and the dye Rhodamine 6G coated onto glass substrates. The surface second harmonic signals induced by a incident pulsed Nd:YAG laser beam (λ = 1064 nm) for each has been investigated. The primary experimental results have demonstrated the wide-applicability of the SSHG technology for biological surfaces / interfaces sensing and analysis.

1. INTRODUCTION

Biological surfaces play important roles in biological, biomedical, environmental, and chemical systems.[1] It is not therefore surprising that biological surface sensing and analysis have attracted increasing attention with many different techniques being developed for application to biologically important surfaces. In the last two decades, the development of optical spectroscopic techniques has led to significant progress in biological surface studies.[2] However, the complexity of the surface phenomena of biological systems combined with the limitations of existing surface analysis techniques ensures that there is a demand for the development of new and improved techniques.

Recent developments in laser and optoelectronics technology has made the use of nonlinear optical techniques much more practical. This has led to the possibility of develop new surface analysis techniques and instruments based on nonlinear optical effects. In particular, surface-second-harmonic generation (SSHG) has received increasing attention in the last a few years, and emerged as a very promising technique for surface and interface sensing and analysis.[3] The surface second harmonic generation technique has a number of features which make it an ideal tool for surface analysis. These include: surface specificity, high sensitivity; ultra-fast response; large anti-Stoke's spectrum shift; orientation sensitivity, simplicity of measurement, etc.

The nature of SSHG means that the technique may potentially be applied to study all kinds of surfaces or interfaces. However, in practice, the technique has been widely reported for studies of semiconductor and metal surfaces, as well as some surfaces and interfaces of certain electro-chemical system.[4-5] Our objective in this work is to investigate the feasibility of using the SSHG technique as a novel biological surface sensing / analysis scheme. In this paper, protein molecules (Bovine Serum Albumin) and dye molecules (Rhodamine 6G) have been used as a model system and been deposited onto glass substrates. The surface second harmonic generation characteristics of these surfaces was then studied, with a view to developing the technique for sensing applications.

2. SURFACE SECOND HARMONIC GENERATION

Optical second harmonic generation is a second order nonlinear optical process in which two incident photons of frequency ω are converted into a single emission photon of frequency 2ω. This nonlinear conversion process, in the electric dipole approximation, requires a noncentrosymmetric media. Since the centrosymmetry of a medium will necessarily be broken at its surface or interface with other medium of different refractive index, the second harmonic signal will be naturally generated at such surfaces (interfaces). The surface second harmonic generation (SSHG) process can only occur in the first a few layers of molecules (atoms) on either side of interface, due to the fact that only the first few layers of molecules on either side of the interface participate in the breaking of the centro- / inversion symmetry of the medium.[5] Therefore, the surface second harmonic (SSH) signal is highly surface selective. Previous research has shown that a submonolayer sensitivity is possible for analysis of metal surface using a SSHG technique,[5] illustrating the potential high sensitivity of the technique. A physical picture of SSHG at a interface between two centrosymmetric mediums of different refractive index, is shown in figure 1.

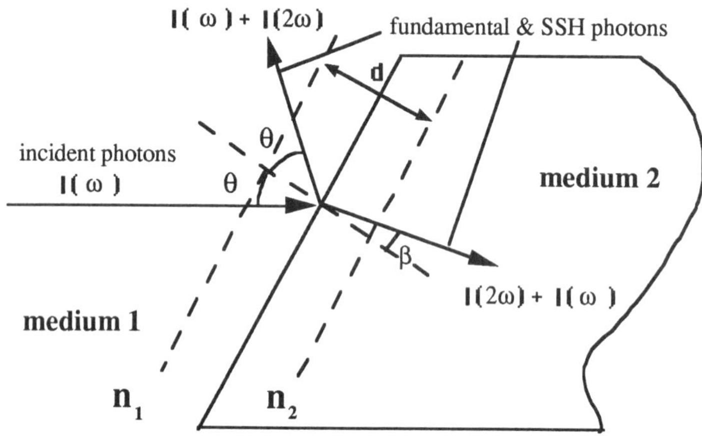

Figure 1 Surface second harmonic generation mechanism

where n_1 and n_2 are the refractive index of the medium 1 and medium 2 respectively, d is the thickness of the surface layer of the interface within which the medium inversion symmetry is broken, and I(ω), I(2ω) are the intensities of the fundamental and SSH beams respectively.

As shown in Fig 1, a SSH polarisation ($P^{(2)}(2\omega)$) will be induced in both the reflected and refracted directions when a intense beam of light is used. The induced polarisation is proportional to square of the incident light intensity ($E^2(\omega)$), as:

$$P^{(2)}(2\omega) = \chi^{(2)} \cdot E^2(\omega) \tag{1}$$

where $\chi^{(2)}$ is the surface second-order nonlinear susceptibility, which is a third-rank tensor, and its independent nonvanishing elements depend on the particular structure and symmetry of the molecules (atoms) in the surface layer(s). The surface second harmonic intensity (I(2ω)) can therefore be derived and can be expressed as follows:[5]

$$I(2\omega) = \frac{32\pi^3\omega^2\sec^2\theta_{2\omega}}{C^3 n_1(\omega) n_1^{1/2}(2\omega)} \; |P(2\omega) \cdot \chi^{(2)} : P(\omega)P(\omega)|^2 \, I^2(\omega) \tag{2}$$

where $\theta_{2\omega}$ is the angle of SSH radiation with respect to surface normal; I(ω) the incident light intensity; P(2ω) the SSH generated; $\chi^{(2)}$, the surface nonlinear susceptibility as defined above; and C the light speed.

3 EXPERIMENT

3.1 Sample preparation

Materials: Rhodamine 6G dye was obtained from the Hopkin & Williams Ltd; Bovine Serum Albumin (BSA) protein (98.99%, fatty acid-free, ELIZA grade) was purchased from the Sigma Chemicals Company; Microscopic cover slides (size: 22x22 mm, glass) manufactured by Chance Proper Ltd; ethanol and other chemicals are the highest grade available from Aldwich or Sigma Chem. Company.

Methodology: To prepare sample coated glass surfaces of the molecules, a Langmuir-Blodgett film deposition method was used. Before the deposition, the glass microscopic cover slides were first cleaned and dried. After the deposition of the Langmuir-Blodgett layer, the coated glass substrates were dried and stored at 4 °C until required.

3.2 SSH Measurement

The surface second harmonic emission characteristics of the sample molecule surfaces prepared above were studied using a surface second harmonic generation system developed in this laboratory and shown schematically in figure 2. As shown in the figure, a pulsed, Nd:YAG laser (Spectron, SL281) was used as the pump. The fundamental laser output (λ = 1064 nm, FWHM = 5 ns) was guided onto the sample surface after passing through a cut-on Schott-glass optical filter (cut-on wavelength: 660 nm).

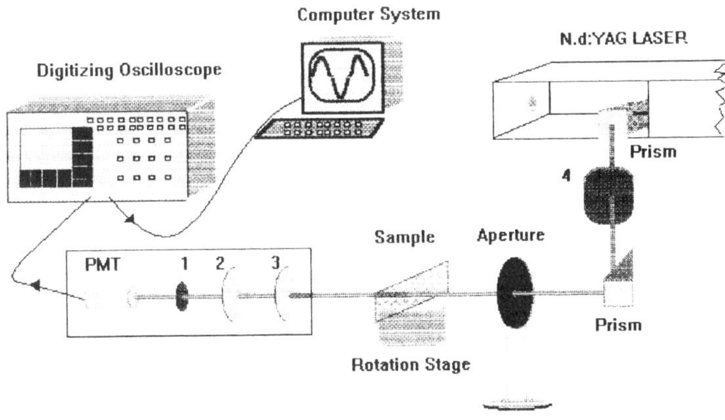

Figure 2 Surface second harmonic generation system. 1, band pass filters (λ=532 nm FWHM=10 nm); 2&3, colour glass band pass filters (λ=532 nm); 4, Schott glass cut-on filters (cut-on λ=660).

To reduce the unwanted second harmonic signals that may arise from the surfaces of the optical components in the experiment, a pair of Colored glass band pass filters (Spinder & Hoyer, BG38-3) were used immediately before the sample surfaces. These were carefully positioned to keep their surface normal parallel with the incident beam direction. After these filters a narrow band-pass interference filter (Ealing Electro Optics, λ = 532 nm, FWHM = 10 nm) was used to select specifically the second harmonic signal generated at the biological sample surface. The surface second harmonic signal radiating along the transmitted beam direction was then detected using a photomultiplier tube, the output of which has connected to a digital storage oscilloscope (Tektronix, TDS620). To reduce the influence of laser output fluctuation and increase signal to noise ratio, each measurement was taken as an average reading over a hundred shots.

4. RESULTS AND DISCUSSION

Typical surface second harmonic results from the sample surfaces are presented in figures 3-5, where figure 3 shows the power dependence of the surface second harmonic signal of a Rhodamine 6G sample surface as a function of pumping laser power. Figures 4 and 5 show the surface concentration dependence of such generated surface second harmonic signals from the dye and protein (BSA) surface samples.

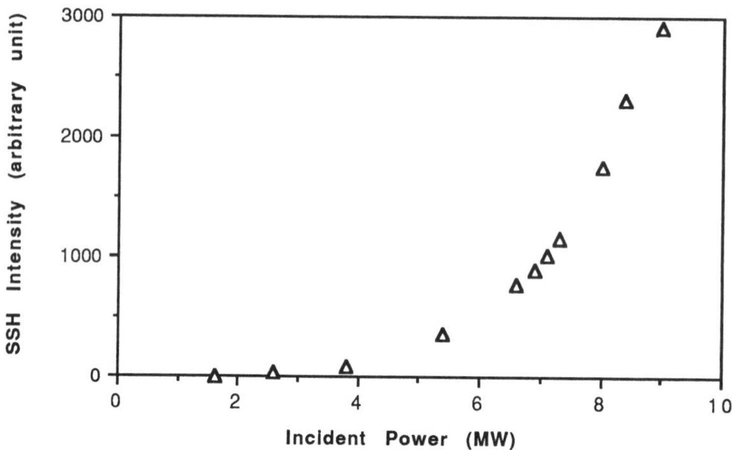

Figure 3 Power dependence of surface second harmonic of Rhodamine 6G sample surface.

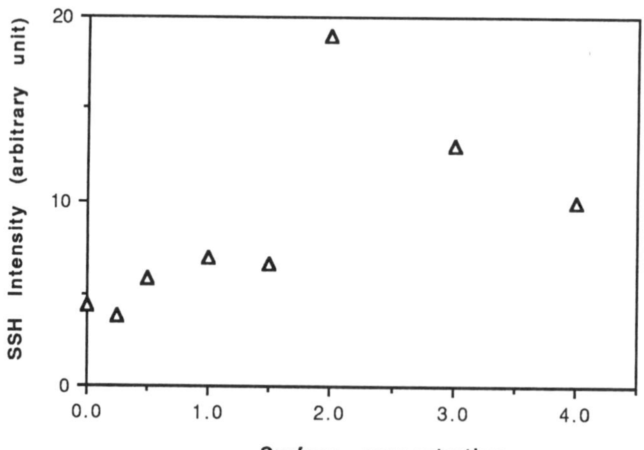

Figure 4 Surface concentration dependence of the Surface second harmonic signal from a Rhodamine 6G dye surface sample.

From figure 3 it can be seen that the surface second harmonic signal generated from sample surfaces vary nonlinearly with the incident laser power. It has been found that the second harmonic signal is proportional to the square of the incident laser power, as would be expected from theory, illustrated in equation 2.

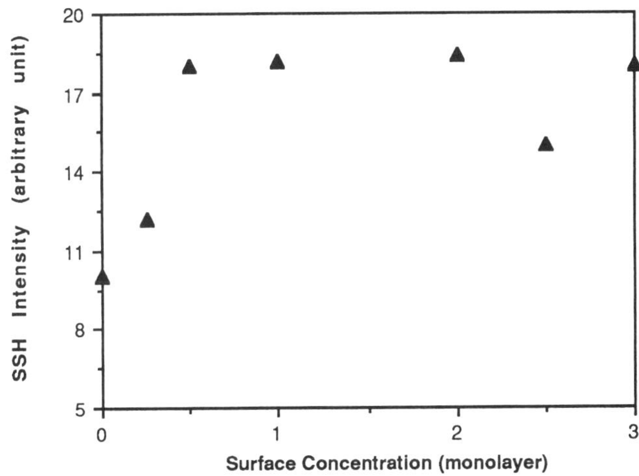

Figure 5 Surface concentration dependence of Surface second harmonic generated from a protein (BSA) on glass.

Figures 4 and 5 record the variation in SSH signal with the surface molecule concentration (surface coverage). As can be seen from the figures the SSH signals generated from both sample surfaces exhibit an obvious dependence on the concentration of the surface molecules. The signal is observed to increase with increasing surface concentration, and reaches its peak values at or around a surface coverage of one - two monolayers of sample molecules. This observation shows that the second harmonic generation technique is very sensitive when the sample concentration is submonolayer, however, when the surface concentration is greater than a single monolayer, the sensitivity is reduced. For a surface coverage of greater than a monolayer the SSH signal is observed to vary nonlinearly with surface coverage, indeed, for some surface coverage the SSH signal is smaller than that observed for sub-monolayer coverage. This is attributed to the complex nature of the interface at such surface coverage, i.e. the interface is no longer glass/dye/air, but dye/dye/air. The $\chi^{(2)}$ value for such an interface may be greater or smaller than that for the submonolayer case, leading to a change in the SSH signal.

Therefore, the technique is especially suitable for ultra low concentration chemical / biological molecules sensing.

5. CONCLUSION

The primary experimental results demonstrate that the SSHG technique can be used to probe and analyse biological surfaces or interfaces. The technique has been shown to have a submonolayer sensitivity. The down shift in wavelength of the SSH signal make it particularly attractive for biological applications, where the intrinsic fluorescence of the biological materials is often a serious limitation in high sensitivity applications.

6. REFERENCES

1. J.D.Andrade (ed.), *Surface and Interfacial Aspects of Biomedical Polymers.* Plenum

Press, New York, 1985.
2. S.S.Perry and C.A.Somoriai, *Anal. Chem.*, **66**(7), 403A, 1994.
3. Y.L.Shen, *Nature*, **337**, 519, 1989.
4. R.M.Corn and D.A.Higgins, *Chem. Rev.*, **94**, 107, 1994.
5. G.L.Rechmond, J.M.Robinson and V.L.Shannon, *Progr. Surf. Sci.*, **28**(1),1, 1988.

A video based system for Food texture analysis

D.McStay and L. Yang
School of Applied Sciences
The Robert Gordon University
St Andrews St, Aberdeen, U.K.

Abstract
A novel system for measuring food texture based on video analysis is reported. The technique involves attaching reflective or fluorescent tags to a human subject and subsequently videoing the subject chewing food samples. The technique has been shown to be effective in distinguishing foods with different textures. The system is simpler, easier to use and less distrubing for the subject than many existing systems. The system also lends itself to use with large numbers of subjects.

Introduction

The food industry is one of the major world industries. It is also highly competitive with food manufacturers investing millions of pounds per year in developing new and improved products. In general there three basic qualities of a food which determine a consumers opinion of it, these are flavour, appearance and texture. The visual perception is known to strongly influence a consumers opinion of a food product. The relative importance of flavour and texture is more difficult to determine but is nontheless important to the overall success of the product. The texture of a food product is therefore important to food manufacturers. The definition of texture with respect to food is, however, difficult but can perhaps best be said to be the perception of a food's physical characteristics by sensations in the mouth, particularly by resistance to mastication [2].

A number of techniques and instruments have been developed to measure food texture. The simplest is a sensory panel comprising volunteers representative of the target population. These panels provide useful information but are prone to a number of volunteer related problems. An alternative approach has been to develop mechanical devices to provide a quantative assessment for the texture. A range of devices based on deforming a food sample and measuring the reaction of the sample to the applied force have been developed [3,4]. These devices are often only applicable to specific types of food [4] and do not address the problem of trying to relate the physical properties of the food to the human response. To overcome this problem a number of methods which incorporate the human factor have been developed. An example of such an approach is the magnetic method of jaw tracking developed by Micheler et al [5]. In this method the subject has a small magnet attached to a front tooth. The motion of the jaw while chewing a food sample is tracked using a series of Hall effect sensors mounted in a headset. This technique provides useful data but is cumbersome, time consuming to calibrate and it is likely that the mood of the subject will be influenced by wearing the headset. Additionally the technique is not suited to the routine testing of a large number of subjects. There is therefore a demand for a food texture measurement which incorporates the human factor and is easy to implement so that a representative number of subjects can be evaluated with the subjects feeling the minimum of discomfort.

In this paper we report initial results from a video based technique for food texture analysis which overcomes many of the problems discussed above.

Video Analysis for food texture analysis

In the new technique volunteers have several small reflective or fluorescent tags attached to their faces. The placing of the tags is not critical, but at least one must be attached to a point on the face, such as the nose and forehead, which is known not to move as part of the chewing process. The subject is then illuminated using a standard light source which incorporates an optical filter, as shown in figure 1. A standard video camera with an optical filter, with the same spectral properties as that on the illumination source for reflective tags and at the fluorescence emission wavelength for fluorescent tags, (mounted in front of the lens) is used to video the subject while they chew a number of replica samples of a food product. The use of the filters helps to reduce the background making subsequent analysis of the images simpler and thus reduces the processing time. After completion of the chewing test the video image is digitised (typically at five frames per second) using a frame grabber (Videologic Captivator) and a 486 computer and stored as a bit map file. The data is then converted into an ASCII file

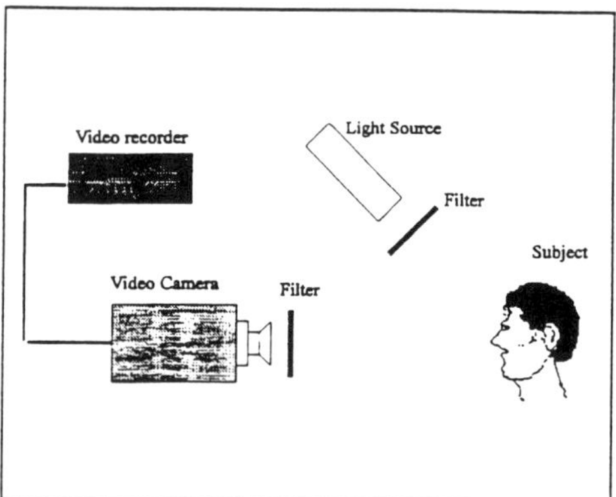

Figure 1: Schematic of the basic video analysis system for food texture analysis.

using specially written software and subsequently imported into a spreadsheet package (Excel) for analysis. The tags are separated form the background on a coordinate plot by imposing a threshold level routine. This rejects all pixels which have values below a preselected number corresponding to different shades of grey on a 0-256 scale (0 being black and 255 white). At this point another specially written software routine is employed to determine the distances between the reference tag and those used to monitor the jaw displacement. In this way it is possible to track the jaw displacement throughout a chewing event. This data can then be used to produce a plot of the jaw displacement with time.

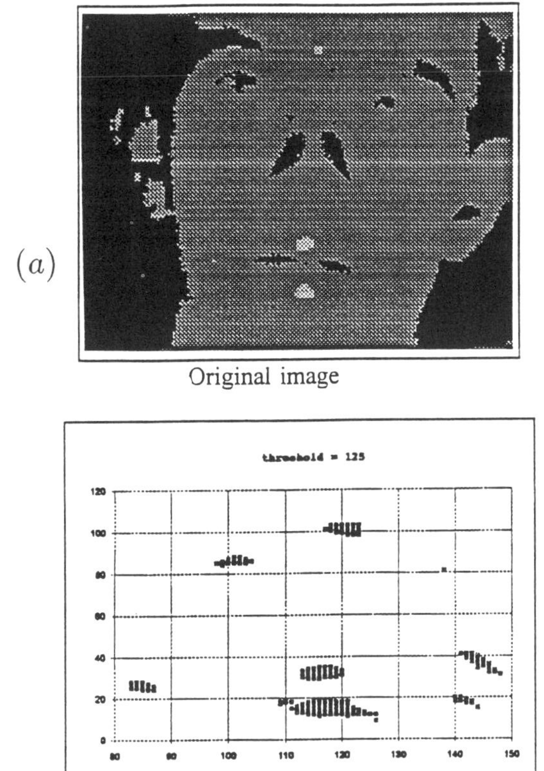

(a) Original image

(b) Threshold too low

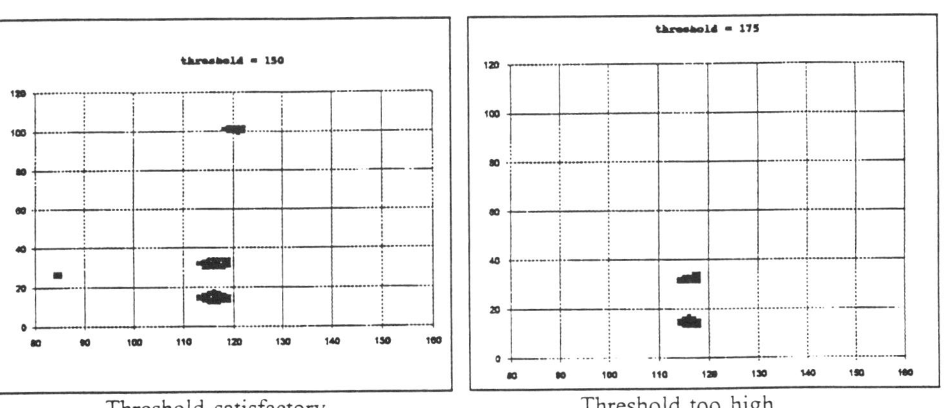

Threshold satisfactory

Threshold too high

Figure 2: a) the original digitised image b) the resultant coordinate plot with varying threshold values.

(a)

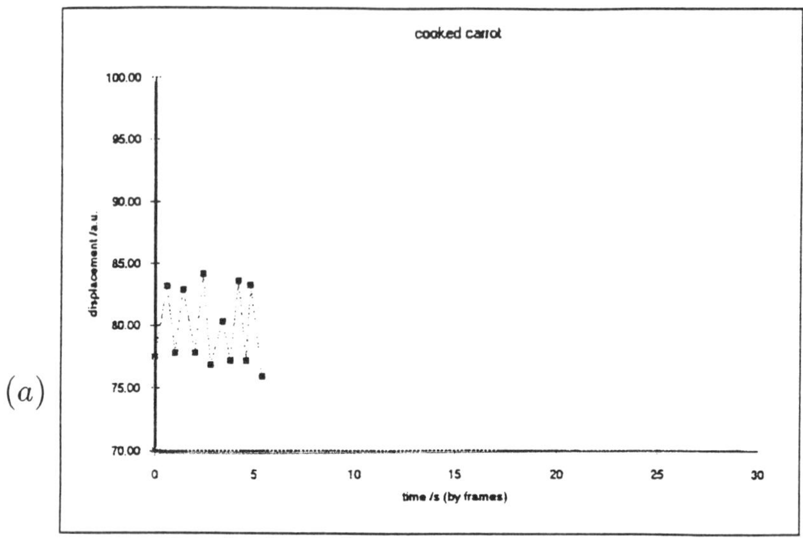

Sample of cooked carrot

(b)

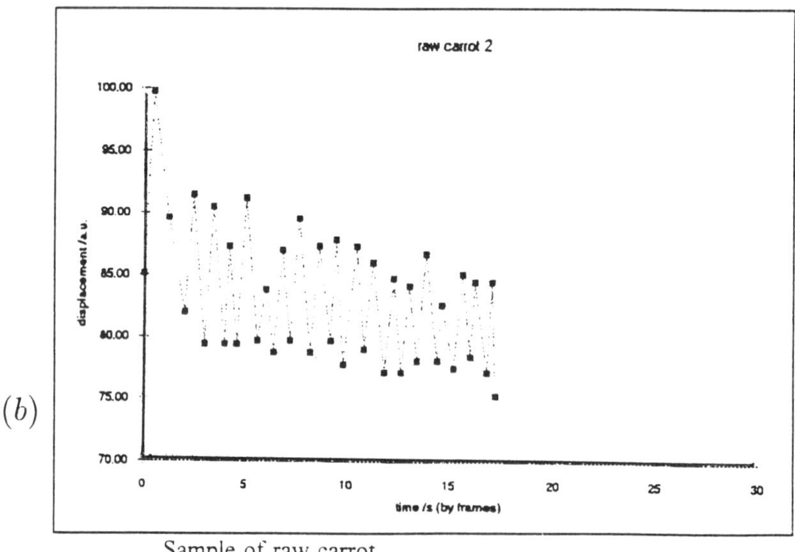

Sample of raw carrot

Figure 3: The maximum and minimum jaw displacements for a subject chewing identical samples of a) cooked carrot and b) raw carrot

Results

Figure 2(a) shows a typical digitised image of a subject with several reflective tags obtained using the video analysis system. Figure 2(b) shows the effect of imposing varying threshold levels. As can be seen it is posssible by optimising both the optical filters and the threshold level to produce an image showing only the tags, on a coordinate plot. Figure 3 shows a typical result obtained for a test in which a subject was given samples of cooked and raw carrot which were all of the same size and all at room temperature. In this figure only the points of maximum and minimum displacement in each chewing cycle have been plotted. It is apparent, even from this limited use of the available data that the food types produce very different jaw displacement patterns. Both the amplitude of the jaw displacements and the duration of the the chewing event are greater for the raw carrot. This is as expected as it has been previously shown that the chewing duration increases with the hardness of the food [6,7].

Conclusions

The use of a simple video technique based on attaching reflective or fluorescent tags to a human subject and subsequently videoing the subject chewing food sample has been shown to be effective in distinguishing foods with different textures. The system includes the human factor i.e. a human subject and provides a quantative result. The system is simpler, easier to use and less distrubing for the subject than many existing systems. The system also lends itself to routine use with large numbers of subjects.

References

1) Matz S.A. (1962) "Food Texture" AVI Publishing
2)Stone H. and Sidel J.L. (1985) "Sensory evaluation practices" Academic Press
3)Voisey P.W. (1976) "Instrumental measurement of food texture" in Rheology and texture in food quality, Ed deMan J.M. et al, AVI Publishing, Westport, Connecticut, U.S.A.
4)Vincent J.F. et al (1991)"The wedge fracture test: a new method for food texture" J. Texture Studies., Vol. 22,45-47
5) Michler L. et al (1987)" Graphic assessment of natural mandibular movements" J. Craniomandibal Disorders: Facial and Oral Pain., Vol. 1, 97-114
6) Chew C.L. et al (1988) "The effect of food texture on the replication of jaw movements in mastication", J. Dentistry., Vol.16,5,210-214
7) Horio T. and Kawamura Y.(1989)"Effects of texture of food on chewing patterns in the human subject" J. Oral rehabilitation., Vol.16, 2, 177-183.

Section G

Sensor Modelling and Interfacing

Applications of a Universal Sensor Interface Chip in Smart Sensor Systems

P D Wilson, S P Hopkins, R S Spraggs and K Robinson

ERA Technology Ltd. , Cleeve Road, Leatherhead, Surrey, KT22 7SA, UK

Abstract. ERA Technology, in conjunction with European partners in the EUREKA project JAMIE, has developed a Universal Sensor Interface Chip (USIC). The USIC represents a highly flexible sensor interfacing block designed to support a wide range of instrumentation and sensor applications. Its combined analogue and digital features enables it to be configured to create many smart sensor systems often using the minimum of external components. The measurement of voltages, resistances, currents, capacitances and frequencies with a resolution of up to 20 bits is combined with local intelligence for configuration, calibration, signal conditioning and interfacing .

1. Introduction

Processing of output signals from all types of sensors is becoming increasingly critical as new sensor technologies are adopted and demands on signal processing, digital communications and local intelligence expand. These requirements make it imperative that the sensor and its associated electronics are viewed as a system and increasingly this system needs to be intelligent or smart. Commonly a smart sensor system is considered to consist of the following elements, but simpler smart sensor systems can be fabricated.[1]

- Sensor elements (primary sensor and appropriate compensating sensors)
- Excitation control
- Analogue signal processing (amplification, filtering etc.)
- Analogue to digital conversion
- Digital information processing (configuration, calibration, signal conditioning etc.)
- Digital communications

To address the need for these smart sensor systems several approaches are available to the sensor manufacturer or system integrator. For high volume applications, specially designed integrated circuits have been developed to meet this demand. The dedicated IC or ASIC approach, however often gives rise to high development costs and requires large volume production to make the manufacture economically viable. Few sensor applications can justify these costs or have the necessary volume demand, this is particularly so when digital as well as analogue functionality is required. As the demand for more complex and accurate mixed signal interfaces grow the number of organisations that can justify this state of the art technology is reducing.

Another approach is to use several proprietary ICs and to make up dedicated circuits for each element of the system. This does not result in the same high setup cost, but normally gives rise to larger size, more expensive production and less robust products.

An alternative solution to this demand is to develop a device that has a high degree of in built flexibility. This IC will not be dedicated to one application or sensor type, but can address the needs of many. The Universal Sensor Interface Chip (USIC) being developed by ERA and its partners in the EUREKA project JAMIE is such a device.

The USIC has been designed as a flexible device and is intended to address the interfacing needs of the many industrial companies who cannot justify the development and production costs of dedicated mixed mode ASICs. The USIC includes all of the processing elements needed to produce many smart sensor systems and gives small and medium sized end users access to this growing market.

2. Background

The USIC is being developed as part of the EUREKA Project JAMIE (Joint Analogue Microsystems Initiative of Europe - EU 579). The objective of the overall JAMIE project is to enable fast, simple and error-free design and manufacture of mixed signal ASICs: The USIC demonstrates and exploits the mixed signal technology developed under JAMIE. ERA has teamed up with the following organisations in the JAMIE consortium.

Company	Country	Project Role
Mikron	Austria	Project Leader & Coordinator of Design
Austria Mikro Systems	Austria	ASIC Fabrication
Joanneum Research	Austria	RISC Processor Design
Dolphin Integration	France	Low Cost CAD Development
Instituto Superior Technico	Portugal	Special Analogue Cell Design

3. User Requirements.

To ensure that the USIC specification meets the needs of industry, the sensor's community have been invited to complete questionnaires on their requirements. Questionnaires from over three hundred companies were returned to ERA from European sensor users and manufacturers. This data was a major factor in determining the USIC specification presented below.

Details of the USIC market requirements have been presented else where[2] therefore only a brief summary is presented here. Eighty-six percent of the companies that responded currently do not use ASICs, the reasons given for this are;

High setup costs	72%
Low adaptability / flexibility	36%
High unit cost	28%
Not considered	17%
Do not know how	12%
Do not need	10%

Measurement accuracy from 8 to 20 bit is required for voltage, current, resistance, capacitance and frequency based sensor inputs. Seventy percent of the applications require at least one compensating or reference input. A wide range of update rates were specified, but more than 80% of those specified were at 1 kHz or lower. Strong interest was shown in signal conditioning with an average of 70% of the responding organisations having interest in these functions; offset, span, linearisation, auto calibration and self test. Both analogue and digital outputs are required from the USIC. Twenty-five types of digital output were specified, but both voltage and current loop outputs are in stronger demand than any one digital output. Size and power consumption are also important factors.

4. USIC-General Description

The USIC represents a complete signal processing capability for data acquisition systems designed to support a wide range of sensor applications. It offers high performance with flexibility and requires only a few external components for many applications. Local intelligence is provided by the integrated RISC processor allowing many smart sensor systems to be created without high development costs being incurred.

The main function blocks of the USIC are shown in figure 1 below.

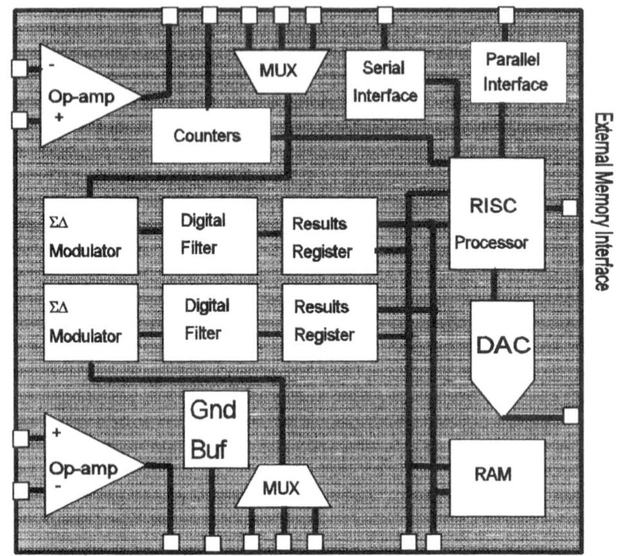

Figure 1 USIC - Block Diagram

The ΣΔ (sigma-delta) technique is used to perform the analogue to digital conversion. Each converter consists of two sections, a modulator producing single bit, highly over sampled output and a FIR digital filter. Twin ΣΔ converters have been included in the design as some applications can achieve better accuracy by employing two ΣΔ converter channels, one for the sensor and one for a reference. The reference channel shadows the effect of, for example, temperature changes on the sensor channel, but is not subject to changes in the measured quantity.

An analogue multiplexer precedes each converter, giving up to three input sources for each converter. This is useful for calibration purposes or for selecting between different sources. However the ΣΔ conversion method is not suited to rapid switching between sources due to the group delay (typically three sample periods) associated with the modulator and the filter.

Two configurable 24 bit counters allow frequency measurement, pulse counting and timing to be performed. The two chopper stabilised, operational amplifiers have many uses including active analogue filters, current sources, integrators or signal summing.

Operating from a single 5V supply, the device includes a 2.5V ground buffer which provides a reference voltage that allows bipolar output from the sensor. Digital output from the device is via the RS485 serial and/or the parallel interface, both are under processor control.

Analogue output may also be generated by the on-chip 16 bit D/A converter. Where analogue output is not required, the D/A converter may be put to other uses. For example, in cases where the analogue input has a significant DC offset, the D/A converter could be used to reduce the effects of the offset. By forming a summing amplifier, the D/A converter output can be subtracted from the analogue input thus increasing the dynamic range achievable in the A/D conversion process.

4.1 RISC Processor and Memory Management

The main features of the RISC processor and the implemented memory management are given below.

- 8 bit high performance CMOS RISC CPU
- 36 single word instructions
- all single cycle instructions except for program branches(two cycles)
- operation speed DC to 23 MHz
- 24 bit wide internal data paths
- 8000 x 8 addressable external program ROM and optional external RAM
- 256 x 8 addressable registers (including 128 byte integrated data RAM)
- 8 levels x 12 bit deep hardware stack
- direct, indirect and relative addressing modes
- hardware interrupt controller
- power saving sleep mode (10 μAmp)

Integration of the RISC processor with the A/D converters is a key element of the design enabling local processing and storage of the sensor data. Sensor data may require linearisation or a block of data may need to be stored before being passed to a central computer. Configuration of the digital filters is controlled by the processor as is the operation of the serial and parallel interface.

The die size of the USIC is approximately 30mm^2 and has 80 pins. It can thus be packaged within an 80-pin Quad Flat Pack(QFP) or used as a bare die in Multi-Chip Module(MCM) solutions. Prototype USICs were realised in the spring of 1995 and fabrication runs of the fully integrated device will yield devices in July 1995. Higher volume fabrication will start by the end of 1995.

5. Smart Sensor Application Examples

5.1. Piezoresistive Bridge

This example, shown in figure 2 is a micromachined pressure sensor based on the piezoresistive bridge principle with an onboard diode for temperature sensing. Op-amp A is configured as a differential amplifier with a single pole low pass filter for anti-aliasing, this provides the multiplexer with an input signal of 1V FSD. Op-amp B generates a C-hebyshev(0.5dB) 2-pole filter that provides a degree of filtering and the gain required so that the input to the multiplexer is typically 4mV/°C.

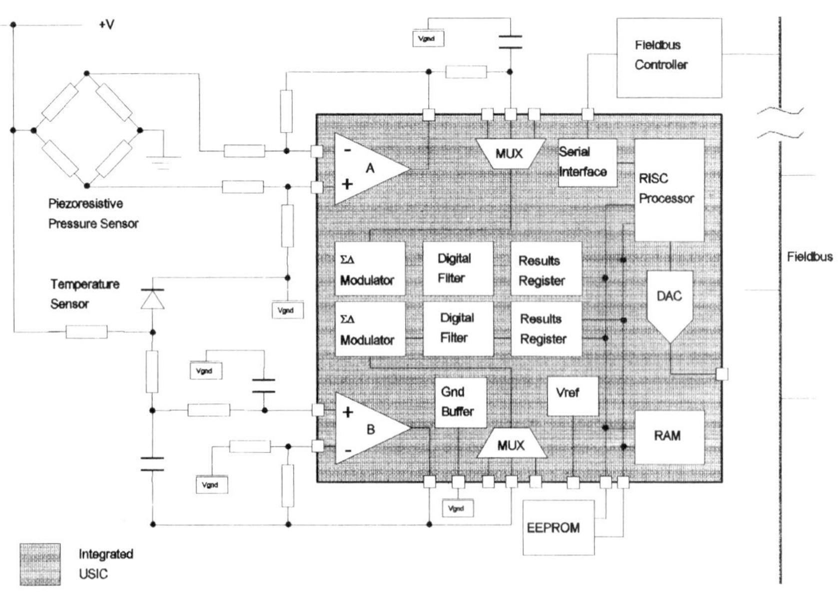

Figure 2 USIC Based - Piezoresistive Bridge Smart Sensor System

The accuracy of the A/D conversion will be limited by CMRR of Op-amp A, configured as a differential amplifier with 0.1% resistors this will be 55dB. For higher accuracy an external differential amplifier could be added.

Calibration, linearisation and offset correction will all be handled in the digital domain by the RISC processor. Off chip EEPROM can be used as a look-up table or to store coefficients for polynomial approximations.

Illustrated is a commercial Fieldbus controller connected to the USIC via its RS485 port. This gives the potential for a Fieldbus based control system, exploiting smart sensor technology to the full.

5.2. Capacitance Measurement

With the growth in microengineered capacitive devices, particularly the emergence of pressure and accelerometers, a capability for the measurement of capacitance has been incorporated into the USIC. The USIC can measure capacitances up to 900 pF at a resolution of 0.01 pf.

Progressive synchronous demodulation has been chosen for use with the USIC, as shown in figure 3 below.

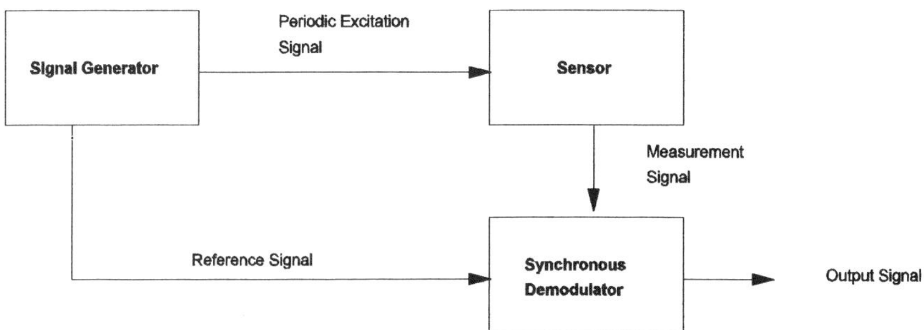

Figure 3 Principal of Capacitance Measurement

The sensor is excited by a periodic voltage signal (100 kHz) and the current flowing through the sensor is detected. The synchronous demodulation is based on the high correlation between the source signal and the sensed signal. Modulation is performed by the multiplication of the source signal with the sensed signal. Low pass filtering removes the high order products.

The methodology is realised with the USIC by connecting an external MDAC (DAC 7801) via the parallel port. One channel is used to generate the 100 kHz source signal with the digital input signal to the MDAC being generated by the USIC. This digital signal is also used for multiplication with the measured signal into the second MDAC channel. For this purpose the resulting current is converted to a voltage signal by one of the available operational amplifiers. The second op-amp is used for the 100 kHz band pass filtering necessary to get a smooth signal. One $\Sigma\Delta$ converter is used to sense the output voltage of the circuit and so gives a representation of the capacitance value of the sensor. A second mode is available to get the real part of the impedance.

This capacitance measurement technique is being used to create smart microsystems in conjunction with ERA's activities in microengineering.

Acknowledgements

ERA's work within the JAMIE project is being supported by the Department of Trade and Industry as part of the EUREKA project EU 579. The work presented in this paper is a result of collaboration between the various partners in the JAMIE project, particular thanks is given to the staff of Mikron who have made a major contribution to the successful realisation of the USIC.

References

1. J Brignell, N White. Intelligent Sensor Systems (Bristol: Institute of Physics Publishing)

2. P D Wilson, J S Brown S P Hopkins. Universal Sensor Interface Chip (USIC): Market Requirements and Specification, IEE Colloquium, Moving and Flexing Microstructures - their Design, Modelling and Production, 15 March 1994

A graphical user interface for intelligent sensor ASIC reconfiguration

A H Taner, J E Brignell

Department of Electronics and Computer Science, University of Southampton, Highfield, Southampton SO17 1BJ, UK.
Tel: _44-703-593580 Fax: _44-703-592901

Abstract.
Reconfigurability of the analogue section of an intelligent sensor under digital control has been shown to be a powerful addition to the amour of intelligent sensing techniques. It introduces, however, a new level of design complexity with a new possibility of introducing errors. The graphical user interface described here simplifies the design process and greatly reduces the risk of error.

1. Introduction

This paper is concerned with the mitigation of problems of complexity that might arise in exploiting the power of ASIC technology for the realisation of intelligent sensors. The relentless progress of this technology in terms of performance and cost suggests that it will become dominant in the future of sensor systems. Digital processing enables many of the defects of primary sensors to be overcome, so that the wide range of potential sensing mechanisms can be revisited for exploitation, resulting in sensors that are accurately targeted for specific sensing tasks. Serial binary communication capacitates robust, versatile and extensive instrumentation systems[1].

There are, however, clear hazards in such an approach. In general more complexity means less reliability. It also poses problems for the designer in appreciating the implications the inter-relationships of all parts of the sensor sub-system, and there is great potential for disastrous errors. The potential implications to fixed costs contributed by the design process are also significant.

We have proposed an approach that enables the greater complexity to be accompanied by an *increase* in reliability, by the introduction of the concept of reconfigurability, which in turn enables processes of self test and self calibration[2]. Costs can be mitigated by adopting a systematic strategy designed to prevent "re-inventing the wheel" and using library elements[1].

A major problem within the design process arises from the need to marshal the information and hence maximise the freedom and control available to the designer. Errors are costly, and their effects can emerge anywhere down the line to final application in the field – the later they reveal themselves the greater the costs. One way of dealing with this problem is to embed a test chip in a standard PC card. Once this is done the full panoply of software aids becomes available. A carefully designed graphical

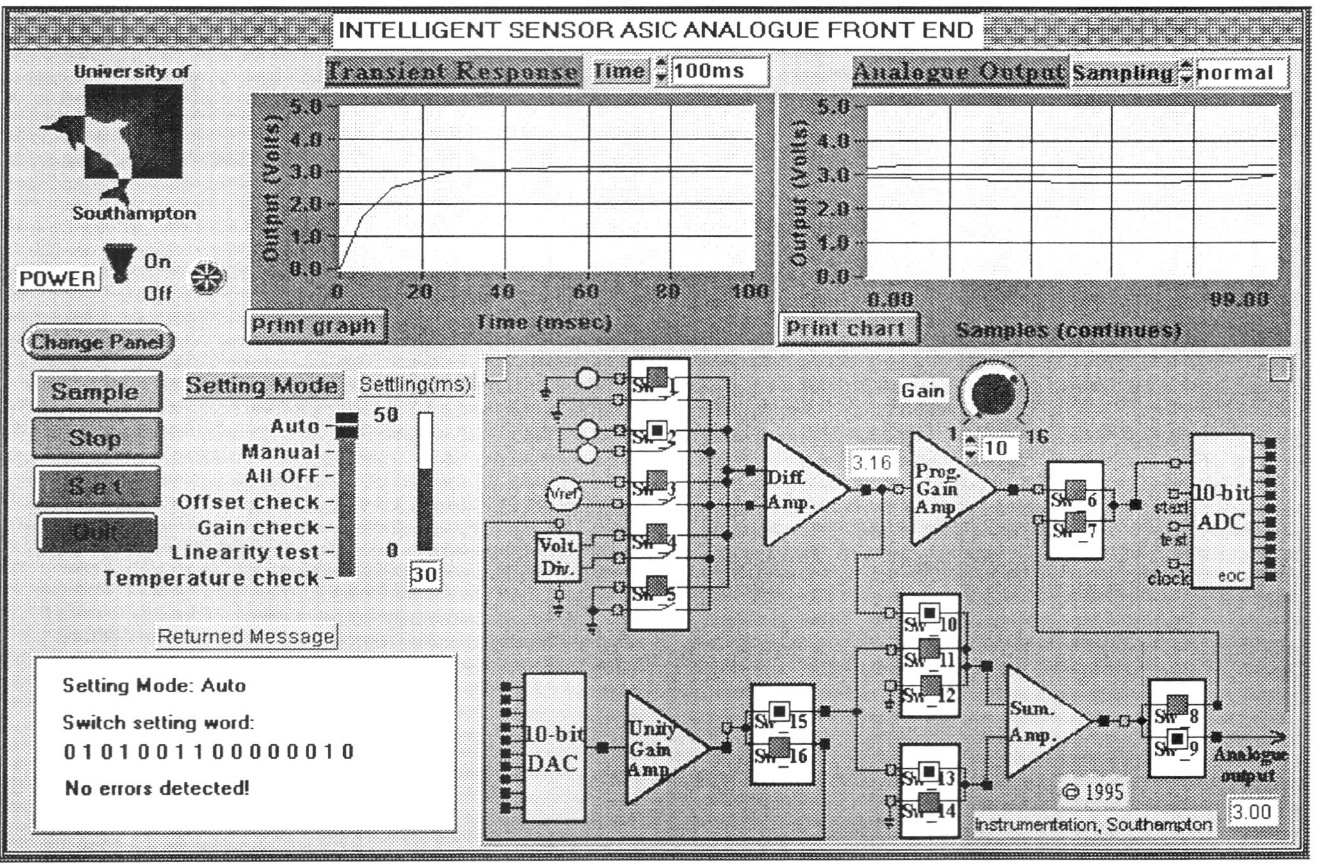

Figure 1. *Graphical user interface for the intelligent sensor analogue front end*

custom user interface then simplifies access to all the information relating to the control of the sensor sub-system, giving the designer sufficient control to avoid the pitfalls, as we hope to establish below. Figure 1 shows the display for such an interface.

It will be seen that, as the control program is stepped through, emphasis is given to maximising the information presented to the designer. In particular, in the bottom left hand corner is the output of the software filters that trap any illegal or destructive configurations. Various standard arrangements may be selected or custom ones implemented. The current configuration is presented in circuit diagram form.

2. Self-test and auto-calibration

We have established that, in order to break the link between complexity and unreliability, self-test and self-calibration are the paramount features of intelligent sensors[2]. A reconfigurable intelligent sensor analogue ASIC front end has been produced which demonstrates the basic principles of self test and self-calibration routines. Its schematic is an essential component of the graphical user interface as shown in Figure 1. Five different inputs are multiplexed to the differential amplifier. The input multiplexer (or changeover switch) and all the other changeover switches are under digital control, as is the gain of the programmable-gain amplifier. In the field the ADC and DAC are connected to the data bus of the microprocessor, but as mentioned above, for the test and development purposes, the microprocessor is replaced by a PC based data acquisition system as described in the next section.

In brief, the operation of the analogue ASIC front end is as follows: The input is sequentially connected to ground, a reference voltage, an internally generated ramp signal and a secondary sensor. The secondary sensor is sensitive to the variable to which main sensor is cross-sensitive, almost invariably temperature. By means of these checks, the necessary coefficients for correction of offset, linearity, drift and cross sensitivity are obtained. Consequently the microprocessor uses these values to carry out the calibration and compensation routines[1,2].

3. Testing problems

It is clear that the sort of system described here cannot be run and tested by conventional means. Therefore a version of the front end has been implemented with maximal pin-out so that it can be embedded in the PC board to provide maximal testability. As mentioned above, some of the thousands of possible configurations are illegal, or even destructive. Other configurations, or combinations of them, may be nonsensical for one reason or another, without being trapped by the software filters.
Examples of illegal combinations are in the implementation of a changeover switch by an array of ON-OFF switches – clearly if more than one switch is set at ON an error has occurred. There are also illegal loop connections – a direct connection between the output and the input of an operational amplifier, for example. These combinations can be detected and eliminated by means of simple logical operations with a table of prescribed masks [2], though it must be remarked that ensuring that such a table is complete is a non-trivial task.

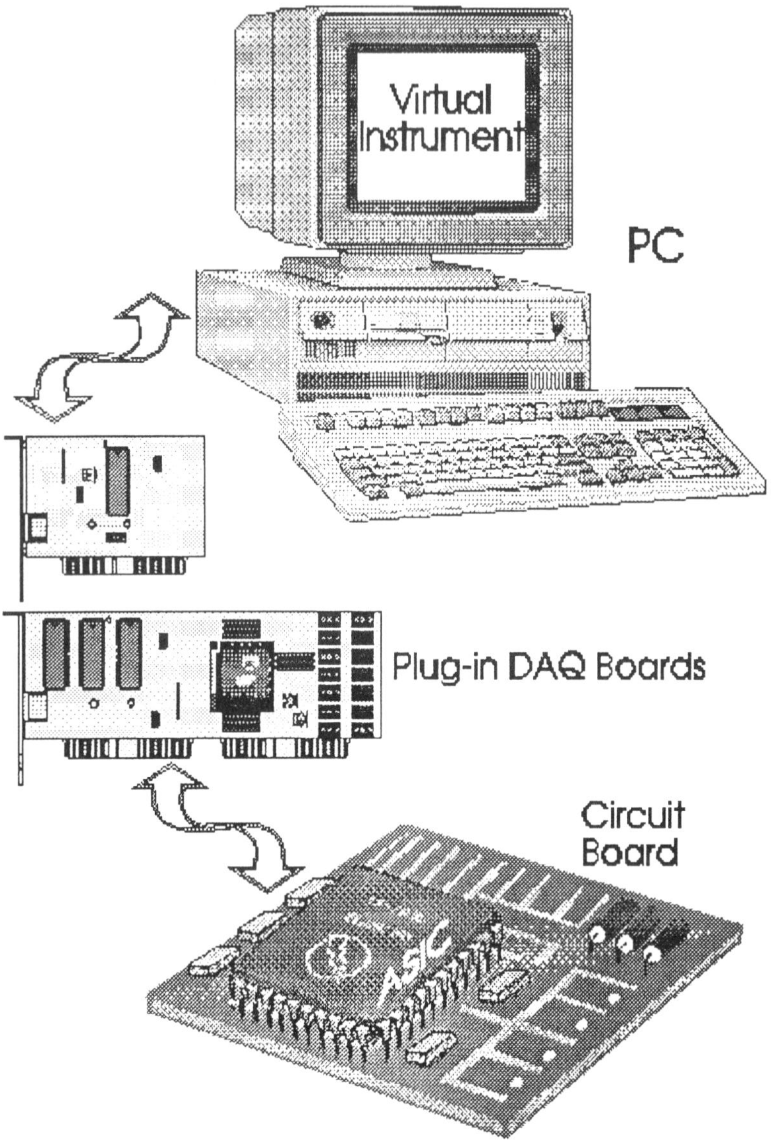

Figure 2. *Hardware set-up of the virtual instrument*

4. Realisation

The hardware structure of the test and development system is shown in Figure 2. The ASIC with maximal pin-out has been embedded in a circuit board. This board is connected to a fast PC via plug-in data acquisition cards. Several software packages are used to control the hardware and implement the user interface. They amount to what has become known as a virtual instrument. In this case the virtual instrument is dedicated to the evaluation of intelligent sensor analogue front end. The configuration allows software development to be carried out on the device itself rather than by simulation, and important feature, since all simulations are to some extent imperfect, especially when they involve connection to the real world.

In addition to some other software such as drivers of the plug-in DAQ card, four main software packages are used during development of the software drivers and the graphical interface : LabWindows. Borland C++ , Corel Draw and Corel Photo-Paint.

The user interface is created by using Labwindows User Interface Libraries. LabWindows is a software development system for BASIC and C programmers. It contains an interactive environment for developing programs and libraries of functions for creating data acquisition and instrument control applications. LabWindows contains a set of software tools for data acquisition, data analysis, and data presentation.

Borland C++ IDE is chosen as the development environment for the complete system. Within Borland's software development kit, a full set of ANSI C is present and by using Borland's object linking facility within the project file, LabWindows libraries can also be utilised.

Corel Draw and Corel Photo-Paint are extensively used to edit, colour, resize and to change the format of images in the graphical user interface. The user interface editor of Labwindows is capable of displaying only PCX images.

5. Discussion and Conclusion

As described above a virtual instrument with a custom graphical user interface has been developed, in this case dedicated to the production of reliable control programs for a reconfigurable analogue front end to the intelligent sensor ASIC. All the functions of the ASIC such as configuration, gain of the amplifier and input output destinations can be controlled and observed via the interface at the touch of the mouse pointer. It is only when one is faced with the task of programming such a versatile device that the value of this sort of support can be fully appreciated.

6. References

[1] J.E. Brignell and N.M. White, Intelligent Sensor Systems, IOP, Bristol, 1994, ISBN 0-7503-0297-6.

[2] A.H. Taner and J.E. Brignell, Aspects of intelligent sensor reconfiguration, Sensors and Actuators A, 1995, to be published.

[3] A H Taner and J E Brignell, *Software drivers for the reconfigurable ASIC analogue front end*, ASICs for Measurement Systems, IEE Digest No: 1994/050, 4/1-4/4.

COATING STRATEGY ASPECTS FOR QCM-SENSOR APPLICATIONS

Carsten Behling, Ralf Lucklum, Peter Hauptmann

Institut für Prozeßmeßtechnik und Elektronik, Otto-von-Guericke-Universität Magdeburg, PF 4120, D-39016 Magdeburg, Germany, Tel.: ++(391) 5592-4622, FAX: ++(391) 561 6358

Abstract: Quartz crystal micro balance sensors are based on interactions between a sensitive layer on the quartz surface and the analyte. The interaction causes a change in the acoustical behaviour of the quartz. This article describes investigations about the sensing effect of QCM-sensors for liquids by help of a KLM model for a viscoelastic sensitive layer.

1. Introduction

The Quartz Crystal Microbalance (QCM) is a mass sensitive principle which can be used for chemical sensing by coating the quartz surface with a sensitive layer. Many papers have been published about quartz crystal gas sensors with a considerable variety of sensitive materials. Recently new applications for under-liquid sensing have been published, too. However, commercial examples are still rare. Many reasons are responsible for this unsatisfactory situation. First of all design rules from the chemical as well as the physical point of view are not available, yet.
The sensing mechanism of a QCM-sensor can be divided formally into two parts. The first step is a chemically controlled interaction between the sensitive coating and the analyte. This process is responsible for the sensitivity and the selectivity of the sensor. It results in a change of the mechanical parameters of the coating. This shift in the material properties leads in a second step to a change of the acoustical behaviour of the system quartz-coating-analyte and finally in a shift of the resonant frequency of the electrical oscillator which drives the quartz.

2. Chemical Design Rules

More or less all kinds of intermolecular interactions can be employed. Simple van der Waals´ forces (dispersion or dipole forces) are from a general type. The sensor selectivity based on this kind of interaction is controlled by thermodynamical parameters. In the typical concentration range, $c_{i,gas}$, of the analyte i Raoult´s law is valid. The mass sensitivity of the coating, $S_{i,mc}$, is given by the distribution coefficient, $K_{i,c}$, which can be expressed by the saturation vapour pressure, $p_{i,0}$, the activity coefficient, $\gamma_{i,0}$, a volume parameter, $V_{w,c}$, the gas constant, R, and the temperature, T.

$$S_{i,mc} = \frac{m_c}{\rho_c} \frac{RT}{p_{i,0}\gamma_{i,0}V_{w,c}} c_{i,gas} \qquad (1)$$

m_c and ρ_c are the mass and the density of the coating, respectively. In cases of low boiling compounds or gases, i.e. high $p_{i,0}$ values, one need a low $\gamma_{i,0}$, i.e. a strong interaction between the coating and the analyte. On the other hand the product $p_{i,0} \cdot \gamma_{i,0}$ compared with other compounds j of the mixture is a sensor selectivity design criterion.
Hydrogen bonding between analyte and coating is much stronger and more selective. This kind of interaction requires electronegative bonding partners like O, N, F in a suited molecular configuration.

Supra molecular cage structures offer a combination of interaction forces and steric effects (host guest interaction) by preorganized cavities. They exhibit molecular recognition properties. Advanced preparation techniques allow a molecular design and therefore a function with defined purposes.

3. Physical Design Rules

3.1 Theory

In the case of an AT-cut quartz, vibrating in the thickness shear mode, the vibration behaviour of the quartz and the wave propagation in the system quartz-coating-medium can be derived from an one dimensional solution of the equation of wave. To calculate the overall system characteristics from the electrical port to the front acoustical port the chain matrix technique of the transmission line model with two acoustical ports and one electrical port (extended KLM model) is used. For a piezoelectric layer like quartz the equivalent circuit is shown in Fig. 1.

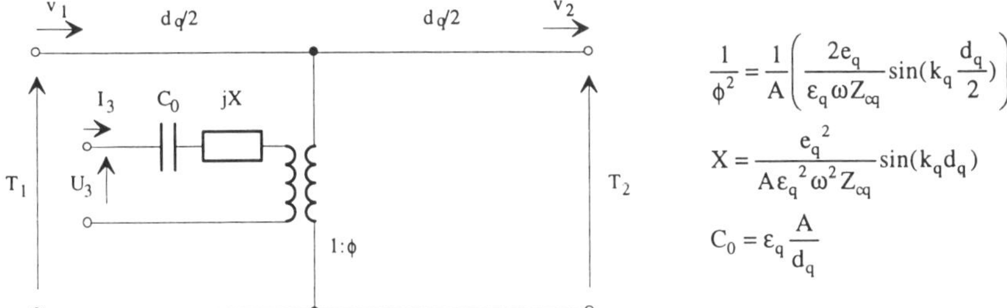

Fig. 1: Three-port KLM model for piezoelectric quartz

The losses during wave propagation in the transmission line are included with a complex wave number k. Non piezoelectric layers are represented by a two-port circuit without transformer and electrical port.

The relation between the properties at both ends of the transmission line (shear stress T and shear velocity v) is as follows:

$$\begin{pmatrix} T_1 \\ v_1 \end{pmatrix} = \begin{pmatrix} \cosh(\gamma_m d_m) & Z_{cm} \sinh(\gamma_m d_m) \\ \frac{1}{Z_{cm}} \sinh(\gamma_m d_m) & \cosh(\gamma_m d_m) \end{pmatrix} \begin{pmatrix} T_2 \\ v_2 \end{pmatrix} \tag{2}$$

Repeated application of this formula for a various number of layers allows an easy treatment for an arrangement of multiple layers by matrix multiplication. The elements of the propagation and the transfer matrices are calculated from materials and geometric parameters. The complex impedances seen at each interface may be calculated but it is not required. Remember, the quartz sensor is in contact with air on one side. This surface is stress-free corresponding to a short-circuit acoustical port. The other surface of the quartz is loaded by the acoustical impedance Z_L which is the resulting impedance from all layers placed on the quartz surface. The following expression for the complex electrical impedance Z can be obtained from the transmission line equivalent circuit:

$$Z = \frac{1}{j\omega C_0}\left\{1 - \frac{K^2}{\alpha} \frac{2\tan\frac{\alpha}{2} - j\frac{Z_L}{Z_{cq}}}{1 - j\frac{Z_L}{Z_{cq}} \cot\alpha}\right\} \tag{3}$$

It is possible to separate the impedance Z into a parallel circuit with a static capacitance C_0 and a motional impedance Z_m of the quartz. With the abbreviation $\zeta = Z_L/Z_{cq}$ follows:

$$Z_m = \frac{1}{j\omega C_0}\left\{\frac{1 - j\cdot\zeta\cdot\cot\alpha}{\frac{K^2}{\alpha}\left(2\tan\frac{\alpha}{2} - j\cdot\zeta\right)} - 1\right\} \tag{4}$$

This motional impedance can be split into two parts. One part stands for the unloaded quartz (ζ=0) and the other one is related to the load:

$$Z_m = \frac{1}{j\omega C_0}\left(\frac{\frac{\alpha}{K^2}}{2\tan\frac{\alpha}{2}} - 1\right) + \frac{1}{\omega C_0}\frac{\alpha}{4K^2}\zeta\frac{1}{1 - \frac{j\cdot\zeta}{2\tan\frac{\alpha}{2}}} = Z_m^0 + Z_m^L \tag{5}$$

Near the resonance of the unloaded quartz approximation leads to a simple expression for the unperturbed part of Z_m:

$$Z_m^0 = R_1 + j\omega L_1 + \frac{1}{j\omega C_1} + \frac{1}{j\omega(-C_0)} = R_1 + j\omega L_1 + \frac{1}{j\omega C_1{}'} \tag{6}$$

For small loads ($\zeta << 2\tan\frac{\alpha}{2}$) the second part of equation (5) simplifies to:

$$Z_m^L = \frac{1}{\omega C_0}\frac{\alpha}{4K^2}\zeta \tag{7}$$

Here the additional impedance is direct proportional to the load ζ. The motional impedance of a quartz near its resonance frequency with a small load only can be described with the following equation:

$$Z_m = R_1 + j\omega L_1 + \frac{1}{j\omega C_1{}'} + \frac{1}{\omega C_0}\frac{\alpha}{4K^2}\zeta \tag{8}$$

This expression can be transformed into a modified motional arm of the known Butterworth-Van Dyke equivalent circuit (BVD) as shown in Fig. 2 and is similar to Martin /1/, nevertheless the elements of the equivalent circuit are defined in another way. However, the differences in the complex impedance near resonance are negligible.

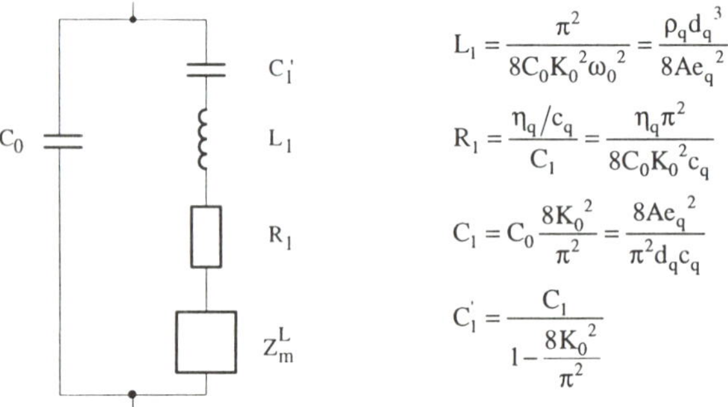

Fig. 2: BVD equivalent circuit of a loaded quartz, the elements R, L, C represent the quartz without load

In the case without any restrictions for the load of the quartz equation (6) can not be applied. The more general equations (3) or (5) are still valid for any loads and can be used to determine the overall complex electrical impedance Z of the quartz depending on the acoustical load Z_L.

3.2 Viscoelastic coating

The acoustical load of a viscoelastic film with finite thickness can be calculated as follows:

$$Z_L = j\sqrt{\rho_f \cdot G_f}\,\tan\left(\omega\sqrt{\frac{\rho_f}{G_f}} \cdot d_f\right) \quad \text{with} \quad G_f = G' + jG'' \tag{9}$$

Substituting this expression into equation (3) leads to the electrical impedance Z of the quartz for a viscoelastic load. A typical oscillator works at a quartz phase angle of zero degree. Thus the investigation of the phase angle of Z provides information about the vibration behaviour of the quartz.
As an example of this analysis the influence of the coating's complex shear modulus is shown in Fig. 3. The phase angle of the complex impedance Z points out a huge frequency shift compared with a typical sensor response with a decrease of the complex shear modulus. Unfortunately the slope of the phase angle decreases dramatically, too. The phase angle even does not reach zero degree near the so called film resonance. A normal oscillator does not work in that case. With some electronical modifications it might work, nevertheless the frequency stability is inadequate.

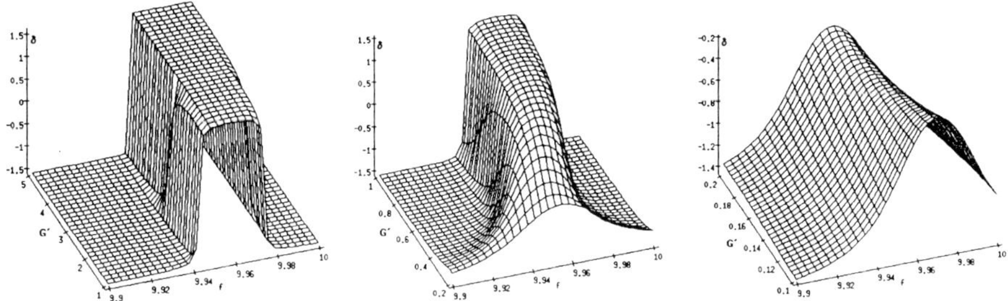

Fig. 3: Impedance phase angle (in rad) of a coated quartz at different shear moduli $d_f = 1\mu m$, $\rho_f = 10^3$ kg/m^3, G´ / 10^7 N/m^2, f / MHz, G´´ = 10^6 N/m^2

3.3 Viscoelastic coating in liquid

Considering a quartz coated with a film (film properties have index f), immersed in a liquid (index l), the load impedance for the quartz can be derived with the matrix technique for two layers:

$$Z_L = Z_{cf}\frac{Z_l + j \cdot Z_{cf}\tan\left(\omega\sqrt{\frac{\rho_f}{G_f}}d_f\right)}{j \cdot Z_l \tan\left(\omega\sqrt{\frac{\rho_f}{G_f}}d_f\right) + Z_{cf}} \tag{10}$$

The acoustical impedance of the liquid alone would be:

$$Z_{liq} = (1+j)\sqrt{\pi \cdot f \cdot \rho_l \eta_l} \tag{11}$$

With these expressions introduced in equation (3) the electrical impedance Z becomes quite complicated.
One example for the phase angle of Z in this case is presented in Fig. 4.

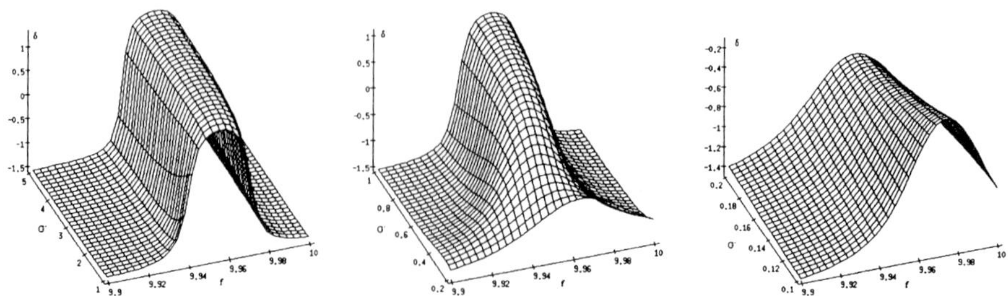

Fig. 4: Impedance phase angle (in rad) of a coated quartz in liquid at different shear moduli $d_f = 1 \mu m$, $\rho_f = 10^3$ kg/m^3, $\rho_l = 10^3$ kg/m^3, $\eta_l = 10^{-3}$ Pa s, G´ / 10^7 N/m^2, f / MHz, G´´ = 10^6 N/m^2

The general behaviour seems to be similar, anyway the values of the phase angle δ are much smaller within G´ =0.5 10^7..5 10^7 N/m^2, the corresponding frequencies are different, too.

3.4 Rules

The optimization of the coating from the physical point of view has to deal with contrary tendencies, the enhancement of the sensor sensitivity and the reduction of the frequency stability with the coating thickness. The limitation is given by the slope of the phase angle of the impedance necessary for the frequency stability of the oscillator. Under Sauerbrey conditions the maximum slope is at about zero degrees near the serial resonance frequency. Under a viscoelastic load it changes to negative angles. It is convenient to add a phase rotating element into the oscillator loop. The optimum phase rotation depends in a complex way on the coating thickness and the complex shear strength of coating and (liquid) load.
A second optimization method is the use of a coupling layer which itself is not sensitive. In that case the sensing mechanism, mass change and/or change in the complex shear modulus, is important. In the simple case where the QCM-sensor works as a viscosimeter the impedance match between quartz and liquid is responsible for sensor sensitivity and quartz damping.

References

1 V.E. Granstaff, S.J. Martin, J. Appl. Phys. 75, 3 (1994), 1319-29

Appendix

quartz constants:

$\rho_q = 2.651 \cdot 10^3$ kg m^{-3}	density
$\varepsilon_q = 3.982 \cdot 10^{-11}$ A^{2}s^{4}kg^{-1}m^{-3}	permittivity
$e_q = 9.53 \cdot 10^{-2}$ A s m^{-2}	piezoelectric constant
$\eta_q = 3.5 \cdot 10^{-4}$ kg m^{-1}s^{-1}	viscosity
$c_q = 2.947 \cdot 10^{10}$ N m^{-2}	piezoelectric stiffened elastic constant
$d_q \approx 1.667 \cdot 10^{-4}$ m	thickness (for 10-MHz-quartz)
$A \approx 1.257 \cdot 10^{-5}$ m^2	electrode area (for 4 mm diameter)

$K_0^2 = \frac{e_q^2}{\varepsilon_q \cdot c_q}$	electromechanical coupling factor for lossless quartz
$K^2 = \frac{e_q^2}{\varepsilon_q \cdot (c_q + j\omega\eta_q)}$	electromechanical coupling factor for lossy quartz
$C_0 = \varepsilon_q \frac{A}{d_q}$	static quartz capacitance
$\omega = 2\pi \cdot f$	angular frequency
$\alpha = \omega \frac{d_q}{v_q}$	wave phase shift in quartz

relations for any medium (Index m):

$v_m = \sqrt{\frac{c_m}{\rho_m}} \qquad k_m = \frac{\omega}{v_m}$	wave velocity and wave number
$Z_{cm} = \rho_m \cdot v_m = \sqrt{\rho_m \cdot c_m}$	specific acoustic impedance
$\gamma_m = j\frac{\omega}{v_m} = j \cdot \omega \sqrt{\frac{\rho_m}{c_m}}$	complex wave propagation constant
$c_m = c + j \cdot \omega \cdot \eta$	complex elastic constant
d_m	thickness of layer
ρ_m	density

A self-balancing capacitance and loss conductance measuring circuit for industrial transducer applications

W.Q.Yang and H.Y.Lam

Department of Electrical Engineering and Electronics, UMIST, P O Box 88, Manchester M60 1QD, UK

Abstract. A capacitance and loss conductance measuring circuit has been developed for industrial transducer applications. By digitally balancing two orthogonal components, the capacitance and loss conductance can be measured simultaneously. Experimental results show that the transducer has good linearity and resolution.

1. Introduction

Although capacitance transducers have been widely used in laboratories and industry for many years, new measurement methods, particularly for small capacitances, are still being reported, including charge/discharge [1], AC impedance measurement based on Op-amps [2, 3], self-balancing capacitance to DC voltage conversion [4], lock-in detection [5] and switched-capacitors [6].

The AC bridge is still recognised as the most accurate and stable method of measuring capacitance [7]. However conventional laboratory AC bridges cannot be used as industrial transducers because they are either unable to give continuous measurement, or are too complex, bulky and expensive. The most common AC bridge used in industrial transducers is the unbalanced transformer-ratio-arm bridge which gives a signal based on the difference between the unknown capacitance and the reference capacitance [2, 7].

For many industrial applications, e.g. measuring mixtures of oil and sea water, it is necessary to measure both capacitance and loss conductance [8]. This paper describes a self-balancing AC measuring circuit which can measure both capacitance and loss conductance at the same time.

2. Principle

In the conventional AC bridge, balance is obtained by adjusting the balance impedance. Alternatively balance can be obtained by varying the amplitudes of the voltages applied to fixed balance impedances. The schematic diagram of such a self-balancing circuit is shown in Fig.1.

A single-chip signal generator provides sine-wave and square-wave signals of 10 KHz. The sine-wave voltage is applied to the unknown capacitance C_x and the parallel loss conductance G_x. The output of the Op-amp in Fig.1 has an in-phase component and a quad-phase component related to the capacitance and the loss conductance respectively. The amplified AC signal passes through two buffers to two phase-sensitive demodulators (PSD) constructed by CMOS switches and Op-amps. To demodulate the two orthogonal components of the AC signal, four reference signals at 90° phase intervals are required. The 0° and 180° reference signals are used for obtaining the information related to C_x and the 90° and 270° for G_x. The demodulated signals comprise DC components and 20 KHz AC components plus harmonics. Two low-pass filters reject the AC signals and the two DC signals representing C_x and G_x pass through a multiplexer to a single ADC in which the analogue voltages are converted into digital signals and input to a micro-controller. Using these signals the micro-controller produces two digital feedback signals which are modulated by the 10 KHz sine wave from the signal generator by means of two multiplying DACs. This produces two analogue AC feedback signals with amplitudes proportional to the digital signals. The feedback voltages are fed back to the Op-amp through a balance capacitor C_b and a balance resistor R_b respectively. In the ideal situation, the feedback signals will balance the signals from C_x and G_x, i.e. the bridge reaches the null point. Therefore it is a self-balancing circuit.

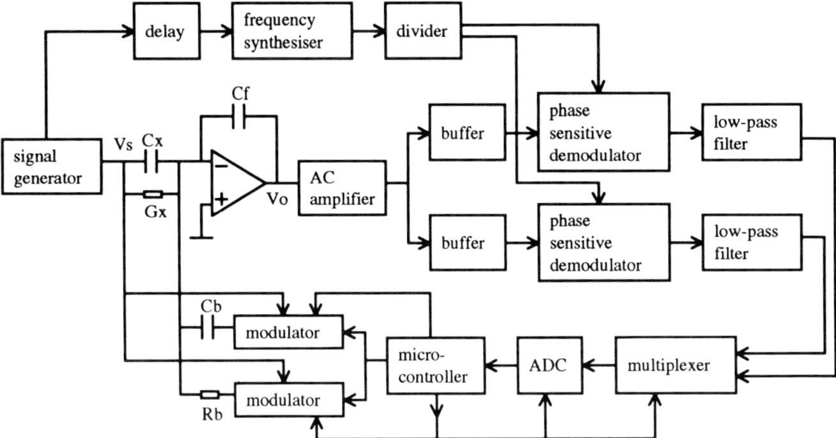

Fig.1 Schematic diagram of self-balanced circuit

To adjust the phase of the reference signals needed by the PSD, the square-wave signal from the signal generator is delayed by a monostable device. The delayed square-wave of 10 KHz is boosted to 40 KHz by a frequency synthesiser constructed by a phase-lock-loop and then divided by four, giving four reference signals at 90° phase intervals.

The output of the first stage, i.e. the Op-amp in Fig.1 is

$$V_o = -\left[\frac{C_x - C_b\dfrac{D_c}{2^n}}{C_f} - j\frac{\dfrac{1}{R_x} - \dfrac{1}{R_b}\dfrac{D_r}{2^n}}{\omega C_f}\right]V_s \tag{1}$$

where, $R_x = \dfrac{1}{G_x}$, V_s and ω are the amplitude and angular frequency of the excitation signal, D_c and D_r are the digital feedback values for capacitance and conductance respectively and n is the number of bits in the multiplying DACs.

When the circuit is balanced, the values of C_x and R_x are related to the digital feedback signals thus

$$C_x = C_b\frac{D_c}{2^n} \tag{2}$$

$$R_x = R_b\frac{2^n}{D_r} \tag{3}$$

The self-balancing circuit utilising feedback measurement has several advantages over conventional circuits using open loop measurement:

(a) The measuring accuracy depends on the feedback loop only. So changes of forward loop gain have no effect on the final measurement.
(b) The feedback measurement provides better linearity which is dependent only on the DACs.
(c) Using a balance capacitor with zero temperature coefficient and a balance resistor with low temperature coefficient, the measurement will have very low drift due to temperature change.

(d) The measurement is not affected by variations of the excitation voltage.
(e) the measurement range is easily set by selecting appropriate values of the balance capacitor and the balance resistor.

3. Results

The following parameters were chosen for experiments: C_b=10 pF, R_b=100 KΩ and n=12. Therefore, the capacitance measurement range is 0 -- 10 pF and the conductance measurement range is 100 KΩ -- 409.6 MΩ (100 KΩ × 2^{12}).

3.1 Successive approximation

The feedback procedure is similar to that used for a 12-bit successive approximation ADC. "1" is set from MSB to LSB successively for each of the multiplying DACs. If the resultant forward signal from one of the low-pass filters remains positive the "1" is kept. Otherwise this bit is reset to "0". After processing the full 12 bits the circuit is balanced to within 1 bit for each of the capacitance and conductance channels. Fig.2 shows a typical balance process.

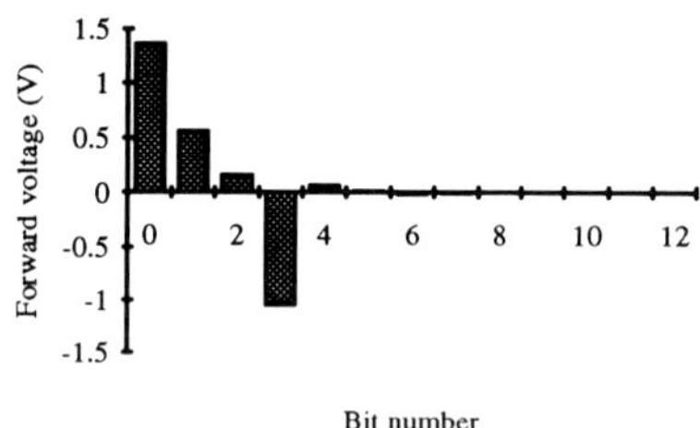

Fig.2 Successive approximation process

3.2 Testing linearity and resolution

Since it is difficult to obtain suitably small capacitors, the circuit linearity and resolution were tested by using a perspex tube with two opposing electrodes. The tube was half filled with oil and a syringe was used to inject small increments of oil. The measured capacitance changes against oil increments and the best straight line fitting the data (using the least square method) are shown in Fig.3. The deviation of the measured data from the best straight line is shown in Fig.4.

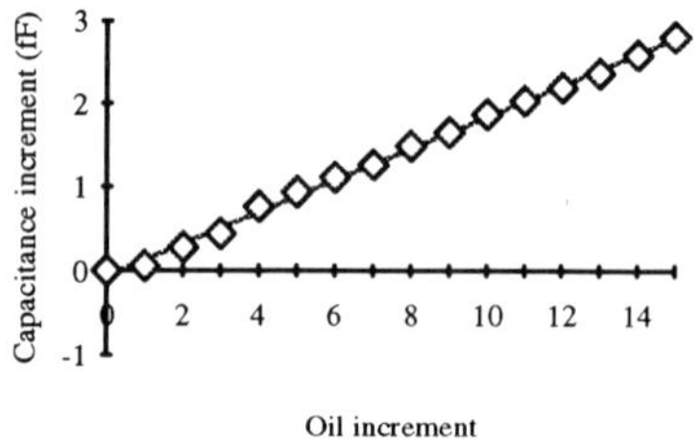

Fig.3 Measured capacitance values and best straight line

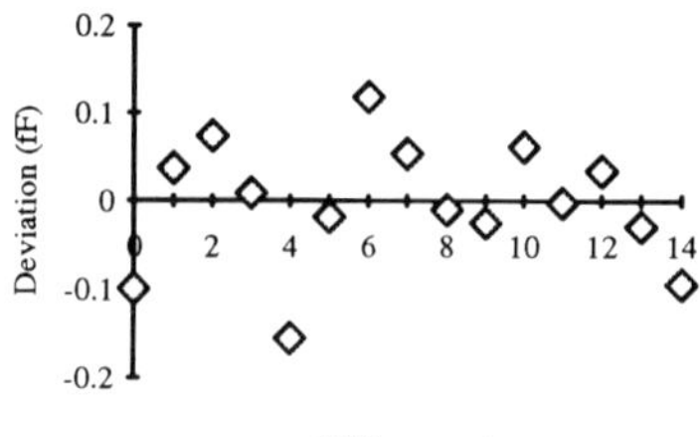

Fig.4 Deviation between measured capacitance values and best straight line

The correlation coefficient between the measured values and the best straight line is 0.998874, showing good linearity. The mean square deviation between the measured values and the best straight line is 0.043 fF. This includes not only the circuit error but experiment method errors, e.g. oil increment volume error, electrode shape error, electrical field edge effect, etc. So the circuit resolution must have been better than 0.043 fF.

3.3 Measuring capacitance and resistance individually

The circuit was tested with capacitance and resistance input individually. The capacitances and resistances were calibrated by an impedance/gain-phase analyser (Solartron-Schlumberger SI 1260). The results are shown in Fig.5 together with the ideal straight line outputs. The mean square errors for capacitance and resistance measurements are 0.72% and 5% respectively.

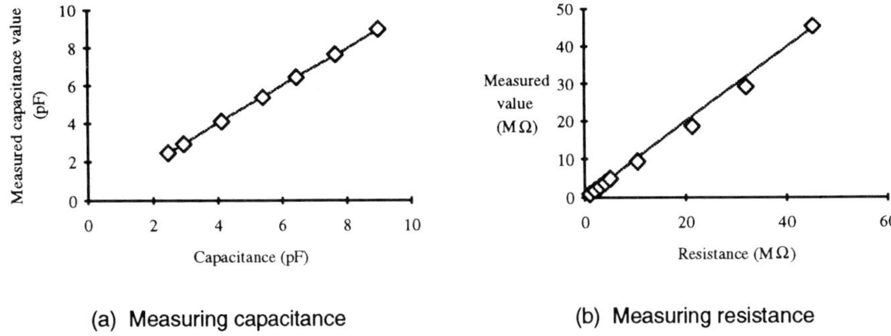

(a) Measuring capacitance (b) Measuring resistance

Fig.5 Measuring capacitance and resistance individually

3.4 Measuring capacitance and resistance simultaneously

Fig.6 shows the results of measuring capacitance and resistance simultaneously over a capacitance range of 2.45 to 8.98 pF) with the resistance constant at 1.01 MΩ.

The results show that the circuit can measure both capacitance and resistance simultaneously and the cross-talk between the two channels was relatively small. The mean square errors for capacitance and resistance measurements are 1.44% and 0.09% respectively.

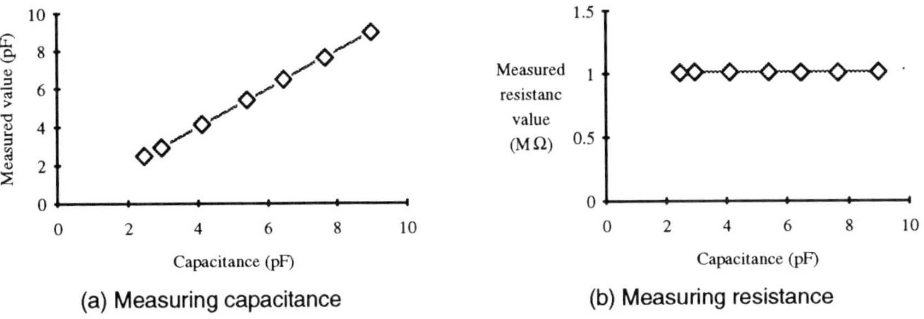

(a) Measuring capacitance (b) Measuring resistance

Fig.6 Measuring capacitance and resistance simultaneously

4. Conclusion

In principle, the self-balancing circuit is similar to the conventional AC bridge. By controlling the digital feedback signals, the circuit can be balanced and the capacitance and loss conductance can be measured simultaneously. The feedback measurement has been shown experimentally to have good linearity and resolution. This circuit is suitable for industrial transducers, especially when both capacitance and loss conductance need to be measured.

References

[1] S.M.Huang, A high frequency stray-immune capacitance transducer based on the charge transfer principle, *IEEE Transactions on Instrumentation and Measurement,* **IM-37** **(3)**, 1988, pp 368-373

[2] A.L.Stott, Correlation and capacitance techniques for solids mass flow measurement, Ph.D. thesis, UMIST, 1991

[3] W.Q.Yang, A.L.Stott, M.S.Beck, High frequency and high resolution capacitance measuring circuit for process tomography, *IEE Proc.-Circuits, Devices Syst.,* **141** **(3)**, 1994

[4] N.Hargiwara, M.Yanase and T.Saegusa, A self-balance-type capacitance-to-DC-voltage converter for measuring small capacitance, *IEEE Transactions on Instrumentation and Measurement,* **IM-36** **(2)**, 1987, pp 385-389

[5] D.Marioli, E.Sardini and A.Taroni, High-accuracy measurement techniques for capacitance transducers, *Measurement Science and Technology*, **4**, 1993, pp 337-343

[6] A.Cichocki and R.Unbehauen, Switched-capacitor transducers with digital or duty-cycle output based on pulse-width modulation technique, *Int. J. Electronics,* **71** **(2)**, 1991, pp 265-278

[7] S.M.Huang, A.L.Stott, R.G.Green and M.S.Beck, Electronic transducers for industrial measurement of low value capacitances, *J. Phys. E: Sci. Instrum.,* **21**, 1988, pp 242-250

[8] G.F.Hewitt, Developments in multiphase metering, *IBC Conference on Development in Production Separation Systems*, London, 4-5 March 1993

Modelling and Computation of Electrostatic Fields in Capacitive Angular Displacement Sensors used in Limited Angle Actuators

S H Khan F Abdullah

Measurement and Instrumentation Centre, Department of Electrical, Electronic, and Information Engineering, City University, Northampton Square, London EC1V 0HB, UK

Abstract - The modelling and computation of electrostatic fields in a robust capacitive angular displacement sensor is carried out using the finite element method (FEM). The sensor, investigated in the paper is used in limited angle actuators, such as the Law's relay type torque motor for the precise sensing of the position of its rotor within a limited angle of actuation of ±5°. Various aspects of finite element (FE) modelling of 2D electrostatic fields in the complex geometry of the sensor is discussed. Results are presented in terms of field plots and the performance curve giving the sensor's output voltage versus angular position of the motor rotor.

1. INTRODUCTION

In recent years cylindrical capacitive sensors are being used in limited angle actuators for position sensing, especially in those applications where robustness, reliability and precise position sensing are of paramount importance. Compared to optical encoders capacitive sensors are simple in construction, robust and highly reliable. The sensor relies on the well-known capacitive technique in which the capacitance/voltage in a system of suitably arranged stationary capacitive electrodes changes owing to the redistribution of electrostatic field caused by some other electrodes moving in between the stationary ones. Fig. 1a shows the cross section of such a 4-electrode capacitive angular displacement sensor used for rotor position sensing in a high performance torque motor. It consists of a compact arrangement of four outer electrodes, one centre electrode and two rotating electrodes. The four segmented outer electrodes, made of stainless steel and insulated from one another by thin insulation strips are mounted symmetrically on the inner surface of the cylindrical electrode body. They are insulated from the electrode body by the outer insulator ring made of epoxy resin bonded glass fabric laminate. The cylindrical centre electrode, also made of stainless steel is positioned concentrically with the outer electrodes by the centre electrode carrier. It is insulated from the centre electrode by the inner insulator ring made of epoxy adhesive. The stainless steel rotating electrodes, attached coaxially to the rotor of the motor rotates with the rotor in the airgap between the inner and outer electrodes. Potentials of the same magnitude but opposite polarities are given to the outer electrodes while the rotating electrodes are kept at zero potential. The electrode body and the

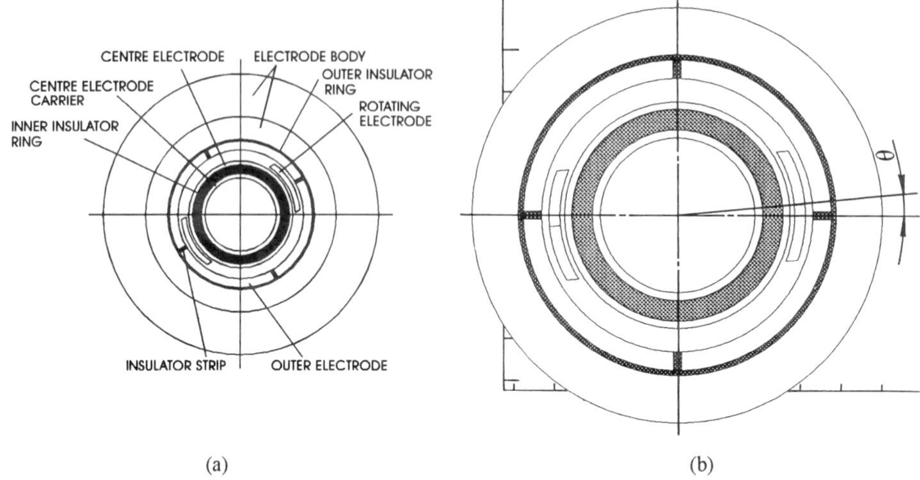

Fig. 1. Cross section of a 4-electrode capacitive angular displacement sensor (not in scale) (a) showing various constructive features, (b) with the rotating electrodes rotated through an angl θ.

centre electrode carrier are earthed and act as a shield against stray fields. The distribution of the resulting electrostatic field and so the overlap between the centre and outer electrodes change as the rotating electrodes rotate with the motor rotor. This results in the change of capacitance between the electrodes and so the output voltage V measured between the centre electrode and the virtual earth. The variation of the output voltage with the position of the rotating electrode θ (Fig. 1b), $V=f(\theta)$ is considered as one one of the most important performance characteristics of the sensor. The linearity of the $V=f(\theta)$ curve determines the effective and reliable performance of the motor to which the sensor is attached. From this, it becomes obvious the importance of accurate field modelling and computation for the CAD, performance evaluation, characterization and prediction of the capacitive angular displacement sensor described above.

2. FINITE ELEMENT MODELLING OF ELECTROSTATIC FIELDS IN CAPACITIVE ANGULAR DISPLACEMENT SENSORS

For modelling and computation of the electrostatic field produced owing to the potential difference between electrodes in the displacement sensor shown in Fig. 1 the following simplifying assumptions are made: (a) The fringing field effects due to the finite axial length of electrodes are negligible; this assumes that the electric field distribution is 2D and remains the same for any plane perpendicular to the lengths of electrodes. (b) The permittivities of dielectric materials (e.g. outer insulator ring) are fixed and do not depend on the field. (c) The dielectric medium in the sensor is piece-wise homogeneous and isotropic. Under these assumptions the electrostatic field in the 2D region Ω of the above sensor is given by the following Laplace's equation:

$$\nabla \cdot [\varepsilon(x, y)\nabla\Phi(x, y)]=0 \qquad \text{in } \Omega \qquad (x, y) \in \Omega \qquad (1)$$

For given geometric and material parameters of the sensor the above equation is numerically solved by FEM [1, 2] in terms of the unknown potential distribution $\Phi(x, y)$ in the 2D region Ω. For this commercial FE software package OPERA-2d [3] running under Unix on a standalone

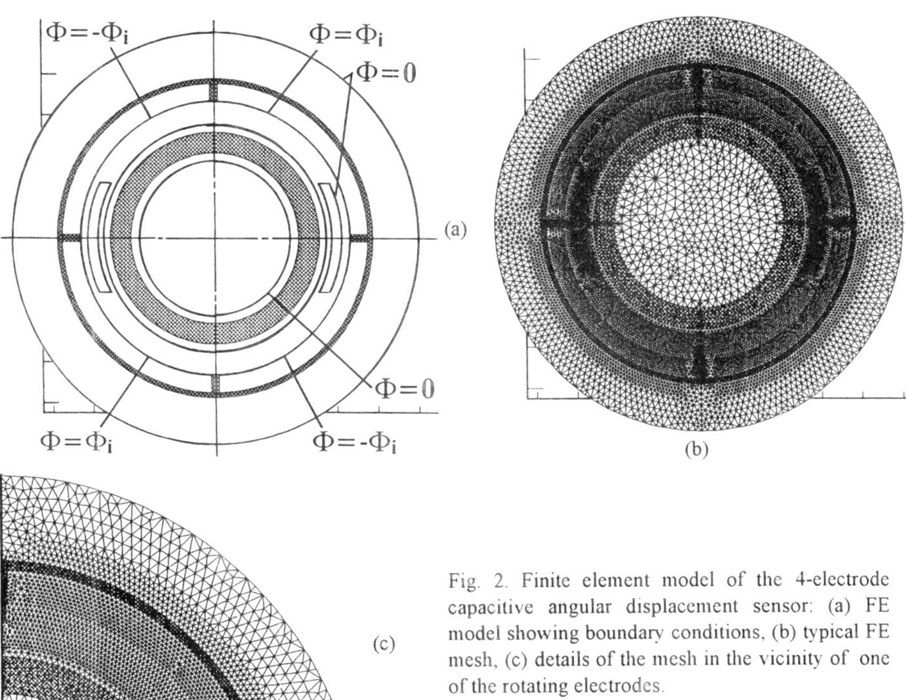

Fig. 2. Finite element model of the 4-electrode capacitive angular displacement sensor: (a) FE model showing boundary conditions, (b) typical FE mesh, (c) details of the mesh in the vicinity of one of the rotating electrodes.

Sun Sparcstation 10/51 has been used. Fig. 2a shows the typical FE model of the sensor shown in Fig.1, and the boundary conditions under which equation 1 is solved. These boundary conditions are mainly dictated by the potentials applied to the electrodes for normal operations of the sensor. It should be noted in this connection that no boundary condition is imposed on the centre electrode. However, as a conductor there must not be any electric field inside this electrode and its surface must be an equipotential surface. This means that the electric field lines (not equipotential lines) must be normal to the equipotential surface of the centre electrode. This condition is satisfied in the FE models of the sensor by considering the permittivity of the centre electrode to be infinitely large. For the discretization of the FE model shown in Fig. 2a first order triangular elements have been used, the number, sizes and the density of which take into account the geometric complexity of the model and the *a priori* knowledge about the distribution of field in the sensor. Furthermore, various modelling aspects mentioned in [4] have also been taken into account in order to minimize preprocessing time and effort, and maximize modelling accuracy and efficiency. Figs. 2b, c give an example of the FE mesh used for various models. The average number of elements used is about 28000 which corresponds to approximately 13000 nodes. No special measures have been taken for remeshing due to the motion of the rotating electrodes. It has been found that the default mesh, provided by OPERA-2d in the 'background' region of the airgap between the centre and outer electrodes for various angle of rotation of the rotating electrodes, is sufficiently consistent and adequate. Following the solution of equation 1 by FEM the required potentials and capacitances are

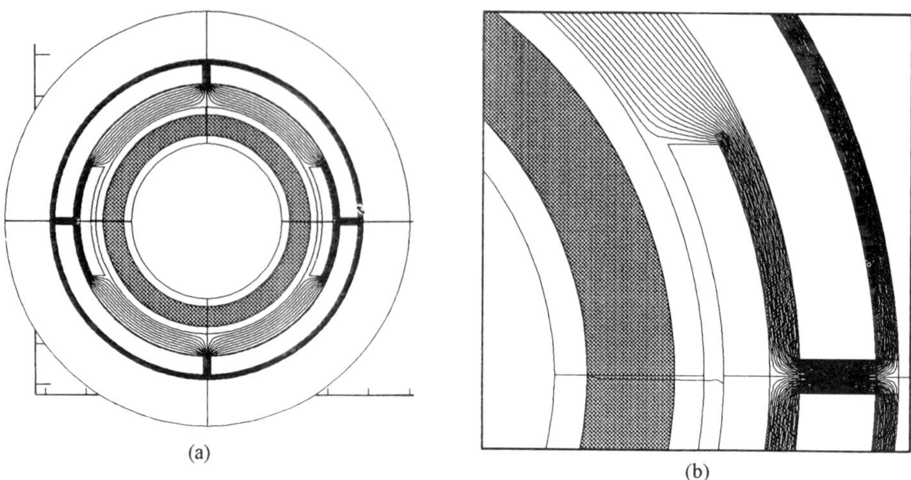

Fig. 3. Equipotential contours in the 4-electrode capacitive angular displacement sensor for the angular position θ=0: (a) equipotential contours in the whole region, (b) equipotential contours in the vicinity of one of the rotating electrodes showing details of potential distribution.

calculated from field computation results. The accuracy of modelling results are verified by special measures [4] and by using the facilities provided by the software used.

3. SOME MODELLING RESULTS AND DISCUSSIONS

Figs. 3-5 show some of the modelling results obtained for the capacitive angular displacement sensor shown in Fig. 1. The equipotential plots in the sensor for two extreme positions of the rotating electrodes are shown in Fig. 3 (θ=0) and in Fig. 4 (θ=5°). It is obvious from Fig. 3 that for θ=0 (symmetrical position) the output voltage V measured between the centre electrode and

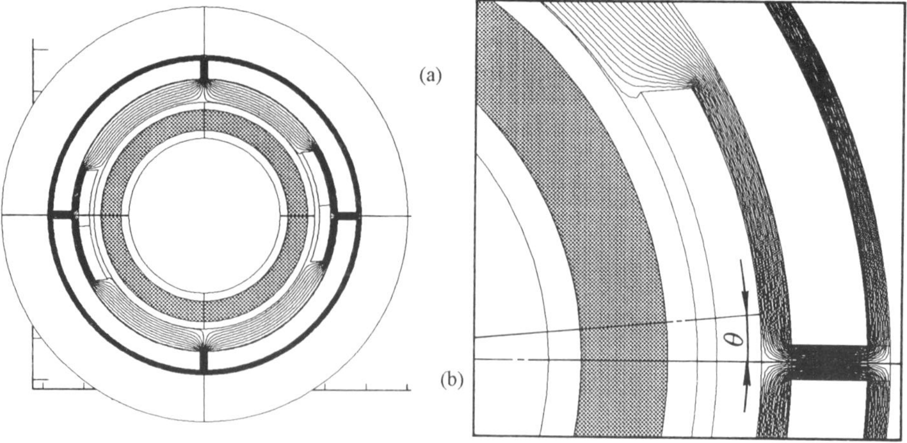

Fig. 4. Equipotential contours in the 4-electrode capacitive angular displacement sensor for the angular position θ=5°: (a) equipotential contours in the whole region, (b) equipotential contours in the vicinity of one of the rotating electrodes showing details of potential distribution at the leading end, and (c) at the trailing end (see next page).

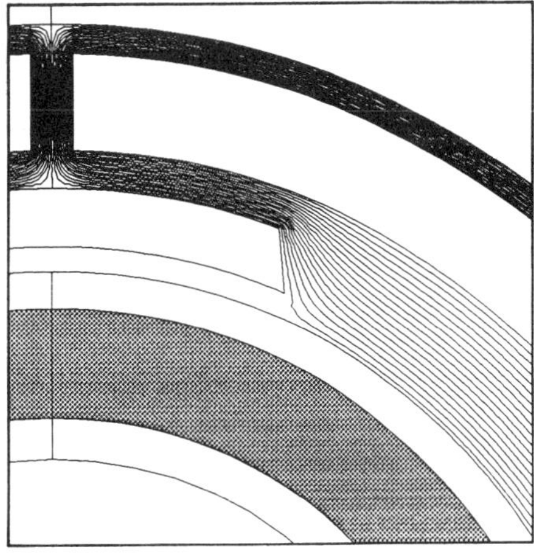

Fig. 4c

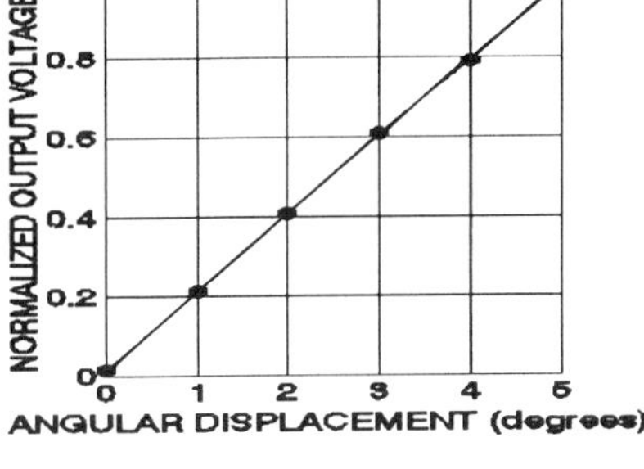

Fig. 5. Variation of output voltage (normalized) with angular position of the rotating electrodes showing linear dependence.

the virtual earth is zero which can be seen from the simulated $V=f(\theta)$ curve shown in Fig. 5. The change in this voltage with angle θ is affected not only by the redistribution of the electrostatic field owing to the variation of the overlap between the centre and outer electrodes, but also the by the change in the field distribution in the vicinity of the rotating electrodes. This can be clearly seen from the equipotential plots presented in Figs. 3b, 4b and 4c. As shown in Figs. 4b and 4c even for a given $\theta \neq 0$ the field distribution is different at the leading (Fig. 4b) and trailing (Fig. 4c) ends of the rotating electrodes (assuming the electrodes to be rotating in the anticlockwise direction). All these factors along with the effects of geometric and material parameters need to be taken into account to ensure the linearity of the performance curve shown in Fig. 5. This constitutes the central part for the CAD, optimization and performance evaluation of the sensor.

4. CONCLUSIONS

A methodology for the FE modelling of robust capacitive angular displacement sensors have been developed. Modelling results show high degree of linearity of the output characteristic $V=f(\theta)$ which is expected from the sensor in its practical exploitation. Appropriate experimental data are being sought to validate the simulation results. The simulation results obtained establishes the importance of field modelling and computation for the CAD of above sensors.

5. ACKNOWLEDGEMENTS

The authors would like to thank EPSRC for financing the work under an EPSRC Grant and Mr. T. J. Collyer and his colleagues at MBM Technology Limited (MBM Electrodynamics), UK for active collaboration and various discussions.

6. REFERENCES

[1] Silvester P. P. and Ferrari R. L.: "Finite elements for electrical engineers" (Cambridge: Cambridge University Press, 1990), 2nd edition.

[2] Binns K. J., Lawrenson P. J. and Trowbridge C. W.: "The analytical and numerical solutions of electric and magnetic fields" (Chichester: John Wiley & Sons Ltd, 1992).
[3] "The OPERA-2d reference manual", Version 2.5, Vector Fields Limited (Oxford: November, 1994).
[4] Khan S. H. and Abdullah F.: "Finite element modelling of multielectrode capacitive systems for flow imaging", IEE Proceedings-G, June 1993, vol. 140, no. 3, pp. 216-222.

Forecasting the Critical State of Deformed Crystal by Analysis of Smart Defect Structure: Fractal Characteristics and Percolation Critical Indexes

Yu G Gordienko[1], M V Karuskevich[2] and E E Zasimchuk[1]

[1]Institute of Metal Physics of the National Academy of Sciences, Kiev, Ukraine
[2]Institute of Civil Aviation, Kiev, Ukraine

Abstract. It is proposed to use specially prepared thin single-crystal plates rigidly fastened to the sample surface area as sensors of the deformation pre-history of multi-phase commertial alloys under fatigue and static loading. Estimation of lifetime and strain of a sample material is based on monitoring band patterns on a sensor surface. Density, direction of deformation bands and fractal dimension of band patterns on surface of sensor are shown to correlate with the number of cycles, maximum applied stress (under fatigue loading) and plastic strain, temperature (under static loading). Simulation of the pattern formation brought to light mechanisms of evolution. On the basis of experiment and simulation appearance of spatial inhomogeneities is considered as a dynamic percolation process and destruction is considered as a static percolation process. Experimental and simulation results are compared and nature of critical percolation indexes are discussed.

1. Introduction

The complex-alloyed multi-phase alloys (MPA) have the very inhomogeneous structure, and their behavior in the mechanical field is determined by a number of simultaneous processes. The plastic shape change promoting the deformation damage of the alloy structure is just one of such processes and it is hardly controllable by conventional techniques, such as direct measuring or the analysis of deformation damage of the structure. As a rule the plastic deformation of MPA is localized within separate microvolumes under any chosen loading conditions. Since it is difficult to predetermine such volumes, the analysis of shape change is complicated. On the definite stage of loading the plastic deformation of single-crystals is localized in elements of band structure; localization of deformation and structural transformations define band stability after loading action is removed (Persistent Slip Bands - PSBs, Shear Bands - SBs, etc.). Localization of deformation in band causes also a forming of surface relief, which is stable to following heating, aggressive environments action, etc. This fact can become a basis of designing single-crystal sensors (in the form of thin plates) of deformation pre-history.

2. Experimental

We used the single-crystal foil [1,2] as a detector to analyze the deformation damage and the lifetime of the D16 alloy samples at the static uniaxial and fatigue loading. The 30 mm long fatigue indicator was made of single-crystal aluminum foil (99.99% wt Al). The indicator was stuck on a flat specimen 1.2 mm thick. The preliminary experiments showed that the presence of detectors in the working part of the sample did not change mechanical properties at any used testing procedures. The TEM study of the D16 alloy structure after the mechanical treatment did not enable to reveal classic symptoms of the deformation damage (an increase of the dislocation density, the appearance of dislocation clews, boundaries, etc.). Only 0.7 - 2.5 mm size inclusions randomly distributed over the area under investigation are seen at the micrographs. Neither direct methods of structure and deformation surface relief study nor indirect ones (i.e., the analysis of hardening and X-ray scattering peculiarities) did not allow to reveal any evident signs of damage (i.e., manifestation of forming the sites of destruction) in the final state of material under the loading. The information about the deformation pre-history of the sample material near detector was derived from the analysis of the detector surface relief. For this purpose the surface of single crystal plate was grinded and electrolitically polished.

3. Results

The following features of the surface relief were observed for fatigue loading:

1) striation density ($k=l/n$, where n is the number of bands crossing the l length segment of line perpendicular to the striation direction) increases with number of loading cycles (N):

$$k \sim N^c \qquad (1),$$

where c = 0.248 for maximum applied stress s = 140 MPa and c = 0.577 for s = 180 MPa [1].

2) the angle between striation inclination and the loading axis is about 80° and does not change with N and s;
3) the striation direction does not correspond to traces of crystal slip (it is "non-crystallographic");
4) the sites of destruction (microcracks) are parallel to the striation direction and they have the "ladder" structure (Fig.1a), which is obviously created by merging the neighbor bands (Fig.1b);

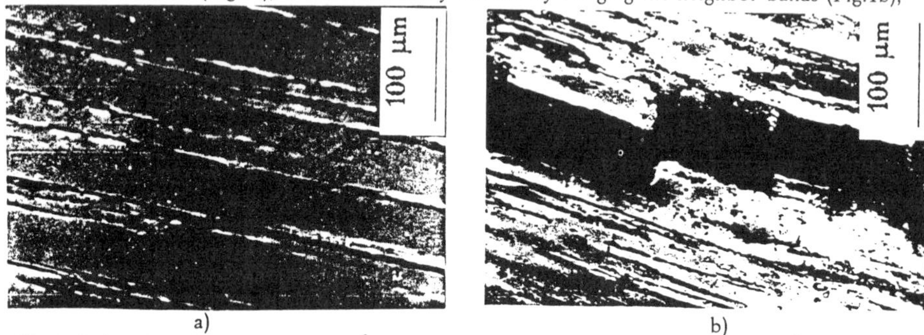

Fig.1. Defect band pattern after 3×10^3 cycles (a) and ladder microcrack after 5.9×10^5 (b) cycles.

5) the fractal dimensions of the band patterns defined by "box" method (Fig.2a) and by "yardstick" method (Fig.2b) changed with loading cycles.

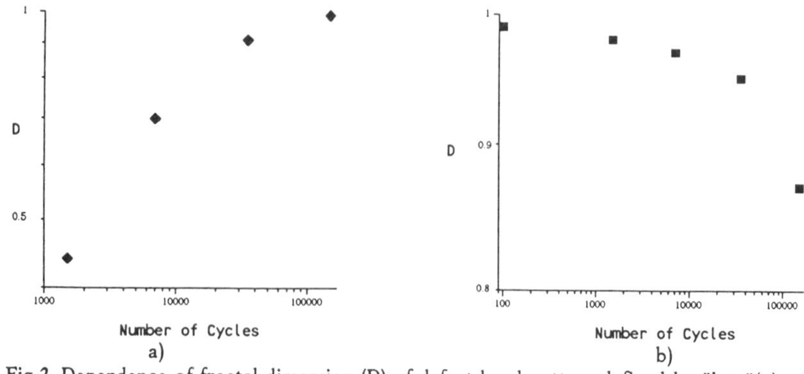

Fig.2. Dependence of fractal dimension (D) of defect band pattern defined by "box"(a) and "yardstick" method (b).

The singly-applied static loading the sensors installed on aforementioned samples brought to light the same features of relief. On the whole the deformation bands had the same properties mentioned in 2)-4). The striation density, k, increases with the applied strain, ε, by:

$$k = B\varepsilon^{b}, \tag{2}$$

where $b = 0.861$, B is the constant[2]. The main differences are existence of pattern of the secondary bands and non-monotonous dependence of its fractal dimension as a function of the applied strain. The direction of bands was 57° relative to the tensile axis. As to the fractal dimension, been initially equal to 1, it sharply falls to 0.23 (up to $\varepsilon = 3$ %) and then slowly increased along with the strain increasing. Moreover, the dependence of fractal dimension as a function of temperature can be observed. This dependence was also observed by other investigators and is known for different materials [3-5]. However, the very careful calibration of the indicators is necessary for their usage.

4. Discussion

Formation of band structures PSBs, SBs and others in crystals under load is well known to be related with defect localization [6-8]. The proposed models [9] are often based on the assumptions about cooperative behavior of dislocations and, in general, are described by the set of non-linear partial equations of reaction-diffusion type. It is known that appearance of instability with regard to spatial inhomogeneous disturbances is possible in the systems. Development of instability leads to spatial structures ("dissipative contrast structures") [10,11], which consist of alternating regions with very different values of defect concentration. This process can be the basis of structure formation of spatial band patterns as PSBs in crystals under fatigue loading and as SBs in crystals under static loading. However, the well-known models [9] cannot explain the non-crystallographic defect patterns observed in single crystals under fatigue loading [2] and single-applied static loading [3-5].

We suppose that at the beginning of loading the surface relief was mainly caused by outlets of single dislocations, which form patterns of randomly distributed bands. Because, fractal dimension D calculated by "yardstick" method [4] is equal to 1 for equidistant (and randomly distributed) bands. However, further ordering in the defect system can be attributed to self-organization effects in ensemble of point-type defects [12], could lead to the very high localization of defects in small region of crystal (non-crystallographic bands). Evolution of the spatial inhomogeneites in crystal lead to higher defect localization. Appearance of the complicated self-similar band patterns can be result of complicated structural evolution. The fractal dimension D calculated by "yardstick" method is allow to determine the band pattern as a self-similar gap set, bum a non-self-similar point set [4]. This method is sensitive to whether the band is considered as a single or a multiple one. The method allowed to observe transformation from single line pattern (with fine bands) to multiple line pattern (with coarse bands). It explains disagreement between Fig.2a and Fig.2b.

For bringing to light the main features of evolution of defect system with randomly distributed point-type defects (vacancies), we carried out computer simulation by cellular automaton technique (described in [13]). System of point-type defects (vacancies) were situated on a square two-dimensional lattice 2050×15 with preferential movement along the longer side. Vacancies can move unidirectionally under gradient of external stresses with probabilities M and aggregate in complexes with probabilities P to abandon complex:

$$M \sim \exp\{-(E_m - A(i,j))/kT\}, \qquad P \sim \exp\{-(\sum E_b - A(i,j))/kT\},$$

where E_m - a vacancy migration energy, A - an external influence, which cause unidirectional movement of vacancies, E_b - a vacancy dissociation energy, local interaction with neighbors is taken into account by summation. As a result we obtained one-dimensional patterns of spatial distribution of free vacancies, complexes and piles of vacancies and complexes. The smoothed plots (for 5 points) of defect density along movement direction are presented on the Fig.3 for different time steps. The plots can be interpreted as cross sections of two-dimensional anisotropic pattern of point-type defect system with creation of the final band structure. After transition period (t~10^3) system reaches asymptotic state. We found two qualitatively different states: "dynamic" state and "static" one. The former is characterized by *homogeneous* distribution of many free vacancies, many small complexes and without piles. The latter is characterized by *inhomogeneous* distribution of few free vacancies, many large complexes, many large piles (Fig.3). We observed the critical character of

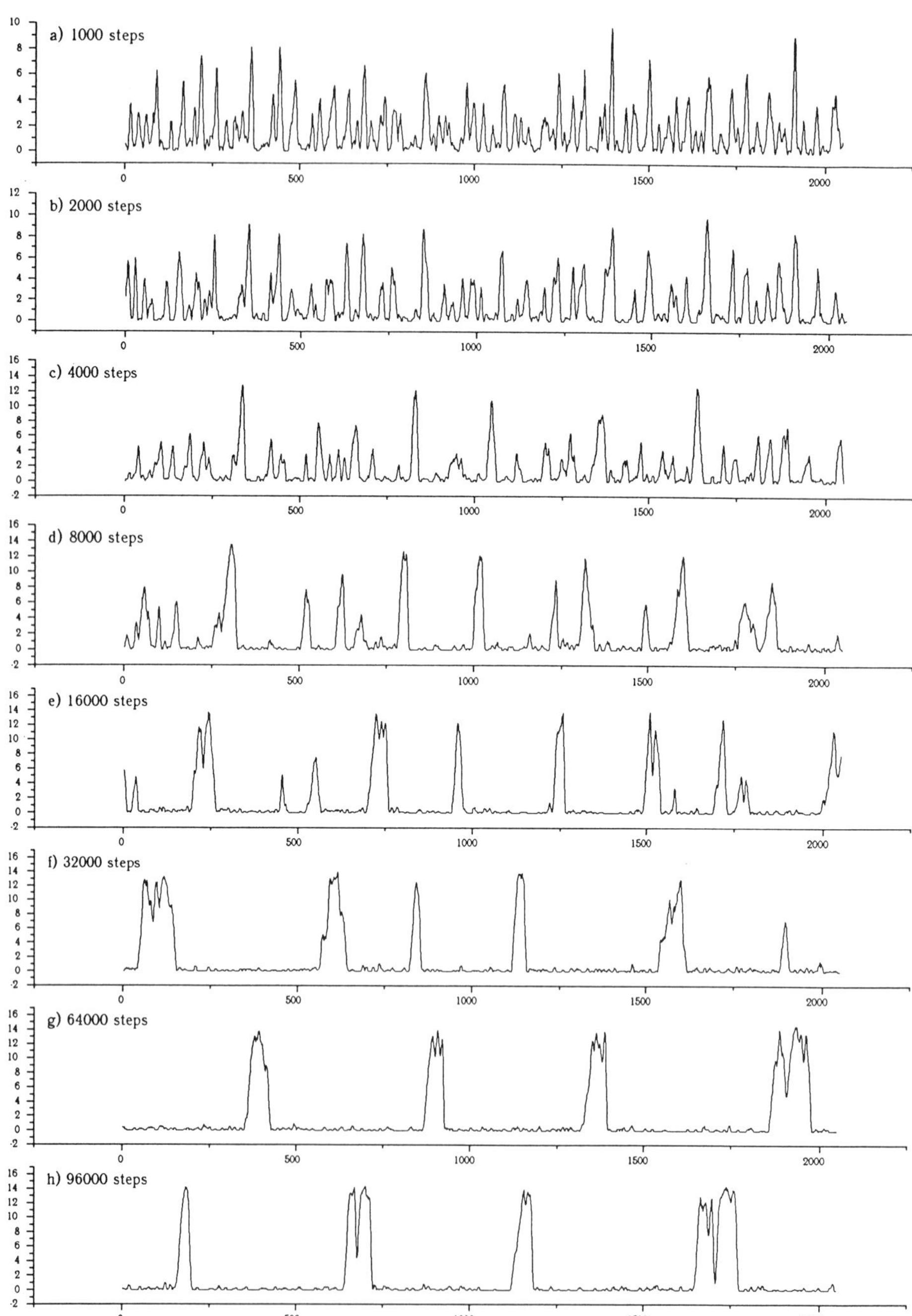

Fig.3. Results of computer simulation of unidirectional movement of point-type particles: defect density plots (in arbitrary units) after some time steps as a function of spatial coordinate.

transition between these two states as a function of R:

$$R = \exp\{(E_b - E_m)/kT\}$$

This transition is known as a dynamic percolation , however, in this case R is used as the control parameter, but not the critical particle concentration observed in traffic jamming transition [14].

Observations of the patterns allowed to define the next mechanisms of evolution (which take place on many scale levels): firstly, by merging neighbor bands and, secondly, by coalescing neighbor bands. The former leads to tightening the band structure with simultaneous dynamic selection of final gaps of gap set. The latter leads to extending the widths of structural elements (piles) with dynamic selection of final piles of pile set. Similarity of the mechanisms of evolution on many scale levels can lead to appearance of similar structures on many scale levels. It is confirmed by availability of self-similar properties of experimental band structures (Fig.1). Both mechanisms correlate with results obtained experimentally, that the band pattern is self-similar *gap* set and single lines tend to be multiple ones.

From quantitative point of view evolution can be described on the basis of observation of behavior of the next characters: free vacancies, vacancy complexes, piles of complexes (inhomogeneites with local defect density higher than average defect density). For example, the number of free vacancies (F), the number of piles (N_p) and the number of complexes (N_c) were interpolated by

$$F = A(R)R^{-\gamma}, \qquad N_p(t,R) = P_0(R)t^{-P(R)} \quad \text{and} \quad N_c(t,R) = C_0(R)t^{-C(R)},$$

where $P(R)$ and $C(R)$ - the non-monotonous functions, which sharply increase in the narrow range $5<R<15$, get maximum near $R_{max} \approx 15 \div 20$ and decrease for $R>R_{max}$ (Fig.4b and Fig.4c), γ sharply change its value in the narrow range $5<R<8$ (Fig.4a) for $10^3<t<10^6$, where t - the number of time steps.

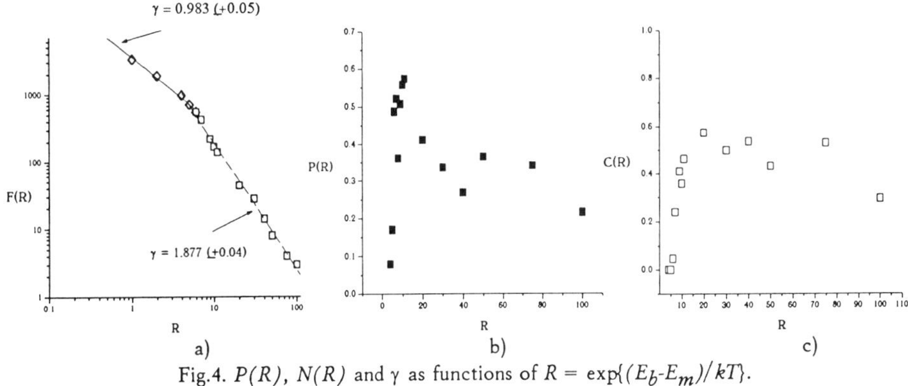

Fig.4. $P(R)$, $N(R)$ and γ as functions of $R = \exp\{(E_b - E_m)/kT\}$.

We interpreted the final inhomogeneous distribution of defects (Fig.3h) as a cross section of the non-crystallographic band patterns observed in experiment. The experimental results (1) and (2) describe appearance of visible structural elements (bands). Rough estimation of the whole number of piles (bands) gives that parameter $c \sim 0.2 \div 0.6$ from equation (1) would be equal to $(1-P(R))$. It is in agreement with results of computer simulation (Fig.4b), where $(1-P(R)) \sim 0.2 \div 1$. In this case change of amplitude of applied stresses can be related with crystal lattice distortion. It is known that the crystal lattice distortions are followed by changes of values of a vacancy migration energy (E_m) and a vacancy dissociation energy (ΣE_b) [15].

It should be noted that specially prepared surface regions of basic materials could be used for monitoring the non-crystallographic "smart" pattern, if the basic material is single crystal or it consists of large single-crystal grains. Then direct observation can allow to estimate directly deformation damage and, moreover, predict critical state of material (destruction or superplasticity). For this purpose critical state can be described in terms of percolation theory: critical concentrations of defect regions and critical percolation indexes [16]. Really, further merging and coalescing the neighbor bands lead to creating defect band clusters with increasing diameter. Finally, defect band clusters will increase and pierce the whole crystal. This can lead to significant macroscopic consequences (i.e., destruction of fatigued samples, high plasticity and destruction of samples under uniaxial tension), which were observed experimentally [1,8]. Such a process is a similar to traditional percolation transition. However, correlation length (average diameter of defect clusters) can be hardly estimated by relationship, which is well-known from theory of phase transformations:

$$L \sim |x - x_c|^{-\nu} \quad (3),$$

where x_c - the critical concentration of defect clusters, ν - the critical percolation index (0.9 for cubic lattice and 1.33 for square lattice [17]). This doubt is caused by two reasons. Firstly, the whole process of band pattern formation consist of two main nested stages: creation of inhomogeneites (dynamic percolation) and creation of global defect cluster (static percolation). Both transition to critical state are ruled by the critical values of control parameters, for example, by parameter R or concentration of vacancies v_c (dynamic percolation) and concentration of piles p_c (static percolation). Secondly, creation of global defect cluster (static percolation) is complicated by drift of piles, that can cause essential decrease of p_c. In this connection, it seems to be very interesting to investigate more thoroughly scaling properties of the pattern formation as a consequence of two nested percolation processes. In addition to this it is not clear whether the critical percolation index (ν) for this regime will be equal to the known theoretical value: 0.9 for cubic lattice and 1.33 for square lattice [17]. Using by scaling dependence (3) we could predict the beginning of the critical state.

5.Conclusions

The method based on monitoring the smart non-crystallographic band pattern allows to obtain the quantitative information about the deformation pre-history in the material during the fatigue and static loading: direction of applied load during fatigue and uniaxial loading; number of applied cycles with high accuracy; maximum applied stress in fatigue; extent of reliability and safety of materials in fatigue. The method is a very cheap and can be applied in the hardly accessible and aggressive mediums, where special expensive devices cannot be used. Moreover, these cheap sensors with smart surface can be distributed on the whole industrial construction with periodical monitoring the smart surfaces by replica technique. Further scanning the set of replica will allow to estimate "health" of the whole industrial construction and catch abnormal and unpredictable overloads of individual elements of construction.

6. References

1.E.E.Zasimchuk, A.I.Radchenko, M.V.Karuskevich, Fatig.Fract.Engng.Mater.Struct. 15, 1281 (1992).
2.Yu.G.Gordienko, E.E.Zasimchuk, Proc. of Second Europ. Conf. on Smart Structures and Materials (Glasgow, UK), ed. by P.Gardiner, A.McDonach, R.S.McEwen, Proc. SPIE 2361, 312-315 (1994).
3.B.Sprusil and F.Hnilica, Czech.J.Phys. B 35, 897 (1985).
4.T.Kleiser and M.Bocek, Z.Metallkde, 77, 582 (1986).
5.I.R.Gobel, Z.Metallkde, 82, 853 (1991).
6.U.Essmann, Phys.Stat.Sol.(a), 12, 723 (1965).
7.A.S.Malin and M.Hatherly, Met.Sci., 13, No.8, 463 (1979).
8.E.E.Zasimchuk and L.I.Markashova,Mater.Sci.Engng.,A127, 33 (1990)
9.L.P.Kubin, G.Martin (editor), Proc. Inter.C.N.R.S. Meeting. Dif. and Def. Data. B3,B4 (1988).
10.V.A.Vasiliev et al., Autowave processes (in Russian), (Nauka, Moscow, 1987).
11.B.S.Kerner and V.V.Osypov, Autosolitons (in Russian), (Nauka,Moscow,1991).
12.Yu.G.Gordienko, E.E.Zasimchuk, Phil.Mag. A70, 1, 99-107 (1994).
13.Yu.G.Gordienko, E.E.Zasimchuk, Proc. of Ukrainian-American Summer School oh Pnysics and Chemistry of Surfaces. 4-10 September 1994, (Kiev, Ukraine), in Functional Materials, 1 (1995).
14.O.Biham, A.A.Middleton and D.Levine, Phys. Rev. A46, R6124 (1992).
15.A.I.Melker, V.V.Sirotinkin, A.A.Vasilyev, Phyzika Tverdogo Tela (in Russian) 29, 3044 (1987)
16.D.Stauffer, Phys. Rep. 54, 1-74 (1979).
17.B.I.Shklovskii and A.L.Efros, Usp. Fiz. Nauk 117, 401 (1975) [Sov. Phys. Usp. 18, 845 (1975)].

Modelling of a recording Head with its shielding

Khan O.R., MIEEE,

Engineering division Matthew Boulton college, Birmingham B5 7DB

Abstract. Perpendicular magnetic recording has been an important research area nowadays. It is being applied to improve and develop both efficient storage and recording Systems, for information storage technology. Thin-film pole heads offer a significant improvement in perpendicular magnetic recording for applications requiring high density and sharp resolution. Experimental verification of the recording process of such a system is an essential feature in evaluating its performance. A large-scale model of a thin-film pole head is described here with its components and its shields. The model is instrumental in evaluating the performance and efficiency of a single-pole recording head. This study highlights the design, the measurement techniques, and the results. Comparison of the magnetic fields under and around an isolated pole and the field under the pole with shielding is made with the results achieved by calculations for both cases. Field strength was found to be optimal for shielding, when placed at a distance equal to the pole thickness.

I. Introduction

Perpendicular magnetic recording is not a new idea. However, interest in it has recently been renewed and it has become a popular and demanding research activity. This method of recording and replay can improve the performance of a system by increasing the density of recording and by achieving sharp magnetic transitions. It is well known that in comparison with a conventional recording system, [1] the magnetic transition width is reduced significantly using a perpendicular recording system, a Co-Cr and Ni-Fe double layer medium [2] and a pole head [3].

Extremely high density (over 100 KFRI) has been reported for a perpendicular recording system with a suitable combination of the head and the recording medium [4]. The interaction between the head and the medium play an important role during the recording and replay process.

To verify the importance of this interaction, an experimental large-scale model of the head was designed and developed. The design outlines are shown in figure 1. The model is used to test the earlier calculated results [5,6] for a thin-film pole head. In addition the model is an excellent practical aid for understanding applications of electromagnetic engineering, it can therefore also be used as a teaching aid. This study highlights various features of the model and many results obtained with it.

The size of the model is ten thousand times the size of actual thin-film pole heads reported [7]. It comprises a pole, its back plate, a plate forming the recording medium and two shielding metal plates, two flanges (large shielding), a Hall probe with electronic circuitry, coil windings and an aluminium supporting system. The two flanges and the shielding poles are optional, to test the effects of shielding the pole head.

Jiro Hokkyo [8] has reported a simple model of a perpendicular-magnetic-recording thin-film pole head. He has evaluated the interrelation parameters between the recording field, the magnetization of the medium, and the interaction between head and the recording medium. A large-scale model of a perpendicular pole head has also been reported [9,10] and the experimental results have been compared with the calculated magnetic field [6].

In this study, the model was tested, and used to verify the results and values (which were plotted in a similar format using a computer program reported earlier [5]). The outcome of the study verifies the results, with advantage that the model can be used to support teaching in the area of applications of electromagnetic engineering. The effect of pole shielding was found to be significant, in terms of improving field strength of the pole and to higher resolution . The results are in close agreement with the theoretical field of the pole with shielding, when the shield is placed at a distance equal to the pole width.

II. MODEL COMPONENTS

With dimensions proportional to the thin-film pole head, a pole plate, a recording medium and a frame plate as a large flux guide with shielding plates were designed (figure 1). The structure of the model includes a frame of five rectangular iron plates of equal cross-sectional area (50x150 mm) with pole shielding (figure 2). Two plates, one used as the recording medium and the other as a large flux guide, are each 500 mm long. The flux-guide plate has two openings of an area of 400 mm long and 10 mm wide openings to enable pole and shielding plates to be mounted at any required position. This area, as shown in figure 2, is used for the bolts to pass through to form joints for supporting the pole piece (third plate) and the shielded poles (fourth and fifth optional plates). Two small openings in the plate used as a recording medium, were provided for further applications. The third plater, 200 mm long, was a pole piece and had holes in the back cross sectional-area to take two bolts used to join it to the large plate.

Two more plates 50x50 mm could be used for pole shielding (figure 2). The 5mm diameter holes in the plates provide strong support to the whole frame set-up, when assembled with nuts and bolts and the supporting iron strips. The media plate is supported vertically by four aluminium bolts, set-in the aluminium stands of the support system. The flux guide (the back pole plate) along with the pole plate and shielding plates rest on the aluminium studs. This makes the support a rigid and strong system, with the flexibility of using verniers.

An amplifier system with a measuring device (a digital voltmeter), is used to display the magnetic field measurements in terms of voltage level. An aluminium vernier for a Hall probe (top in figure 2) and another vernier, to change the spacing between the pole face and the medium (bottom right in figure 2), with the base of the support system are employed. The setup is useful both for diversity applications and the accuracy of the measurements. The whole setup is placed on a thin piece of polyethylene to avoid any variation in the parameters due to any environmental vibrations.

A copper coil of twenty turns, shown in figure 2, over a PVC cavity around the pole plate facing the media plate, is used to generate the magnetomotive force to activate the pole plate. A combination of two more coils of 50 and 100 turns respectively was also fabricated, and placed at an appropriate position for producing higher magnetic fields. The cross-sectional diameter of the copper wire is 0.1 mm, which was estimated to produce enough magnetomotive force for the model system.

The Hall probe, with the required specification [11] (and its additional electronic circuitry for signal amplification), is used to detect the magnetic field at any point along the recording medium plate. The signal detected by the Hall effect sensor is amplified and processed for measurements on a designed electronic circuit.

III. EXPERIMENTAL METHODS

The medium plate can be placed at an adjustable distance (an adjustable air gap/spacing), along the y-axis, facing the pole plate, with the use of a vernier. The Hall-effect sensor detects the field intensity at any point/position along the x-axis (along the medium plate) using the top vernier during the measurements. Fine adjustment up to 0.1mm could be made for the air gap and for the Hall-probe displacement using these verniers.

The whole setup is mounted on two parts, (a) a stand platform system, (consisting of large bottom plate holding the medium plate and the Hall-probe vernier) and (b) a moveable system (upper bottom platform, holding the pole plate and its

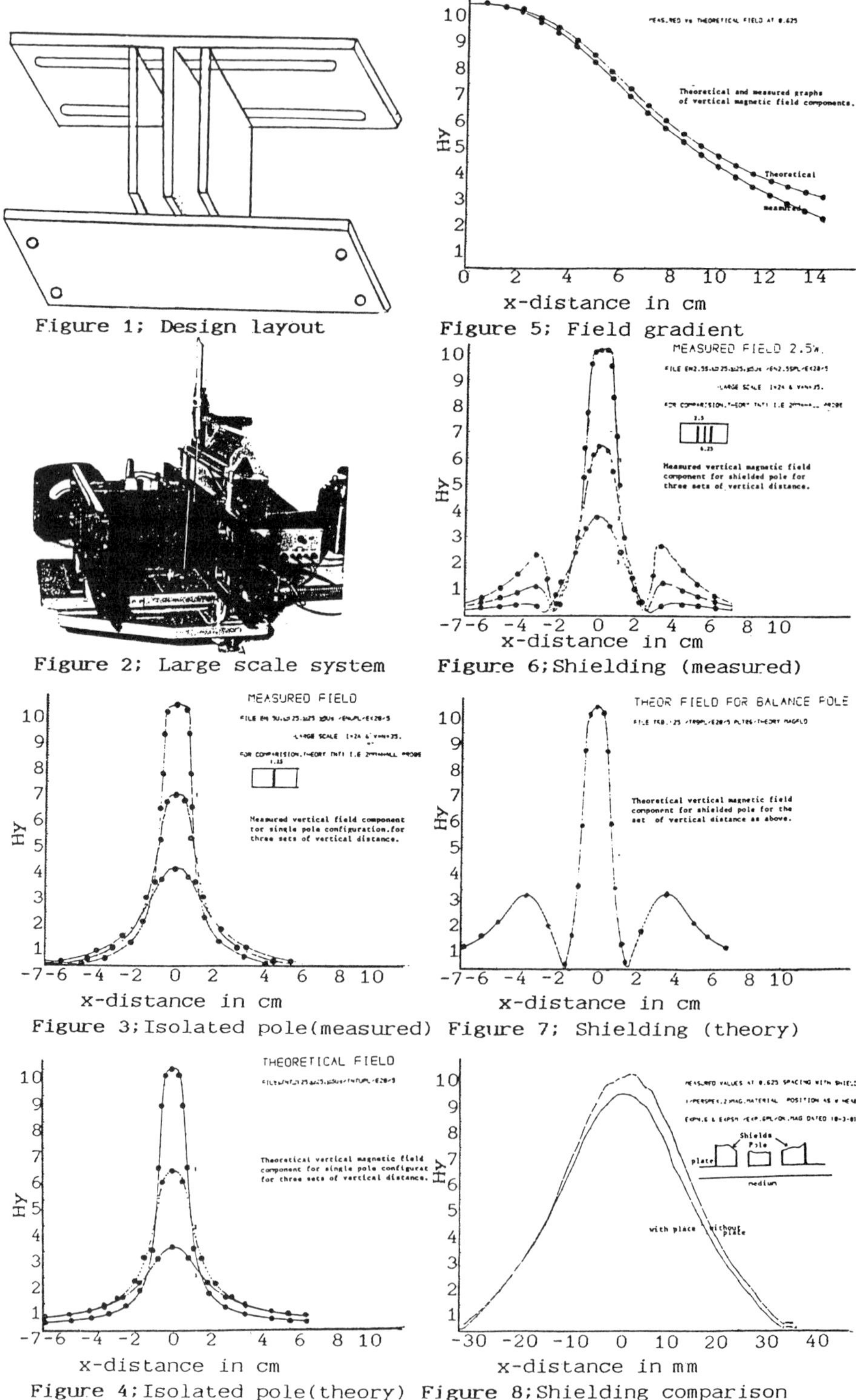

Figure 1; Design layout

Figure 2; Large scale system

Figure 3; Isolated pole(measured)

Figure 4; Isolated pole(theory)

Figure 5; Field gradient

Figure 6; Shielding (measured)

Figure 7; Shielding (theory)

Figure 8; Shielding comparison

back). Thus one part is the holder of the medium plate, facing the pole plate, and the other part is moveable in the opposite direction, the other components of the frame are rigidly and firmly supported. The field intensity for a position of the Hall-effect sensor is noted in steps of 5mm, starting from some initial position on the left, for both setups, with and without shielding.

IV. FIELD MEASUREMENTS

In each run the field intensity was measured along the media plate, using the Hall probe. The initial measurement was on the left, and the Hall probe was progressively moved to the right through the zero position (i.e. probe faces the centre of the pole plate) and further to the right.

All other parameters were kept constant, so as in the theoretical calculation of the magnetic field to enable comparison with the run. Field plots for normalized values of both experimental and theoretical values were obtained by using a Fortran computer program. The data values were kept in an input file to plot the graph of the field strength of a pole head. Three values, proportional to the spacing values (the head to medium gap) used in the calculations, 5, 2.5 and 1.25 um, were kept fixed in the measurements with and without the pole shielding.

V. RESULTS AND CONCLUSION

The plots which were obtained using the above methods, are shown in figure 3. Keeping all the parameters constant, the magnetic fields were also calculated [5,6] using a computer program for the equivalent spacing (the media to head gap) and the plots were obtained in the same format, to compare with the experimental results. These plots are shown in figure 4 for an isolated pole.

The field gradient plot was also plotted for a value (air gap) for both calculated and experimental methods, for comparison purposes. It is shown in figure 5. The comparison shows that the theoretical results are getting higher than the experimental values increasingly at the region away from the pole (from 2 cm ≈ 2 um to 14 cm). But it shows a very close agreement at the centre and under the pole regions (for about 2 cms).

The results of measured and calculated fields for the effects of pole shielding are shown in figure 6 and 7 respectively , which are in close agreement except at the side peaks (low frequency flux components), which, in general improve the field to a higher resolution because of the shunting effect. The arrangement of shielding the pole leads to increase the field at the trailing edge of the peak as shown in figure 6.

Measured results of the field under the pole length (10mm ≈1 um) for larger size shielding (with a thickness, several times to that of the pole thickness) with the shielding of pole thickness are shown in figure 8, which shows some better results for the shielding of the same thickness to that of the pole head when placed at the pole size distance.

This concludes that the theoretical evaluation [5 and 6] of the actual pole head is meaningful to allow for the design of the thin film pole head of this configuration. The study also verifies the dimensions, designing methods, parameters and assumptions used in the theoretical evaluation.

VI. ACKNOWLEDGMENT

The author would like to acknowledge the moral support of the colleagues at the Physics department and thanks to Miss Lubna Obaid, who typed the manuscript. Financial support was given by SERC UK.

VII. REFERENCES

[1] V. Zieren, et. al.,International conference on magnetics, 1986, AB-5.S.

[2] Iwasaki and Y. Nakamura, IEEE Trans. Magn., Vol. MAG-13, no.5, 1272-1277, 1977.

[3] K.Hokkyo, K.Hayakawa, S. Stake and T.Simamura, Proceedings of international magnetics conference 1978, pp 26-9.

[4] Iwasaki et. al., IEEE transactions on magnetics 17(6), p 2535, 1981.

[5] Khan O.R., Thin film MR read/write devices for computer memory application technology, department of electrical and electronic engineering, Plymouth Polytechnic, 1988.

[6] Wilton D.T. and Mapps D.J.," An analysis of a shielded magnetic Pole for perpendicular recording", IEEE trans. on Magnetics, Vol 29 (6), p 4182-4193, Nov 1993.

[7] Khan O.R., Thin Film read/write devices, IEE VADR eighth international conference, Birmingham university Birmingham 1990, p 35.

[8] Jiro Hokkyo, et. al., IEE Transactions on magnetics, MAG-16, No.5, p 887, 1980.

[9] Szczech T.J. et. al., IEEE Transactions on magnetics, MAG-18 (6), p 1176-78, 1982.

[10]Zaho, Y.H.,Chen Y.X., and Zhang, " A study on character istics of improved perpendicular recording write head with a large scale model", IEEE Trans. Magn., Vol. Mag-27, no 4, pp 2488-2490, September 1987.

[11]Guide to Installation, UGN-3503U and UGS-3503U, "Ratiomet eric, Linear Hall Effect Sensors"

Section H

Sensor Applications

SENSORS AND ACTUATORS FOR ACTIVE VIBRATION CONTROL OF SMALL CANTILEVERS.

D F L Jenkins, M J Cunningham and W W Clegg

Division of Electrical Engineering, Manchester School of Engineering,

University of Manchester, Oxford Road, Manchester M13 9PL, England.

The deflection of very small cantilevers is at the heart of scanning probe instruments such as atomic force and magnetic force instruments. Such instruments are extremely sensitive to building and air borne vibrations and great care is normally required to isolate the instruments from such sources. Nevertheless, mechanical vibration of the cantilever limits the resolution of the instruments. The position is further complicated by the fact that in magnetic force instruments, the cantilever is vibrated deliberately near its resonant frequency in order to increase instrument sensitivity. In a magnetic force instrument the forces of interest are long range and the cantilever tip must be maintained at a constant distance from the sample in order to build up an image of the spatial distribution of the magnetic field over the area of the sample. Typically the tip-sample distance is around 40nm.

Scanning force microscopy has been used to investigate a wide range of materials which produce spatially varying forces on a suitable tip attached to a cantilever. Various techniques have been used to measure the deflection of the cantilever, such as optical beam deflection [1], interferometry [2] and capacitance [3]. However, Franks [4] has observed how induced unwanted cantilever vibrations limit the performance of atomic force microscopes. It is therefore of interest to investigate the use of active vibration control methods for such cantilevers. The employment of a suitable actuator may permit a substantial reduction in instrument size and cost whilst maintaining the ability to isolate and control the cantilever, since massive air bearing benches are the conventional pre-requisite for acceptable images.

In this application it is necessary to implement active vibration control whilst retaining the ability to position the cantilever statically or dynamically using a suitable actuation method. The experimental arrangement used is schematically depicted in Figure 1. To investigate the application of vibration control to scanning force microscopes a small glass cantilever was initially used, of somewhat larger dimensions than used in such instruments. This cantilever was fabricated from 100μm thick glass and was 1mm wide and 30mm long. The cantilever was coated with an 80nm aluminium film to provide a good optical reflector. The dynamic behaviour of the cantilever was monitored using optical beam deflection [5]. In optical beam deflection, light from a diode laser, in this case at 670nm, is focused onto the end of the cantilever and the reflected light detected by a position sensing quadrant photodiode. In order to align the detector, and also permit simultaneous observation of bending and torsional modes of the cantilever, if required, the detector is operated as switchable lateral detector. For small displacements, as in this case, the response of the position sensing detector can be assumed to be linear. In order to induce cantilever motion, a small loudspeaker was placed in close proximity and driven by a train of 2ms pulses. These pulses exited the cantilever at its natural frequency, the amplitude of which subsequently decays before the arrival of the next pulse. To effect electrostatic vibration control the cantilever was placed at ground potential and a small aluminium plate (electrode) was placed below the free end. The separation of the electrodes was adjusted to bring them almost into contact, and parallel, with each other. To effect active vibration control the signal from the optical sensor was filtered, amplified and summed with a DC bias voltage such that the feedback signal did not cross 0V line. If the signal is not DC biased the electrostatic feedback is effected at twice the drive frequency. Figure 2 shows the open loop response of the cantilever when it is excited by the acoustic pulse train from the loudspeaker. When the electrostatic feedback signal is applied to the lower plate, the cantilever vibration is damped, as shown in Figure 3.

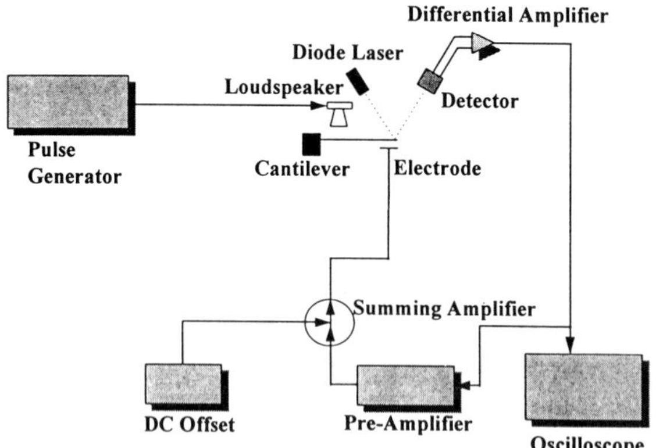

Fig.1 Schematic experimental arrangement

This technique is now extended to the application of active vibration control to a scanning force microscope. The force microscope used here is a Differential Interferometric Scanning Force Microscope constructed by the authors [6], the construction of this instrument is depicted in block form in Figure 4.

The technique of electrostatic feedback was applied to micro-cantilevers used in this force microscope. The experimental arrangement used is depicted schematically in Figure 5. A PZT element attached to the cantilever holder was driven by a train pulses to excite cantilever motion. Figure 6 (upper trace) shows the open loop response of the cantilever when it is excited by the acoustic pulse train from the loudspeaker. This signal was then filtered and amplified before being summed with a DC bias voltage. This signal is then applied to the sample. The cantilever itself is at ground potential. Figure 6 (lower trace) shows that the vibratory motion of the cantilever is damped when electrostatic feedback is applied. The residual noise here is due to processing electronics and not cantilever motion.

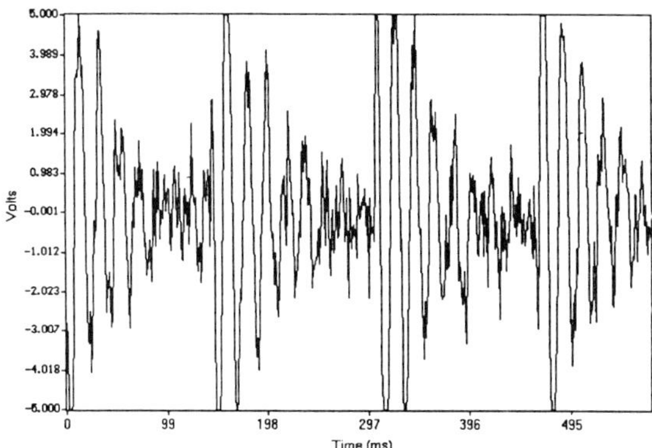

Fig. 2 Open Loop Response

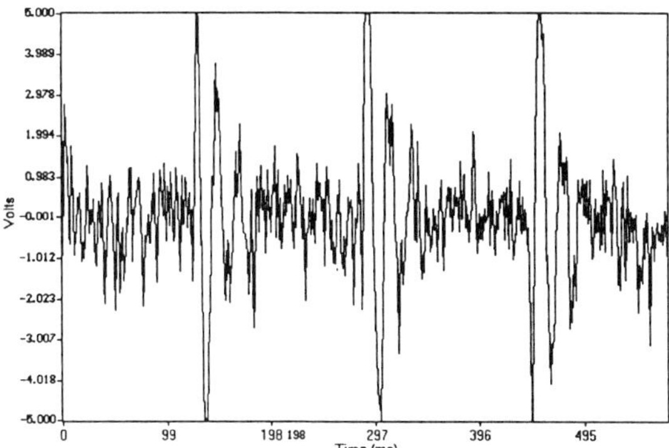

Fig. 3 Closed Loop Response

An active vibration control and actuation system has been developed which enables simultaneous active vibration control and micro-actuation using electrostatic forces. Using this technique, small, lightweight, structures can be controlled and manipulated by low voltages, typically less than 15V.

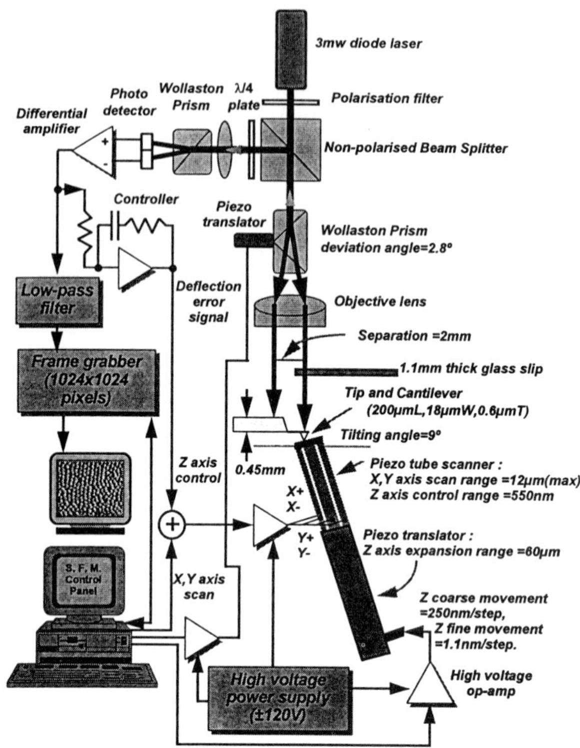

Fig. 4 Block diagram of the scanning force microscope system incorporating interferometric deflection sensing.

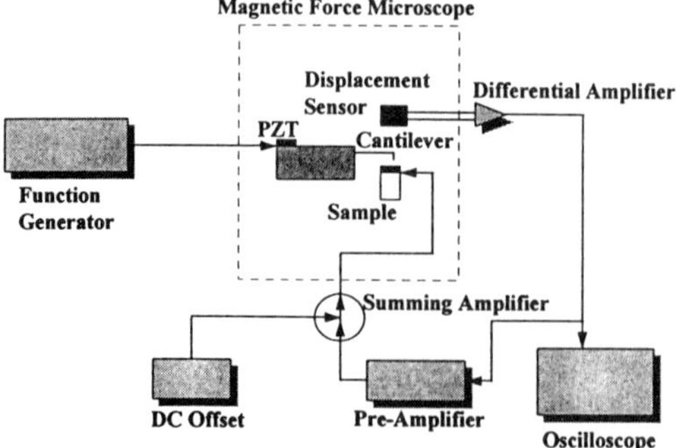

Fig. 5 Experimental Arrangement Used In Electrostatic Active Vibration Control

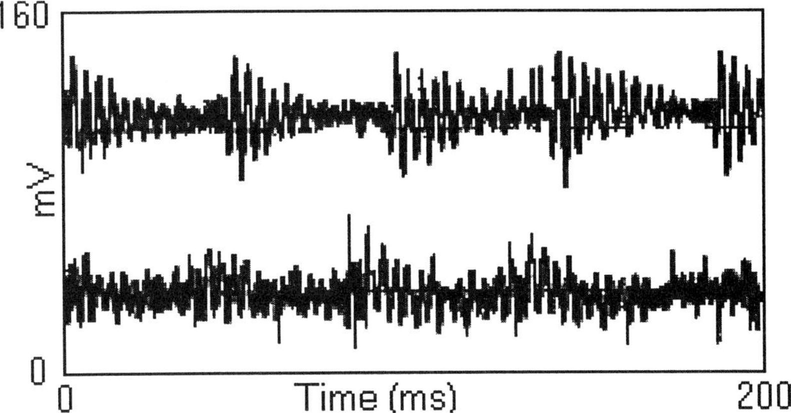

Fig. 6 Open loop response (upper trace) and closed response (lower trace)

References

1. C Schonenberger and S F Alvarodo. A differential interferometer for force microscopy. Rev. Sci. Instrum. 60, 3131-4.

2. G Meyer and N M Amer. Erratum: Novel optical approach to atomic force microscopy. Appl. Phys. Lett. 2400-2, 1988.

3. T Goddenhenrich, H Lemke, U Hartman and C Heiden. Force microscope with capacitive displacement detection. J. Vac. Sci. Technol. A8, 383.

4. Franks. Progress towards traceable nanometric surface metrology. Nanotechnology, Vol. 4, pp200-205, 1993.

5. D F L Jenkins, M J Cunningham, W W Clegg and M M Bakush. Measurement of the modal shapes of inhomogeneous cantilevers using optical beam deflection. Meas. Sci. and Technol, Vol. 6(1995), 160-166 .

6. M J Cunningham, S T Cheng. A differential interferometer for scanning force microscopy. Meas. Sci. Technol. 5(1994), 1350-1354.

Optical sensing technique for Nd:YAG laser welding process monitoring, using light returned through the cladding of the delivery fibre optic

†D. P. Hand, ‡C. Peters, †J. D. C. Jones

†Physics Department, Heriot-Watt University, Riccarton, Edinburgh EH14 4AS

‡Lumonics Ltd., Cosford Lane, Swift Valley, Rugby CV21 1QN

Abstract. A non-intrusive optical sensing technique for Nd:YAG laser welding monitoring is described. The Cladding Power Monitor (CPM) monitors the light returned through the cladding of the delivery fibre optic, and appropriate optical filtering isolates the light radiated from the plume of hot gas which forms above the weld. The successful detection of welding faults using this optical signal is described, including laser focus errors and shield gas interruption.

1. Introduction

Laser welding is becoming a widely used jointing technique with many different applications, but as with all welding techniques, the relationship between process parameters and ultimate weld quality is complex. In addition to problems caused by changes in material composition and texture, particularly at the surface, alignment errors can be significant, especially when dealing with large structures. Such errors typically affect the resultant weld quality, and may even lead to total failure to weld. Furthermore, before commencing welding on new workpieces (of different composition, or geometry) it is at present necessary to conduct a series of experiments followed by destructive analysis of the welds produced, which typically have compositional, microstructure and stress changes as compared to the original material, in order to determine the optimum processing conditions.

Despite these difficulties, the industrial application of laser welding continues to increase. In the case of Nd:YAG laser welding systems, the applications are stimulated partly by the commercial availability of optical fibre beam delivery systems, thus facilitating the integration of laser welding with robotised operation, and allowing operation on three-dimensional work-pieces.

The objective of this paper is to investigate the feasibility of non-invasive in-process monitoring of Nd:YAG laser welding, with the ultimate intention of closed-loop control of the process. An in-process monitoring system would allow errors to be detected as they occur. Corrective action can then be taken, reducing the quantity of scrap generated and avoiding the possibility of weld failure, particularly important in applications such as aerospace where weld integrity is paramount. A further step would be to close the process monitoring loop, giving real-time process control of parameters known to affect the weld quality, such as the laser intensity, scan speed, focal position, or spot size. Such process control would also assist greatly in optimising parameters for different materials, weld types, or workpiece dimensions. At present, highly time-consuming experiments are required for such evaluation because of the considerable number of parameters to be optimised.

A number of different potential sensors for weld monitoring have been previously reported, primarily with CO_2 laser systems. These include optical sensors, detecting either the UV/visible light emitted from the plume which forms above the surface, or infrared black-body emission from the melt pool [1]. An electrode to detect the flow of charge within the plasma has also been used [2], and acoustic sensors to detect shock waves generated by the plasma, or fluctuations in the light returned from the weld pool [3].

The fundamental design principle in the present work was to devise monitoring techniques that did not intrude on the process, and which were compatible with standard laser welding and beam delivery systems. In

This paper has been reproduced from Hand *et al* 1995 *Meas. Sci. Technol.* **6** 1387–1392.

our previous work, we have developed an optical sensor for fibre beam delivery systems, based on the detection of laser light propagating in the cladding of the fibre. A European patent (0 507 483) has been issued for this cladding power monitor (CPM), and we have exploited it for a number of applications including launch efficiency monitoring [4], workpiece position sensing (at low optical powers) [5], and drilling penetration monitoring [6].

In this paper the CPM is investigated as a weld monitor, based on the detection of light radiated by the plume of hot gas above the surface of the weld, using filters to block the Nd:YAG laser light and longer wavelengths. As an example, one particular type of process fault, occurring when the work-piece is out of the focus of the laser beam, is explored. This type of error is common and significant in robotised systems, particularly when welding large objects where errors in surface form are sufficient to cause a loss of focus. Signals were recorded for a wide range of focal errors for both steel and zinc-coated steel samples. A clear correlation was observed with out-of-focus error. The optical signals show considerable intensity modulation, due to the unsteadiness of the plume. Subsequent frequency analysis of the intensity modulation revealed spectral content changes which were correlated with theoretically-predicted weld pool oscillations. The monitoring technique can also, in principle, be extended to other types of weld error. For example, we were also able to detect the presence or absence of shield gas in the experiments, and to observe the effect of coating vaporisation in the zinc-coated steel samples.

2. Experimental

The basis of operation of the CPM technique is as follows. The CPM is installed in the fibre delivery system used with the Nd:YAG welding laser, close to the output optics. The CPM is analogous to a conventional optical fibre directional coupler, in that it couples light from the delivery fibre to a monitor fibre. However, it is designed to couple light from the *cladding* of the delivery fibre to the core of the monitor fibre. Hence, a fraction of the light radiated from or scattered in the region of the workpiece is collected by the output optics of the delivery system, and some of that light enters the cladding of the delivery fibre where it is guided (by internal reflection at the cladding-buffer interface) to the CPM, coupled into the monitor fibre, and thence travels to a photodetector.

A schematic diagram of the CPM is shown in fig. 1. Optical coupling between the cladding of the delivery fibre and the core of the monitor fibre is provided by an index matching adhesive, which is optimised to give maximum power transfer. The monitor fibre guides the light to the detection system, where optical filtering is used to detect the appropriate wavelength range. In these experiments, the UV/ visible light emitted from the plume was detected using a UV-enhanced silicon photodiode with a KG-5 glass filter, giving a wavelength detection range from ~ 0.3 to ~ 0.75 μm. It was known that radiation from the surface of the work-piece occurs predominantly at longer wavelengths.

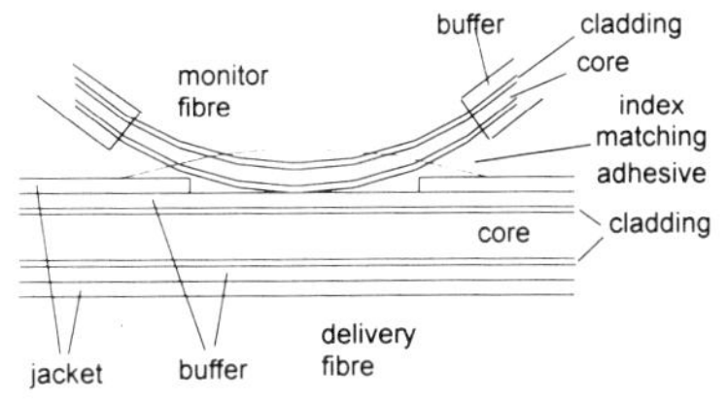

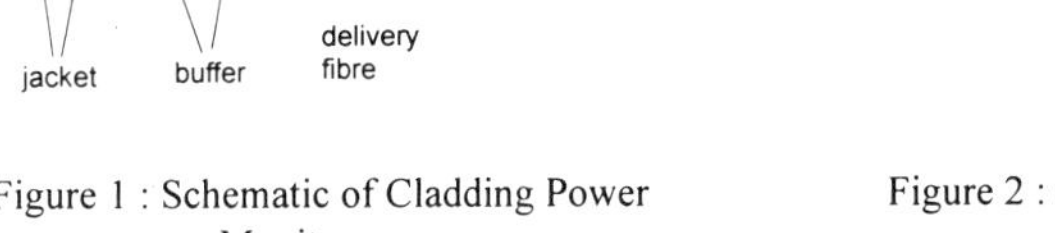

Figure 1 : Schematic of Cladding Power Monitor

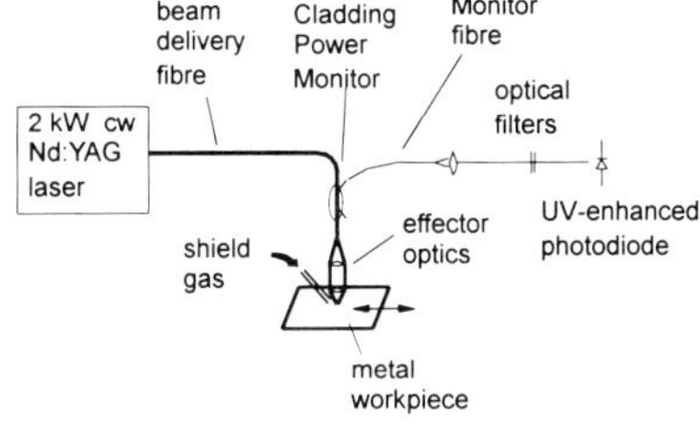

Figure 2 : schematic of welding arrangement

The laser source used was a Lumonics MultiWave™ 2000 Nd:YAG laser, operating cw at 1.06 μm and producing an average power of 2 kW. This was delivered to the workpiece through a silica fibre optic with a core diameter of 1mm. The CPM was incorporated into this fibre ~ 0.5 m from the effector (output) optics, which collimate and focus the light onto the workpiece, giving a 0.5 mm diameter spot.

A schematic diagram of the welding arrangement is shown in fig. 2. The metal plates to be welded were clamped onto a motorised stage which traversed underneath the effector optics to produce 'lap' welds. An inert 'shield' gas, in this case Argon, was introduced at the leading edge of the effector optics, flowing across the metal surface to prevent oxidation in the heated region. To investigate the out-of-focus problem, the metal plates were mounted at an angle, producing a continuous change in focal position relative to the surface as the stage was traversed, going from ~ 3 mm above the surface to ~ 4 mm below.

Experiments were performed with both single and double sheets of 1 mm thick mild steel, using a suitable traverse rate to give full weld penetration (producing a weld bead on the rear surface) in the region of optimum focus. In addition, the effect of interrupting the shield gas was investigated. Welds were also produced in zinc-coated steel. The low vaporisation temperature of zinc compared to steel can cause problems, particularly when lap welding two sheets, as the zinc vapour bubbles through the molten steel, causing spatter and pores within the resultant weld. This can be avoided by ensuring a small gap between the two sheets which allows the zinc vapour to escape. CPM measurements were therefore made with and without such a gap, in order to determine whether 'spattering' could be detected. A number of the welds were subsequently sectioned to determine melt pool volume and cross-sectional areas.

3. Results

A typical signal obtained from the CPM UV/ visible detector when welding across an angled steel plate is plotted in fig. 3. This signal profile was found to be highly repeatable, reaching a maximum to one side of the optimum focus position, when the focus was below the surface of the workpiece. Optimum focus is defined as the centre of the full penetration region, denoted by the vertical dotted lines in the graph. In addition there is an increase in signal modulation around the optimum focus point. Also marked on this graph (by the vertical dashed lines) is the start and end of the 'enhanced coupling' region. This corresponds to the start and end of metal vaporisation at the surface. A clear change in CPM signal level is seen at this point.

The increased modulation when at focus led us to try two forms of further analysis in an attempt to arrive at a simpler descriptor of the process, not reliant on the absolute level of the signal intensity. Both analysis techniques were based on measures of the average frequency of the intensity modulation of the signal, calculated either from FFTs or from the standard deviation, and are plotted in this figure; each show a peak corresponding to the optimum focus.

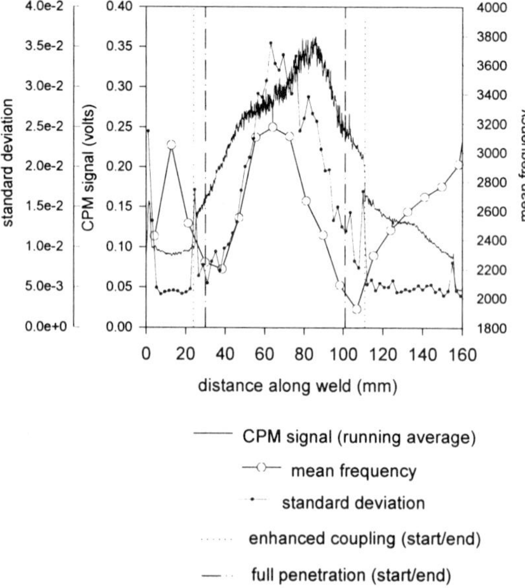

Figure 3 : signal from UV/ visible CPM detector when welding across an angled mild steel plate

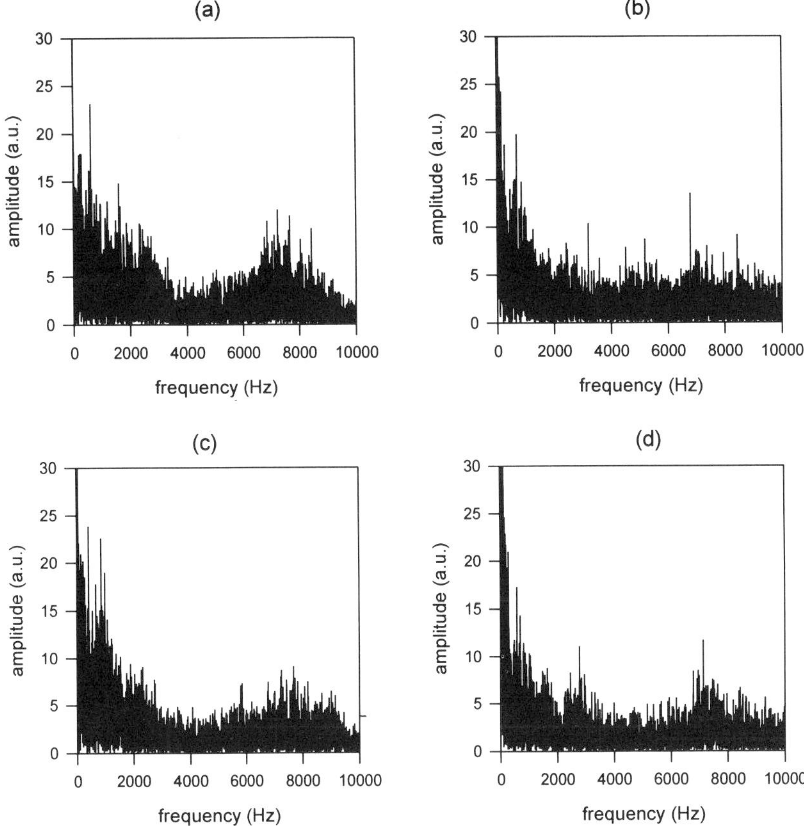

Figure 4 : normalised frequency spectra calculated from UV/ visible CPM detector when welding across an angled mild steel plate at various focal positions, measured as distance of the focus above the surface:

(a) = +1.1 mm (b) = 0 mm
(c) = -1.1 mm (d) = -1.5 mm

The frequency spectra used to calculate the average frequency are plotted (normalised) in figure 4, and it can be seen that the spectra in each case are very broad. This type of weld was repeated for a double sheet, with a suitable reduction in welding speed to ensure full penetration, and the signal detected is plotted in fig. 5, together with the standard deviation and average frequency. The traces are very similar to those obtained with the single sheet; and the lower average frequency may be accounted for by the lower sampling rate used in this case (due to the longer sampling interval), and the broadband nature of the frequency spectra.

The effect of blocking the shield gas is shown in figure 6, where the CPM signals are plotted for two welds made in single sheets of steel mounted horizontally, with identical welding parameters, but (a) with shield gas, and (b) without.

With a double sheet of zinc-coated steel mounted at an angle, and an 80 μm gap between the sheets, the trace shown in fig. 7 was obtained. A much larger modulation is observed than with the uncoated mild steel, but similar trends are still observed with both the average frequency and standard deviation signals. When welding was attempted with no gap, the modulation became more extreme.

Subsequent measurements of weld cross-sections were made for a number of the different angled steel samples. A clear narrowing of the melt pool is observed in the centre of the full penetration region, corresponding to the improved focus of the beam.

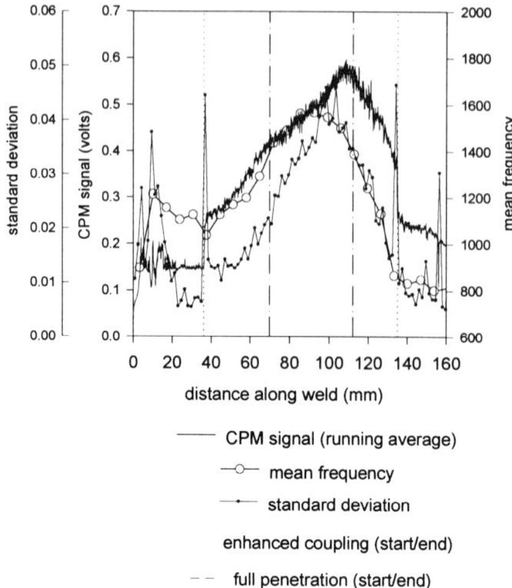

Figure 5 : CPM signal when welding across two sheets of mild steel plate mounted at an angle

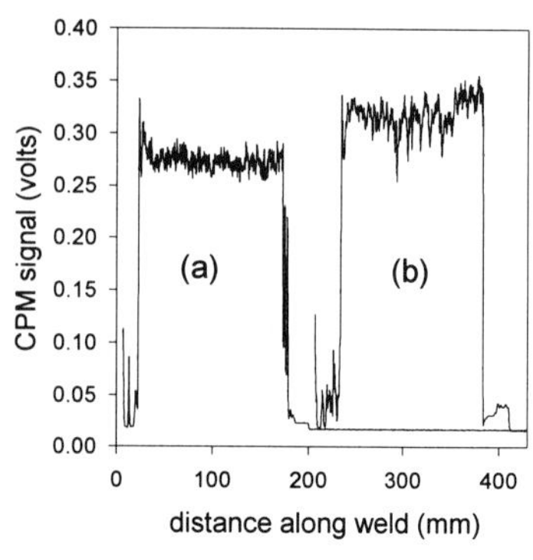

Figure 6 : CPM signal when welding single sheets of horizontally-mounted mild steel (a) with shield gas, and (b) without.

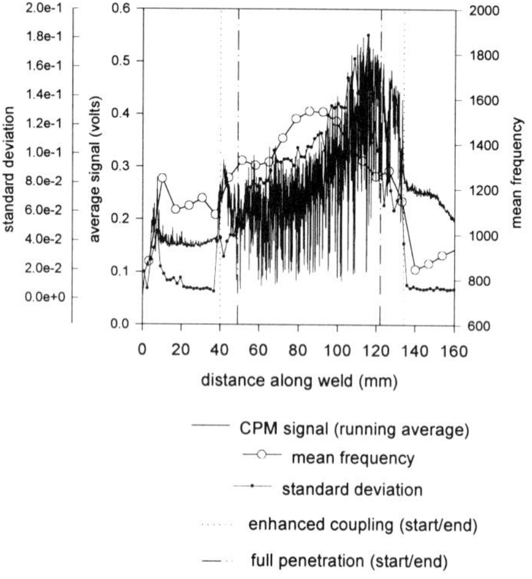

Figure 7 : CPM signal when welding across two sheets of zinc-coated steel plate mounted at an angle

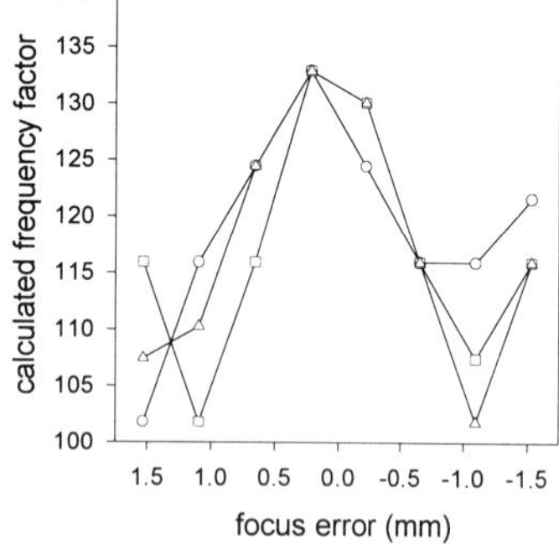

Figure 8 : frequency factor F for three different weld samples plotted as a function of focus error

4. Discussion

The results demonstrate that the optical signal from the plume detected with the CPM has a clear relationship with out-of-focus error. Most importantly, the asymmetry of this signal would allow the out-of-focus direction to be determined, particularly if combined with the symmetrical standard deviation or FFT calculation. This signal could therefore be used as the basis of a process control system.

The sharp increase in signal level observed at the start and end of the enhanced coupling region is easily explained by the abrupt increase in metal vaporisation, giving rise to a significant plume.

The variation of the signal mean frequency with focal error corresponds with previous theoretical work on weld pool oscillations [7], where the frequency of these oscillations was found to be dependent on weld pool dimensions, associated with the fundamental modes of the melt pool. A number of different modes may be excited, and the oscillation frequency of each is described in terms of a dimensionless frequency ω', independent of melt pool dimensions. Provided that the melt pool depth is greater than its surface width, d, then

$$\omega = F\omega', \quad \text{where} \qquad F = \left(\frac{\Lambda_0}{d^3\rho}\right)^{\frac{1}{2}} \qquad (1)$$

where Λ_o is the average surface tension of melt, and ρ the melt density.

Plotting F for the values of d measured by the sectioning described above, for three different weld samples, the graph shown in fig. 8 is obtained. A peak in frequency is observed, similar to that measured with the CPM. The CPM, of course, looks at the plume of hot metal gas instead of the surface. As the surface oscillates, however, it is reasonable to expect the gas which produces the plume is ejected from the surface at correspondingly fluctuating angles. It is therefore likely that light emitted from this plume would experience oscillatory coupling with the fibre cladding.

5. Conclusions

A non-intrusive optical sensing technique for Nd:YAG laser welding monitoring has been demonstrated. This sensor is incorporated into the delivery fibre optic, detecting the light returned through its cladding, and so is compatible with existing beam delivery systems. We have demonstrated the potential of this sensor for process control applications, with successful detection of a number of welding faults, including out-of-focus errors and shield gas interruption.

6. Acknowledgements

This work was partially supported by the CDP committee of the EPSRC. One of the authors (CP) is partially supported by a Fellowship from the Royal Commission for the 1851 Exhibition, in association with the University of Liverpool.

6. References

[1] H. B. Chen, L. Li, D. J. Brookfield, W. M. Steen, 'Multi-frequency fibre optic sensors for in-process laser welding quality monitoring', NDT&E International, **26** (1993) 67-73.

[2] L. Li, N. Qi, D. J. Brookfield, W. M. Steen, 'Laser weld quality monitoring and fault diagnosis', in proc. of Conf. on Laser Systems Application in Industry, Torino, Italy (1990) paper 3.7

[3] L. Li, W. M. Steen, 'Non-contact Acoustic Emission Monitoring During Laser Processing', in proc. of ICALEO, Orlando, U.S.A. **75** (1992) 719-728

[4] A. A. P. Boechat, D. Su, J. D. C. Jones, 'Bi-directional cladding power monitor for fibre-optic beam delivery systems', Meas. Sci. Technol. **3** (1992) 897-901

[5] D. Su, D. R. Hall, J. D. C. Jones, 'Workpiece position sensing by means of a fiber optical beam delivery system', Optical Engineering **32** (1993) 1823-1826

[6] D. Su, I. Norris, C. Peters, D. R. Hall, J. D. C. Jones, 'In-situ laser material process monitoring using a cladding power detection technique', Optics and Lasers in Engineering **18** (1993) 371-376

[7] N. Postacioglu, P. Kapadia, J. Dowden, 'Capillary waves on the weld pool in penetration welding with a laser', J. Phys. D **22** (1989) 1050-1061

A Distributed Optical Fibre Sensor for Offshore Applications

B. P. Ludden, J. E. Carroll and C. J. Burgoyne

University of Cambridge, Engineering Department, Trumpington Street, Cambridge. CB2 1PZ.

Abstract. There is an urgent need in the offshore industry to monitor damage to Parallel-lay ropes that are widely used in that hostile environment. We will present an account of ideas and experiments that show how to assess the damage rate to parallel-lay ropes of up to a few kilometres length. It is envisaged that the information collected by the sensor will be stored and displayed as a histogram of localised damage accumulation verses rope position.

1. Introduction

The idea of 'smart structures' emerged some eight years ago from America. Since then there has been increasing interest and research in incorporating sensors in a variety of structures in the aerospace and civil engineering industries[1].

One of the most promising candidates as a sensor is the optical fibre[2,3]. The advantages of using an optical fibre as the sensing element have been well discussed. Amongst these advantages their immunity to electromagnetic interference, their electrical passivity, high sensitivity and high bandwidth are probably the most attractive. However, their ability to measure a diverse range of measurands and robustness also makes them adaptable to many applications even in the most hostile of environments.

Here we propose a distributed optical fibre sensor to measure the structural damage of Parafil ropes used in the offshore industry. It is envisaged that such ropes will play an important part in the realisation of ambitious structural projects in the marine environment such as deep-sea oil platforms and submerged-tube bridges.

2. Parafil Ropes

Parafil ropes[4] made from synthetic fibres were developed in the 1960's by Linear Composites.

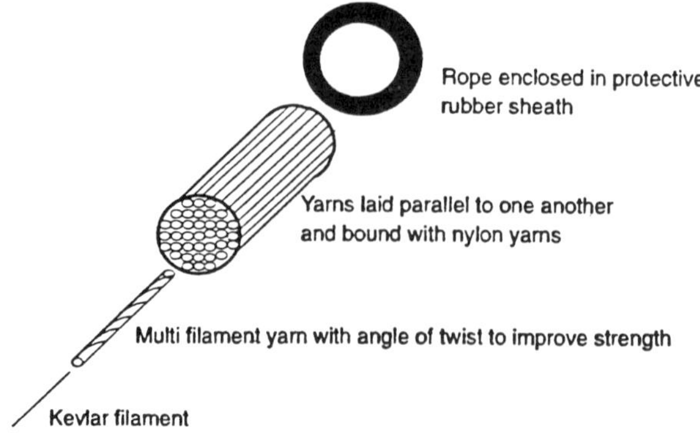

Figure 2.1 Architecture of Parafil ropes

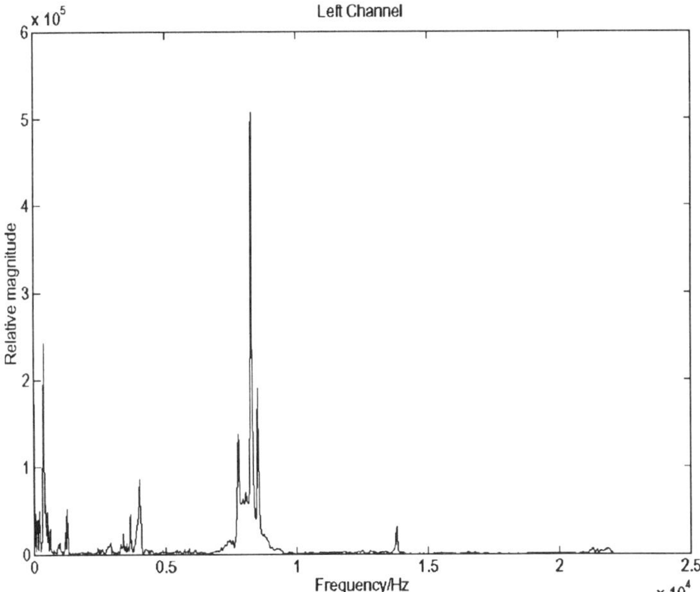

Figure 2.2 The frequency spectrum of snapping yarns in aramid Parafil ropes. The peak centred at 8kHz is characteristic of the snapping yarn. The other two peaks represent the resonant frequencies of the test equipment and the rope itself ('guitar string' effect) respectively.

They are basically a parallel array of 'yarns' bundled together and encased in a polymeric sheath. Each yarn consists of about a thousand micron diameter fibres laid together and then slightly twisted to acheive optimum strength. They have very attractive properties for the construction industry including high tensile strength, high elastic modulus, low weight and high corrosion resistance. A variety of yarns can be used, but for structural applications polyester or aramid yarns are the most common.

Research into the failure of Parafil ropes has yielded some important facts for any attempt to develop a sensor to monitor their integrity. When a Parafil rope sustains damage it occurs as complete individual yarns snapping. The failure of the yarns is characterised by clearly audible acoustic signal (figure 2.2). A yarn which has failed will recoil back into the rope in both directions until the friction forces between it and its neighbours cause it to take up the load again. The length that the yarn recoils along the rope is called the characteristic length and is typically a few metres. Thus, the damage that a rope accumulates will be local and will only weaken the rope in a region equal to twice the characteristic length.

It is therefore desirable to be able to detect yarn snapping as it occurs and to locate where each yarn has snapped. A histogram could then be generated giving a simple display of the condition of the rope along its entire length.

3. Principles of Sensor Operation

3.1. Detection of Rope Damage

It is widely known that the properties of light propagating along a fibre, such as polarisation, mode or intensity, can be affected by external fields[5,6,7]. For example, we have demonstrated that a small acoustic signal will cause light to be coupled between modes in a two moded fibre at the frequency of the acoustic field (figure 3.1.1).

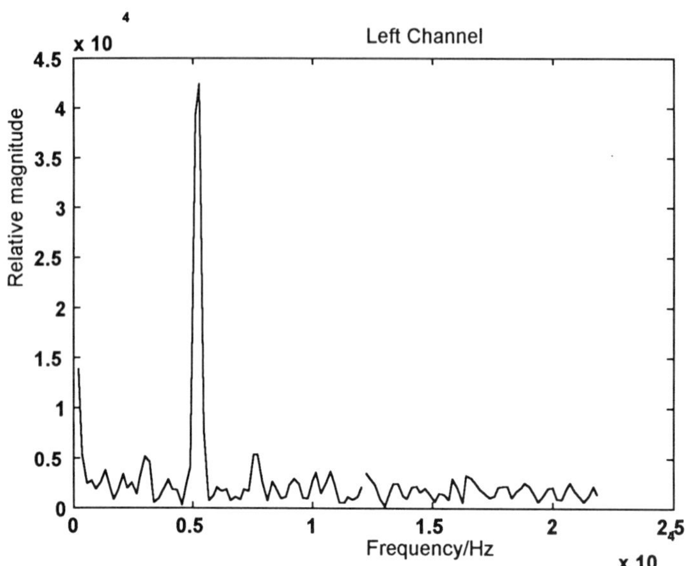

Figure 3.1.1 Response of a 'modal coupling' optical fibre sensor to a 5kHz acoustic wave

If an optical fibre is incorporated into a Parafil rope then a snapping yarn will change the properties of the light propagating along the optical fibre. The changes in, say, the amount of light in the zero order mode will carry information about the frequency of the detected signal. It is therefore possible to detect yarns snapping, and to distinguish the snaps from other noise in the sensing environment.

The principles of using this technique have been demonstrated by helically winding a small length of an 800m long standard telecommunications fibre around a 'bare' (polymeric sheath removed) six ton Parafil rope of 3m length and loading the rope to failure.

The laser source was a He-Ne 633nm laser and light was launched into the communications fibre (which is multimoded at 633nm) from a single moded fibre. The light was reflected by a mercury mirror at the end of the sensing fibre and the returning signal was mode stripped before detection, to monitor the amplitude modulation of the light in the zero order mode. The signal was then amplified and recorded on a dat-recorder.

The signals from the breaking yarns were easily distinguishable from other noise in the workshop environment, but were dominated by the 'guitar string' effect of the 3m long rope vibrating (this would not be a factor in the large area, long ropes employed in offshore anchoring) making frequency analysis difficult. However, the robustness of the optical fibre

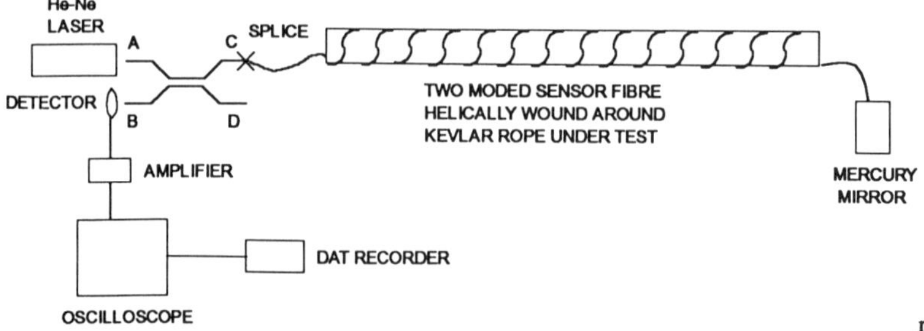

r

Figure 3.2.2 Detection of yarns snapping in an aramid Parafil rope using an optical fibre sensor was highlighted by this experiment. The unprotected optical fibre remained undamaged by the snapping yarns in all tests except one, where the fibre only broke after complete failure of the Parafil rope. This is an important indication of the compatibility of the optical fibre with this sensing environment.

3.2. Location of Rope Damage

A modification to the above techniques enables distributed sensing. If a number of broad band partial reflectors are distributed along the length of the optical fibre, it would be possible to locate the position of snapping yarns with a resolution dependent on the reflector spacing.

A narrow pulse of light (typically 10-100ns) is launched into the optical fibre. The small percentage of reflected light sent to the detector from each reflector will carry information about the properties of the light passing that point. By analysis of the pulse signals arriving from each reflector it is possible to determine between which two reflectors a signal has occurred.

We have demonstrated the principles of this technique by launching 100ns pulses of highly polarised 1300nm laser light into a sensor fibre with 5 partial reflectors spaced at 100m intervals. The returning pulses of light were passed through a polariser before detection at a photo diode. An oscilloscope was used to display the pulses and it was possible to locate along which 100m section of the sensing fibre a test signal occurred.

An example of a distributed sensor for monitoring large Parafil ropes is shown in figure 3.2. The details of the laser source, fibre type, detector technology, signal processing and light property to be measured have been omitted. It is the subject of further work to determine the optimum configuration of the system.

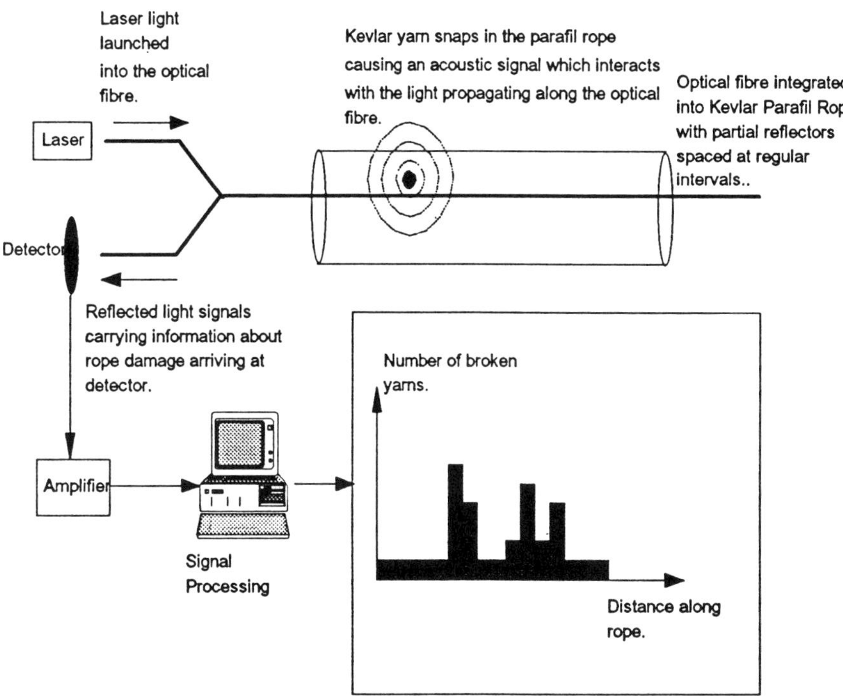

Histogram of Rope Damage.

4 Conclusion

We have demonstrated the principles of a distributed optical fibre sensor for monitoring the integrity of large Parafil ropes in offshore applications. It has been shown that by introducing a number of partial reflectors along the fibre, and by interrogation of one particular property of the light , it is possible to detect and locate a short transient signal such as the acoustic emission caused by a yarn snapping.

It is envisaged that this type of sensor could be applied to any application where it is desirable to locate and/or identify a transient signal, such as a security perimeter fence or to measure traffic speed.

5 References

[1] D. Uttamchandani, "Fibre-optic sensors and smart structures: Developements and prospects", *Electronics & Communication Emgineering Journal,* Vol. 6 No. 5, 1994.
[2] T. G. Giallorenzi et al, "Optical Fibre Sensor Technology", *IEE Journal of Quantum Electronics,* Vol. QE-18 No. 4, 1982.
[3] M. J. Dill & I. L. Curtis, "Monitoring of concrete structures using optical fibre sensors", *Published by LTG,* 1993.
[4] C. J. Burgoyne (editor), "Symposium on Engineering applications of Parafil ropes", *Imperial College of Science and Technology Department of Civil Engineering,* 1988.
[5] J. F. Sevic & D. B. Patterson, "Noninvasive polarization eigenmode coupling in elliptical-core optical fibre based on bulk-shear acoustic waves", *Optics Letters,* Vol. 18 No. 23, 1993.
[6] S. J. Garth, "Intermodal coupling in an optical fiber using periodic stress", *Applied Optics,* Vol. 28 No. 3, 1989.
[7] Herga information sheet on "Fibre optic sensing safety products".

Sensing the Cleanliness of Teats by Optical Inspection

C R Bull, N J B McFarlane & R Zwiggelaar

Silsoe Research Institute, Wrest Park, Silsoe, Bedford, MK45 4HS, UK

Abstract

In order to implement robotic milking it is necessary to automate the inspection and cleaning of the cow teats prior to milking. This paper investigates the feasibility of producing a practical sensor of teat cleanliness based on the optical reflectivity of the teat. A fibre optic sensor and fast scanning monochromator were used to make a series of measurements of the visible and the near infra red reflectance properties of teats and the major surface contaminants. From this data spectral features which uniquely correspond to manure were identified. This paper then assesses the potential of detecting these spectral differences using a standard colour CCD camera.

1. Introduction

The first priority for automatic milking has been the development of a robotic technique to locate the teats and attach teat cups (Frost, 1991). However, an important element of any milking systems is preparation of the teats prior to the attachment of the teat cups. One option would be to clean all teats, irrespective of their state of cleanliness. However, an increase in the washing and drying of teats has been shown to increase the incidence of skin sores and chaps (Phillips et al., 1981). Consequently there is much to be gained from washing only those teats which require contamination to be removed. This indicates a requirement for an automatic inspection procedure.

In this paper we seek to identify absorption features in the visible to near infra-red spectral range (380 - 1100 nm), which uniquely correspond to manure or the clean teats themselves, with a view to identifying a discrete number of measurement wavelengths that can be used to assess teat cleanliness. This data is presented in more detail in Bull et al., (1995). In the second part of the study we investigate the possibility of detecting the spectral variations in reflectivity using a colour CCD camera.

2. Equipment and method

The spectral measurement system consisted of a light source, a bifurcated light guide and a monochromator (Rees Instruments Ltd.). Light from the source was guided onto the teat surface along one arm of the bifurcated light guide whilst the reflected light from the teat was transmitted by the second set of fibres into the spectrum analyzer. The fibre optic light guides were 500 mm long, allowing the sensor heads to be held close to the teat. The optical spectrum analyzer took a spectral scan in the wavelength range of 380 - 1100 nm. Spectral scans were taken on a series of clean and contaminated teats.

To obtain images of the cow teats the cow was admitted to a milking stall. In the stall she stood with her front legs on a 15 cm raised platform. This helped to move the udder forward from between the rear legs, hence improving the view of the teats. The camera was placed approximately 1 m in front of the cow at an angle of 45°. This usually gave an adequate image of the udder and the four teats. The scene was illuminated by 2 x 100 W halogen floodlights placed on a beam 500 mm each side of the camera. This gave reasonably diffused lighting to the scene although the rear teats were often shaded by the legs and the surfaces of the teats closest to the light source were brighter than those further away. The camera used in this investigation was a Sony camera (DXC-151AP) which took a 256 x 256 pixel image. Each pixel of information corresponded to approximately 1 mm^2 of the teat surface. When a suitable image was displayed on the monitor the image was stored. For the investigation 10 cows were selected. A series of images at various contamination levels were taken for each cow. Once the images had been obtained the parts of the image corresponding to the teats were manually identified. Data for each teat was separately stored and analysed.

3. Results

Typical absolute reflectance scans for clean and contaminated teats are shown in fig. 1.
The clean white teat has a number of prominent spectral features. In contrast the clean teat with black pigmentation shows few spectral features with the reflectance varying only slightly across a spectrum. The reflectivity spectra of manure is very different from that of the white or black teat (fig. 1) with a major absorption feature at 670 nm, which corresponds to the chlorophyll absorption band.

A colour analysis programme has been used to relate these spectral responses to the colour coordinates that would be given by a colour camera (using the PAL system). The RGB values given by each channel of the camera depend on the intensity of the image as well as the colour. This is affected by variations in the lighting and also in shading caused by, for example, changes in stance of the cow. To overcome this we can look solely at the colour information by

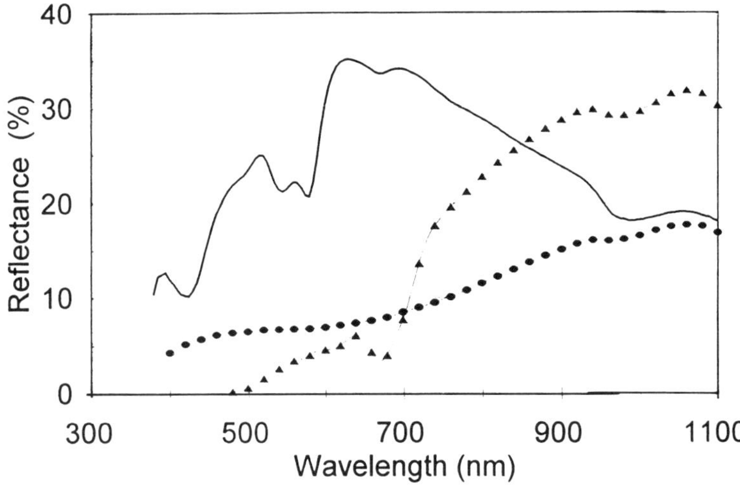

Fig. 1. The reflection spectra of clean white (——) black (......) and manure contaminated (▲▲▲▲) teats.

removing the intensity component and examining the normalised colour coefficients. For example, the normalised red coordinate r_n is given by $r_n = R/(R+G+B)$. The r_n and g_n coordinates of the spectral scans under white light illumination are shown in fig. 2.

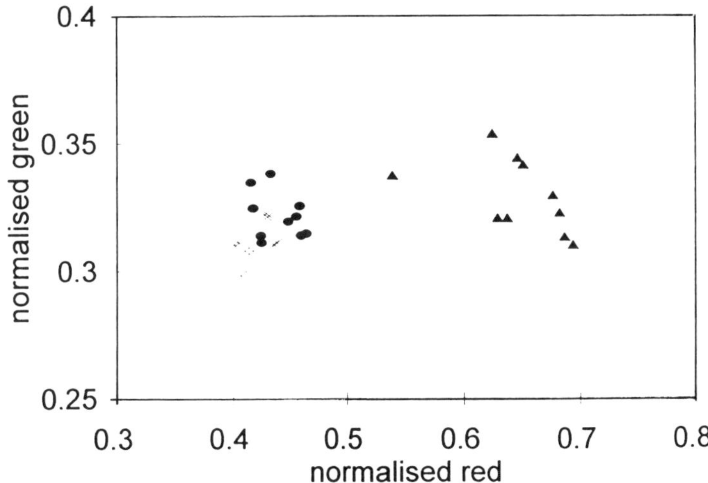

Fig. 2. The colour coordinates of spectral scans corresponding to clean white (×) and black (●) teats and manure (▲)

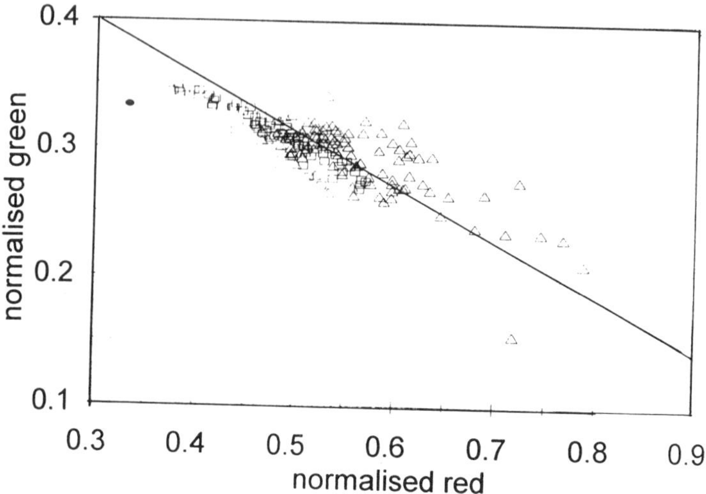

Fig. 3. The colour coordinates of selected pixels from an image of a white teat. Pixels corresponding to manure (▵) and clean flesh (□).

The colour coordinates in normalised colour space of selected pixels from an image of a contaminated white teat are shown in fig. 3. There is considerable overlap between the groups of clean and contaminated teats. Nevertheless the pixels corresponding to manure tend to have higher $r_{n,}g_n$ coordinates.

4. Interpretation

Results from the spectral scans show that the white pigmented teat is more reflective than both the black and contaminated teats. However, it is not possible

to discriminate between white and black or contaminated teats on the basis of the magnitude of the reflection because the reflection signal changes markedly with distance of the sensor from the teat. In order to distinguish between the different scans we must consequently look for unique spectral features. Examination of Fig. 1 shows that manure is strongly absorbing at 670 nm due to the presence of the chlorophyll pigment. By taking the ratio of the reflectance at this chlorophyll absorption band and at a closely adjacent reference band it is possible to identify scans in which chlorophyll is present. The ratio of reflectances at 670 and 690 nm has been calculated for each of the scans and the values are presented in fig. 4, plotted against the teat contamination level. It can be seen from this figure that it is possible to select a threshold ratio which can be used to distinguish between the dirty and clean teats.

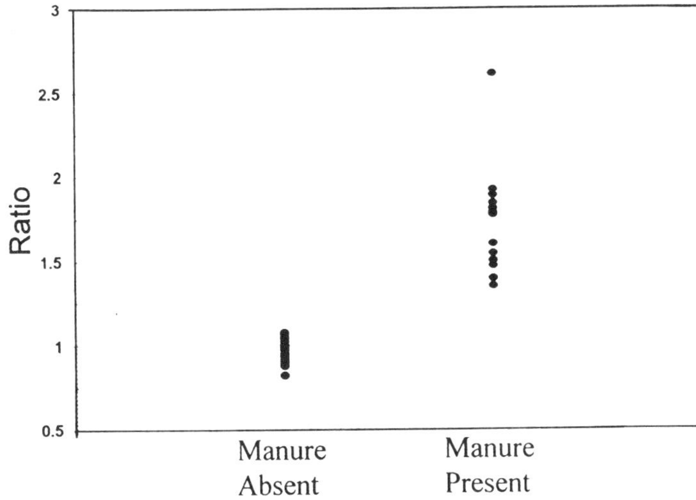

Fig. 4. Ratio of the reflectance at 690/670 nm as a function of the teat condition.

The conversion of these spectral scans into equivalent RGB signals using the PAL conversion system (fig. 2) suggests that the spectral differences are sufficient to give colour differences in the normalised colour space available to a colour camera. However, if we look at figure 3 we can see that there is considerable overlap in the colour pixel coordinates of a clean white teat and manure. There are several factors that affect our perception of colour.

The first is intensity. This component does not appear in the normalised colour coordinate. The second is saturation. This corresponds to the fraction of white light (coordinates $r_n=0.33, g_n=0.33$)in the signal. Finally there is the hue, which is the dominant spectral content of the signal. For example the colour pink has a red hue but is
pale (or of low saturation) because of the addition of white light, which brings the colour coordinates closer to the 0.33, 0.33 coordinate.

The colour coordinates of the reflection spectra illustrated in figure 2 show that the hue of each of the colour clusters is very similar and that the major variation in colour between clean and contaminated teats is the degree of saturation. In the image (fig. 3) the colour coordinates of pixels corresponding to the manure and flesh appear to merge along a line which can be extrapolated back to the white light colour coordinates ($r_n = 0.33$ $g_n = 0.33$). In other words the saturation of the pixels corresponding to the two colour regions varies considerably. This variation in saturation is due to the variations in specular reflectance. Fig. 2 shows the reflectance properties of manure and clean skin due to diffused reflectance alone. In the camera images some of the light from the floodlights will be specularly reflected (mirror-like reflection) back to the camera from the teat. This will affect the coordinates of the pixel by decreasing its saturation although the hue will remain invariant.

Despite the merging of the colour coordinates of pixels corresponding to clean and dirty teats there is some degree of separation in fig. 3 with the manure pixels having slightly higher r_n,g_n values. Unfortunately if we compare these colour coordinates with those taken from another image we find that they have shifted slightly in saturation and hue and much of the discrimination in colour space between clean and dirty areas of the teat is lost. This is probably due to slight changes in the colour of the illuminating light. In fig. 3 a line is drawn which optimally divides the colour clusters associated with manure and clean flesh. The equation of this line is g_n=-0.43+0.529. If we isolate pixels with high r_n,g_n values (above the line drawn in fig. 3) such that g_n>(-0.43+0.529) then we still detect pixels with a green hue corresponding to manure. If we define a teat with more than 4% of its pixels lying in this green hue region as dirty, then all the dirty teats (12 images) are correctly identified whereas some (22%) of the clean teats are assessed as dirty.

Conclusion

It is possible to identify the presence of manure on teats by comparing the ratio of reflectance at 670 to that at 690 nm or by looking at the colour coordinates of the diffusely reflected light. However, if we move from the controlled situation of the sensor being close to the teat the returning signal becomes harder to interpret as we have less control of the direction and colour content of the illuminating light, and consequently there are difficulties in interpreting the information from a colour image. In this investigation we attempted to classify pixels of an image of a cow teat as clean or dirty from the basis of their hue, saturation and intensity. It was found however that hue, saturation and intensity of the manure or clean flesh varied between images. Furthermore without knowing which, if any, part of the scene corresponds to clean flesh it is not possible to identify pixels as dirty by comparison. Consequently an accurate assessment of the teat contamination on a pixel by pixel basis is not possible. However, it has been shown that patches of manure on the skin display areas that have a greener hue than flesh, and by looking for these outlying points and colour space it is possible to identify grossly contaminated teats even when the lighting cannot be strictly controlled.

References

Frost, A.R., 1991. Robotic milking: A review. Robotica, 8: 311-318.

Bull, C.R., Mottram, T. and Wheeler, H. 1995. Optical inspection for automatic milking systems. Accepted by Computing and Electronics in Agriculture.

Phillips, D.S.M., Malcolm, D.B. and Copemen, P.J.A., 1981, Milking preparation methods - their effect and implications. Proceedings of Ruakura Farmers Conference, Ruakura Animal Research Station, Hamilton, New Zealand.

Application of optical fibre temperature sensors to large machine tools for quality improvement

J S McKenzie, Y Lu and C Butler

Department of Manufacturing and Engineering Systems, Brunel University, Uxbridge, Middlesex, UB8 3PH, UK.
Tel: 0895-274000 ext 2954 Fax: 0895-812556

Abstract This paper reports on a multi-channel optical fibre temperature sensor system developed to allow multi-point temperature measurement on a 5 meter bed milling machine. The temperature sensor is based on the band-edge shift of a semiconductor element, Gallium Arsenide (GaAs). Light from a single LED light source is launched into an optical fibre and then passes through a GaAs element, the light transmitted is collected by a return fibre and conveyed to the detectors. These are two element photodiodes with optical filters placed over each. Two wavelength bands of light are selected using the interference filters. One wavelength band is used as the sensing wavelength and the other is the reference. Thus changes in the attenuation of the optical path can be largely cancelled out. The range of the system is 0-100 °C with a resolution of 0.1 °C and an accuracy of +/- 1.8 °C. The complete system with 24 channels is connected to a PC which allows the results to be monitored and logged.

1.Introduction

Modern machine tools are capable of manufacturing components to fine tolerances and at high speed. Control of the output is usually effected by means of post process inspection using statistical methods and the effects are usually fed-back into the process. This approach falls short of the ideal for advanced machines because of the time delay in the feed-back loop and the omission of other important process parameters such as temperatures, forces and vibrations in the machine. To achieve more effective control, higher productivity and improved piece quality, it is necessary to provide sensors for the important machine and piece parameters and to process this data very rapidly to detect machine problems and act accordingly. This approach requires a number of sensor types including:-

- Temperature - In contact with various parts of the machine
- Vibration - In contact with the machine
- Force - Incorporated into the machine
- Dimensional - Non-contact displacement, surface finish/roughness

Due to the presence of high power electrical motors the machine tool environment is extremely noisy. Conventional electrical sensors require a great deal of shielding to perform in such environments. It is well known that optical fibre sensors can offer a number of key advantages (Jones 1991). The main ones for this kind of environment being immunity of these sensors from electromagnetic interference (EMI). Although optical fibre sensors have undergone substantial development relatively few are commercially available.

For the 5 metre bed milling machine under study, thermal deformations of the machine due to changes in ambient temperature and localised heating of parts of the structure by the operation of the motors have been found to introduce large errors in the dimensions of the parts being machined. It is expected that by monitoring the temperature at various points on the machine and correcting for these deformations that these errors can substantially reduced and the quality of the parts improved.

The main requirements for the temperature sensors are to measure the temperatures inside the structure of the machine with an operating range of 0 -100 °C and a resolution of at least 0.1 °C. As these sensors are to be used to measure the temperature of the structure of the machine they must be easy to insert into the machine. In addition, as the machine vibrates during operation the sensor head should desirably have no moving parts. The speed of response of the sensor must be 5 seconds to allow significant changes in the machine temperature to be detected.

A number of single point optical fibre temperature sensors have been produced including a number based on fluorescent materials (Grattan and Palmer 1987) these tend not to offer the speed of response required and use quite exotic materials. A number of sensors are based on refractive index change of modified fibres (Scheggi et al 1983 Soares and Dantas 1992), but these cannot be referenced with respect to changes in the optical path due to microbending, changes in coupling efficiencies etc.

A number of interferometric temperature sensors have been developed. These are based mainly on the Mach-Zender configuration and Fabry-Perot cavities. The Mach-Zender interferometer offers extremely high sensitivity, but has the disadvantage of requiring a reference arm that must not be disturbed and typically the associated optics and signal processing are relatively expensive.

With a Fabry-Perot cavity the wavelengths that are either transmitted through or reflected back from the cavity vary according to the thickness of the cavity or its refractive index. A number of temperature sensors based on these cavities have been produced (Farahi et al 1987, Akhavan-Leilabady 1987 Beheim et al 1987). The complexity and cost of the detectors (spectrum analyzers, reference cavities etc) and the susceptability of the cavity itself to vibration under certain circumstances make this kind of technique impractical for this application.

Another technique is to use a semiconductor material as the sensing element (Kyuma et al 1981). The band-edge of semiconductors changes with temperature and with it the wavelength of the light that can pass through it (see Figure 1). Using a single wide band LED centred at 0.88 µm a temperature range of -10 to 300 °C could be covered using a GaAs crystal connected in line with two optical fibres with an overall accuracy of +/- 3 °C (Kyuma et al 1981). A further refinement of this system was to use a second LED source at 1.27 µm as a reference wavelength (Kyuma et al 1982).The LEDs were alternately pulsed. The two wavelengths were combined using a coupler and passed through the GaAs sensing element and detected using a single photodiode. The system had a reported accuracy of +/- 1 °C over the range -10 to 300 °C. Both of these systems used in-line coupling of the fibres and a GaAs crystal. A commercial version (Mitsubitshi Electric Corp) of this device used an additional coupler and a single fibre connected to an anti-reflection coated GaAs crystal. This system operated over the range -20 to 150 °C with an accuracy of +/- 2 °C (Dakin and Culshaw 1989). This kind of system requires a number of expensive optical couplers.

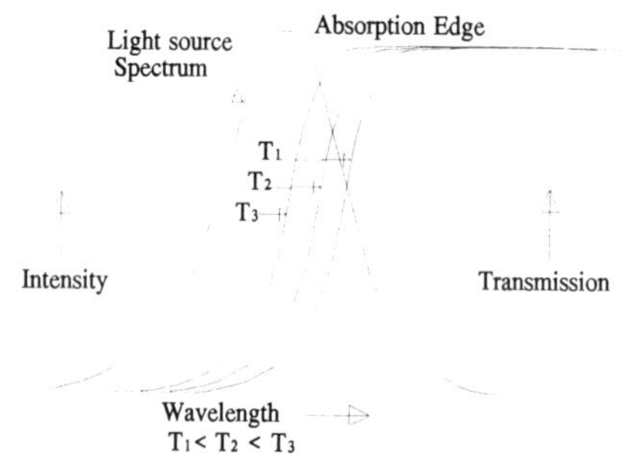

Figure 1 Semiconductor absorption edge principle

The system described in this paper is based on the shift in band edge of a GaAs element with temperature and does not require the use of any expensive couplers.

Initial experimentation

Using a white light source, an in-line probe and an optical spectrum analyzer the magnitude of the change in band gap was investigated (see Figure 2). The position of the band gap at two temperatures is shown in Figure 3 and Figure 4. A water bath was used to ensure the temperature remained stable. As can be seen, the position of the band gap increases with increasing temperature and is estimated to be 0.35 nm/°C as found by other authors (Kyuma et al 1981).

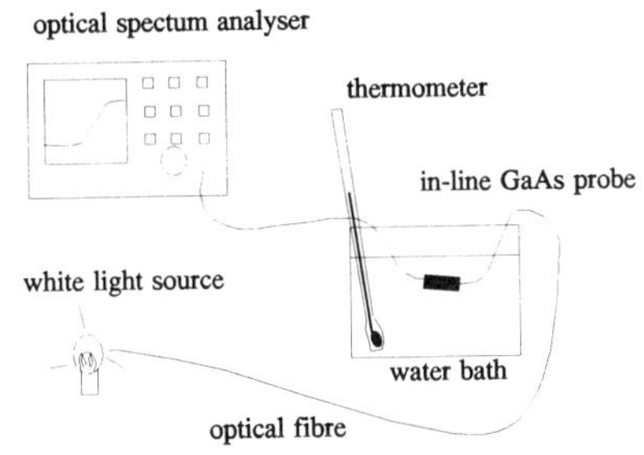

Figure 2 Schematic of test apparatus

The in-line probe is not very applicable for insertion into the body of the machine and a number of other designs were tried all based on the one fibre in and one fibre out design in order to eliminate the need for a coupler. One used a prism of GaAs to reflect the light back into the return fibre, see Figure 6 type 2 (Christiansen and Vaguine 1987), this proved difficult to manufacture and was replaced with a probe with two parallel fibres with a GaAs platelet between the fibre ends, with the fibre ends polished at 45 ° to allow the light coupling between the fibre ends (Schoener et al 1984), see Figure 6 type 1. The transmission loss of light through this type of probe is at best 8 dB (Schoener et al 1984).

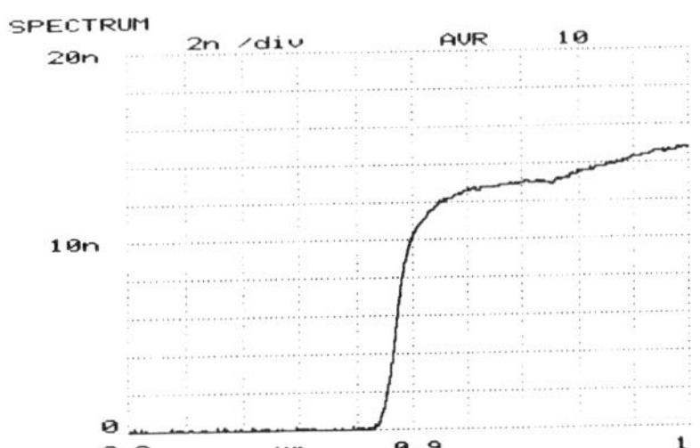

Figure 3 Optical transmission spectrum of GaAs at 21 °C illuminated with white light.

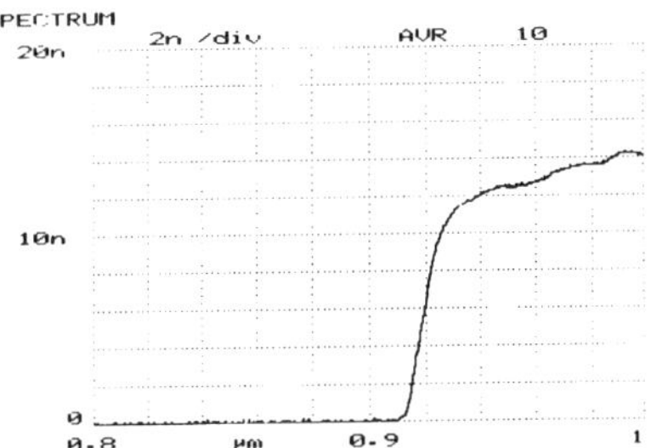

Figure 4 Optical transmission spectrum of GaAs at 98 °C illuminated with white light.

System development

The absorption edge of gallium arsenide is in the wavelength region where inexpensive high powered wide band LED light sources are available thus allowing for a single LED source and optical filters to be used to select a sensing and a reference wavelength. The filters used are a narrow bandpass 890 nm filter and a >940 nm bandpass filter mounted on a 2 element photodiode with each photodiode separated by a small distance such that light from each filter falls on only one of the photodiodes with the advantage that the photodiodes have the same characteristics as they are manufactured on the same substrate. The photodiode and filters are mounted in an single SMA housing. Thus the cost of producing these sensors can be reduced, through the use of fewer expensive optical components like couplers and spitters. The sensing wavelength is selected by the 890 nm narrow band pass filter and the reference wavelength is largely unaffected by changes in the temperature of the Gallium Arsenide (GaAs) sensing crystal and thus changes in the transmission properties of the fibres due to microbending and souce intensity variations can be referenced out. A schematic of this system is shown in Figure 5.

A number of possible light emitting diodes (LEDs) were tested to find one with a suitable wavelength profile and suitable packaging to allow large amounts of power to be coupled into the optical fibre. The Siemens SFH485P (centre wavelength 0.89 μm) was finally chosen. A number of type 1 probes (see Figure 6) were built and tested using HCP-M200T 200μm core glass optical fibre by Ensign Bickford.

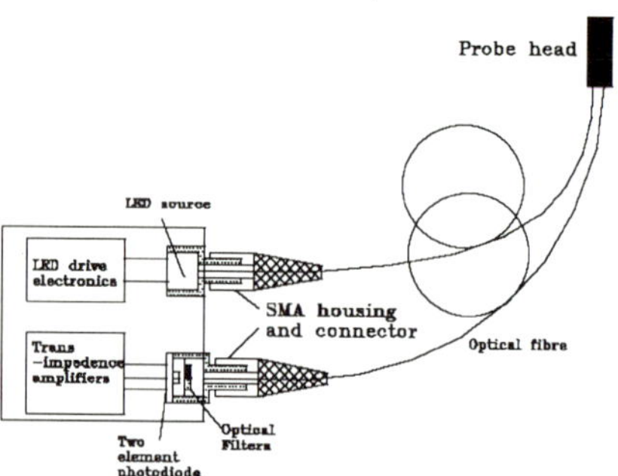

Figure 5 Schematic of the referenced temperature system

One of the probes was used in a sensor test with the LED coupling 21 μW of optical power into the optical fibre and a Photodyne optical power meter (a non-referenced sensor). The probe temperature was varied between 0 and 100 °C, a graph of the results is shown in Figure 7. This graph shows that the amount of light transmitted by the probe decreases with increasing temperature. The poor linearity is attributed to the low accuracy of temperature measurement. Microbending of the fibre was found to be significant affecting the readings by up to 10 % and source power variations were even more significant, highlighting the need for referencing techniques.

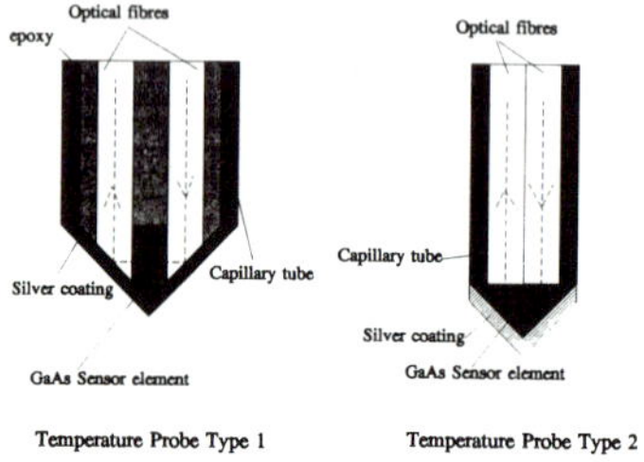

Figure 6 Schematics of prototype temperature probes

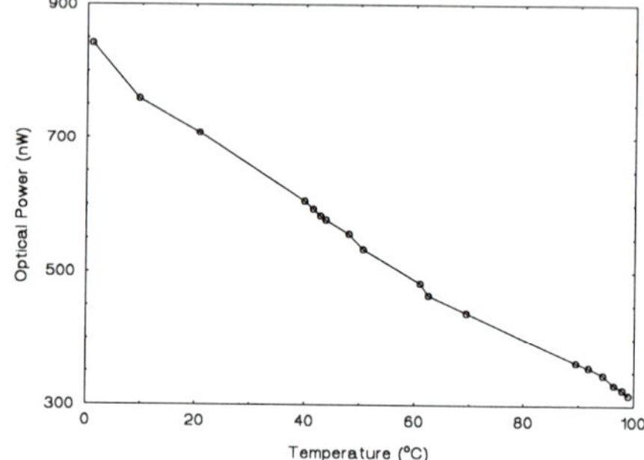

Figure 7 Graph showing initial GaAs temperature sensor results (non-referenced system)

Experimental system

A number of prototypes were produced using the split photodiode and the optical filters approach see Figure 5. Results were promising and as a Eurocard Rack system with 24 sensor channels was produced. The readings from the sensors (both sensing and reference channels) were multiplexed and read using a 16-bit Analogue to Digital Converter (ADC) and sent through a serial link to a PC.

Software was written to allow the sensors to be calibrated from a PT100 precision thermometer with a resolution of 0.01 °C and an accuracy of +/- 0.03 °C. This software logs the raw signals from both the reference and sensing channels of each sensor, the ratio of these two signals and the temperature from the precision PT100 probe. The results of one of the calibrations for a single temperature probe is shown in Figure 8 and Figure 9. Figure 8 shows that the sensing and to a lesser extent, the reference channels are affected by the temperature of the probe. The reference channel changes with temperature because of a small amount of cross talk between the channels and the fact that the optical filter on the reference channel transmits a small amount of light below 940 nm. However the ratio is still sufficient to determine the temperature, as is shown in Figure 9.

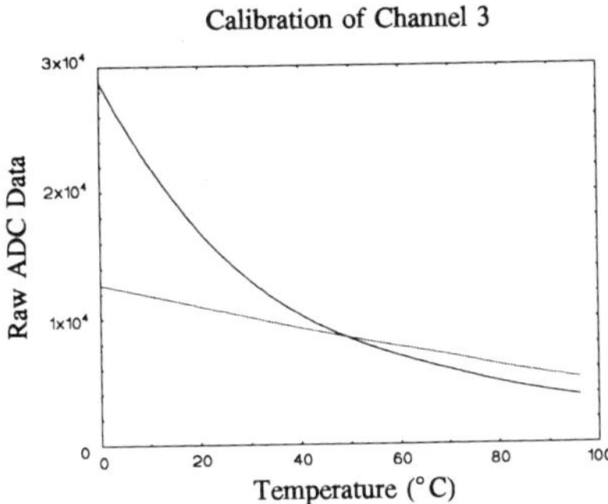

Figure 8 Graph showing the variation of the reference and sensing channels with temperature

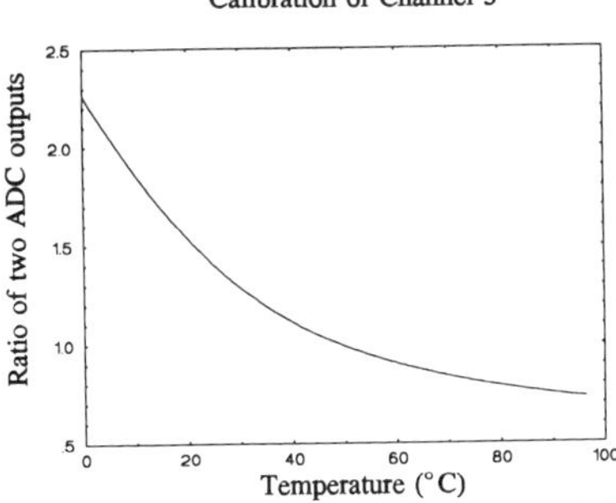

Figure 9 Graph showing the ratio of the sensing and reference channels and the variation with temperature

Figure 10 Photograph of the Eurocard Rack 16 Channel Optical Temperature Measuring System developed

The 24 channel rack optical temperature sensor system with 16 probes attached is shown in Figure 10. This rack, when linked to a PC running a display and loggig program running under Microsoft Windows, allows the raw temperature data coming from the Rack to be compared to the calibration data for each channel and converted into actual temperature values for display on the screen.

However, with all the temperature probes at the same temperature there was a large variation in the displayed temperatures. The spread in values was of the order of 5 - 6 °C and the stability of the rack system was poor. Drifts in the temperature recorded were found to correlate with changes in ambient temperature. The principle causes of the poor stability were identified as:-

- LED temperature - wavelength shift with temperature adjusts the ratio and thus the displayed temperature
- Photodiode dark current variation - dark current roughly doubles every 10 °C
- Optical filter temperature stability - transmission characteristics change with temperature
- The different degree of cross talk between sensing and reference channel for each sensor

One solution to the ambient temperature dependence of the system is to control the temperature of the LEDs, photodiodes and optical filters. A metal block containing the photodiodes and LED sources was produced and its temperature was regulated using a Peltier element. To reduce the power consumption of the Peltier element and improve the temperature stabilisation, the block was insulated. The block temperature can be stabilised to +/- 0.4 °C for changes in the ambient temperature of 15 to 30 °C. The system with the temperature stabilised optical components performed much better than the unstabilised system. The resolution remained unchanged at 0.1 °C and the accuracy was found to be +/- 1.8 °C (95 % confidence limits).

To test the referencing technique the amount of light from the LED was varied and the displayed temperature was noted. For a 3 dB change in light power the displayed temperature changed from 17.6 °C to 17.4 °C (0.2 °C change) and for a 6 dB change the displayed temperature changed from 17.6 °C to 16.9 °C (0.7 °C change) Thus the referencing is sufficient to overcome any normal variations in the fibres transmission characteristics, microbending of the fibre and source / fibre coupling changes.

Conclusions and future work

A 24 channel optical fibre temperature sensor system has been built which contains relatively few expensive optical / optical fibre components making possible production of the system at a reasonable cost.

The accuracy of the system is limited by the temperature sensitivity of the optical components and cross talk between the reference and the sensing channels. Optical filters with a lower transmission at the cut off wavelengths, less cross talk and a more uniform degree of cross talk between channels would improve the accuracy of the system. The stability can be further increased by more accurate temperature control of the optical components.

Acknowledgements

The authors would like to thank the EU for funding the work through a Brite-Euram project. The help given by the other partners in defining the specifications is acknowledged as is the work by another of the of the partners, 3D Digital Design and Development, Southgate, London in building the electronics and providing the software for this work.

References

Akhavan Leilabady P 1987 *Optical fibre point temperature sensor* SPIE Vol 838 Fibre optics and laser sensors V 231-237

Beheim G Fritsch K Anthan D J 1987 *Fibre optic temperature sensor using a spectrum modulating semiconductor etalon* SPIE Vol 838 Fibre optic and Laser sensors V 238-246

Christensen D A and Vaguine V A 1987 *A fiberoptic temperature sensor using wavelength-dependant detection* SPIE Vol 838 Fibre optic and Laser sensors V 252-256

Dakin J P and Culshaw B 1989 *Chapter 17 Physical and Chemical Sensors for Process Control, Optical Fibre Sensors: Systems and Applications Volume 2* Artech House Inc; Norwood, USA p664

Farahi F Jones J D C Jackson D A 1987 *High speed thermometry utilising multiplexed fibre Fabry-Perot interferometers* SPIE Vol 838 Fibre optics and laser sensors V 216-222

Grattan K T V and Palmer A W 1987 *Fibre-optic-addressed temperature transducers using solid-state fluorescent materials* Sensors and Actuators 12 375-387

Jones B E 1991 *Optical fiber sensors for industrial measurement* Revue Francaise de Mecanique no 1991-2 ISSN 0373-6601 133-138.

Kyuma K Tai S Sawada T Matsui T and Nunoshita M 1981 *Fibre optic sensor for temperature* Proceeding of 1st Sensor Symposium 241-244

Kyuma K Tai S Sawada T matsui T and Nunoshita M 1982 *Fibre optic instrument for temperature measurement* IEEE J Quantum electronics Vol QE-18 No 4 pp676-679

Scheggi A M, Brenci M, Conforti G, Falcial R and Preti G P *Optical fibre thermometer for medical use* OFS 1983 IEE Conf Publication No 221 13-16

Schoener G Bechtel J H and Salour M M 1984 *Novel fiber coupler for optical fiber temperature sensor* Proc 2nd Optical Fibre Sensors Conf. Stuttgart 203-6

Soares E A and Dantas T M 1992 *Simple theory for a temperature sensor* SPIE Vol 1795 Fibre optic and Laser Sensors X 236-240

Low noise compact gradiometric Induction magnetometer system.

R.J. Prance, T.D. Clark and C. Watkins.

ENGG, University of Sussex, Falmer, Brighton, BN1 9QH.

Abstract. Induction magnetometers are both simple in principle, and robust in operation as sensors for measuring weak extensive magnetic fields. Traditionally large (~1m) coils have found applications in geophysical surveying [1] while more compact versions are widely applied to motion and position sensing [2]. In this paper we describe a radically new design for compact sensors (~10cm) with sensitivity comparable to that obtained with large coils.

1. Introduction

The compact induction magnetometer system described here has been created as a result of two new developments. Optimisation of the coil design and incorporation of high permeability core materials has allowed us to reduce the physical size of the coils, whilst maintaining the required very high inductance (>700H). The second development concerns the electronics used to amplify the coil current, and to produce an output voltage proportional to this current. Two alternative approaches are presented and the relative merits of each in terms of noise performance and operating bandwidth is discussed. We conclude that the choice of electronics provides the options of either the lowest noise performance (<1nT/$\sqrt{Hz}$), or the largest operating bandwidth (DC-50kHz). Anticipated improvements in performance will put these induction magnetometers in a position to challenge cryogenic SQUID magnetometers in most areas of application. We have already observed considerable reductions in the noise (x3) by cooling our system to liquid nitrogen temperatures (77K). Future work will be centred around achieving this level of performance at room temperature.

2. Coil Design.

The main aim of this work has been to produce a compact induction magnetometer system with a magnetic field sensitivity which is comparable with conventional large induction systems. The major considerations when designing high inductance pick up coils are to maximise the number of turns, the effective area, and, the length of the coil in order to produce the maximum induction from the applied field. However, our constraint on the physical size of the system requires that we minimise these same quantities as far as possible. A compromise solution has been found using small diameter wire (0.06mm) and high permeability magnetic core materials. The dimensions of a single coil are shown in figure 1. The magnetic core consists of annealed laminations of super mumetal strips (0.065mm thick), to form a 10mm by 10mm cross section core, 150mm long. The effective permeability of such a core depends not only on the intrinsic permeability of the magnetic material, but also on its aspect ratio. For the case we are considering here, the length/diameter ratio is approximately 15, giving an effective permeability[3] of approximately 25. The minimum practicable wire diameter (0.06mm) gives a maximum of 10^5 turns (n) on a coil of these dimensions. Typical measured values for such a coil are:

Inductance (L) 780H; Resistance (R) 23kΩ; Corner Frequency (R/L) 29Hz

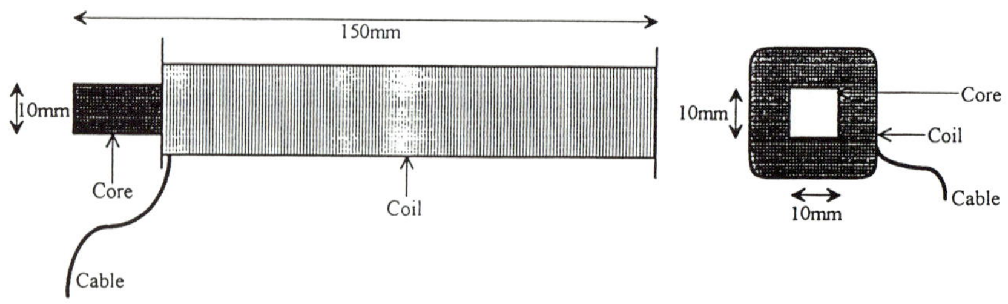

Figure1

3. Current/Voltage Conversion.

Two methods have been employed in order to produce an output voltage proportional to the applied field. In the first case we use the fact that the current (I) which flows in the coil is proportional to the applied magnetic flux (Φ), which is independent of frequency above R/L.

$$\Phi = LIn$$

The inherent simplicity of a single op amp current/voltage converter means that in principle we may produce a very low noise system. The fundamental limits on the performance of this configuration of magnetometer are associated with the current noise in both the op amp and the feedback resistor. It should be noted that component values for this circuit are such that the resistor contributes little to the overall noise; in addition the bandwidth is restricted to a range from 29Hz (R/L) to the maximum frequency of operation for the coils (approximately 50kHz).

The second method employed is to use an analogue electronic integrator, since by Faraday's law the induced voltage is given by the derivative of the applied magnetic flux.

$$V = -\dot{\Phi} = -L\dot{I}n$$

For a simple electronic integrator the output voltage is given by:

$$V_{out} = \frac{1}{RC}\int V_{in}\,.dt$$

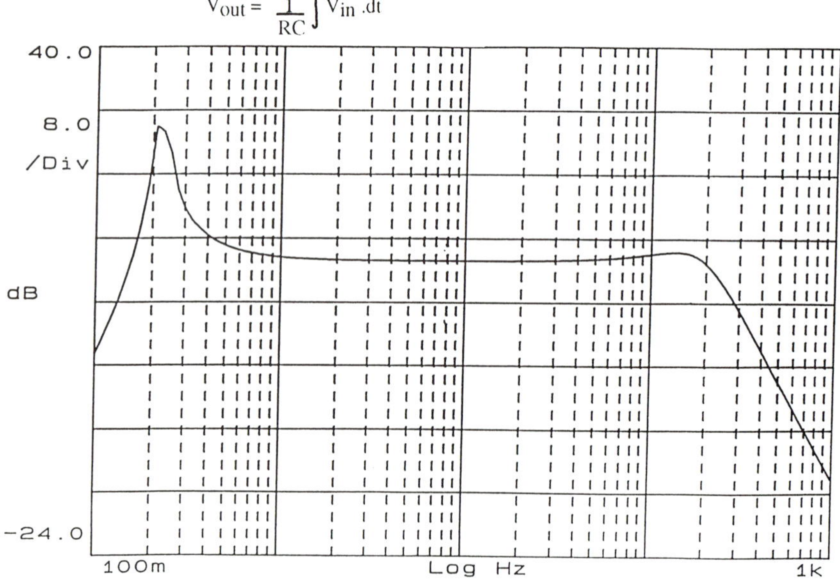

Figure2

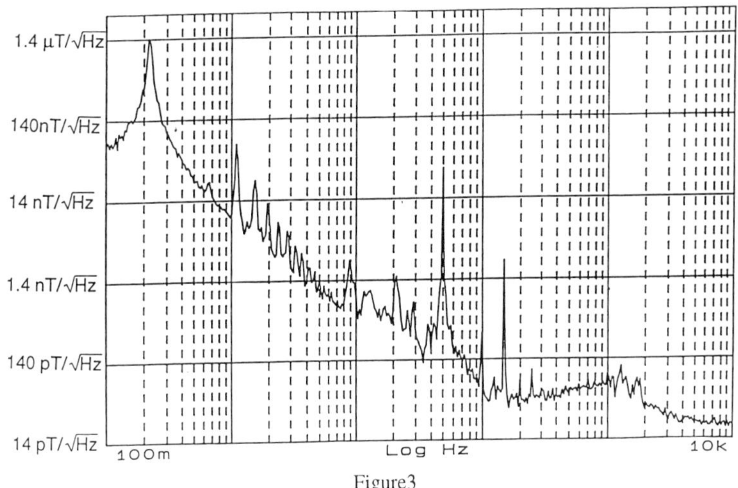

Figure3

By appropriate choice of the resistor and capacitor values we may set the gain of the integrator. One problem which exists with this type of system is that the gain is essentially infinite at DC. This is the penalty for operating an induction magnetometer which in principle responds to DC signals. Obviously this is an unstable situation and the op amp will saturate very quickly. Automatically resetting integrators have been constructed but these have proved to be an unsatisfactory solution, since they cause further problems including limiting the high frequency bandwidth and introducing more noise. We have solved the problem in a different way, by incorporating a low pass filter circuit in the feedback network, thus limiting the low frequency gain of the integrator to a large but finite number at DC. This effectively stabilises the system with the limitation that the lowest frequency at which the magnetometer now responds is set by the low pass filter. However this can be made arbitrarily close to DC. In figure 2 we show the combined frequency response for the coil and integrator system from 0.1Hz to 100Hz. For this example we have chosen 0.2Hz as the lower cut off frequency, and it should be noted that the peak in the response may be suppressed using suitable compensation components.

4. Results.

For use as a measurement magnetometer a gradiometric configuration must be used in order to reduce the effect of external noise sources. With the system described here it is possible to balance electronically gradiometric systems by providing each coil with its own integrator and trimming the relative gain in order to achieve balance. This allows differences between coils to be nulled with ease. We note that this simple procedure represents a significant improvement over conventional techniques of gradiometer balancing by adding extra turns or shimming with magnetic material. In figure 3, where we present preliminary measurements on a first order gradiometric magnetometer, the magnetic field noise is plotted as a function of frequency. These were conducted in an electrostatically screened enclosure (no magnetic screening), and calibrated for field sensitivity using a commercial Hall probe.

5. Conclusions.

We have demonstrated the feasibility of a compact induction magnetometer system exhibiting a frequency response from close to DC to up to 50kHz, and with a noise performance better than 100pT/√Hz at100Hz. From the results presented above we conclude that further improvements in the noise performance can be achieved by using much lower noise op amps (particularly in the low frequency 1/f noise region) for the integrator circuit of this room temperature magnetometer. In addition we have chosen magnetic core materials with properties which are essentially temperature independent. This will allow us to cool the sensor coils to liquid nitrogen temperature (77K), or lower, in order to reduce the thermal noise from the coils.

6. References.

1. W.W.L. Taylor, B.K. Parady, P.B. Lewis, R.L. Arnoldy and L.J. Cahill, "Initial results from the search coil magnetometer at Siple Antarctica", J. GeoPhys. Res. 80, No.34, pp4762-4769.
2. "Sensors A Comprehensive survey", Edited by, W. Gopel, J. Hesse and J.N. Zemel, Published by, VCH, Weinham, 1989, chapter 7.
3. M.F.Demaw, "Ferromagnetic core design and applications handbook", Published by Prentice-Hall Inc., 1981.

Review of Data Acquisition System Developed for the MIRAM Project (BRITE-Euram BREU-CT91-04623)

P.R. Drake, R.I. Grosvenor and A.D. Jennings

University of Wales Cardiff, Cardiff School of Engineering, PO Box 917, Newport Road, Cardiff CF2 1XH.

Abstract. A computerised data acquisition system (DAS) was developed for use (i) to collect field failure data from machine tools in use in industry (ii) to provide information to the overall maintenance management system. The latter, developed by project partners, integrated the DAS, machine tool models for normal and fault conditions, fault classification algorithms, user interface designs and inputs into wider maintenance management modules. Key areas of the DAS development were the auditing of the machine tool, the selection of signals and sensors, the rationalisation of the sensors and their signal conditioning, the method of data collection from remote sites and the management of the vast amount of data collected. The use of a database concept to integrate the systematic data capture and manipulation is reported. The DAS design is seen to provide an economic and powerful solution to machine tool condition monitoring and is applicable to other machines and processes.

1. Introduction

A data acquisition system (DAS) was developed, in the Cardiff School of Engineering, as part of a European Union BRITE/Euram project entitled 'Machine Management for Increasing Reliability, Availability and Maintainability (MIRAM)'. Other partners in the project were responsible for aspects of the overall system. The system operation was such that a fault classifier compared signals obtained from a machine tool with the outputs of mathematical models which were provided to the same format as the DAS signals for both normal and fault conditions. The classifier essentially then provided guidance to an operator to assist with fault diagnosis and the faster return of the machine to operational status. The classifier further provided an interface to maintenance management modules. The partnership included the German machine tool manufacturer Heckler and Koch GmbH (H&K) whose machine tools, located at various factories in Germany, were used as test beds. The work of the various partners was integrated via the use of a central data repository for data exchange.

The role of Cardiff in the project was to provide the data acquisition facilities for two distinct stages of the system development. Firstly, DAS units were required to collect extensive field data for the Heckler and Koch BA25 machines. Real data was to be recorded from machines working in industrial environments with the aim of identifying which signals were most sensitive to the machine faults which were expected to occur during the data collection period. A refined version of the DAS, drawing on the experiences of the field data collection, was then to be provided for use with the overall MIRAM system, whose operation was summarised above. This DAS was know as the research DAS.

This paper aims to describe the key stages in the development of the DAS units. The designs are considered to offer more than just data logging facilities linked to sensors on machine tools. The DAS is a flexible, generically applicable (to many other machines and processes than machine tools) and inexpensive solution to sensor based condition monitoring applications. It included a methodology for analysing data logging requirements and a relational database that supported that methodology. The database concept allowed simple reconfiguration of the DAS to implement new monitoring requirements and to customise it to the needs of different machines.

2. Machine Tool Audit

During the early stages of the project all partners agreed on the definitions of user requirements and an initial specification. Further the general philosophies, such as monitoring during non-cutting periods of machining cycles, of the research were agreed along with details such as preferred platforms, software packages and data formats for example. A vital area of work was the highly detailed audit of the H&K BA25 machine tool [1]. A potential set of some 300 signals (analogue and digital) were identified. Consultations with partners then refined the list of signals to be monitored. This included existing signals and also defined the transducer requirements for additional signals to be monitored. The list was formulated to suit the needs of the partners involved in modelling activities and drew on experience as to those signals likely to give good fault detection coverage. With both the existing and additional signals there was a prime requirement that the monitoring of the signals should not affect in any way the normal operation of the machine tool. A full failure mode and effect analysis (FMEA) [2] was completed along with a sensitivity analysis. The latter was used to reinforce the signal selection which needed to balance fault coverage against transducer and computer costs and timing overheads.

A review of condition monitoring techniques was performed and this also provided an input to the selection of the signals to be monitored. A review of commercially available data acquisition systems was also performed, with the conclusion that such systems either exceeded an agreed cost criteria (maximum cost was to be less than 10 % of the cost of the BA 25 machine tool) or did not fully meet the technical specification requirements.

3. DAS Design

The decision was therefore made to base the DAS design on available PC components. A detailed costing exercise along with a weighted decision making technique (analytical hierarchical process) were then employed to defined the major components such as processor type, mother board, data storage and display devices [3].The DAS design also included the selection and or design and manufacture of suitable interface cards and signal condition cards. For both the field DAS and the research DAS the remote installation and communication aspects were also fully considered. There was a need to guarantee operation of the DAS in the remote sites. This was achieved via the use of a watchdog card designed and manufactured at UWCC. This and the other cards which were designed and manufactured at UWCC were done so to professional standards. This in-house work was carried out when no suitable commercially available cards were found despite detailed enquiries.

The transducers required to provided the additional signals were sourced and work was done to ensure a standardisation of the output format of the different types of transducer (pressure, flow, electrical current, electrical power, vibration, temperature etc.). The format selected was 4 - 20 mA. This was done in order to minimise the number of different signal conditioning and interface cards required.

3.1. DAS Specification

The hardware configuration of the research DAS unit and the associated signal conditioning rack are shown schematically in Figure 1. The main components are summarised below (firstly for the field DAS):

Motherboard included: a 486SX processor running at 20 MHz, a backplane allowing connection of 8 daughter boards, 2 MBytes RAM

Other Cards included: Multifunction I/O card (to control - 3.5", 1.44 MByte Floppy Disk Drive (external), 40 MByte Hard Disk, Jumo Tape Streamer (40 MByte per tape, use to transfer field data back to UWCC)), VGA graphics card, 3 analogue input cards (Bede Technology PC-ADH24) - 24 channels per card, 2 configured for DMA operation and 1 for interrupt operation, maximum sampling rate of 60 kHz, 12 bit converters and ± 10 V input range. Up to 3 channels could be sampled 'simultaneously', within a limit of 1 ms, 2 digital input cards (Bede Technology PC-DIP24) - 24 channels per card, 2500 volt isolation, Watchdog card (UWCC) - wide range programmable timer (100 ms - 5 min). Nominal times were

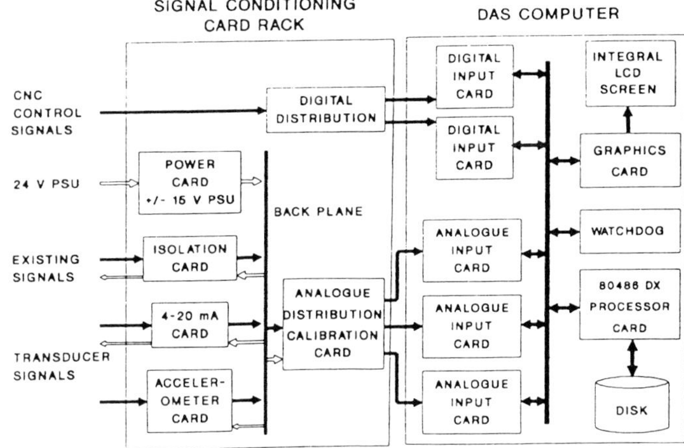

Figure 1. Schematic of DAS Hardware

measured for all software routines. Timer set at start of routine and reset at end of routine for normal operation. If fault in routine caused timer to exceed set value a reset was initiated causing DAS to re-boot.

Signal conditioning cards (all UWCC): For the field DAS units these consisted of four 4 - 20 mA transducer receiver cards (each with 4 channels and 6 software selectable filters), three isolation cards (each with 6 channels and 6 selectable filters), an accelerometer interface card (2 channels, programmable filter), a power distribution card, a digital converter card (240 V to 24 V) and a temperature interface card (8 channels and with calibration facilities)

Case: A tower case was used to house the PC components. This was modified to better accommodate all of the components within the available space. A back plate fitted with suitable connectors was added to the back of the unit to provide easy connection of the signal cables from the machine tool. An extra cooling fan was added and one of the temperature transducers was used to monitor conditions within the DAS case.

Research DAS Specification: The design of the research DAS drew heavily on the experience gained with the field DAS units. Many of the elements were indeed common to the two designs with the main differences for the research DAS being the use of a rack mountable industrial computer with integral monitor, the use of a separate rack for backplane mounting of all signal conditioning cards and the use of a communications port in place of the tape streamer. The main items of the research DAS were as follows: 19" rack mountable industrial computer with 80486DX processor, 14 slot passive backplane and 4 MBytes RAM, Data storage via 40 MByte hard disk and 3.5", 1.44 MByte Floppy disk drive, Signal conditioning: four 4 - 20 mA cards, 1 power supply card, 5 isolation cards, 1 accelerometer card, 1 digital termination card (48 channels) and 1 analogue distribution and calibration card.

Transducers and existing signals: For the field machines the monitored signals were as follows: TRANSDUCER SIGNALS: Pressure (6), Flow (5), Power (5), Accelerometer (2),Temperature (3); EXISTING SIGNALS: X Axis (3 i.e. Demand, Current, Tachogenerator), Y Axis (3), Z Axis (3), Spindle (3), Drives (2), Digital (46)

4. DAS Database and Software

The controlling software was implemented in a generic, object-oriented fashion and was based on the use of a database. The database contained the following: test definitions (analogue signal to be measured, digital control signal, sampling times etc.), data processing method (DPM) definitions (cross referenced to signals), and interface details (card number, signal channel, transducer details and calibration etc.). The tables of data in the database were addresses by software written in turbo pascal. A generic BA25 database was constructed and populated with data and was then tailored to each of the 3 field installations. A database for the Wadkin machine tool in UWCC's lab facility was also constructed and populated. Fuller details of the database appear in reference [2]. The structure of the database is summarised in Figure 2.

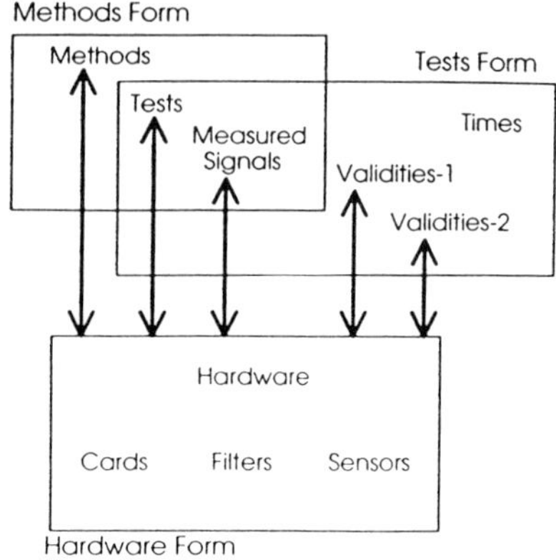

Figure 2. DAS Database Structure

The DAS control software was used to extract (with suitable checking) data from the database and to create all of the output datafiles. These included a header (to a format suggested by UWCC and agreed with all partners) containing all essential information about any test and the actual data in one of two agreed formats. The DAS software monitored all the digital signals (used as triggers and validity conditions) and then selected an appropriate test to perform (allowing for different periodicity with different signals as required). For the selected test the database provides the sampling rate and duration and the DPM to be employed. The software checked on validity condition during the data acquisition. The DAS software also controlled the logging of data for the field machines. The user inserting a tape into the drive initiated the automatic back up of all the relevant files. On completion the user then mailed the tape back to UWCC. The available hard disk space on the DAS was also intelligently managed by the DAS software. For the research DAS the data was instead sent via a communication port to the MIRAM computer. Evaluation tests at UWCC helped to define a radio modem as the means of achieving this communication of data. The DAS software was able to monitor the link and supply data on request from the MIRAM computer.

The data processing methods referred to above were required in order to reduce the considerable volumes of data encountered when using raw signal data and to facilitate the comparison to modelled data for normal and faulty machine states. The algorithms were drawn from recognised sources and were coded into the required software routines largely by UWCC. The information was then made available to partners to ensure consistency of the methods when use to compare and classify faults. The DPM methods implemented were then as follows: Raw Data (used during initial testing and familiarisation), Statistical (Number of points, mean, rms, standard deviation, skew, kurtosis), Linear regression, Polynomial (2 parameters), Polynomial (3 parameters), FFT (Hanning window), FFT (Parzan window), FFT (Welch window), FFT (Octave filtered output), FFT(1/3 Octave filtered output), Auto Correlation, Cross Correlation, Correlation coefficient

5. DAS Installations and Data Collection

During the project 3 field DAS units were installed in industrial sites in Germany. The actual installations of the transducers and the DAS units were supported by and co-ordinated with service personnel of H&K. The routine collection of initially raw data and later processed data from the field machines was planned and the industrial users were briefed on the procedures. By the end of the project a total of some 500 MBytes of data had been collected from the field sites and distributed to all project partners.

6. Discussion

The DAS units proved to be a powerful solution to the condition monitoring data acquisition tasks. The units had a high technical specification and yet were inexpensive. The total cost of the DAS hardware, including I/O devices, but excluding labour, software development costs and transducer costs was less then 5500 pounds.

The number of additional sensors required for a final commercial condition monitoring system would need to be reduced. The main signals of use in fault diagnosis would be retained whilst others would be considered as not being justified to monitor. During the field data collection exercise the inherent high reliability of the H&K BA25 machine was demonstrated. Very few faults developed during the period and thus further data gathering is on-going before any decision can be made on the refined list of signals can be attempted. The DAS data did detect faults ranging from empty coolant tanks to progressive damage to spindle bearings.

It was interesting to note the need to design and fabricate the signal conditioning cards in-house. Further the careful sourcing of transducers, from the wide range with diversity of outputs, enabled the standardisation on a 4-20 mA output format and a reduced number of required signal conditioning cards.

The technology should be easily transferable to other machines and processes. The need for a detailed machine/process audit has been demonstrated. This then leads easily, through the programming of the DAS database, to the setting-up of the DAS hardware.

7. Acknowledgement

The financial support of the European Union via BRITE/Euram contract BREU CT91-04623 is fully acknowledged along with the contributions of the other project partners.

8. References

1. "Identification of the BA25 Machine Tool Signals", 1992, Jennings A.D. & Whittleton D., Technical Note 194, UWCC, Cardiff (MIRAM document no MIRAM-UWC-2.2-EXT-ADJ&DW-3-A).
2. "A Data Acquisition System for Machine Tool Condition Monitoring", 1995, Drake P.R., Jennings A.D., Grosvenor R.I. & Whittleton D., Quality & Rel. Eng. Intl., Vol 11, pp 15-26.
3. "A Computer Based Data Acquisition System for Monitoring a CNC Machine", 1992, Nicholson P.I., M.Sc Dissertation, UWCC, Cardiff.

Theoretical Modelling of Chemical and Biological Prism-Coupled Surface Plasmon Resonance Sensors

CR Lavers

BRNC, Dartmouth, Devon, UK, TQ6 OHJ

Abstract. Initial theoretical modelling for several proposed prism-coupled surface plasmon resonance (SPR) sensors is presented and the results compared with earlier theoretical modelling for non-prism waveguide sensors (i.e. planar ion-exchange waveguide sensors which may support surface plasmon resonance) [1]. The results indicate that such sensing techniques are in fact complimentary and allows for flexibility in real device design.

1. Introduction

Optical techniques are at the leading edge of modern analytical methods in chemistry due to their inherent high sensitivity to specific measurands. Sensors which allow for rapid detection of minute refractive index changes in thin sensing films, or the adsorption of a thin layer of chemical species are highly desirable. Several techniques are currently of interest, including; absorption spectroscopy [2], light emission from labelled fluorophores [3] and surface plasmon resonance (SPR) [4]. The latter of these, SPR, has received considerable attention recently and has lead to several commercial devices [5]. SPR is usually accomplished with the conventional prism-coupling technique although other methods such as grating coupling, [6] and fibre coupling [7] have been demonstrated.

2. Theory of Prism-Coupling and Resonant Modes

Prism-coupling as a means of probing thin layers to obtain information on the optical dielectric tensor parameters has been known for a long time. The Otto [8] and Kretschmann-Raether surface plasmon configurations [9] have been well documented as reliable methods for examining surface parameters of thin films. As well as the surface plasmon resonance mode, sharp Fabry-Perot resonant modes may also be seen in the Kretschmann-Raether geometry [10].

To analyse the optical dielectric parameters of a multilayered structure it is fundamental to measure the reflection properties of the system. The prism-coupling technique is used where optical guided modes and surface waves may be propagated. The important consideration is that a guided mode may only be excited when the momentum of the incident radiation and the particular guided mode are matched and this is then observed as a sharp dip in the reflectivity spectra

The surface plasmon resonance or SPR as it is more commonly known, is a charge density electromagnetic field coupled oscillation which can be supported at an interface between media of opposite sign of the real part of the dielectric constant such that

$$\varepsilon_{1real}+\varepsilon_{2real}<0$$

where ε_{1real} and ε_{2real} are the respective real parts of the dielectric constants of the metal and the dielectric films respectively. The SPR is highly localised to the interface with exponentially decaying fields into each medium. It is excited with p-polarised light only and is the TM_0 mode of the system. Thus the TE mode may be used as an accurate reference sensor for the system under study.

2.1 Theory of Modelling Method

Theoretical reflectivities are calculated as a function of angle of incidence for a multilayered isotropic medium. The multilayered system is considered as a series of uniform slabs of dielectric material, where each layer thickness is much less than the wavelength of the incident light. A scattering matrix formalism is used to overcome instabilities in the conventional transfer matrix method developed by Azzam and Bashara [11]. The method takes account of the reflection and transmission coefficients at each interface between successive media and uses Fresnel's equations. The scattering matrix method, developed by Ko and Sambles [12], couples the incoming fields at any given interface to the outgoing fields by a matrix which is implicitly stable for all incident angles.

3. Device Modelling

Two devices are considered, one a sensor for the detection of chlorine in aqueous solution, the other a sensor for monitoring refractive index changes due to the adsorption of charged chemical species, both based upon similar planar waveguide sensors which have already been demonstrated elsewhere [1].

3.1 Device A SPR Electrochemical Sensor

For the first SPR device, configured as a chlorine sensor the device was modelled as shown in figure 1.

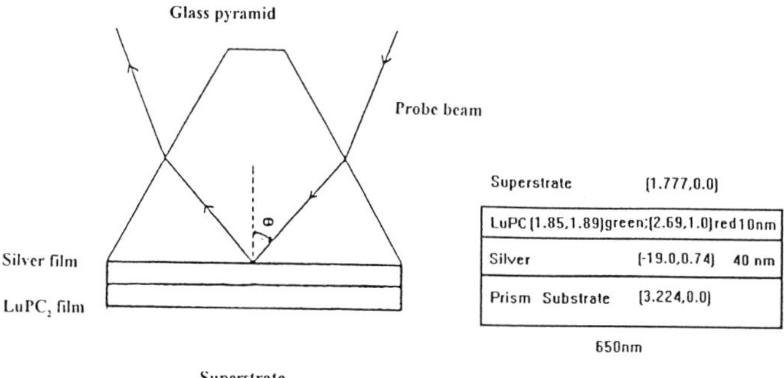

Figure 1

A 10 nm thick layer of lutetium biphthalocyanine ($Lu(PC_2)$ initially in a reduced green state (state a figure 2), was deposited on a 40 nm thick layer of silver evaporated onto a glass prism. Values for the metal dielectric constants at a probe wavelength of 650 nm were taken from Johnson and Christy [13]. The conductive metal layer may also act as an electrode so that the device may be used in a resettable mode [1]. In the planar case Indium tin oxide was used and therefore did not support SPR. Figure 2 shows theoretical predicted reflectivity against incident angle and reveals high optical sensitivity to changes in the dielectric constants of the lutetium film caused by the presence of chlorine. Upon complete oxidation by aqueous chlorine the dielectric constant for the lutetium film changes significantly, ($\delta\varepsilon_{real}$= 0.84, $\delta\varepsilon_{imaginary}$ = 0.89) giving a new set of prism-coupling conditions for the SPR mode and thus a new plasmon curve as shown (state b figure 2). The maximum change in reflectivity possible is ~220% at θ_{SPR}=53.0^0 compared with a planar transmission change of ~90%

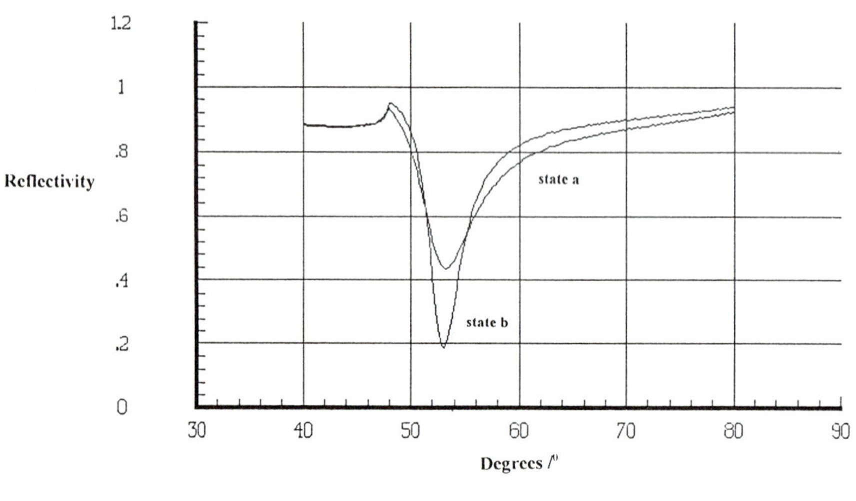

Figure 2

From earlier work using the planar waveguide-coupled chlorine sensor at 950 nm offered an optimum "window" on high specific sensitivity to chlorine detection [1], with a transmission change between the initial green reduced state and red oxidised states of 225 %. The maximum reflectivity change between the reduced (state a) and oxidized (state b) prism system occurs at θ_{SPR}=53.5^0 giving a favourable 200 % change in reflectivity, figure 3.

The planar waveguides with which the prism-coupled data is compared may be fabricated by potassium ion-exchange in soda-lime silicate glass slides or more expensive BK7 glass and is shown in figure 4. Channels are defined by conventional photolithography, in an aluminium mask, and then immersed in a molten salt bath for a specified duration of time to produce a region of modified refractive index that will act as a light guide. This work is reviewed extensively elsewhere [14]

In terms of electrochemical stability silver is a poor choice of electrode, with gold or platinum as the usual preferred choices. Taking a gold surface of 40 nm we may repeat monitoring of the reduced green (state a) to red oxidised (state b) process at 650 nm for the TM_0 mode (figure 5).

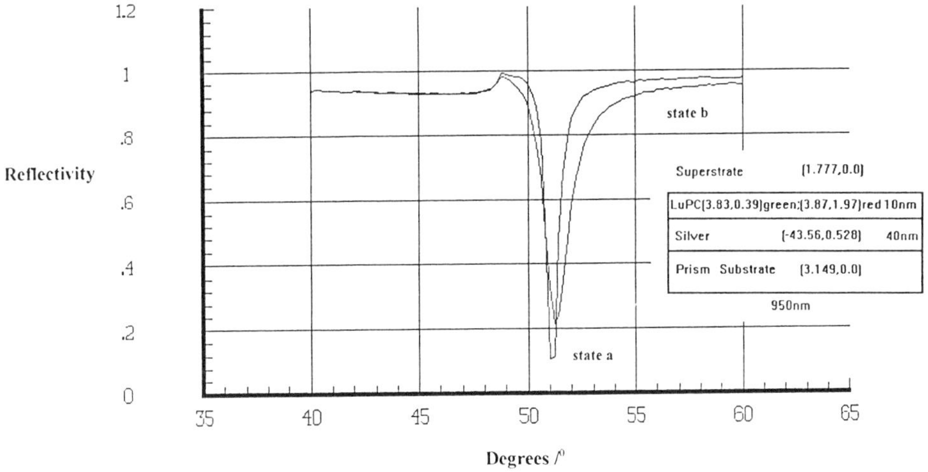

Figure 3

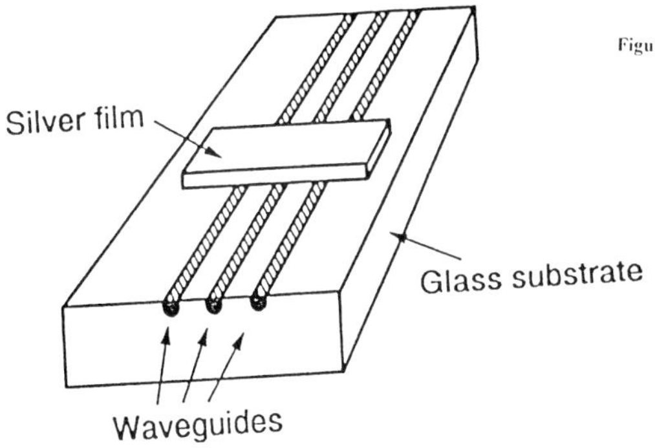

Figure 4

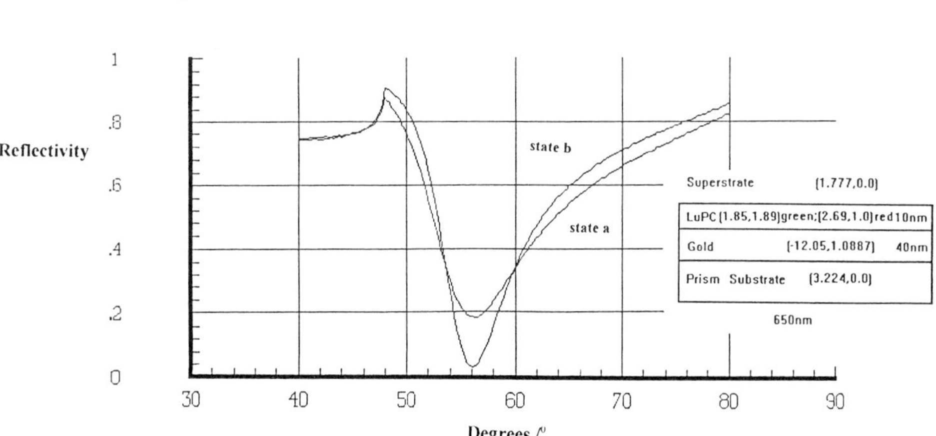

Figure 5

3.2 Device B A SPR Adsorption Sensor

Here a prism-coupled system was compared with a recently demonstrated planar waveguide-coupled surface-plasmon sensor for monitoring monolayer adsorption in an aqueous environment [1]. Monitored optical changes were observed due to preferential monolayer adsorption under the application of an electric field (figure 6).

A 40 nm thick silver layer of silver was taken, identical with both previous modelling and experimental work. SPR reflectivity changes were considered at a probe wavelength of 632.8 nm. Previous planar optical results suggested that an initial monolayer of thickness 0.5 nm (Nreal ~1.58) on the waveguide surface was replaced by a second monolayer 0.8 nm thick having Nreal~1.51. Dielectric constants for this system are shown in figure 6. Inserting these values into a prism-coupled model the reflectivity at the optimum coupling angle (minimum reflectivity) was observed to change by only 0.015 % and compares poorly with the planar waveguide-coupled sensor having an attenuation ($\frac{Power_{out}}{Power_{in}}$) change of 0.263 %. This can be improved upon significantly by choosing the point of maximum slope on the reflectivity curve $\theta_{spr} = 50.8^0$ and therefore $\Delta R_{pp} = 0.07\%$, 27 % of the predicted planar waveguide value. The prism-coupled system may be optimised further, e.g. by correct choice of silver film thickness, to reduce this deficit.

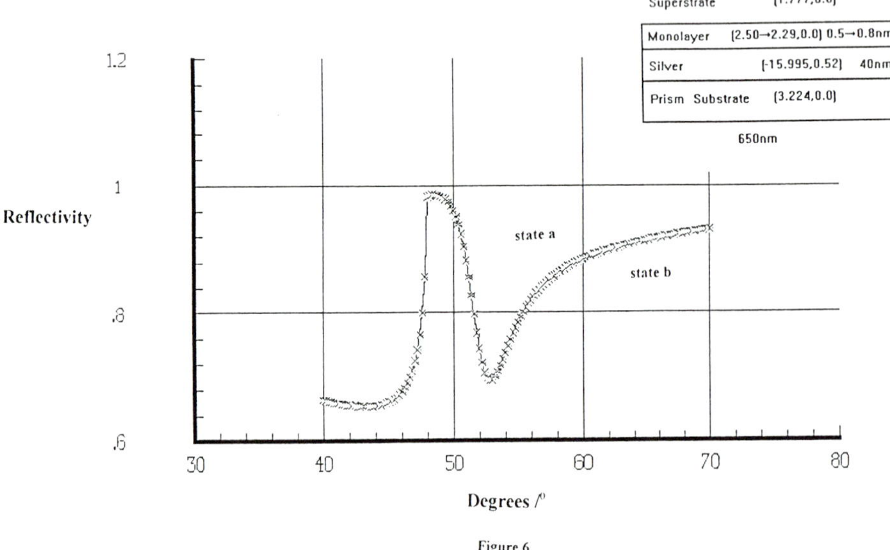

Figure 6

4. Conclusions

A simple theoretical comparison between a prism-coupled surface plasmon resonance sensor and a planar waveguide-coupled surface plasmon sensor has been demonstrated. A scattering matrix formalism allowed a numerical model to predict the prism reflectivities as a function of both incident angle and superstrate index. Work is now in progress to evaluate more rigorously the two waveguide systems' performances for these particular configurations and other electrochromic systems. Such evaluation is necessary so as to design the optimum device performance in the best experimental configuration. For monolayer detection the prism-coupling system studied here does not have as great a sensitivity as a planar waveguide system having long optical path length.

5. Acknowledgements

The author would like to thank Dr AW Machin and Mr TR Mason of BRNC for their support of this research, Dr AC Walton (BRNC) for her help with computational modelling of planar waveguides, and Professor JR Sambles and Dr CR Lawrence of the Physics Department at Exeter University for theoretical modelling using the scattering matrix formalism.

6. References

[1] CR Lavers, C Piraud, JS Wilkinson, M Brust, K O'Dwyer and D J Schiffrin, Proc. Opt. Fibre Sensors 9 Firenze, Italy, p.193, 1993.
[2]K Nishizawa, E Sudo, M Maeda and T Yamasaki, Proc. European Conf. Opt. Commun., pp.99, 1987.
[3] AN Sloper, JK Deacon and MT Flanagan, Sensors and Actuators, B1, p.589, 1990.
[4] U Jonsson, L Fagerstam, B Ivarsson, B Johnsson, R Karlsson, K Lundh, S Lofas, B Persson, H Roos, I Ronnberg, S Sjolander, E Stenberg, R Stahlberg, C Urbaniczky, H Ostlin and M Malmqvist, BioTechniques, 11, p.620, (1991).
[5] J Hodgson, Bio/Technology, 12, p.31 (1994).
[6] MC Hutley and Maystre D, Opt. Comm. 19, p.431, 1976.
[7] S Barcelos, P St J Russell, MN Zervas and CR Lavers, Proc. LEOS USA. Nov 11-14th (1993).
[8] A Otto, Z Phys, 216, p.368 (1968).
[9] E Kretschmann and H Raether, Z Naturforsch, Teil A, 23, p.2135 (1968).
[10] CR Lavers and JR Sambles, Liquid Crystals, 8, 577 (1990).
[11] RMA Azzam and NM Bashara, Ellipsometry and Polarised Light, North-Holland, Amsterdam, (1986).
[12] DYK Ko and JR Sambles, J.Opt. Soc. Am, A, 5, p.1863, (1988).
[13] PB Johnson and RW Christy, Phys. Rev. B, 6, p.4370 (1972).
[14] RV Ramaswamy and R Srivastava, J. Lightwave Technology, 6, p.984, (1988).

Index